PROCEEDINGS

OF THE

XIV INTERNATIONAL BOTANICAL CONGRESS

BERLIN 1987

PROCEEDINGS

OF THE

XIV INTERNATIONAL BOTANICAL CONGRESS

BERLIN, 24 JULY – 1 AUGUST 1987

EDITED BY

WERNER GREUTER & BRIGITTE ZIMMER

KOELTZ SCIENTIFIC BOOKS
KÖNIGSTEIN, 1988

ISBN 3-87429-282-7

PREFACE

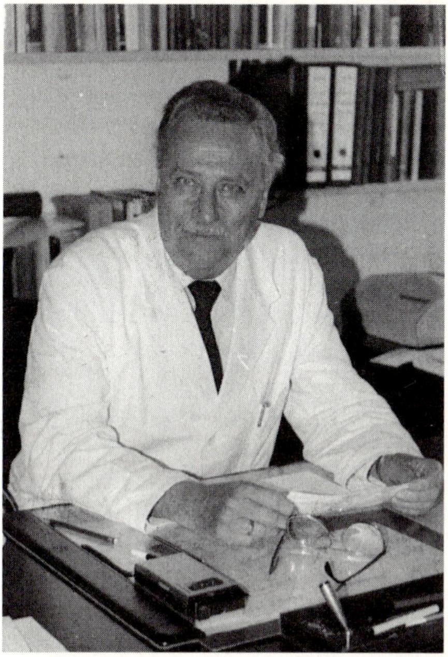

Everything comes to an end. Congresses are no exception to that rule. We are now looking back to the XIV International Botanical Congress from a distance of a few months, and take pleasure in remembering its main events and in recapitulating its happenings.

This volume is the final document of the Congress, and is sent to all its members. It is divided in two main portions, the first of which comprises the proceedings proper, including an overview of the scientific programme, due mention of the main social events and corollary manifestations, and full accounts of the Opening and Closing Ceremonies.

The second part of the volume comprises the full texts of the General Congress Lectures if submitted by their authors, plus an English translation of the text of one of the two Public Lectures that had been given in German language.

Even now when the euphoria of the happening has settled we still believe that the Berlin Congress was a great success. Many letters that we have received, and many opinions expressed to us, reinforce this judgement. We are well aware of the fact that the credit for this success goes in equal shares to the organizers and to the participants. It is to you, members of the XIV IBC, that we as organizers owe the feeling that our efforts have not been vain.

This is the moment to thank once more all who have taken part in the Berlin Congress. They have kept the tradition of International Botanical Congresses well alive and have ensured that it will last for many years to come. The XIV IBC belongs to the past. Our cheers go to the XVth IBC and to the following ones:

Vivant, crescant, floreant!

Karl Esser

XIV INTERNATIONAL BOTANICAL CONGRESS – BERLIN (WEST)

Under the auspices of the International Union of Biological Sciences
Sponsored by the Senate of Berlin and by the Deutsche Forschungsgemeinschaft

Congress Officers

K. Esser – President of the Congress
W. Greuter – Secretary General
H. Sukopp – Treasurer
Brigitte Zimmer – Scientific Secretary
H.-D. Behnke W. Haupt O. Kandler

Honorary Congress Officers

F. A. Stafleu – Honorary President

Honorary Vice-Presidents:

K. Faegri, Norway
M. Furuya, Japan
E. I. Gabrielian, Soviet Union
N. Grobbelaar, South Africa
J. Heslop-Harrison, United Kingdom
C. C. Heyn, Israel
J. Kornas, Poland
A. Lang, United States of America
G. Melchers, Federal Republic of Germany
F.-H. Wang, People's Republic of China
J. H. Weil, France
D. von Wettstein, Denmark

Past Presidents:

R. Robertson A. Takhtajan

Programme Committees

1. Metabolic Botany

O. Kandler – Convener
E. Elstner – Secretary
E. Kessler
U. Lüttge
H. Marschner
P. Matile
B. Parthier
W. Tanner
A. Trebst
M. H. Zenk
H. Ziegler

2. Developmental Botany

W. Haupt, Convener
N. Amrhein, Secretary
K. Hahlbrock
H. Meinhardt
H. Mohr
W. Nultsch
A. Sievers
M. H. Weisenseel

3. Genetics and Plant Breeding

K. Esser – Convener
U. Stahl – Secretary
R. Herrmann
W. Nagl
G. Röbbelen
G. Wenzel

4. Structural Botany

H.-D. Behnke – Convener
P. Leins – Secretary
W. Hagemann
R. Kollmann
W. Liese
E. Schnepf
P. Sitte

5. Systematic and Evolutionary Botany

W. Greuter – Convener
U. Jensen – Secretary
A. Bresinsky
F. Ehrendorfer
W. Frey
L. Kies
K. U. Kramer
J. Poelt
F. Schaarschmidt
J. Stöckigt
G. Wagenitz

6. Environmental Botany

H. Sukopp – Convener
R. Bornkamm – Secretary
H. Haeupler
W. Holzner
F. Klötzli
G. Lang
O. L. Lange
M. Runge
Lore Steubing
Otti Wilmanns

XIV INTERNATIONAL BOTANICAL CONGRESS – BERLIN (WEST)

Special Committees of the Congress

Film Committee

Brigitte Zimmer (Convener), B. Hock (Secretary), W. Dewitz, S. Franz, H.-K. Galle, G. Werz

Exhibitions Committee

H.-W. Lack (Convener), T. Brandis, W.-D. Dube, A. Germann, Eva Potztal, H. Sontag, W. Stegmann

Excursions Committee

Brigitte Zimmer (Convener), W. Greuter (Secretary), W. Barthlott, R. Y. Berg, R. Bornkamm, K. Browicz, A. Charpin, Viera Feráková, C. Gómez-Campo, H. Kürschner, E. Landolt, K. Larsen, Maria Lawrynowicz, E. Mayer, G. Moggi, H. Niklfeld, D. Phitos, S. E. Snogerup, Y. Vasari

Committee for the History of Botany

H. Lorenzen (Convener), H.-J. Küster, W. Plarre, H. Scholz, H. Sukopp

Bureau of Nomenclature

F. A. Stafleu (President), W. Greuter (Rapporteur-Général), J. McNeill (Vice-Rapporteur), P. Hiepko (Recorder)

The Task Force in Berlin

Scientific Congress Secretariat

W. Greuter, Brigitte Zimmer, Rosemarie Ziegler, Suzyon Wandrey

Poster Area Manager: J. Hinrichs

Slide Check Coordinators: D. Höner, T. Kersten

Grant Accounting: U. Brühe, Ilse Kowalewsky, Christel Maubach

EDP Exhibition: P. Schallock, W. Schmitt-Rennekamp, B. Hein

Botanical Tours Coordinator: R. Böcker

ICC Congress Manager: Franziska Heiduk

Congress Bureau and Registration Management: Ingrid Rexrodt, R. Mikisch

Planning and Coordination of the Opening and Closing Ceremonies, Congress Dinner and Press Information: K. Esser

Daily Bulletin Editor: H.-D. Behnke

The Congress was placed under the general Motto: "Forests of the World"

XIV INTERNATIONAL BOTANICAL CONGRESS – BERLIN (WEST)

Sponsors and Donors

Sponsors

Berlin, Senator für Wirtschaft und Arbeit
Berlin, Senator für Wissenschaft und Forschung
Bundesministerium für Forschung und Technologie
Bundesministerium für Innerdeutsche Beziehungen
Bundesministerium für Umwelt, Naturschutz & Reaktorsicherheit
Deutsche Botanische Gesellschaft
Deutsche Forschungsgemeinschaft
Deutsche Stiftung für Internationale Entwicklung
Deutscher Akademischer Austauschdienst
International Phycological Society
International Union of Biological Sciences
KLM Royal Dutch Airlines
Krupp von Bohlen und Halbach-Stiftung
Merck'sche Gesellschaft für Kunst und Wissenschaft
Schloßstraße, Arbeitsgemeinschaft
United Nations Development Programme
United Nations Educational, Scientific & Cultural Organization
United Nations Environment Programme
Vereinigung für Angewandte Botanik

Donors

Bradford Exchange, Frankfurt
Gelsenwasser, Gelsenkirchen
Schering, Berlin
Johnson & Johnson, Austin
Schöller Lebensmittel, Nürnberg
Sparkasse der Stadt Berlin West
Hoechst, Frankfurt
Badische Anilin- und Sodafabrik, Ludswigshafen
Daimler-Benz, Stuttgart
Brauerei Schultheiß, Berlin
Bayer, Leverkusen
Berliner Commerzbank, Berlin
Nicolaische Verlagsbuchhandlung Beuermann, Berlin
Boehringer Ingelheim International
Boehringer Mannheim
Franz Enning, Recklinghausen
Gustav Fischer Verlag, Stuttgart
Heidelberg Instruments
Madaus, Köln
Jung, Nußloch
Balzers Hochvakuum, Wiesbaden
Coca Cola, Berlin

PART I

A REPORT OF THE CONGRESS

by Werner Greuter and Brigitte Zimmer

Introduction: Preparing the XIV IBC

The International Botanical Congresses are an institution with a venerable tradition. A first series of twenty such Congresses, which were partly botanical and partly horticultural, took place between 1864 and 1892 in various European cities (details are given by F. A. Stafleu in the Proceedings of the XI IBC). The 19th Century Congresses, which were held almost annually between 1864 and 1885, came to an end when the 21st of them, being the first one to be organized outside Europe (at Madison in 1893), declared itself to be non-international.

The present series of International Botanical Congresses started in Vienna in 1900 and had a normal periodicity of five years – severely disrupted by the two World Wars – up to 1969, and of six years since that date. As decided by the VII IBC in Stockholm (1950), Congresses have, from the VIII IBC in Paris onward, been alternating in their venue between Europe and extra-European countries (some data on the previous International Botanical Congresses can be found under the next heading, "Statistics of the Congress").

International Botanical Congresses have long become the major global event in plant sciences. They relate to all groups of plants and to all scientific disciplines that are partly or totally concerned with the study of plants, in the field of both pure and applied research. By tradition, the rules of botanical nomenclature are discussed and updated during a meeting of the Congress's Nomenclature Section that takes place in the days preceding the Congress proper.

At the final Plenary Session of the XIII IBC in Sydney in 1981, the unanimous decision was taken to accept the invitation, by the German Botanical Society and the Governing Mayor of Berlin, to hold the next Congress in Berlin (West). The XIV IBC therefore took place in Berlin's International Congress Centre from July 24 to August 1, 1987. This was the first of the 20th Century IBCs to be held in Germany.

On November 14, 1981, the Council of the German Botanical Society appointed an organizing committee of seven, to plan and carry out the Congress. On that same date, the seven organizers (five of whom had previously constituted the German Botanical Society's special task force that had prepared the invitation) decided to incorporate themselves as an association under German law and henceforth acted independently to achieve their mandate. This small team, which remained unaltered throughout the six years of its functioning, kept full control of all organizational Congress matters. The German Botanical Society as the inviting party was present through a delegate at all important meetings of the organizing committee, and also sanctioned the nominations of the members of the scientific programme committees. At its 1984 general assembly in Vienna, the GBS officially designated K. Esser, the chairman of the organizing committee, as the Congress President.

The secretary of the organizing committee, W. Greuter, immediately undertook to set up a congress secretariat at the Botanical Garden and Botanical Museum Berlin-Dahlem. In September 1982, a second member of the organizing committee, Brigitte Zimmer, joined the Museum's scientific staff and considerably reinforced the secretariat. From the beginning of 1983, a half-time position of secretary was available for Congress purposes which, after some fluctuation, was held by Mrs. Rosemarie Ziegler from October 1983 onward. Two years later, a second half-time secretary, Mrs. Suzyon Wandrey, completed the number of those who were most immediately active in the implementation of the Congress. This was still a very small group of people to bear a workload that became unexpectedly heavy in the two or three years preceding the Congress, but thanks to good teamwork, to a dedicated effort by everyone and, also, to modern equipment (word processor and later personal computer) its task could be successfully completed.

After thorough negociations, the organizing committee signed a contract with a professional congress agency, DER Congress, to assist in the more technical aspects of congress organization. Most prominent among these were the accounting and, later on, the registration of participants as well as the booking of accommodation. DER Congress provided advance funding on a credit base for the whole pre-Congress period – indeed a far from negligible assistance.

Finance was an early if not a lasting concern of the organizers. The first drafts of a global Congress budget date back to January 1982. A complete and detailed preliminary budget was finalized in March the same year, and was submitted along with a grant request to the Senate of Berlin. This resulted in a most generous grant promise which, as early as May 1982, enabled to establish the Congress on a sound financial basis. The budget had, of course, to be revised and updated annually, but its general structure remained unaltered throughout.

A preliminary booking of the International Congress Centre in Berlin for the dates that were later to become the actual Congress dates had been made already prior to the XIII IBC. Both the booking and the dates were confirmed in a convention between the organizing committee and the society that is running the Congress Centre (Ausstellungs-Messe-Kongress GmbH, AMK Berlin), which was signed in July 1982.

Perhaps not surprisingly for those who have an understanding of human psychology, the most disputed issue during the early months of organizational work was the choice of the Congress emblem. It took twelve months of deliberations and the circulation of several sets of drafts to reach a decision. In December 1982 the balance finally tilted, by a narrow majority, in favour of Mrs. Haupt's design of *Hepatica nobilis*.

The general basis of the scientific Congress programme, and the main principles underlying it, were soon firmly established. The organizers felt that the justification of such a big international congress was primarily in the fact that it permitted an open exchange between specialized disciplines, so that subjects at the crossroads between the traditional fields of research were to be given prominence. In order to achieve this, the numerous specialized sections of former congresses were abolished and replaced by six major programme divisions. In addition, every effort was made to promote fields on the borderlines of those divisions. Some subjects of very general scope were not left to the individual programme divisions, but were placed under the direct responsibility of the whole organizing committee. Such was the case of the history and philosophy of botany, of the presentation (if possible by demonstration) of new methods of research, and in particular of the ever more important field of electronic data processing techniques and their applications in plant biology.

Without regrets, the organizers dismissed the traditional concept of contributed paper sessions. Contributions could be presented either orally in a number of symposia on well defined subjects, or in the form of poster demonstrations. Oral contributions to symposia were to be in part upon direct invitation, but for the other part they were to be selected from among offered contributions. Such selection should not, however, discriminate against the poster exhibits which were to be placed in the very centre of the Congress, being in many respects, and for many subjects, the most appropriate and most modern form of presentation of scientific results.

The responsibility for the scientific programme was confided to six programme committees, each convened by one of the organizing committee members and appointed, on his proposal, by the German Botanical Society. (The seventh organizer undertook to convene the excursion committee and the film committee.) The programme committees were appointed in September 1982 and held their constituant meeting in Berlin in June 1983, when, by an overwhelming majority, they sanctioned the organizers' suggestion that English was to serve as the official Congress language. A similar decision had been taken for the XIII IBC in Sydney, but choosing English as the single official language was nevertheless a noteworthy novelty for a Congress held in a non-English-speaking country.

The programme committees set to work immediately, in conformity with detailed guidelines that had been prepared to their intent. By the time they gathered in a second common meeting, in September 1984 in Vienna, the first Congress circular had been distributed and provisional lists of symposia had been set up. The first circular, together with a preliminary announcement for display, had been widely circulated by direct mailing and through scientific societies, mostly between November 1983 and February 1984. The requested feedback by means of a preliminary reply form gave valuable indications, among others, on the interests of the prospective participants. The number of symposia allotted to each programme committee (and, concurrently, the grant money share of which it could dispose) was established for one half on the basis of the show of interest evidenced by the preliminary replies. The corresponding figures are interesting enough: In mid-June 1984, when c. 3700 replies had been received, Systematic and Evolutionary Botany had taken the lead (30.5% of the expressed preferences), followed in order by Environmental Botany (24.5%), Metabolic Botany (13.8%), Developmental Botany (13.2%), Structural Botany (10.0%) and Genetics and Plant Breeding (8.0%).

The winter 1984/85 was largely devoted to the implementation of what the programme committees had decided at the Vienna meeting. The list of scientific symposia was finalized, the organizers of each symposium – one usually resident in the German language area, the second from abroad – were selected and appointed, and the procedures for organizing the symposia and inviting the speakers were defined. By mid-1985, when the final numbering of the symposia was agreed, the first half of the speakers (those directly selected by the symposium organizers) had been invited. At the same date, a final selection was made of the subjects of the 24 general lectures.

The mailing of the second Congress circular to the c. 9000 persons and institutions who had expressed their interest took place in October 1985. It included a call, to all but the previously invited speakers, to offer a scientific contribution (symposium paper or poster) for presentation at the Congress. (Since one could easily foresee that a majority at least of those attending would wish or need to present a scientific contribution of their own, the organizers had decided at an early stage that each participant would be allowed to present one and only one such contribution – a rule that was adhered to without a single exception!) The deadline for offering contributions had been set on June 30, 1986, and in the weeks preceding and following that date the first big tidal wave of mail swept over the Berlin secretariat. By mid-July, when the organizing committee met in order to assign each offered contribution to the appropriate programme division, no less than 1589 prospective participants had sent in such an offer (621 of them expressing preference for a poster presentation), to which the 954 who had previously been invited to present a paper on a given subject are to be added.

The following months were a busy time for all involved, including the programme committees and symposium organizers. The latter had to select the additional speakers from among the offered contributions, whereas the former had to define the subjects of the poster sessions. It was a difficult but rewarding task to group the posters into thematically coherent groups of a handy size (normally between five and ten) so that they could be presented in a satisfactory and attractive manner. An additional duty was the selection and invitation of competent and able persons, to serve as moderators of the poster sessions. From the autumn 1986 until the end of January 1987, invitations to scientific contributors – first to additional speakers then to poster authors – were leaving the Berlin secretariat in a steady flow. This operation was completed scarcely more than a fortnight before the end of the regular registration period, on February 15, 1987!

Meanwhile, the third Congress circular with the registration and abstract forms had been mailed to all prospective participants in December 1986. The deadline for submission of the offset-ready abstract text had deliberately been set very late, on 31 May 1987, so as to enable the authors to provide informative abstracts, including their latest results and faithfully reflecting the contents of their contribution. Even so a number of abstracts were received after the deadline, and a major effort was necessary to enable their inclusion in the printed volume. The actual cut-off date was June 29, but after that date – and partly even during the Congress itself – 74 supplementary abstracts were submitted which were included as an appendix in the post-Congress library edition of the abstract volume.

The first half of April 1987 was devoted to an operation that can be compared to a gigantic jigsaw puzzle. Hundreds of symposia and poster sessions had to be fitted together into a coherent pattern of time and space within the narrow boundaries of which the Congress could dispose. An invaluable statistical material was available for that purpose, consisting of the stated programme wishes of c. 2100 prospective participants that had been numerically digested with the aid of the computer. Tailoring the programme to the needs and wishes of the audience was, however, brainwork in which the computer could not assist. Eventually, the 14 lecture halls at our disposal in the ICC were made to fit the programme as tightly as a glove. We were pleased to note that, as a result of this tedious exercise, little if any complaints about thematic overlaps and clashes were received.

The scientific contributions, with their authors and titles, had been recorded in the form of electronic files all along. With the symposia and poster sessions gradually acquiring their definite shape and contents, these files were refined and updated and provided the basis on which the symposium organizers and session moderators were regularly informed and asked to comment. As the Congress approached, the files underwent a final checking, editing and indexing process which resulted directly in the printed Congress programme. The cut-off date for changes that were incorporated in the programme volume was barely 4-5 weeks ahead of the opening of the registration counters. Programme modifications notified after that date were announced in the daily Congress Bulletins.

In retrospect it is hard to believe that, eventually, everything ran smoothly and was finished on time. During the six months preceding the Congress, our schedule was so tightly packed that any major unforeseen event might have threatened the whole enterprise. Yet some such events happened, as they always do – not fortunately in the form of a major computer break-down in a crucial moment! A nice example is the disappearance of the first 400 printed copies of the programme book, which were either mis-directed or stolen, so that a new run-off had to be ordered at a very short notice. It was delivered at the registration counter half an hour before the last volume of the previous stock had gone.

We have stated in our introduction to the programme book to how many persons and institutions we owe thanks for their help and assistance in preparing the XIV International Botanical Congress. It is appropriate here to add one last, major acknowledgement: We thank a propitious fate that has kept us going and has spared us from major catastrophies.

Statistics of the Congress

Attendance

Participation in the XIV International Botanical Congress was largely in excess of the budgeted figures, although it might have been considerably larger but for the drastic decline of the dollar rates that had taken place in 1986. 3546 scientific members from 79 different countries (Table 1) were in attendance, of which 702 were registered as students. The number of accompanying persons was 449. To these figures 409 one-day registrants may be added, of which 243 were students. A list of scientific members, with addresses, is enclosed in microform as an annex to the present volume.

The perseverant efforts of the organizers to attract young people to the Congress were not vain. The Berlin Congress was the youngest ever, with close to 20% student members. A special grant scheme for young botanists as well as generous financial arrangements (low Congress fee in spite of full membership rights, later deadline for normal registration) made this Congress particularly attractive for post- and undergraduate students, from Germany as well as from abroad.

Special efforts were also made to bring botanists from the Eastern European countries and the German Democratic Republic to Berlin. The results were noticeable, although the figures are still very much short of the actual number of interested colleagues from those countries. The sizeable delegations from, e.g., Poland and the GDR support the claim that Berlin is ideally suited as a meeting point between East and West.

Table 1: Participants of the XIV International Botanical Congress 1987

(number of scientific members, including students, in attendance + number of accompanying persons)

Argentina	9	+	1		Korea	10			
Australia	127	+	20		Kuwait	1			
Austria	67	+	2		Lesotho	1			
Barbados	1				Liechtenstein	1			
Belgium	46	+	1		Luxemburg	1			
Bolivia	1				Malaysia	3			
Botswana	1	+	1		Mexico	5			
Brazil	23	+	2		Morocco	1			
Bulgaria	4				Mozambique	2			
Canada	114	+	21		Nepal	1			
Chile	3				Netherlands	115	+	15	
China (P.R.)	59	+	1		New Zealand	15			
China (Taiwan)	11	+	1		Nigeria	3			
Cyprus	1	+	1		Norway	12	+	2	
Czechoslowakia	9				Pakistan	3			
Denmark	36	+	2		Panama	3	+	1	
Ecuador	1				Papua New Guinea	1			
Egypt	10	+	1		Paraguay	1			
Ethiopia	3				Peru	1			
Finland	38	+	2		Philippines	7			
France	121	+	21		Poland	50	+	2	
Germany, GDR	42				Portugal	10			
Germany, FRG	935	+	86		Qatar	1			
Great Britain	235	+	13		Singapore	2	+	1	
Greece	8	+	1		South Africa	30	+	3	
Guinea	1				Soviet Union	20			
Honduras	1				Spain	38	+	3	
Hongkong	1				Sri Lanka	3			
Hungary	25				Sweden	92	+	15	
Iceland	2				Switzerland	102	+	11	
India	101	+	4		Tanzania	2	+	1	
Indonesia	1				Thailand	3			
Iran	4	+	1		Turkey	2			
Iraq	1				Uganda	1			
Ireland	6	+	2		USA	666	+	155	
Israel	41	+	6		Venezuela	1			
Italy	95	+	26		Yugoslavia	9	+	2	
Jamaica	1				Zaire	1			
Japan	135	+	22		Zimbabwe	3			
Kenya	3								
					total	3546	+	449	

Table 2: The 20th Century International Botanical Congresses

IBC	Place	Year	Presidents	Secretaries	Scientific Attend-ance	Foreign Scientific Members	Student Members (if known)
I	Paris	1900	J. de Seynes	E. Perrot [1]	233	64	
II	Vienna	1905	R. von Wettstein J. Wiesner	A. Zahlbruckner [1]	504	222 [6]	
III	Bruxelles	1910	Baron de Moreau Th. Durand	E. de Wildeman [1] M. G. Hegh	305	230	
IV	Ithaca	1926	L. H. Bailey	B. M. Duggar	912	93 [7]	
V	Cambridge	1930	A. C. Seward	F. T. Brooks T. F. Chipps	1175	688	
VI	Amsterdam	1935	J. C. Schoute	M. J. Sirks [1] H. J. Lam	963	711	
VII	Stockholm	1950	C. Skottsberg	E. Åberg [1] A. Nygren	1521	1243	
VIII	Paris	1954	R. Heim	P. Chouard [1] R. de Vilmorin	1805	1377	135
IX	Montreal	1959	W. P. Thompson	C. Frankton [1] R. Pomerleau M. L. Berlyn	2124	1640	
X	Edinburgh	1964	H. Godwin G. Taylor [2]	H. Fletcher [1] G. E. Fogg A. Brook	2583	1646	56 [8]
XI	Seattle	1969	K. Thimann K. B. Raper [2]	R. S. Cowan [1] G. Fischer [3]	3861	1220	521
XII	Leningrad	1975	A. L. Takhtajan	O. V. Zalensky [1] N. S. Snigirevskaya [4]	3688	1870	
XIII	Sydney	1981	R. Robertson	W. J. Cram	2798	1568	487
XIV	Berlin	1986	K. Esser F. A. Stafleu [5]	W. Greuter [1] B. Zimmer [4]	3546	2611	702

1) secretary general
2) chairman organizing committee
3) executive director
4) scientific secretary
5) honorary president
6) from outside Austria and Hungary
7) from outside the US and Canada
8) undergraduates only

If there was a major deficit in Congress participation, it was the meagre attendance of botanists from developing countries. The needs of prospective participants from those areas were disproportionate when compared to the available funds – which proved, moreover, exceedingly difficult to obtain. It is to be hoped that, in the future, international solidarity will materialize in a more generous support of scientists from the Third World.

We were interested to compare the Berlin figures to those of previous International Botanical Congresses. We have therefore verified and updated the overview that had been provided in tabular form by F. A. Stafleu in the Proceedings of the Seattle Congress (Table 2). In so doing, we have recalculated the earlier membership figures, to make them cover only scientific members in attendance (figures previously quoted for some Congresses did include accompanying persons and non-attending members as well). We have also quoted separately the number of participants from outside the host country and, when it was known, the number of student members.

The figures in Table 2 show that the Berlin Congress was the third in size of this century's International Botanical Congresses, following the Seattle and the Leningrad Congress. It was by far the largest in terms of foreign participation. In other words, International Botanical Congresses have by no means lost their attractiveness for botanists throughout the world, students in particular. The Berlin Congress has proved that the IBCs are as vital as ever, and that their future is secured in the long run under the only proviso that persons and institutions can be found who are willing to organize them.

Scientific programme

The *Congress Timetable* is graphically represented on page 12 of the programme book. The main scientific features of the Congress are briefly outlined below.

The *Nomenclature Section* of the Congress deliberated as usual during four and a half days prior to the Congress proper. 336 proposals to modify the "International Code of Botanical Nomenclature" had been placed before the Section, whose commented synopsis had been previously published. The results of the preliminary mail ballot on these proposals had been tabulated by J. McNeill, Vice-Rapporteur, and were made available to all members of the Section. The deliberations were recorded on tape and will be published separately. All registered members of the Congress were entitled to enroll themselves as voting members of the Nomenclature Section, whose actual membership was 157 and included official delegates of 115 botanical research institutes and collections from all over the world.

Section members were invited to take part in the inauguration of the new herbarium and library wing of the Botanical Museum Berlin-Dahlem, with subsequent reception, on Wednesday, July 23. Those familiar with history will note the remarkable analogy with the inauguration of the Botanical Museum Berlin-Dahlem by Engler in 1910, that was attended by a delegation of 70 botanists on their way home from the III International Botanical Congress in Brussels.

There were 24 *General Lectures* on the Programme, four on each full Congress day (two in sequence and two in parallel). Following the example of the Sydney Congress, they were all held in the main lecture theatres during the lunchtime breaks. Except for the Congress films, some society meetings and possible individual needs for food uptake, there was no competing event at the same time. The subjects of the general lectures had been chosen most carefully on the basis of proposals made by the six programme committees (3 by each of them, and an additional 3 for the programme divisions 5 and 6).

230 *Scientific Symposia* and round-table discussions were held during 7 morning sessions, 6 afternoon sessions and 4 late evening sessions, 13 or 14 of them running in parallel all along. Each normally comprised 3 hours of oral presentations and discussions plus half an hour's break. The number of symposium speakers on record in the programme book is 1678. Supposing that all symposia were exactly on schedule, their total duration was 674 hours, and each speaker could on average dispose of 24 minutes (minus the time for discussion and for introductory and closing remarks from the chair).

The prominence given to *Poster Presentations* was perhaps the most successful innovative feature of the Berlin Congress. For the first time, posters were shown in the very centre of a congress, on everyone's way, and for the first time also, the poster sessions were moderated by renowned scientists specialized in the field. The quality of the posters was, on average, quite remarkable and justified the interest that the audience took in them.

Figure 1: A view of the posters in the main lobby ("Hauptfoyer") of the ICC, where posters belonging to Metabolic and Developmental Botany were on display.

Figure 2: Presentation of a poster during one of the sessions on Developmental Botany to an attentive audience which includes the convener of that programme division, W. Haupt (first row on the left).

Posters as such are not of course an invention of the Berlin Congress. To quote from the second circular of the Sydney Congress: "The Organizing Committee wishes to encourage authors of contributed papers to present their material as a poster. The poster session was first introduced at scientific meetings about 10 years ago: in those early days the poster sessions were small and often tucked away in rooms some distance from the main meeting area. This is not the intention for the XIII International Botanical Congress. Instead, poster sessions will be given prominence ..."

The origin of posters in IBCs is, in fact, much older. Their invention is probably to be credited to the IX IBC in Montreal, where they were part of what was then newly promoted under the designation "demonstration papers". Again to quote from the corresponding second circular: "The demonstration method permits personal contact between the author and his audience. The audience is able to examine material closely and to ask questions freely. Most of the subjects dealt with by the Congress can be presented as demonstrations provided there is careful, imaginative planning. Wherever possible members of the Congress are urged to present their papers as demonstrations rather than as lectures. Demonstrations will be set up in laboratories: bench tops will be used for specimens, microscopes or other equipment, and vertical panels for charts, diagrams and photographs. The author will be asked to be in attendance at his demonstration for a period of either two or three hours. Demonstrations on related subjects will be grouped together ..." These so-called demonstrations have steadily increased in numbers and importance from Montreal in 1959 (75 presentations) to Leningrad in 1975 (326 presentations) when the term "poster" was first used for a particular kind of demonstration. In Sydney in 1981 the number of posters again showed a marked increase, to reach 520.

The programme of the Berlin Congress lists no less than 1191 posters, which means that more than 40% of the scientific contributions were presented in this form. In Metabolic Botany, there were even significantly more posters than oral papers (58% against 42%). The posters were grouped in 123 poster sessions (up to 6 being held in parallel) for which a total time of 156 hours was available according to schedule (a far from rigid schedule, however, since opportunity was provided to continue discussions either formally or informally beyond the imparted time limit). The average time available for each poster presentation was 8 minutes.

Table 3 provides a *numerical overview* of the scientific programme items that were placed under the care of the six programme committees and (for general subjects, No. 7) of the organizing committee. The general lectures, symposia and poster sessions, which constituted the core of the scientific Congress programme, lasted 815 hours altogether, corresponding to a time span of 34 days around the clock. More than 80% of the registered Congress members presented a scientific contribution of their own.

Table 3: A quantitative digest of the Scientific Congress Programme
(Figures are based on the Programme and Abstract Books)

Progr. Division	*General Lectures*	*Scientific Symposia*	*Symposium Papers*	*Poster Sessions*	*Posters*	*Total Contributions*
1	3	31	219	31	311	533
2	3	28	168	18	155	326
3	3	30	192	11	138	333
4	3	32	255	7	72	330
5	6	50	382	24	213	601
6	6	51	403	28	269	678
7	–	8	59	4	33	92
Total	24	230	1678	123	1191	2893
Published Abstracts	21	–	1514	–	1074	2609

43 *Scientific Films* were grouped under 10 different headings and shown during 6 sessions totalling slightly less than 20 hours. The film sessions were moderated, and the audience was given opportunity to ask questions to the author of the film or to his representative. In addition, a popular film on the Berlin Botanical Garden and a sequence of educational films placed under the Congress motto "Forests of the World" were shown during each of the lunchtime breaks.

A most successful innovation of the Berlin Congress were the *Special Interest Group Meetings,* small informal meetings organized independently by a single person or a group of persons, for which the Congress merely provided a room and programme space. There were 37 such meetings, of very diverse nature and attendance, including one that was announced during the Congress in one of the daily bulletins (by G. L. Stebbins on "The future of biosystematics in the molecular age"). Up to four such meetings were running in parallel simultaneously with the symposia and poster sessions.

Botanical Software Demonstrations took place on 5 days (July 26-27 and 29-31) from 11 to 15 hrs in a special room equipped with personal computers and modems, also granting access to a selection of database services (Vicieae Database, Commonwealth Agricultural Bureau Database). Details were announced in the daily Congress Bulletins.

Society Meetings are by tradition an important feature of International Botanical Congresses. During the Berlin Congress, no less than 26 international scientific associations and research groups held one or more meetings of their members and/or governing bodies. The Congress provided them room facilities, space in the printed programme and daily bulletin, and opportunities of posting information materials on apposite boards in the resting area.

The "Buchhandlung Ziegan" organized a sizeable and representative *Book Exhibit* at the Congress, in a centrally situated area of the ICC. 80 exhibitors (61 science publishers and 19 non-commercial institutions or societies) were represented. The catalogue published for this exhibit lists 822 titles, but adding those that were received after the cut-off date the actual number was, presumably, in excess of 900.

5 *Satellite Meetings* took place just before or just after the Congress, dealing with special topics. They were fully independent, organizationally, from the International Botanical Congress, but they were announced in the second and third Congress circulars so as to enable participants to make early arrangements for attending. These additional meetings were: The 5th International Conference on Mediterranean-Climate Ecosystems in Montpellier (France) on July 15-21; a Bryological Methods Workshop in Mainz (FRG) on July 17-23; the 10th International Meeting and Conference of the International Association of Botanic Gardens in Frankfurt a.M. (FRG) on August 2-7; the 2nd International Chrysophyte Symposium in Berlin (West) on August 3-5; and a Symposium on the Expanding Realm of Yeast-Like Fungi in Amersfoort (The Netherlands) on August 3-6.

An Outline of the Scientific Congress Programme

General Lectures

1 J. Coombs – Bioenergetics and plant productivity.
2 D. Boulter – Biochemical evolution in plants.
3 G. Schatz – Formation of organelles and the organization of the Eukaryotic cytoplasm.
4 R. Goldberg – Molecular biology of seed development.
5 E. J. Klekowski – Mechanisms that maintain the genetic integrity of plants.
6 P. H. Quail – Molecular biology of phytochrome.
7 J. Webster – Botany and mycology.
8 D. von Wettstein – Botany and biotechnology.
9 L. Bogorad – Botany and genetic engineering.
10 T. Cavalier-Smith – Eukaryote cell evolution.
11 P. K. Hepler – Calcium and development.
12 T. Sachs – Internal controls of plant morphogenesis.
13 H. W. Lack – Berlin and the world of botany.
14 V. H. Heywood – Rarity, a privilege and a threat.
15 N. Myers – The plant world of the Tropics: green hell or paradise lost?
16 W. G. Chaloner – Early land plants – the saga of a great conquest.
17 F. Ehrendorfer – Stability versus change, or how to explain evolution.
18 D. B. McKey – Promising new directions in plant/ant interactions.
19 J. White – Demography of plants.
20 O. L. Lange – Ecophysiology of photosynthesis.
21 M. M. Caldwell – Plant root systems and competition.
22 P. D. Manion – Pollution and forest ecosystems.
23 H. Ziegler – Deterioration of forests in Central Europe.
24 D. Mueller-Dombois – Canopy dieback and ecosystem processes in the Pacific area.

Programme Division 1: Metabolic Botany

Symposia

1-01 Biosynthetic assimilations (CO_2, NO_3^-, $SO_4^{2=}$, etc.) (Org. A. Trebst)
1-02 Biochemistry of ecological adaptations in photosynthesis (Org. C. B. Osmond, M. Kluge)

1-03 Plant bioenergetics (Org. B. A. Melandri, G. Hauska)
1-04 Nucleic acid metabolism (Org. B. Parthier, H. Kössel)
1-05 Nitrogen nutrition (Org. A. Quispel, H. Bothe)
1-06 Plant lipids and steroids (Org. P. Benveniste, E. Heinz)
1-07 Secondary plant products and biotechnology (Org. M. Tabata, M. H. Zenk)
1-08 Biosynthesis and functions of glycoproteins (Org. M. Chrispeels, W. Tanner)
1-09 Metabolic regulations (Org. B. Buchanan, H. W. Heldt)
1-10 Calcium in plant metabolism (Org. A. M. Boudet, H. Kauss)
1-11 Intracellular transport (Org. G. Schatz, U. Heber)
1-12 Molecular mechanisms of membrane transport (Org. A. Goffeau, E. Komor)
1-13 Biochemistry of host parasite relationships (Org. J. A. Bailey, J. Ebel)
1-14 Mode of action of herbicides and fungicides (Org. C. J. Arntzen, J. Dekker, P. Böger)
1-16 Stomatal action (Org. W. H. Outlaw, H. Schnabl)
1-17 Gland physiology (including carnivorous plants) (Org. Y. Heslop-Harrison, G. Heinrich)
1-18 Translocation of assimilates (Org. J. S. Pate, W. Eschrich)
1-19 Physiology of fruits (Org. C. F. Jenner, F. Bangerth, F.)
1-20 Phytochemical aspects of coevolution between plants and animals (Org. J. B. Harborne, U. Jensen)
1-21 The evolution and ecophysiology of vascular plants as epiphytes (Org. D. H. Benzing, U. Lüttge)
1-22a Physiology of halophytes and salt tolerance I: Mechanisms of salt tolerance (Org. R. L. Jefferies, R. Albert)
1-22b Physiology of halophytes and salt tolerance II: The biology and utilization of halophytes (Org. R. L. Jefferies, R. Albert)
1-23 Physiology of aquatic macrophytes (Org. J. A. Raven, A. Melzer)
1-24 Nutrients and rhizosphere (Org. P. B. Tinker, A. Jungk)
1-25a Root responses to environmental factors I (Org. D. T. Clarkson, H. Marschner)
1-25b Root responses to environmental factors II (Org. D. T. Clarkson, H. Marschner)
1-26 Plant responses to air pollution (Org. P. R. Miller, H. Ziegler)
1-27 New plants for food and resources (Org. D. O. Hall, G. Röbbelen)

1-28 New methods in histochemistry and microlocalization (Org. A. Läuchli, R. Hampp)
1-29 System analysis in plant physiology (Org. W. J. Cram, K. Brinkmann)
1-30 Bioenergetics of Cyanobacteria (round-table conference) (Org. W. Lockau)

Poster Sessions

1-101 Biosynthetic assimilations (CO_2, NO_3^-, $SO_4^{2=}$, etc.) (Moderated by H. W. Heldt)
1-102a Biochemistry of ecological adaptations in photosynthesis I (Moderated by M. Kluge)
1-102b Biochemistry of ecological adaptations in photosynthesis II (Moderated by B. Osmond)
1-103 Plant bioenergetics (Moderated by G. Hauska)
1-105 Nitrogen nutrition (Moderated by H. Bothe)
1-106 Plant lipids and steroids (Moderated by P. Benveniste)
1-107a Secondary plant products I (Moderated by M. H. Zenk)
1-107b Secondary plant products II (Moderated by M. Tabata)
1-107c Secondary plant products III (Moderated by F. Constabel)
1-109a Metabolism and its regulation I (Moderated by H. W. Heldt)
1-109b Metabolism and its regulation II (Moderated by T. ap Rees)
1-109c Metabolism and its regulation III (Moderated by R. Scheibe)
1-109d Metabolism and its regulation IV (Moderated by H. W. Heldt)
1-110 Calcium in plant physiology (Moderated by H. Kauss, M. Tazawa)
1-112a Membrane transport I (Moderated by U. Heber, E. Komor)
1-112b Membrane transport II (Moderated by U. Heber, E. Komor)
1-113a Host-parasite relationships I (Moderated by J. Ebel, H. J. Reisener)
1-113b Host-parasite relationships II (Moderatored by J. Ebel, H. J. Reisener)
1-113c Host-parasite relationships III (Moderated by J. Ebel, H. J. Reisener)
1-114 Mode of action of herbicides and fungicides (Moderated by P. Böger)
1-118 Translocation of assimilates (Moderated by W. Eschrich)
1-122a Physiology of halophytes and salt tolerance I: General physiology of salt tolerance, with emphasis on crop plants (Moderated by R. G. Wyn Jones)

1-122b Physiology of halophytes and salt tolerance II: Ecophysiology and microalgae (Moderated by R. G. Wyn Jones)
1-123 Physiology of aquatic macrophytes (Moderated by A. Melzer)
1-125 Root responses to environmental factors (Moderated by S. A. Barber, D. Clarkson)
1-126a Plant responses to air pollution I: Cultivated plants (Moderated by T. A. Mansfield)
1-126b Plant responses to air pollution II: forest trees (Moderated by T. A. Mansfield)
1-126c Plant responses to air pollution III: Cryptogams and general topics (Moderated by T. A. Mansfield)
1-130 Physiology of Cyanobacteria (Moderated by W. Lockau)
1-131 Primary reaction in photosynthesis (Moderated by G. Hauska)
1-132 Stomatal action and gland physiology (Moderated by H. Schnabl, G. Heinrich)

Programme Division 2: Developmental Botany

Symposia

2-01 The molecular basis of plant development (Org. R. B. Flavell, G. Feix)
2-02 Significance of cell and tissue culture in developmental research (Org. Z. R. Sung, H. Lörz)
2-04a Spatial and temporal patterns in development I (Org. W. K. Silk, H. Meinhardt)
2-04b Spatial and temporal patterns in development II (incl. round-table) (Org. W. K. Silk, H. Meinhardt)
2-05 Control of polar development by ionic currents (Org. L. F. Jaffe, M. H. Weisenseel)
2-06a Calcium in development and movements I (Org. S. J. Roux, G. Wagner)
2-06b Calcium in development and movements II (incl. round-table) (Org. S. J. Roux, G. Wagner)
2-07 Strategies in plant development (Org. T. Sachs, H. Mohr)
2-08a Photomorphogenesis I (Org. M. Furuya, E. Schäfer)
2-08b Photomorphogenesis II (incl. round-table) (Org. M. Furuya, E. Schäfer)
2-09a Stress factors in development I (Org. F. Larher, B. Hock)
2-09b Stress factors in development II (incl. round-table) (Org. F. Larher, B. Hock)

2-10 Developmental biology of non-vascular plants (Org. W. Jacobs, M. Bopp)

2-11a Control points in the generation cycle of vascular plants (Org. J. Heslop-Harrison, E. Heberle-Bors)

2-11b New approaches to the photoperiodic control of flowering (Org. D. Vince-Prue, G. Bernier)

2-12 Regulation of cell growth (Org. R. Cleland, P. Schopfer)

2-13 Phytohormones in development (Org. H. Kende, E. W. Weiler)

2-14 Movements based on differential flank growth (Org. P. E. Pilet, J. Bruinsma)

2-15a Senescence in higher plants I (Org. W. H. Woolhouse, H. Thomas, P. Matile)

2-15b Senescence in higher plants II (Org. W. H. Woolhouse, H. Thomas)

2-16 Development and germination of seeds and spores (Org. M. Black, H. Schraudolf)

2-17a Structural and molecular basis of cell motility I (Org. P. Satir, M. Melkonian)

2-17b Structural and molecular basis of cell motility II (Round-table conference) (Org. P. Satir, M. Melkonian)

2-18 Intracellular movements (Org. R. Williamson, W. Haupt)

2-19 Control of cell movement by external signals (Org. M. J. Doughty, D. P. Häder)

2-20 Perception and transduction of gravitationl stimuli (Org. M. B. Wilkins, A. Sievers)

2-21 Chemical communication and recognition between cells (Org. R. C. Starr, L. Jaenicke)

2-22 Plant space biology – present and future (round-table conference) (Org. T. K. Scott, D. Volkmann)

Poster Sessions

2-101 Molecular aspects of development (Moderated by G. Feix)

2-102a Cell and tissue cultures I: External factors controlling development (Moderated by P. E. Read)

2-102b Cell and tissue cultures II: Regeneration (Moderated by H. Lörz, D. Durzan)

2-102c Cell and tissue cultures III: Aspects of differentiation (Moderated by A. Kumar)

2-108 Photomorphogenesis (Moderated by L. O. Björn, M. Furuya)

2-109 Stress factors in development (Moderated by K. Dörffling, F. Larher)

2-110 Developmental biology of non-vascular plants (Moderated by M. M. Johri, M. Bopp)
2-111b Regulation of transition to flowering (Moderated by D. Vince-Prue)
2-112 Regulation of cell growth (Moderated by D. J. Cosgrove)
2-113a Phytohormone metabolism (Moderated by P. Schopfer)
2-113b Phytohormone metabolism (Moderated by J. A. D. Zeevaart)
2-113c Phytohormone physiology (Moderated by E. W. Weiler)
2-116a Seed and spore germination I: Light and temperature as regulating factors (Moderated by R. W. Scheuerlein)
2-116b Seed and spore germination II: Role of internal factors (Moderated by M. Delseny)
2-116c Seed and spore germination III: Regulation by chemical signals (Moderated by K.-H. Köhler)
2-120 Perception and transduction of gravitationl stimuli (Moderated by D. Volkmann)
2-131 Cell motility and intracellular movement (Moderated by W. Nultsch)
2-132 Biorhythms in development (Moderated by D. Vince-Prue)

Programme Division 3: Genetics and Plants Breeding

Symposia

3-01 Genetics of nitrogen fixation (Org. P. M. Gresshoff, A. Pühler)
3-02 Biotechnology and agriculture (Org. G. Goma, H. Katinger)
3-03 In vitro recombination (Org. L. E. Fish, U. Kück)
3-04 Vector development in Eukaryotes (Org. P. A. Lemke, U. Stahl)
3-05 Fungal genetics (Org. N. Gunge, F. Meinhardt)
3-06 Genome and gene structure (Org. C. A. Cullis, V. Hemleben)
3-07 Chromatin organization (Org. M. L. Moreno, V. Hemleben)
3-08 Physical aspects of gene regulation (Org. T. Cavalier-Smith, W. Nagl)
3-09 Differential DNA replication (Org. M. J. Olszewska, K.-H. Neumann)
3-10 Mechanisms of karyotype evolution (Org. M. D. Bennett, D. Schweizer)
3-11a The plastome (plastid genome) I (Org. L. Bogorad, R. Herrmann)
3-11b The plastome (plastid genome) II (Org. L. Bogorad, R. Herrmann)

3-12 The chondriome (mitochondrial genome) (Org. C. J. Leaver, P. Tudzynski)
3-13 Viruses, viroids, plasmids, tumorigenesis (Org. L. van Vloten-Doting, R. A. Schilperoort)
3-14a Nucleo-cytoplasmic interactions I (Org. D. von Wettstein, G. Michaelis)
3-14b Nucleo-cytoplasmic interactions II (Org. D. von Wettstein, G. Michaelis)
3-16 Molecular markers in genetics and plant breeding (Org. S. D. Tanksley, G. Wricke)
3-17 Triticale – a new crop (Org. T. Lelley)
3-18 Utilization of non-chromosomal diversity in plant breeding (Org. G. Pelletier, W. Odenbach)
3-19 Genetic resources for future plant breeding (Org. J. T. Williams, G. Röbbelen)
3-20 Breeding of long-living plants (Org. J. V. Possingham, G. Alleweldt)
3-21 Breeding strategies (Org. A. Gallais, P. Ruckenbauer)
3-22 Gene organization (Org. J. R. S. Fincham, D. von Wettstein)
3-23 Somatic recombination (in vitro techniques) (Org. D. Dudits, O. Schieder)
3-24 Resistance against stresses and diseases (Org. P. R. Day, G. Wenzel)
3-25a Early screening procedures in plant breeding (Org. H. Stegemann)
3-25b Tropical and Subtropical crops (Org. M. S. Swaminathan, G. Röbbelen)
3-26 Evolution of sexual dimorphism in plants (Org. K. S. Bawa, H.-R. Gregorius)
3-27 Gene flow (Org. S. K. Jain, K. Wöhrmann)
3-29 Genetics and physiology of secondary metabolism (Org. J. W. Bennett, W. Roos)

Poster Sessions

3-106 Genome and gene structure (Moderated by V. Hemleben)
3-110 Mechanisms of karyotype evolution (Moderated by D. Schweizer)
3-111 The plastome (Moderated by R. Herrmann, J. H. Weil)
3-112 The chondriome (Moderated by G. Michaelis, P. Tudzynski)
3-113 Viruses, viroids, plasmids, tumorigenesis (Moderated by J. Schröder)

3-123a Somatic recombination I (Moderated by O. Schieder, D. Dudits)

3-123b Somatic recombination II (Moderated by D. Dudits)

3-131 Recombination genetics (Moderated by F. Meinhardt, P. Lemke)

3-132 Utilization of natural variation for plant breeding (Moderated by J. T. Williams)

3-133 New methods and crops for plant breeding (Moderated by G. Röbbelen)

3-134 In vitro selection and early screening (Moderated by P. R. Day, H. Stegemann)

Programme Division 4: Structural Botany

Symposia

4-01 Membrane flow and membrane transformation (Org. D. J. Morrè, D. G. Robinson)

4-02 Cell walls: structure, chemistry, formation and degradation I (Org. R. M. Brown Jr., W. Herth)

4-03 Cell walls: structure, chemistry, formation and degradation II (Org. J. G. H. Wessels, H. Meier)

4-05 Mitosis, meiosis, cytokinesis (Org. J. D. Pickett-Heaps, W. Herth)

4-06a Organelles and cell differentiation I (Org. M. W. Steer, U. Kristen)

4-06b Organelles and cell differentiation II (Org. M. W. Steer, U. Kristen)

4-07 Ultrastructure of fertilization (Org. R. B. Knox, M. M. A. Sassen)

4-09a Cytoskeleton and morphogenesis I (Org. B. E. S. Gunning, E. Schnepf, C. W. Lloyd, P. B. Green, O. Kiermayer)

4-09b Cytoskeleton and morphogenesis II (Org. B. E. S. Gunning, E. Schnepf, C. W. Lloyd, P. B. Green, O. Kiermayer)

4-09c Cytoskeleton and morphogenesis III (Org. B. E. S. Gunning, E. Schnepf, C. W. Lloyd, P. B. Green, O. Kiermayer)

4-10 Cytosymbiosis (Org. R. Trench, P. Sitte)

4-11 Recognition and interrelationships between fungi and green plants (Org. J. R. Aist, H. R. Hohl)

4-12 Parasitic higher plants (Org. B. A. Fineran, J. H. Visser, I. Dörr)

4-13 Intercellular connections (Org. A. W. Robards, R. Kollmann)
4-14 Meristems (Org. J. D. Mauseth, W. Hagemann)
4-15a Floral development I: evolutionary aspects (Org. S. C. Tucker, P. K. Endress, P. Leins)
4-15b Floral development II: special topics (Org. S. C. Tucker, P. K. Endress, P. Leins)
4-16 Organogenesis in higher plants (Org. D. R. Kaplan, W. Hagemann)
4-17a Ovule and seed development I (Org. W. A. Jensen, F. Bouman)
4-17b Ovule and seed development II (Org. W. A. Jensen, F. Bouman)
4-18 Growth forms of higher plants (Org. T. I. Serebryakova (deceased), F. Weberling)
4-19 Inflorescences (Org. A. Weber, F. Weberling)
4-20 Structures and organs concerning interactions between animals and higher plants (Org. F. R. Rickson, S. Vogel)
4-21 Metabolic peculiarities and their cytological and morphological implications (Org. W. Wiessner)
4-22 Sieve elements: the state of our knowledge 150 years after Hartig's discovery (Org. R. D. Sjolund, H.-D. Behnke)
4-23 Development and structure of xylem elements (Org. P. Baas, J. J. Sauter)
4-24 Modern approaches to wood anatomical identification (Org. R. B. Miller, H. G. Richter)
4-25 Recent progress in wood anatomy (IAWA Symposium) (Org. B. J. H. ter Welle, J. Bauch)
4-26 The integrative function of the vascular system in the plant body (Org. R. F. Evert, W. Eschrich)
4-27 Restitution in higher plants (Org. R. Kollmann)
4-28 Structure and systematics of vascular plants: the characters and their application to higher categories (Org. M. F. Danilova, H.-D. Behnke)
4-29 Anthropogenous effects on plant structure (Org. S. Huttunen, W. Liese)

Poster Sessions

4-102 Cell walls (Moderated by J. G. K. Wessels)
4-106 Organelles and cell differentiation (Moderated by M. W. Steer)

4-117 Ovule and seed development (Moderated by B. M. Johri)
4-118 Growth forms and inflorescences (Moderated by D. Müller-Doblies, H. A. Froebe)
4-128 Structure and systematics of vascular plants (Moderated by D. F. Cutler)
4-132 Meristems, Organogenesis (Moderated by W. Hagemann)
4-133 Vascular anatomy (Moderated by J. J. Sauter)

Programme Division 5: Systematic and Evolutionary Botany

Symposia

5-01 Classification of higher categories of lichenized and non-lichenized Ascomycetes (Org. O. Eriksson, A. Henssen)
5-02a Systematics of lichens at the generic and species I (Org. T. Ahti, J. Hafellner)
5-02b Systematics of lichens at the generic and species II (Org. T. Ahti, J. Hafellner)
5-03 Systematics of non-lichenized Ascomycetes (Org. R. A. Shoemaker, E. Müller)
5-04 Chemosystematics of lichenized fungi (Org. J. A. Elix, S. Huneck)
5-05 Fungi as symbionts and parasites on archegoniates and lower plants (Org. L. Holm, J. Poelt)
5-06 Structure and systematics of the Heterobasidiomycetes (Org. K. Wells, F. Oberwinkler)
5-07 The Prokaryotes, fossil record and evolutionary trends (Org. W. J. Schopf, S. Golubic)
5-08 The species concept in algae (incl. blue-green algae) (Org. Anagnostidis, D. Mollenhauer)
5-09 Ultrastructure and systematic relationships in algae (Org. S. P. Gibbs, M. Melkonian)
5-10 Systematics of green algae with special attention to the ancestry of land plants (Org. Ø. Moestrup, L. Kies)
5-11 Life histories of marine algae (Org. M. D. Guiry, D. G. Müller)
5-12 Life strategies in Bryophytes (Org. S. R. Gradstein, M. C. F. Proctor, J.-P. Frahm, W. Frey)
5-14 Taxonomy of higher categories and generic concepts in Bryophytes (Org. D. H. Vitt, R. M. Schuster)
5-15 Evolution in Bryophytes: chemosystematic approach (Org. S. Huneck, R. Mues)

5-16 Cytotaxonomy, ultrastructure and systematic relationships in Bryophytes (Org. B. E. Lemmon, A. E. J. Smith, B. Crandall-Stotler, W. Frey)

5-18 Classification in the higher leptosporangiate ferns (Org. W. H. Wagner, E. Hennipman)

5-19 How to correlate the systematics of fossil and recent Pteridophytes (Org. C. R. Hill, K. U. Kramer)

5-20 Chemosystematics of the Pteridophytes (Org. G. Cooper-Driver, E. Wollenweber)

5-21 Multidisciplinary approaches to the systematics of the *Papaveraceae* and *Annonaceae* (Org. P. J. M. Maas)

5-22 Multidisciplinary approaches to the systematics of the *Compositae* (Org. T. J. Mabry, G. Wagenitz)

5-23 Multidisciplinary approaches to the systematics of the *Gramineae* and *Cyperaceae* (Org. P. Van der Veken, H. Scholz)

5-24 Multidisciplinary approaches to the systematics of the *Araceae* (Org. P. B. Tomlinson, J. Bogner)

5-25 Multidisciplinary approaches to the systematics of the *Orchidaceae* (Org. P. B. Tomlinson, J. Bogner)

5-26 A multidisciplinary study of glucosinolate plants (Org. R. M. T. Dahlgren (deceased), M. G. Ettlinger)

5-27 Steps toward the natural system of the Dicotyledons – Rolf Dahlgren memorial symposium (Org. R. F. Thorne, H. Huber)

5-28 Systematics, evolution and ecology of early land plants (Org. D. Edwards, H.-J. Schweitzer)

5-29 Evolution of early seed plants (Org. S. E. Scheckler, H. Pfefferkorn)

5-30 Early fossil Angiosperms: reproductive structures and evolution (Org. D. L. Dilcher, F. Schaarschmidt)

5-31 Evolution and the fossil plant record (Org. W. Chaloner, H. Walther)

5-32 Southern hemisphere fossil flora (Org. T. N. Taylor, K. U. Leistikow)

5-33 Evolution in crops and weeds (Org. G. Müller)

5-34 Problems of classification in cultivated plants (Org. P. Hanelt)

5-35 Medicinal and poisonous plants of the Tropics (Org. I. Hedberg, A. J. M. Leeuwenberg)

5-36a Spores of Pteridophytes and pollen grains: development, function, comparative morphology and evolution I: Systematic aspects (Org. I. K. Ferguson, Michael Hesse)

5-36b Spores of Pteridophytes and pollen grains: development, function, comparative morphology and evolution II: Developmental aspects (Org. I. K. Ferguson, Michael Hesse)
5-37a Theory and practice of botanical classification I: Theories of botanical classification (Org. C. J. Humphries, F. Ehrendorfer)
5-37b Theory and practice of botanical classification II: Cladistic case studies (Org. C. Humphries, F. Ehrendorfer)
5-38 Macromolecular data in plant systematics: nucleic acids (Org. G. E. Fox, E. Stackebrandt)
5-39 Macromolecular data in plant systematics: proteins (Org. D. Boulter, B. Parthier)
5-40 Polyclonal and monoclonal serological data in plant systematics (Org. G. Cristofolini, U. Jensen)
5-41 Alkaloids in plant systematics (Org. D. S. Seigler, Manfred Hesse)
5-42 Speciation in Angiosperms: changes in chromosome structure and repetitive DNA (Org. R. B. Flavell, D. Schweizer)
5-43 Biosynthetic pathways and the evolution of higher plants (Org. O. R. Gottlieb, K. Kubitzki)
5-44 Isozymes, DNA differentiation and population biology (Org. D. Gottlieb, H. Hurka)
5-45 Establishment and demographic patterns in plant populations (Org. J. L. Harper, H. Runemark)
5-46 Modes of speciation in annual vs. perennial plants (Org. T. F. Stuessy, J. Grau)
5-47a Modes of reproduction and evolution of woody Angiosperms in tropical I (Org. G. T. Prance, G. K. Gottsberger)
5-47b Modes of reproduction and evolution of woody Angiosperms in tropical II (Org. G. T. Prance, G. K. Gottsberger)
5-48 How to apply biochemical and molecular evidence at various levels of phylogenetic divergence and taxonomic rank (round-table conference) (Org. D. E. Fairbrothers)

Poster Sessions

5-122 *Compositae* (Moderated by T. F. Stuessy)
5-123 *Gramineae* and *Cyperaceae* (Moderated by P. Van der Veken, P. Goetghebeur)
5-124 *Araceae* (Moderated by P. B. Tomlinson)
5-134 Cultivated plants (Moderated by B. Pickersgill)
5-135 Medicinal plants (Moderated by I. Hedberg)

5-136 Palynology (Moderated by Michael Hesse)
5-151 Homobasidiomycetes (Moderated by A. Bresinsky)
5-152 Ascomycetes and lower fungi (Moderated by A. Bresinsky)
5-153 Lichen studies (Moderated by H. Hertel, J. Poelt)
5-155 Blue-green and red algae (Moderated by J. Seckbach)
5-156 Systematics and structure of green algae (Moderated by J. F. Gerrath)
5-157 Algae (various groups) (Moderated by P. Silva)
5-158 Bryophytes (Moderated by J.-P. Frahm)
5-159 Pteridophytes (Moderated by K. U. Kramer, C. Jermy)
5-161 Flowering plants: morphology of seeds, fruits and (nectaries) (Moderated by W. Barthlott)
5-162a Flowering plants: dicotyledonous groups I (Moderated by A. Cronquist)
5-162b Flowering plants: dicotyledonous groups II (Moderated by A. Cronquist)
5-162c Flowering plants: dicotyledonous groups III (Moderated by A. Cronquist)
5-163 Flowering plants: monocotyledonous groups (Moderated by F. N. Rasmussen)
5-164 Fossil plants: gymnosperms (Moderated by T. Kimura)
5-165 Fossil plants: various groups (Moderated by A. Traverse)
5-166 Chemosystematics of flowering plants (Moderated by J. Stökkigt)
5-167 Biosystematics of flowering plants (Moderated by G. L. Stebbins)
5-168 Cytotaxonomy of flowering plants (Moderated by F. Ehrendorfer)

Programme Division 6: Environmental Botany

Symposia

6-02 Changes of plant cover and environment during the Tertiary (Org. G. Dolph, D. Mai)
6-03 Quaternary vegetation history (Org. Y. Vasari, W. A. Watts, G. Lang, E. Grüger)
6-05 European temperate zone – actual problems in maintaining natural and semi-natural vegetation (Org. R. Neuhäusl, U. Bohn)
6-06 Comparative ecology of high mountain and arctic areas (Org. G. Nahutsrishvili, O. Hegg)

6-07 Vegetation and environment of Central Asian high mountain areas (Org. Weilie Chen, W. Holzner)
6-08 Plant life under extreme environmental conditions in Antarctica (Org. E. Friedmann, L. Kappen)
6-09 Ecology and evolution of the Mediterranean vegetation and flora (Org. A. Shmida, S.-W. Breckle)
6-10 Vegetation and environment in deserts (Org. K. A. Batanouny, K. Müller-Hohenstein)
6-11 Tropical forests: vegetation and environment (Org. A. Miyawaki, E. Bruenig)
6-12a Stand-level dieback and ecosystem processes: a global perspective – forest dieback in the Atlantic region (Org. S. V. Krupa, U. Arndt)
6-12b Stand-level dieback and ecosystem processes: a global perspective – forest dieback in the Pacific region (Org. C. H. Gimingham, D. Mueller-Dombois)
6-13 Freshwater vegetation and environment (Org. C. den Hartog, G. Wiegleb)
6-14 Ecology of seagrasses (Org. P. H. Nienhuis, R. J. Orth)
6-15 Biogeography of marine benthic algae (Org. C. van den Hoek, K. Lüning)
6-16 Quantitative aspects of plant distribution (Org. P. L. Nimis, H. Haeupler)
6-17a Plant demography I: Models and life history traits (Org. P. A. Werner, W. H. O. Ernst)
6-17b Plant demography II: Models and demography (Org. P. A. Werner, W. H. O. Ernst)
6-18 Population genetics and population biology (Org. P. Jacquard, K. M. Urbanska)
6-19a Vegetation dynamics I (Org. J. Miles, W. Schmidt)
6-19b Vegetation dynamics II (Org. J. Miles, W. Schmidt)
6-20 Vegetation dynamics under isolation: phytocoenological aspects of island biogeography (Org. E. van der Maarel, H. Haeupler)
6-21 Plant architecture and resource acquisition (Org. M. M. Caldwell, M. Küppers)
6-22 Symbiosis and parasitism in higher plants (Org. Y. Hiratsuka, F. Oberwinkler)
6-23 Interference (competition, cooperation, allelopathy and coexistence) (Org. F. Berendse, A. Gigon)
6-24 Synecology of pollination: functioning of flowers (Org. R. Heithaus, S. Vogel)

6-25 Pollination ecology: flowers as food sources and costs of reproduction (Org. B. Heinrich, A. Bertsch)
6-27 Ecology of plant dispersal (Org. R. Y. Berg, D. Podlech)
6-28 Seed and germination ecology (Org. J. Silvertown, J. van Andel)
6-29 Water relations and photosynthetic production of cryptogams (Lichens, Bryophytes) (Org. T. H. Nash, L. Kappen)
6-30 Plant growth and water relations (Org. H. A. Mooney, E. D. Schulze)
6-31 Ecophysiology of leaf gas exchange (Org. J. R. Ehleringer, C. Körner)
6-32 Modelling of photosynthetic response to environmental conditions (Org. G. D. Farquhar, J. D. Tenhunen)
6-33 Ecological implications of respiration (Org. D. T. Canvin, H. P. Fock)
6-34 Mineral nutrition of forest trees (Org. G. M. Will, H. W. Zöttl)
6-35 Distributional relationships between higher plants and larger fungi (Org. A. Bujakiewicz, E. J. M. Arnolds)
6-37 Structural and functional responses to environmental stresses: low and high temperature (Org. W. L. Berry, A. Bogenrieder)
6-38 Structural and functional responses to environmental stresses: water shortage (Org. J. S. Boyer, K. H. Kreeb)
6-39 Mycorrhizas (Org. J. Webster, F. Schönbeck)
6-40 Life cycle: allocation of assimilates (Org. F. Bazzaz, G. Geisler)
6-41 Life cycle: relationship between flowers, fruits and seeds (Org. R. B. Primack, F. Lenz)
6-42 Ecological effects of air pollution on plants (Org. K. Grodzinska, H. J. Jäger)
6-43 Plant responses to enriched ambient CO_2 concentration (including round-table) (Org. B. R. Strain, H. Lieth)
6-45 Distribution of plants and plant communities: recent anthropogenic changes (Org. E. Hadač, E. Jäger, H. Sukopp)
6-46 Changes of vegetation patterns in agrophytocoenoses (Org. H. Haas, E.-G. Mahn)
6-47a Reconstitution of heavily disturbed and endangered plant communities I (Org. A. D. Bradshaw, F. Klötzli, H. Schmeisky)
6-47b Reconstitution of heavily disturbed and endangered plant communities II (Org. A. D. Bradshaw, F. Klötzli, H. Schmeisky)
6-48 Degradation of land cover in arid areas of temperate zones (Org. P. L. Gorchakovsky, H. Freitag)

6-49 Plants and plant communities in the urban environment (Org. S. Hejny, H. Sukopp)
6-50 Conservation of threatened plants (Org. G. L. Lucas, J. Kornas)
6-52 Alien plants in natural and semi-natural plant communities (Org. M. Numata, H. Sukopp)
6-53 Plant conservation (round-table conference) (Org. P. H. Raven)

Poster Sessions

6-102 Palaeoecology (Moderated by G. Lang, D. Mai)
6-112 Forest dieback and ecosystem processes (Moderated by U. Arndt, S. V. Krupa)
6-113a Freshwater flora and vegetation I: Macrophytes (Moderated by C. den Hartog, G. Wiegleb)
6-113b Freshwater flora and vegetation II: Microphytes (Moderated by U. Geissler)
6-119 Vegetation dynamics (Moderated by J. Miles, W. Schmidt)
6-121 Plant architecture, resource acquisition (Moderated by M. Caldwell, M. Küppers)
6-122 Symbiosis and parasitism in higher plants (Moderated by F. Oberwinkler)
6-123a Interference I: Competition and coexistence (Moderated by A. Gigon)
6-123b Interference II: Allelopathy (Moderated by S. J. H. Rizvi)
6-130 Plant growth and water relations (Moderated by E.-D. Schulze)
6-137 Plant responses to temperature stress (Moderated by W. L. Berry, A. Bogenrieder)
6-138 Plant responses to water stress (Moderated by K. H. Kreeb)
6-139 Mycorrhizas (Moderated by J. Webster, F. Schönbeck)
6-142 Ecological effects of air pollution on plants (Moderated by K. Grodzinska)
6-147 Anthropogenic effects on plant communities (Moderated by A. D. Bradshaw, H. Schmeisky)
6-149 Plants and plant communities in the urban environment (Moderated by S. Hejny, H. Sukopp)
6-152 Weeds and alien plants (Moderated by M. Numata, H. Sukopp)
6-153 Plant conservation (Moderated by C. Jermy)
6-160a Flora and vegetation of temperate zones I (Moderated by U. Bohn)

6-160b Flora and vegetation of temperate zones II (Moderated by U. Bohn, B. Markert)

6-161 Flora and vegetation of China (Moderated by W. Holzner)

6-162 Flora and vegetation of mountain and arctic areas (Moderated by O. Hegg, G. Nahutsrishvili)

6-164a Flora, vegetation and plant ecology in tropical and subtropical areas I: Growth and management (Moderated by E. F. Bruenig)

6-164b Flora, vegetation and plant ecology in tropical and subtropical areas II: Vegetation structure, productivity, types (Moderated by E. F. Bruenig)

6-165 Marine flora and vegetation (Moderated by C. van den Hoek)

6-166 Plant population studies (Moderated by S. Kawano)

6-169 Ecophysiology of gas exchange (Moderated by C. Körner)

6-170 Biomass yield, resource allocation and reproductive effort (Moderated by R. B. Primack, F. Bazzaz)

General Symposia and Sessions (7)

Symposia

7-01 History of botany (Org. F. A. Stafleu, H. W. Lack)

7-02 Ways in biological thinking (Org. R. Sattler, H. Mohr)

7-03 EDP in botany: information systems and data bases (Org. V. H. Heywood, J. A. Diment, G. Pflug)

7-04 EDP in botany: the computer as a tool in research (Org. T. J. Crovello, E. van der Maarel, W. Schmitt-Rennekamp)

7-05 Methods and techniques in structural botany (Org. R. P. C. Johnson)

7-06 Advances in ultrastructural technology (Org. G. T. Cole)

7-07 Novel methods at the molecular and infrastructural levels and their application in systematic botany (Org. D. Crawford, P. Gerstberger)

7-08 New biophysical methods in plant physiology (Org. L. O. Björn, L. Fukshansky)

Poster Sessions

7-104 EDP in botany (Moderated by T. J. Crovello)

7-111 History of botany, principles of taxonomy and nomenclature (Moderated by H. W. Lack)

7-112 Floras and floristic databases (Moderated by A. Strid)

7-113 Education and research in botany (Moderated by G. Schaefer)

Films

1 Fungi (Moderator B. Hock)
2a Vascular plant biology (Moderator B. Hock)
2b Palaeobotany (Moderator B. Hock)
3a Physiology, cytology, embryology (Moderator B. Hock)
3b EDP in botany (Moderator B. Hock)
4a Algae (Moderator C. Sautter)
4b Aquatic plants (Moderator C. Sautter)
5a Plant resources (Moderator C. Sautter)
5b Botanical gardens and research stations (Moderator C. Sautter)
6 Vegetation (Moderator C. Sautter)

Special Interest Group Meetings

1 Structure and function(s) of calcium oxalate in plants and fungi (Org. H. J. Arnott)
2 Proposed International Space Year 1992: objectives and tasks of the botanical sciences (Org. D. F. Baer)
3 Regulation of secondary metabolism (Joint Org. J. W. Bennett, P. Häggblom)
4 Plants of the amphibious habitats (Joint Org. T. Bhardwaja, W. Hagemann)
5 International Legume Database & Information Service General Meeting (Org. F. A. Bisby)
6 Seed coat structure and its application to taxonomy (Org. L. H. Bragg)
7 Nomenclature of cultivated plants in relation to the International Code of Nomenclature for Cultivated Plants (Org. C. D. Brickell)
8 Plasmalemma redox functions in plants (Joint Org. F. Crane, I. M. Møller)
9 Artificial intelligence and expert systems in biological research (Org. T. J. Crovello)
11 Natural stable isotope variation analysis (Org. H. Ziegler)
12 Floras and floristic publications (Org. D. G. Frodin)
13 Quantitative botanical cytochemistry (Org. P. B. Gahan)
14 Bryological bibliographic databases: directory project (Org. S. W. Greene)
15 Space biology – results and outlook of German microgravity missions (Joint Org. H. Binnenbruck, K. Kreuzberg)

18 Botanic gardens in conservation (Org. V. H. Heywood)
19 Mycorrhizas: their relevance in plant nutrient uptake, their physiology and biocontrol (Org. B. L. Jalali)
20 Palaeofloristic and palaeoclimatic changes in the Cretaceous and Tertiary (Org. E. Knobloch)
21 Starting a new international journal on pollination biology (Joint Org. K. U. Kramer, G. Gottsberger)
22 *Arabidopsis* – model and tool of molecular and classical plant biology (Org. Kranz, A. R.)
23 Publication problems of journals in plant physiology (Org. A. Kylin)
24 Direct seeding methods for woody plants in harsh environments (Org. C. V. Malcolm)
25 10th IUFRO (International Union of Forestry Research Organizations) Mycoplasma Group Meeting (Org. R. Marwitz)
27 Tropical bryophyte taxonomy and floristics of the Pacific (Joint Org. D. H. Vitt, H. O. Whittier, R. M. del Rosario, H. A. Miller)
28 Inflorescences of Monocots, and the "Families and genera of vascular plants" project (Org. D. Müller-Doblies)
29 Problems and methods in phytogeography (Org. P. L. Nimis)
31 Applied frontiers of allelopathy (Org. S. J. H. Rizvi)
32 International trends in biological education (Coordinator G. Schaefer)
32a The computer in biological education (Org. T. J. Crovello)
32b Modern trends in botanical education (Org. R. H. Saigo)
33 Problems in the study of the Gesneriaceae (Org. L. E. Skog)
34 Experimental cell physiology and its use to detect genetic variation (Org. E. J. Stadelmann)
35 Joint Meeting of the ISHS Working Group for Natural Resources and the IUBS Working Group for Medicinal Plants (Org. P. Tétényi)
36 Morphogenesis of the meristem of root tips (Org. H. Toriyama)
37 6th Meeting of the International Group for the Study of Mimosoideae (Org. J. Vassal)
38 Genetics of self-incompatibility (Org. K. Bawa)
39 Discussion on flowering (Org. G. Bernier)
40 Studies on Cyanidium (Org. H. Seckbach)
41 The future of biosystematics in the moelcular age (Org. G. L. Stebbins)

Society Meetings

The following international scientific societies or groups held a meeting or meetings during the Congress.

1 Advisory Board of the Flora of the Guianas
2 Council on Botanical and Horticultural Libraries
3 International Association of Aquatic Vascular Plant Biologists
4 International Association of Bryologists
5 International Association of Botanical Gardens
6 International Association of Botanical and Mycological Societies
7 International Association for Ecology
8 International Association for Lichenology
9 International Association for Plant Physiology
10 International Association for Plant Taxomomy
11 International Association of Pteridologists
12 International Association for Vegetation Science
13 International Association of Wood Anatomists
14 International Commission for the Nomenclature of Cultivated Plants
15 International Group for the Study of Mimosoideae
16 International Legume Database & Information Service
17 International Mycological Association
18 International Organization of Palaeobotany
19 International Organization of Plant Biosystematists
20 International Palynological Congress
21 International Phycologial Society
22 International Society of Plant Morphologists
23 International Society of Plant Population Biologists
24 International Union for Conservation of Nature and Natural Resources, Plant Advisory Group
25 Linnean Society of London
26 Systematics Association

Special Congress Features

The Emblem

The Hepatica (*Hepatica nobilis* Schreber), a modest spring flower with pretty, star-like blue blossoms and characteristic liver-shaped leaves, was chosen as a fitting symbol for a gathering of adepts of the "amiable science". It is a native of Central Europe, where it grows in beech and oak woods. Just as the forest, whose very existence is presently under threat in many parts of the world, *Hepatica nobilis* is endangered as a wild plant on the territory of Berlin (West), although it is widely grown in gardens. Close relatives of the European *Hepatica nobilis* are found round the globe, from North America to Japan, thus aptly symbolizing the world-wide connections of the botanical sciences.

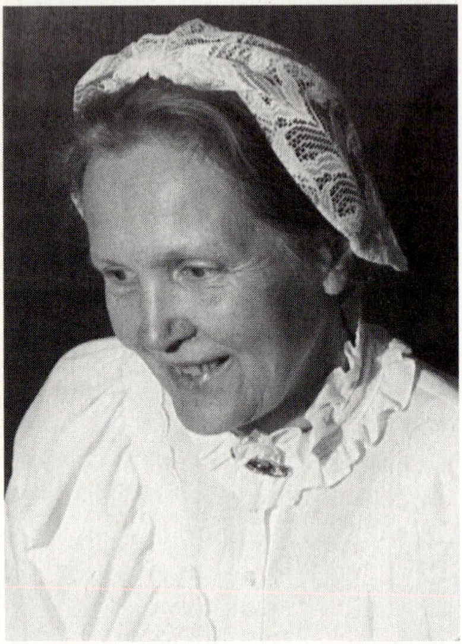

Figure 3: The artist, Gerda Rohde-Haupt, facing her original drawing of the hepatica that served as model for the Congress emblem.

The Congress and the World at large

The newspapers and broadcasting agencies gave ample space to the Congress and to some of the results and hypotheses that were discussed at its meetings. On July 24, immediately before the official opening of the Congress, the Organizing Committee held a press conference in order to give as accurate and balanced information as possible on the aims and prospects of the Congress. Throughout the following days, the Congress President and many others provided statements and comments to the representatives of press, radio and television. To our concern, but not to our surprise, not all such statements were correctly and completely rendered, which caused some stir at and around the Congress. The problems of forest decline and canopy dieback, being of special concern to the general public, soon gained a kind of monopoly in the reporters' view of the Congress. Although this was indeed one of the main focuses, it did obviously not do justice to the Congress as a whole.

Figure 4: The Congress Organizers at the inaugural press conference, duly labelled (H.-D. Behnke sitting to the left, W. Haupt to the right).

Public Lectures

Two lectures given in German language in the big lecture hall of the Technical University, by two of Berlin's leading botanists, were another of the Congress's windows to the public. Dealing with subjects of great general interest and scientific actuality, they managed to attract their audience in spite of the fact that Berlin is rather depopulated in holiday times.

On July 27, U. Stahl spoke on "Biotechnology, yesterday, today, tomorrow", whereas on July 31 R. Bornkamm spoke on "Plants in the city – doormats or pampered kids?". An English translation of the latter lecture is published at the end of the present volume.

Trees and Forests

Being aware of the value of trees and forests as the foundation of all life on earth, and mindful of the potential role of botanists and botanical institutions in research on and in the conservation, monitoring, and promotion of the understanding of forest ecosystems, the Organizing Committee decided at an early date to place the whole Congress under the general motto "Forests of the World". This was the first time to our knowledge that a general theme was given such prominence at an International Botanical Congress.

As a consequence, the Programme Committees were instructed to pay special attention to the inclusion of aspects relevant to forests and trees in the scientific congress programme. The second circular listed no less than 25 different symposia that were purely or in part concerned with such topics. No less than 3 general lectures (nos. 22-24) were to deal with the problems of forest ecosystems. Their full text is reprinted elsewhere in these pages. One of the Congress excursions (no. 13) was devoted entirely to field and laboratory studies of forest decline phenomena in Central Europe. O. Schulz-Kampfhenkel's educational film series "The forests must live!", produced specifically on this occasion, was another prominent feature illustrating the Congress motto.

Biotechnology

For the first time in a Botanical Congress, emphasis was given to biotechnology, in which domain a considerable number of special symposia were held. They, as well as individual lectures on biotechnological subjects at other symposia, were placed under the special sponsorship of the German Federal Ministry of Research and Technology. The integration of biotechnology into botany was further highlighted by two of the General Lectures (nos. 8 and 9), the first of which is reprinted here in full.

Applied Botany

Advice was sought and received from the German Union for Applied Botany on the best ways in which applied aspects could be given prominence in the scientific Congress programme. It was decided not to relegate applied botany to a section of its own, but to fully integrate it into the scientific programme by giving it due consideration and weight within each and every programme division. By consequence, applied botany was virtually omnipresent, no less than 75 symposia (listed in the second Congress circular) having their main emphasis or at least a significant side interest in its realm.

Congress Films

Two general-purpose films had been produced on the occasion of the Congress and were shown during the lunch-time breaks.

Bearing in mind the general Congress motto "Forests of the World", the Institut für Weltkunde in Bildung und Forschung (WBF) in Hamburg prepared a series of educational films under the common title *The forests must live!* (idea and coordination: Otto Schulz-Kampfhenkel and Karl A. Belgardt). This series, comprised of interesting and informative reports from eleven regions all over the world, was intended primarily for young people and schools. It was divided into two parts of c. 45 minutes each, shown on alternate days, and both together rehearsed on August 1 in one of the main lecture theatres.

With these films the WBF and Dr. Schulz-Kampfhenkel provided, on their own initiative and at their expense, a manyfold modern approach to teaching the notion of forest to the younger generation and to foster our children's awareness of the value of our green heritage and of the need of safeguarding it for the future.

Not a paradise, but still a Garden of Eden: the Botanical Garden of Berlin (presented by Peter Baumann; camera: Klaus Noack) is a 45-minutes video produced by the second channel of German television on the occasion of the Congress and of the 750th anniversary of Berlin, featuring the past and present of Berlin's botanical garden. The English version of this film had been prepared specifically for the Congress.

Leisure at the Congress?

No doubt many of those attending will have left with the feeling of having had a very busy time, and will be enclined to deny that the programme gave them any leisure at all. Nevertheless it did.

The evening of July 27, just about half-way through the Congress, was set aside for the official *Congress Dinner.* No less than 1109 persons took part in the banquet and delighted themselves with dancing at the pleasant sounds of the Ronny Peller Band, many – young and old alike – carrying on enthusiastically well beyond midnight.

The following day, July 28, was the free day of the Congress, on which several *Local Tours* of cultural and touristic interest had been organized. Most of them were limited to one half of the day, but three, leading to Potsdam, Dresden and other parts of the German Democratic Republic, were scheduled for the whole day. Similar local tours were offered, in smaller numbers, all along the Congress, being mainly intended for the benefit of accompanying persons.

Society Dinners that took place on various evenings provided other leisurely opportunities of turning one's back to the constraints of the Congress programme and spending some social hours with one's own peers. To mention but two of them, the traditional dinner of the International Association for Plant Taxonomy took place on August 23 in the Forsthaus Paulsborn by the Grunewald hunting castle, and the equally traditional soirée of the Linnean Society of London found its fitting ambience in the Humboldt castle at Tegel where it was hosted by the von Heinz family, its present owners and last descendants of Wilhelm von Humboldt.

Exhibits outside the ICC

Some public exhibitions, featuring botanical themes, were organized in close contact with the Congress, mostly on the initiative of its Exhibition Committee and under the care of its members.

A special exhibition "*100 Botanical Jewels*", displaying one hundred selected items of outstanding botanical interest, including medieval manuscripts, early printed herbals, famous folio editions of works by, e.g., Jacquin and Humboldt, and a newly discovered collection of drawings depicting plants in the private gardens of Franz I Emperor of Austria, was arranged at the "Staatsbibliothek". Berlin's State Library with its over 3.1 million volumes, 31500 current periodicals, numerous manuscripts, musical scores, etc. is one of the most important scientific libraries of Central Europe.

The exhibition "*Science in Berlin*", giving an overview of the role of science in Berlin from the 16th Century to the present, had been mounted in Berlin's "Kongreßhalle". (The famous "pregnant oyster", a symbol of America's contribution to Berlin's post-war reconstruction, was reopened for that purpose, having had its partly collapsed roof repaired.) In view of the Congress, botany had been given special prominence. A selection of plants described by leading Berlin botanists, as well as plants with names commemorating them, was shown in a glasshouse. Additional documents such as portraits and biographical items were also on display.

An exhibition featuring *Schweinfurth,* the well-known Berlin botanist and explorer of Africa, had been set up at the Botanical Museum in Berlin-Dahlem. Behind the public scene, the Museum's herbarium and library stood ready to welcome the numerous specialists who took advantage of their stay in Berlin to check and solve some of their own research problems. Berlin's Botanical Garden, belonging with the Museum, also attracted many by the riches of its living collections.

The Schloßstraße in Steglitz had assumed a special relation to the Botanical Congress. In more than fifty of its shop-windows it was displaying botany in a show commemorating the Congress and organized by the nearby Botanical Garden, under the motto "*Plants and flowers from all over the world*". The Schloßstraße stores had, on that account, sponsored a shuttle bus service with stops at all the afore-mentioned places plus the Berlin Zoo which, by a special arrangement, granted free access to the Congress members.

The year 1987 also marked the *750th anniversary of Berlin* with its corollary of festivities that were unprecedented in the annals of the city: a wide variety of cultural, sporting and musical events to demonstrate and celebrate Berlin's unique history. Some of the major performances happened to coincide in time with the Congress, the most noticeable being the Historical Fair, the Ship Pageant, and the giant Rock Concert in front of the "Reichstag" building. Several non-botanical public exhibitions also took place concurrently.

Figure 5: Congress stamps with special Congress obliteration and first-day obliteration.

Congress Stamp and other Congress Items

The Landespostdirektion Berlin issued a 60 Pfennig stamp on the occasion of the Congress, one of the nine stamps that Berlin was entitled to produce that year. Its subject (designed by R. Gerstetter) reminds the Congress motto: A single oak tree is encircled by a colourful ring of two curved arrows, to symbolize either the turn of the seasons, or a sequence of decline and (hopefully) recovery, or (as the official version has it) the environmental stresses threatening the tree – make your choice.

At the post office counter in the International Congress Centre two special obliterations were available, one relating to the first day of issue (July 16), the other special to the Congress. A limited edition of 500 of the official Congress envelopes with a block of four of the stamp and first-day obliteration was soon sold out at the Congress shop.

Other items that had been produced especially for the Congress, bearing its emblem, and that were on sale at that shop include a tie and a scarf in marine blue and bordeaux red as well as a silver coin showing a portrait of Alexander von Humboldt on the reverse. They will forever identify their owners and bearers as members of the XIV IBC confraternity.

Congress Medal

The Congress Medal was awarded at the closing ceremony to 26 persons in recognition of their official functions at this and earlier botanical congresses: H.-Dietmar Behnke (organizing committee), W. John Cram (past secretary general), Karl Esser (president), Knut Faegri (honorary vice-president), Wolfgang Franke (president, German Union for Applied Botany), Masaki Furuya (honorary vice-president), Eleonora Gabrielian (honorary vice-president), Werner Greuter (secretary general), Nathanael Grobbelaar (honorary vice-president), Wolfgang Haupt (organizing committee), Jack Heslop-Harrison (honorary vice-president), Clara C. Heyn (honorary vice-president), Otto Kandler (organizing committeee), Jan Kornaś (honorary vice-president), Anton Lang (honorary vice-president), John McNeill (Secretary, International Association of Botanical and Mycological Societies), Georg Melchers (honorary vice-president), Wilhelm Nultsch (president, German Botanical Society), Sir Rutherford Robertson (past president), Frans A. Stafleu (honorary president), Herbert Sukopp (treasurer), Armen Takhtajan (past president), Fu-hsjung Wang (honorary vice-president), Jacques H. Weil (honorary vice-president), Diter von Wettstein (honorary vice-president), and Brigitte Zimmer (scientific secretary). The medal consisted of a silver coin showing the Congress emblem, individually engraved on the plain reverse.

Excursions and Botanical Tours

The Congress Excursions

31 Congress Excursions – of the 62 that were initially announced – did actually take place: 14 before and 17 after the Congress. They had 487 participants altogether, somewhat less than 16 on average. These figures are disappointing when compared to the declared intentions as expressed on the preliminary reply forms, when almost two thirds of the prospective Congress members had hoped to take part in one or other of the field trips. The reasons for low participation are many, ranging from the relatively high cost and the unfortunate fall of the dollar rates to budgetary difficulties of many research institutions and some phobic reactions consequent to Libyan terroristic threats and the Tchernobyl accident. North American participation, in particular, was severely affected.

The original choice of options was doubtless the richest and most varied that had ever been planned for an International Botanical Congress. Even when it had to be severed by one half, what remained was remarkably manyfold. Congress excursions led into 15 different European countries and to areas as far apart as the Canary Islands and the arctic coast of Norway, or Crete and the Faroes. They covered a large number of special topics, including palaeobotany, marine botany, lichenology, bryology, pteridology, botanic garden management, forest decline and the visit of scientific institutions. It is worth noticing that the most popular subject was, however, sheer touristic sightseeing, as evidenced by the 48 participants to excursion no. 1!

Whatever feedback we had from excursionists was enthusiastic and testifies to the competent and devoted work of the excursion organizers. Many of them gave proof of almost unbelievable flexibility, in particular when asked to restructure their programme to fit a small group of people rather than the full bus-load that had originally been planned – without increasing the cost if at all possible. Nevertheless some excursions unavoidably ran into a deficit, but small surpluses of others were happily available to balance the account.

It was particularly unfortunate that two excursions, no. 34 to Greenland and no. 40 to Spain, had to be cancelled at the very last moment, having at first been confirmed, due to the defection or failure to react of several of those who had committed themselves to take part. Both tour leaders had already completed the text of the excursion guide booklet (it was actually published in spite of the cancellation), and both had made prospections in the field. Alas, they had to be disappointed.

Descriptive material relating to the Congress excursions was issued at three stages. In the second circular, a fairly full account of each of the 62 original offers was given (on pp. 35-76), with their itineraries and main subjects and goals. Upon payment of the dues, each participant received individually an information sheet relating to his or her excursion, giving all such useful details as might be needed to join and to get prepared, mostly including a selected bibliography. Finally, at the beginning of each pre-Congress excursion, and upon registration in Berlin for the post-Congress ones, guide booklets were handed out with ample details on the areas visited and on the tour itself. These booklets (listed in the bilbliography below) were also on sale at the Congress bookshop.

The following enumeration recapitulates the 31 Congress excursions and their basic data.

01: A smiling image of Germany: its landscapes and monuments (FRG) – 2 to 8 August (Organizer: Renate Mack, Deutsches Reisebüro, Frankfurt; 48 participants). Route: Frankfurt – Koblenz – Frankfurt – Heidelberg – München.

02: Visiting Berlin (West) as a scientist – 20 to 23 July (Organizers: O. Schieder and Heike Schieder, Freie Universität, Berlin; 30 participants). Base: Berlin (West).

07: Fossil floras of the Tertiary (FRG and Netherlands) – 17 to 23 July (Organizer: H.-J. Gregor, Gröbenzell, FRG; 12 participants). Route: Berlin (West) – Tegelen (Netherlands) – Darmstadt – Günzburg – München – Landshut – Schwandorf – Berlin (West).

08: The ecology and evolution of fossil floras in W Germany (FRG) – 2 to 8 August (Organizer: V. Mosbrugger, University of Bonn; 5 participants). Route: Berlin (West) – Hannover – Soest – Bonn – Darmstadt – Bamberg – Nürnberg – Berlin (West).

11: The vegetation and landscapes of SW Germany (FRG) – 2 to 9 August (Organizers: G. Philippi, Landessammlungen für Naturkunde, Karlsruhe; Otti Wilmanns, University of Freiburg; 3 participants). Route: Frankfurt – Tauberbischofsheim – Freiburg – Frankfurt.

13: Forest decline in C Europe: a study of the phenomenon and of its possible causes (Switzerland, FRG, ČSSR) – 7 to 23 July (Organizer: F. Rebele, Technical University, Berlin; 19 participants). Route: Zürich – Davos – Freudenstadt – München – Bayerisch Eisenstein – Špindlerův Mlýn – Fichtelgebirge – Essen – Neuhaus – Berlin (West).

14: The bryophytes and bryogeographical regions of Switzerland – 3 to 9 August (Organizer: Patricia Geissler, Conservatoire botanique, Genève; 6 participants). Route: Genève – Leukerbad – Claro – Brütten – Zürich.

15: Cutting across the Alps of eastern Switzerland – a comparative study of their vegetation types – 2 to 11 August (Organizer: E. Landolt, Swiss Federal Institute of Technology, Zürich; 12 participants). Route: Zürich – Davos-Schatzalp – Pontresina – Zürich.

16: The vegetation of the northern, central and southern Alps of Switzerland, a comparative overview – 2 to 11 August (Organizer: H. Zoller, University of Basel; 15 participants). Route: Zürich – Engelberg – Melchsee-Frutt – Meiringen – Airolo – Locarno-Muralto – Zürich.

17: The flora and vegetation of the mountains of western Switzerland – 2 to 9 August (Organizer: P. Hainard, University of Lausanne; 10 participants). Route: Genève – Nant – Derborence – Sion – Zermatt – Genève.

18: High alpine flora and vegetation of the Tyrolean Alps (W Austria) – 17 to 23 July (Organizer: G. Grabherr, University of Wien; 14 participants). Route: München – Innsbruck – Obergurgl – Bregenz – Bezau – München.

19: The vegetation of the Land of Salzburg (Alps of Austria) – 3 to 9 August (Organizer: H. Wagner, University of Salzburg; 17 participants). Route: Salzburg – Wiesmayerhaus (Radstädter Tauern) – Glocknerhaus (Hohe Tauern) – Rudolfshütte (Hohe Tauern) – Salzburg.

22: Bohemia, Moravia and Slovakia: the lowlands, hills and high mountains of Czechoslovakia – 2 to 12 August (Organizers: Viera Feráková, University of Bratislava; J. Jeník, Czechoslovak Academy of Sciences, Třeboň; 30 participants). Route: Berlin (West) – Špindlerův Mlýn – Praha – Pieštany – Rožňava – Tatranská Lomnica – Praha – Berlin (West).

23: National parks and nature reserves in southern Poland – 2 to 10 August (Organizer: K. Browicz, Polish Academy of Sciences, Poznań; 32 participants). Route: Berlin (West) – (Warsaw) – Kraków – Zakopane – Kraków – (Warsaw) – Berlin (West).

24: From the Jurassic to the Holocene: the palaeoflora and palaeoecology of W and S Poland – 2 to 11 August (Organizers: L. Stuchlik, Polish Academy of Sciences, Kraków; Maria Reymanówna, Palaeobotanical Section of the Polish Botanical Society, Kraków; 13 participants). Route: Berlin (West) – Poznań – Wrocław – Kraków – Warsaw – Berlin (West).

25: The phanerogamic and cryptogamic flora and vegetation of NE Poland – 15 to 24 July (Organizer: Maria Ławrynowicz, University of Łódź; 12 participants). Route: Berlin (West) – Warsaw – Grajewo – Białowieża – Chełm – (Warsaw) – Berlin (West).

29: Taiga, tundra and North Atlantic coast: the vegetation of N Scandinavia (Sweden and Norway) – 2 to 8 August (Organizer: M. Sonesson, Abisko Scientific Research Station; 14 participants). Route: Kiruna – Abisko – Kiruna.

30: The high mountain flora and vegetation of central Norway – 4 to 10 August (Organizer: O. Gjaerevoll, University of Trondheim; 11 participants). Route: Oslo – Kongsvoll (Dovre Mts.) – Oslo.

31: Botany, nature and Man in western Sweden – 18 to 23 July (Organizer: G. Weimarck, Botanical Garden, Göteborg; 17 participants). Base: Göteborg.

33: The vegetation and bryophyte flora of the Faroe Islands (Denmark) – 16 to 23 July (Organizers: Jette Lewinsky, Botanical Museum, Copenhagen; J. Jóhansen, Museum of Natural History, Tórshavn; 7 participants). Route: Copenhagen – Tórshavn – Sørvágur – Copenhagen.

35: The phanerogamic and cryptogamic flora and vegetation of Tenerife (Canary Islands, Spain) – 15 to 22 July (Organizer: W. Wildpret de la Torre, University of La Laguna, Tenerife; 19 participants). Base: Santa Cruz de Tenerife.

39: The systematics and ecology of the pteridophytes of northern Spain – 3 to 11 August (Organizers: R. Viane, University of Gent; C. Jermy, British Museum, Natural History, London; M. Mayor López, University of Oviedo; 10 participants). Route: Oviedo – Cangas de Onis – Santillana – Santander.

41: The vegetation and endemic flora of the Spanish Pyrenees – 2 to 10 August (Organizers: P. Montserrat and L. Villar, Instituto Pirenaico de Ecología, Jaca; 27 participants). Route: Barcelona – Sort – Cerler – Panticosa – Barcelona.

44: The botany of the Picardie (N France) – 3 to 10 August (Organizer: Josiane Paré, Université de Picardie, Amiens; 8 participants). Route: Paris – Amiens – Paris.

47: Marine botany in NW Corsica (France) – 2 to 10 August (Organizer: V. Demoulin, University of Liège, Belgium; 10 participants). Base: Calvi

49: The lichen flora of Sardinia (Italy) – 13 to 22 July (Organizers: P. L. Nimis, University of Trieste; J. Poelt, University of Graz, Austria; 18 participants). Route: Cagliari – Lanusei – Alghero – Olbia.

56: Illyrian forest vegetation and Balkan endemics in western and central Yugoslavia – 14 to 23 July (Organizers: I. Puncer,M. Zupančić and Mr. A. Seliškar, Slovenian Academy of Sciences and Arts, Ljubljana; 14 participants). Route: Ljubljana – Plitvice – Sarajevo – Višegrad – Tjentište – Dubrovnik.

57: From Belgrade to the Iron Gate: the vegetation and relic flora of E Serbia (Yugoslavia) – 11 to 19 July (Organizer: Olga Vasić, Museum of Natural History, Belgrade; 9 participants). Route: Belgrade – Golubac – Donji Milanovac – Borsko jezero – Belgrade.

58: The mountain flora and vegetation of northern Greece – 10 to 23 July (Organizer: A. Strid, University of Copenhagen, Denmark; 7 participants). Route: Thessaloniki – Kavala – Mt. Vrondous (Lailia) – Metsovo – Litohoro – Mt. Olympus – Litohoro – Thessaloniki.

59: The flora, vegetation and monuments of classical Greece – 2 to 10 August (Organizers: S. Diamantoglou, University of Athens; U. Kull, University of Stuttgart, FRG; 19 participants). Route: Athens – Delphi – Olympia – Tripolis – Athens.

60: Endemism and island botany: the flora and vegetation of Crete (Greece) – 16 to 23 July (Organizer: M. Damanakis, University of Crete, Iraklio; 19 participants). Route: Iraklio – Ajios Nikolaos – Hania – Hora Sfakion – Hania – Iraklio.

Statement of the participants of Excursion no. 13

The participants to this excursion, that had focused on forest problems in Switzerland, Czechoslovakia and Germany, issued a common final statement upon their arrival in Berlin. It was published in the second Congress bulletin under the heading "*Forest decline in Central Europe*", and was also made available to the press. It is reprinted here.

"The following statement is based on our 17 day (4,200 km) excursion in which we visited 10 research groups, had discussions with more than 40 research workers and visited 12 mountain ranges in 4 countries.

"Ours is not an in depth scientific critique of the many competing hypotheses for the cause of forest decline, but a group evaluation based on discussions with a majority of the primary investigators of the decline phenomenon at their laboratories and field sites in Central Europe. We are neither forest practitioners nor specialists in air pollution, however, we do draw on extensive experience in the physiology, ecology, environmental chemistry, anatomy, pathology, mycology, reproduction and evolution of forest trees.

Figure 6: F. H. Evers (Freiburg i. Br.) demonstrating symptoms of forest damage to participants of excursion no. 13, in the northern Black Forest (picture by R. Bornkamm).

"In our visits to sites in the mountainous areas of Central Europe we have observed many forest stands which appeared healthy. However, in spruce forests of all ages, we saw widespread symptoms of tree disorder including needle yellowing and needle loss. In some locations we observed extreme dieback in old stands. From the expositions given to us, the discussions which followed and our own observations, it was clear that no one single cause

existed for decline at all sites. It did appear, however, that a diverse array of seemingly unrelated factors including past forest management, climatic stress and aerial inputs of pollutants were associated with tree decline and the deterioration of forest soils.

"Research groups we have visited are elucidating some of the complex processes involved. However, we feel that a greater integration of different approaches is needed to unambiguously establish the causes and the course of forest decline in Central Europe. An increase in cooperative research efforts and greater standardization of evidence is essential if we are to identify a sufficiently compelling explanation of decline to stimulate judicious political response by the European nations, both East and West. We therefore call for the formation of a multinational working group to address these problems."

Botanical Tours at the Congress

On July 28, the "free day" of the Congress, 15 botanical half-day excursions were organized, to compete with 23 other tours whose scope was touristic or cultural. The botanical tours were planned and organized by the Institute for Ecology of the Technical University (R. Böcker), to which all tour leaders, unless otherwise specified, were associated. They were a real success and attracted no less than 615 participants, 342 on the morning and 273 on the afternoon. They are described in the third Congress circular (pp. 44-47) and are again briefly enumerated below.

01: The Botanical Garden Berlin-Dahlem (Leaders: H. Ern and B. E. Leuenberger, Botanical Garden and Museum; A. Bley, T. Engel, M. Jeske, Christiane Köhler, Brigit Lehmann, M. Menzel, Anne Schäpermeier, C. Weiglin, students; 229 participants).

02: The urban fringe of the northern Grunewald forest (Leader: R. Böcker; 23 participants).

03: The "Pfaueninsel" (Peacock Island) (Leaders: H. Scholz, Botanical Garden and Museum, M. Seiler, Pfaueninsel; 25 participants).

04: The "Tiergarten" and the "Diplomatenviertel" (Leaders: I. Kowarik, Ulrike Sachse; 7 participants).

05: Amphibian aid programme for the Spandau forest (Leader: A. Auhagen; 19 participants).

06: The reed-banks of the Havel lakes (Leaders: Kerstin Wöbbecke, L. Trepl; 13 participants).

07: The cemeteries of Berlin as vascular plant and bryophyte habitats (Leaders: Annerose Graf, Annemarie Schaepe; 15 participants).

08: The "Karolinenhöhe" sewage farm area in Gatow (Leader: Carol Salt; 11 participants).

09: The Tegel creek valley (Leader: R. Böcker; 41 participants).

10: The Tegel forest and the Humboldt Castle (Leaders: N. Schacht, W. Seidling, U. Starfinger, U. Sukopp; 121 participants).

11: The Glienicke park and the Düppel museum village (Leaders: A. Brande, W. Tigges; 16 participants).

12: Anhalt goods station: vegetation of waste land (Leader: I. Kowarik; 18 participants).

13: Berlin's villages and the "Spandauer Burgwall" (Leaders: Hanna Köstler, D. L. Patterson; 23 participants).

14: Plant ecology at the "Kehler Weg" (Leader: R. Bornkamm; 38 participants).

15: Algae in urban waters (Leaders: Ursula Geißler, Regine Jahn, Susanne Wendker, Institute of Systematic Botany and Plant Geography, Free University; 16 participants).

A Selected Bibliography of the Congress

Official Congress Publications

XIV International Botanical Congress. Preliminary announcement. August 1983. [1 sheet in colour, for display; printing: 10,000.]

XIV International Botanical Congress. First Circular. September 1983. [folded prospectus with 8 printed faces, including preliminary reply form; printing: 54,000.]

XIV International Botanical Congress. Second circular and preliminary programme. October 1985. [Brochure of 80 pages plus coloured cover, with forms A (offer of contributions), B (excursion registration) and C (programme wishes) as loose enclosures; printing: 10,170.]

XIV International Botanical Congress. Third and final circular. December 1986. [Brochure of 52 pages plus coloured cover, with forms B (updated), D (final registration) and E (abstract) as loose enclosures; printing: 10,200.]

XIV International Botanical Congress. Opening Ceremony. Programme. July 1987. [Brochure of 4 pages plus coloured cover (including text on musical programme and address of G. Turner, reprinted hereafter); printing: 4,050.]

W. Greuter & B. Zimmer (eds.): *Programme of the XIV International Botanical Congress, Berlin, July 24 to August 1, 1987.* July 1987. [432 page book with flexible, coloured cover and cover flaps, with folded tabular programme as loose enclosure; printing: 4,200.]

[H.-D. Behnke (ed.):] *XIV International Botanical Congress Berlin. Bulletin.* No. 1, 25 July 1987 (9 pages); No. 2, 26 July 1987 (4 pages); No. 3, 27 July 1987 (4 pages); No. 4, 29 July 1987 (4 pages); No. 5, 30 July 1987 (4 pages); No. 6, 31 July 1987 (4 pages); No. 7, 1 August 1987 (4 pages). [stapled sheets; printing: 4,000 each.]

W. Greuter, B. Zimmer & H.-D. Behnke (eds.): *Abstracts of the general lectures, symposium papers and posters presented at the XIV International Botanical Congress, Berlin, July 24 to August 1, 1987.* July 1987. [480 page book with flexible, coloured cover; printing: 3,700.]

W. Greuter, B. Zimmer & H.-D. Behnke (eds.): *Abstracts of the general lectures, symposium papers and posters presented at the XIV International Botanical Congress, Berlin, July 24 to August 1, 1987 [Library edition].* August 1987. [495 page book with plastified, coloured hard cover; printing: 432; the additional pp. 481-495 also exist as a reprint.]

[NB.: the present "*Proceedings*" book is itself the last official publication of the Congress.]

Excursion Guide Booklets *

W. Welss: XIV International Botanical Congress. Excursion no. 01. *A smiling image of Germany: its landscapes and monuments.* Berlin, [July] 1987. [32 pages, 4 fig., 1 tab.]

* Two of the 31 Congress excursions that actually took place had no guide booklet of their own: No. 11 (for which the same text as for No. 01 was used) and No. 19 for which no corresponding text was submitted. In compensation, the texts for two excursions (Nos. 34 and 40) that had to be cancelled at a late date were nevertheless printed. The participants of excursion No. 49 were given a book rather than a guide booklet; while belonging to the present series it has in the same time been published as vol. 7 of the series "Studia Geobotanica" and has, therefore, a format and cover of its own.

O. Schieder & H. Schieder: XIV International Botanical Congress. Excursion no. 02. *Visiting Berlin (West) as a scientist.* Berlin, [July] 1987. [46 pages, 1 fig.]

H.-J. Gregor: XIV International Botanical Congress. Excursion no. 07. *Fossil floras of the Tertiary (FRG and Netherlands).* Berlin, [July] 1987. [60 pages, 76 fig., 3 tab.]

J. van der Burgh, K. Helmerich, W. Jung, H. Kaiser, V. Mosbrugger, G. Pelzer, W. Riegel, F. Schaarschmidt, J. Schweitzer & V. Wilde: XIV International Botanical Congress. Excursion no. 08. *The ecology and evolution of fossil floras in W Germany.* Berlin, [July] 1987. [81 pages, 33 fig., 2 tab.]

F. Rebele (ed.): XIV International Botanical Congress. Excursion no. 13. *Forest decline in C Europe: a study of the phenomenon and of its possible causes.* Berlin, [July] 1987. [52 pages, 18 fig., 5 tab.]

P. Geissler: XIV International Botanical Congress. Excursion no. 14. *The bryophytes and bryogeographical regions of Switzerland.* Berlin, [July] 1987. [24 pages, 15 fig.]

E. Landolt: XIV International Botanical Congress. Excursion no. 15. *Cutting across the Alps of eastern Switzerland – a comparative study of their vegetation types.* Berlin, [July] 1987. [39 pages, 9 fig.]

H. Zoller: XIV International Botanical Congress. Excursion no. 16. *The vegetation of the northern, central and southern Alps of Switzerland – a comparative overview.* Berlin, [July] 1987. [55 pages, 26 fig., 3 tab.]

P. Hainard: XIV International Botanical Congress. Excursion no. 17. *The flora and vegetation of the mountains of western Switzerland.* Berlin, [July] 1987. [18 pages, 2 fig.]

G. Grabherr: XIV International Botanical Congress. Excursion no. 18. *High alpine flora and vegetation of the Tyrolean Alps (W Austria).* Berlin, [July] 1987. [82 pages, 31 fig., 3 tab.]

D. Bernátová, Z. Dúbravcová, V. Feráková, V. Grulich, I. Háberová, J. Jeník, M. Krizo, D. Magic, Š. Maglocký, L. Mucina & A. Petrík: XIV International Botanical Congress. Excursion no. 22. *Bohemia, Moravia, and Slovakia: the lowlands, hills, and high mountains of Czechoslovakia.* Berlin, [July] 1987. [51 pages, 8 fig.]

K. Browicz, A. Bujakiewicz, J. Kornaś, A. Medwecka-Kornaś, Z. Mirek, H. Piekos-Mirkowa & K. Zarzycki: XIV International Botanical Congress. Excursion no. 23. *National parks and nature reserves in southern Poland.* Berlin, [July] 1987. [43 pages, 11 fig., 2 tab.]

K. Birkenmajer, S. Dyjor, E. Dzięciołowski, J. Ichas-Ziaja, T. Kuszell, A. Obidowicz, M. Reymanówna, A. Sadowska, L. Stuchlik, K. Tobolski & E. Wcisło-Luraniec: XIV International Botanical Congress. Excursion no. 24. *From the Jurassic to the Holocene: the palaeoflora and palaeoecology of W and S Poland.* Berlin, [July] 1987. [54 pages, 11 fig., 1 tab.]

J. B. Faliński, R. Olaczek, A. Pałczyński & B. Sałata: XIV International Botanical Congress. Excursion no. 25. *The phanerogamic and cryptogamic flora of NE Poland.* Berlin, [July] 1987. [65 pages, 43 fig.]

M. Sonesson: XIV International Botanical Congress. Excursion no. 29. *Taiga, tundra and North Atlantic coast: the vegetation of N Scandinavia.* Berlin, [July] 1987. [11 pages, 1 fig.]

S. Britten & O. Gjaerevoll: XIV International Botanical Congress. Excursion no. 30. *The high mountain flora and vegetation of central Norway.* Berlin, [July] 1987. [22 pages, 10 fig.]

G. Weimarck: XIV International Botanical Congress. Excursion no. 31. *Botany, nature and Man in western Sweden.* Berlin, [July] 1987. [30 pages, 5 fig.]

J. Lewinsky & J. Jóhansen: XIV International Botanical Congress. Excursion no. 33. *The vegetation and bryophyte flora of the Faroe Islands (Denmark).* Berlin, [July] 1987. [52 pages, 8 fig.]

S. Laegaard: XIV International Botanical Congress. Excursion no. 34. *The subarctic flora and vegetation of S Greenland (Denmark).* Berlin, [July] 1987. [29 pages, 1 fig., 1 tab.]

W. Wildpret de la Torre, W. Greuter & B. Zimmer: XIV International Botanical Congress. Excursion no. 35. *The phanerogamic and cryptogamic flora and vegetation of Tenerife (Canary Islands).* Berlin, [July] 1987. [54 pages.]

R. Viane, M. Mayor López & C. Jermy: XIV International Botanical Congress. Excursion no. 39. *The systematics and ecology of the Pteridophytes of northern Spain.* Berlin, [July] 1987. [54 pages, 28 fig., 3 tab.]

S.-W. Breckle, J. Vigo, J. M. Montserrat & J. Cortina: XIV International Botanical Congress. Excursion no. 40. *The flora and vegetation of Catalonia, from the seashores to the heights of the Pyrenees (NE Spain).* Berlin, [July] 1987. [62 pages, 56 fig.]

P. Montserrat & L. Villar: XIV International Botanical Congress. Excursion no. 41. *The vegetation and endemic flora of the Spanish Pyrenees.* Berlin, [July] 1987. [66 pages, 65 fig.]

J. Paré: XIV International Botanical Congress. Excursion no. 44. *The botany of the Picardie (N France).* Berlin, [July] 1987. [26 pages, 3 fig.]

V. Demoulin: XIV International Botanical Congress. Excursion no. 47. *Marine botany in NW Corsica (France).* Berlin, [July] 1987. [33 pages, 17 fig.]

P. L. Nimis & J. Poelt: [XIV International Botanical Congress. Excursion no. 49.] *The lichens and lichenicolous fungi of Sardinia (Italy), an annotated list.* Trieste, 1987. [(4) + 269 pages, 10 fig. (Studia Geobotanica, vol. 7).]

Ž. Bjelčić, J. Puncer, A. Seliškar, V. Stefanović & M. Zupančić: XIV International Botanical Congress. Excursion no. 56. *Illyrian forest vegetation and Balkan endemics in western and central Yugoslavia.* Berlin, [July] 1987. [46 pages, 29 fig.]

O. Vasić: XIV International Botanical Congress. Excursion no. 57. *From Belgrade to the Iron Gate: the vegetation and relic flora of E Serbia (Yugoslavia).* Berlin, [July] 1987. [24 pages, 2 fig.]

A. Strid: XIV International Botanical Congress. Excursion no. 58. *The mountain flora and vegetation of northern Greece.* Berlin, [July] 1987. [43 pages, 1 fig.]

U. Kull & S. Diamantoglou: XIV International Botanical Congress. Excursion no. 59. *The flora, vegetation and monuments of classical Greece.* Berlin, [July] 1987. [66 pages, 28 fig., 1 tab.]

M. Damanakis & U. Matthäs: XIV International Botanical Congress. Excursion no. 60. *Endemism and island botany: the flora and vegetation of Crete (Greece).* Berlin, [July] 1987. [21 pages, 2 fig., 3 tab.]

Botanical Tours' Guide Leaflets

R. Böcker: *Urban fringe, northern Grunewald.* International Botanical Congress Berlin (West) 1987, Local Tours, 28. July, No. 2. [12 pages, 11 figures, 1 table.]

H. Scholz & M. Seiler: *Pfaueninsel.* International Botanical Congress Berlin (West) 1987, Local Tours, 28. July, No. 3. [6 pages, 3 figures, 2 tables.]

I. Kowarik & U. Sachse: *Tiergarten, Diplomatenviertel.* International Botanical Congress Berlin (West) 1987, Local Tours, 28. July, No. 4. [11 pages, 13 figures, 5 tables.]

A. Auhagen: *Forest Spandau, amphibian-aid-programme.* International Botanical Congress Berlin (West) 1987, Local Tours, 28. July, No. 5. [7 pages, 6 figures.]

K. Wöbbecke & W. Ripl: *Havel reed banks.* International Botanical Congress Berlin (West) 1987, Local Tours, 28. July, No. 6. [19 pages, 9 figures, 22 tables.]

A. Graf & A. Schaepe: *Cemeteries of Berlin.* International Botanical Congress Berlin (West) 1987, Local Tours, 28. July, No. 7. [11 pages, 10 figures, 1 table.]

C. Salt: *Sewage farm, Karolinenhöhe.* International Botanical Congress Berlin (West) 1987, Local Tours, 28. July, No. 8. [14 pages, 4 figures, 4 tables.]

R. Böcker: *Tegel Creek Valley.* International Botanical Congress Berlin (West) 1987, Local Tours, 28. July, No. 9. [10 pages, 14 figures, 1 table; 3 folded maps in colour.]

W. Seidling, N. Schacht, U. Starfinger & U. Sukopp: *Tegel forest, Humboldt Castle.* International Botanical Congress Berlin (West) 1987, Local Tours, 28. July, No. 10. [12 pages, 10 figures, 3 tables.]

A. Brande & W. Tigges: *Düppel Museum Village + Glienicker Park.* International Botanical Congress Berlin (West) 1987, Local Tours, 28. July, No. 11. [9 pages, 6 figures, 1 table.]

I. Kowarik: *Anhalt goods station.* International Botanical Congress Berlin (West) 1987, Local Tours, 28. July, No. 12. [9 pages, 9 figures, 2 tables.]

H. Köstler & D. L. Patterson: *Villages in Berlin.* International Botanical Congress Berlin (West) 1987, Local Tours, 28. July, No. 13. [9 pages, 11 figures, 5 tables.]

R. Bornkamm: *Plant ecology, Kehler Weg.* International Botanical Congress Berlin (West) 1987, Local Tours, 28. July, No. 14. [8 pages, 5 figures.]

U. Geißler, R. Jahn & S. Wendker: *Algae in urban waters.* International Botanical Congress Berlin (West) 1987, Local Tours, 28. July, No. 15. [9 pages, 3 figures.]

Publications Issued on the Occasion of the Congress

H. W. Lack: *Botanik und Zoologie.* Berlin, June 1987. [8 pages; separately paged reprint from T. Buddensieg, K. Düwell & K.-J. Sembach (eds.):

Wissenschaften in Berlin (the official, 3-volume guide to the homonymous exhibit); offered to all Congress participants.]

H. W. Lack, P. J. Becker & T. Brandis: *100 botanische Juwelen. 100 botanical jewels.* Staatsbibliothek Preußischer Kullturbesitz, Berlin, July 1987. [216 pages, 100 plates in colour or black-and-white, flexible cover; the official guide book to the homonymous exhibit.]

[K. Ziegan]: *Katalog 318. Botanik.* Buchhandlung Ziegan, Berlin, July 1987. [80 unnumbered pages; this is the catalogue of the book exhibit at the XIV IBC, available there.]

Botanica acta. Berichte der Deutschen Botanischen Gesellschaft. Journal of the German Botanical Society. Volume 101, No. 0. Thieme, Stuttgart, July 1987 [vi + 56 + 6 pages; the "zero" issue of the new series of this traditional journal, under a new title, cover and editorship, includes 9 papers later re-published, with a different pagination and layout, in issue No. 1 (together with 5 new ones); it was available free of cost at the book exhibit of the Congress.]

K. Noack & P. Baumann: *Der Botanische Garten Berlin.* Nicolai, Berlin, July 1987. [62 pages, 24 figures, 54 colour photographs, hard cover; the volume to accompany the film "Not a paradise, but still a Garden of Eden", shown at the Congress.]

H. Scholz (ed.): *Botany in Berlin.* Botanical Garden & Museum Berlin-Dahlem, July 1987 (Englera vol. 7). [288 pages, paper; produced under the auspices of the Committee for the History of Botany of the XIV IBC; offered to all Congress participants.]

H. Lorenzen (ed.): *Beiträge zur neueren Geschichte der Botanik.* G. Fischer, Stuttgart, "July" [November] 1987 (Berichte der Deutschen Botanischen Gesellschaft, vol. 100). [x + 441 pages, paper; produced under the auspices of the Committee for the History of Botany of the XIV IBC; announced as being available at the Congress in the programme book (p. 415), but publication delayed (subscription offer included in the daily Congress bulletin no. 1, p. 8).]

Caroli Ludovici Willdenow Florae Berolinensis Prodromus. Reprint of the original 1787 edition, Koeltz, Königstein, "1 July" [mid-December] 1987 (Verhandlungen des Berliner Botanischen Vereins, special volume). [(4), xvi, 440 pages, 7 plates of drawings, 1 folded map in colour, paper with coloured dust-jacket; announced as being available at the Congress in the programme book (p. 415), but publication delayed.]

Proceedings of Symposia, Sessions, Meetings and Excursions *

2-14. Proceedings to be included in P. Barlow (ed.): *Differential growth.* (Environmental and Experimental Botany, special issue.)

4-10. Papers of this symposium are being published in the journal *Symbiosis.*

4-15. P. Leins, S. C. Tucker & P. K. Endress (eds.): *Aspects of floral development.* Proceedings of the Symposium "Floral development: evolutionary aspects and special topics", held at the XIV International Botanical Congress, Berlin (West), Germany, July 24 to August 1, 1987. Cramer, Stuttgart, 1988 (in press).

4-22. H.-D. Behnke & D. Sjolund (eds.): *Comparative structure of the sieve elements.* Springer, Berlin etc., 1989 (in prep.). [A contribution from Symposium **4-27** and some additional chapters are included.]

5-12, 5-14, 5-15, 5-16. W. Frey & S. Hattori (eds.): *Proceedings of the bryological symposia of the XIV International Botanical Congress* Berlin (West) 24 July to 1 August 1987. (Journal of the Hattori Botanical Laboratory, vol. 64, 269 pages.) Nichinan, 1988.

5-30, 6-02, SIGM-20. D. W. Dilcher, E. Knobloch, D. H. Mai & F. Schaarschmidt (eds.): *Proceedings* [exact title unknown]. Review of Paleobotany and Palynology, special issue. Elsevier, Amsterdam & New York (in prep.).

5-34. Publication of the proceedings is expected to take place in *Biologisches Zentralblatt,* G. Fischer, Jena.

5-35. A. J. M. Leeuwenberg (ed.): *Medicinal and poisonous plants of the tropics.* Proceedings of Symposium 5-35 of the 14th International Botanical Congress, Berlin, 24 July – 1 August 1987. Pudoc, Wageningen, [November] 1987 (152 pages).

5-36. The invited lectures of this symposium are to be published as a special volume of the journal *Plant Systematics and Evolution.*

[5-45, see 6-19.]

---

* This list cannot claim to be either complete or accurate. Few of the proceedings announced to us have yet been published, of some we may not have been notified, and some that were planned may fail to materialize. Nevertheless we felt that our attempt to provide at least a rough overview was worth while. No record has been kept of papers published individually. Numbers in bold-face are those used in the scientific programme.

5-47. G. T. Prance & G. K. Gottsberger (eds.): *Modes of reproduction and evolution of woody plants in tropical environments.* New York Botanical Garden (special publication), 1989 (in prep.).

[6-02, see 5-30.]

6-05, 6-160. U. Bohn & R. Neuhäusl (eds.): *Vegetation and flora of temperate zones.* Proceedings of the XIV International Botanical Congress: symposium "European temperate zone – actual problems in maintaining natural and semi-natural vegetation" and poster presentation "Flora and vegetation of temperate zones". SPB Academic Publishing, The Hague, 1988 (in press).

6-12. D. Mueller-Dombois (ed.): *Stand-level dieback and ecosystem processes: a global perspective.* GeoJournal vol. 17, No. 2, September 1988.

6-15. Proceedings to be published in volume 42 (3) of the journal *Helgoländer Meeresuntersuchungen,* 1988.

6-18. P. Jacquard & K. Urbanska (eds.): *Population genetics and population biology.* Proceedings of the symposium held during the 14th International Botanical Congress, Berlin, July 26 – 1987. (Oecologia Plantarum, special issue 1, 1988). Gauthier-Villars, Montrouge, 1988.

6-19, 6-20, 6-119. J. Miles, W. Schmidt & E. van der Maarel (eds.): *Vegetation dynamics.* (Vegetatio, special volume). Kluwer, Dordrecht, 1988. [Includes some contributions from Symposium 5-45.]

6-38, 6-138. K. H. Kreeb (ed.) *Proceedings* [exact title unknown]. SPB Academic Publishing, The Hague, 1988 (in press). [To include some contributions from Symposium **6-30** and Poster Session **6-130**.]

6-43. Proceedings to be published as a volume of the series *Tasks in Vegetation Science* (T: VS) by Kluwer, Dordrecht (in prep.).

6-49. H. Sukopp & S. Hejny (eds.): *Plants and plant communities in the urban environment.* SPB Academic Publishing, The Hague, 1988 (in prep.).

SIGM-8. F. L. Crane & I. M. Møller (eds.): *Proceedings of the Special Interest Group Meeting Plasmalemma redox functions in plants* at the XIV International Botanical Congress, 24 July – 1 August, 1987, Berlin, FRG. Issued as a separate fascicle [reprinted from Physiologia Plantarum, vol. 73: 161-200], Copenhagen, May 1988.

[SIGM-20, see 5-30.]

SIGM-22. A. R. Kranz (ed.): *Proceedings of the 22nd Special Interest Group Meeting* (XIVth International Botanical Congress) Berlin (West), July 29. 1987. (Arabidopsis Information Service, No. 25.) Frankfurt a.M., "December 1987" [1988], viii + 142 pages.

Exc. 44. J. Paré: XIV International Botanical Congress - Berlin 1987 - *Botany in Picardie,* 16th - 23rd July 1987. Amiens, [1988], 91 pages and folded insert in colour, numerous figures.

Congress Reports *

K. Esser: Bericht über die Tätigkeit des Organisationskomitees für den XIV. International Botanical Congress Berlin über den Zeitraum von Sept. 1982 bis Sept. 1984, vorgelegt vom Vorsitzenden. – Ber. Deutsch. Bot. Ges. 98: 487-488, December 1985.

K. Esser & W. Greuter: Aktuelles vom XIV. Internationalen Botanischen Kongreß. – Ber. Deutsch. Bot. Ges. 99: 1-2, May 1986.

W. Nultsch: IBC und "Waldsterben". – Bot. Acta 101: A6-A7, February 1988.

K. Esser: Zum Thema "Wald". – Bot. Acta 101: A8, February 1988.

K. Esser & W. Greuter: Abschließender Bericht über den XIV International Botanical Congress Berlin 1987. – Bot. Acta 101 (in press).

* Official reports by the Congress and its sponsors only. No attempt has been made to list the countless reports published by individual Congress members on parts or all of the Congress or its excursions.

Figure 7. The Opening Ceremony in the huge lecture theatre 1 of the International Congress Centre: A glimpse at the audience.

Figure 8. The front row at the Opening Ceremony, with (from the right) F. A. Stafleu, K. Esser, Mrs. Esser, G. Turner, D. Heckelmann (President of the Free University), W. Nultsch, W. Greuter, O. Kandler, Mrs. Kandler, H.-D. Behnke, Mrs. Behnke, A. Takhtajan, Mrs. Takhtajan, R. Robertson, Mrs. Robertson, and J. Cram.

PART II

OPENING CEREMONY OF THE CONGRESS

Opening of the Congress – by K. Esser, President of the Congress

Herr Senator, Dr. Ride, Herr Vorsitzender der Deutschen Botanischen Gesellschaft, Mr. Honorary President of the Congress, Ladies and Gentlemen,

On behalf of the Organizing Committee, I have the honour to welcome you to the XIV International Botanical Congress. We are very appreciative and proud of the fact that so many of you have accepted our invitation. We hope that the next few days will enable us to exchange ideas and experimental data, irrespective of geographical distances and of political barriers, in a free and open atmosphere. We also hope that this Congress, and especially our social events, will help to renew old friendships and to create new ones.

While we, the members of the Organizing Committee, were busy making preparations for this Congress, we repeatedly asked ourselves: Is such a big Congress really effective now, as we approach the end of the twentieth century? We were also aware of the fact that this question is being discussed all over the world, and that there are many people who say: "Scientific knowledge can really be circulated and communicated only at smaller meetings, in the framework of conferences, where specialists can meet and exchange ideas, in an atmosphere where everyone knows his colleagues personally and has an understanding of the details of their work".

This view has its merits, but we are of the opinion that a large Congress is needed and even more effective, not only on the grounds of purely technical merit, but also because of valid scientific reasons as well. First of all, we have organized the Congress in such a way that the main emphasis is placed on the symposia and on special interest groups, having attached posters to the symposia, so as to enable the advocates of specialized symposia to meet while at our Congress just as they do in isolated conferences. What is more, each participant will have the opportunity to expand his or her own area of expertise and to be informed of advances in other fields, resulting in a teaching and learning process during this Congress, a sort of intellectual symbiosis, which is the main advantage of any large Congress. This process of affiliation becomes evident during the General Lectures, which are designed to give someone unfamiliar with a certain topic a clear and concise introduction to the essential points of that special subject area.

Our second reason for favouring a large Congress is justified by the nature of botanical science per se: *scientia amabilis* has been rapidly developing over the last decades, a fact which I can attest to by merely comparing the VIII Botanical Congress of 1954 in Paris, which was the first in which I participated as a young man, with the present Congress. In the space of 33 years, which corresponds to only one generation, enormous developments have taken place in the field of molecular biology, and these developments have also been influencing the field of botany. More and more – not only in genetics but also in taxonomy – molecular biological methods are being used to trace the science of life from marginal descriptions on the morphological or biochemical level to the basis of genetic information underlying structure and function.

In addition to this burgeoning interest in molecular biology, yet another important development has occurred in recent decades. Because of external political impulses, interest has shifted to problems of environmental biology, that is, ecological questions. This finds its expression during our Congress in the numerous excursions showing the broad spectrum of European vegetation, ranging from marine algae to virgin forests and therewith emphasize the concern for environmental protection. In order to deal with these questions, one must have a profound knowledge of anatomy and morphology; i.e., of classical botany. I must admit that education in this field has been neglected at many universities in recent years, which was probably a result of the lack of interest on the part of young botanists, who were principally, if not exclusively, interested in molecular biology. Luckily, a reverse in this trend is presently being observed, which will hopefully lead to wellbalanced education and research in the field of botany, from classical botany via ecology and physiology to molecular genetics.

There is currently a further development, which has been established over the past few years, namely biotechnology. That is to say, the deliberate and concerted exploitation of the physiological activities of cells to produce certain substances or to convert certain substances into others. Although biotechnology basically deals with microorganisms and the biomedical industry, it is presently a known fact that higher organisms, especially plants, can also be systematically included in biotechnological processes and will undoubtedly revolutionize agriculture.

These concerns and considerations have promoted us to give this Congress two crucial points of emphasis, one being from the field of ecology, namely "Forests of the World", and the other being from the field of genetics, namely "Biotechnology". These subjects will also be reflected in the General Lectures which hopefully will focus on the connection between classical and modern botany.

In this manner, we strive to portray the full and broad spectrum of botany, which affects not only science but society in general. We believe, that, in the course of this international Congress, we are in a position to inform a broad international audience of scientists and to give them food for thought with regard to various fields of science in which they are not personally active. We however also believe that, thus nourished, we can demonstrate to the external public that botanists are capable of actively dealing with current problems which affect all of us, namely, the preservation of our environment as well as the application of special capabilities of living matter through biotechnology, and thereby continually improve human welfare. I hope that you share with me, that these considerations justify the integrative character and interest of our Botanical Congress.

Let me use this occasion to express our sincerest feelings of gratitude to our predecessors, represented by Sir Rutherford Robertson and Prof. Cram, who, as president and secretary general respectively, were responsible for the XIII International Botanical Congress at Sydney, for their valuable support and advice.

In addition to my English welcome to each of you, allow me to express representatively for all people not of English tongue some special greetings to various groups of people of different geographical origin, namely the Francophone, the Spanish-speaking, the Slavonic, those from the Far East and, last but not least, those speaking my mother tongue.

Mesdames et Messieurs,

C'est très chaleureusement que je souhaite la bienvenu à Berlin à tous nos invités des pays francophones. Je rappellerai à cette occasion que le premier congrès de botanique a eu lieu à Paris en 1900. C'était également le cas du huitième congrès de 1954, il y a deja plus de 30 ans, qui a donné une forte impulsion à la botanique.

Bienvenu à Berlin!

Muy estimados señoras y señores:

Tengo el honor de saludar a los numerosos botánicos de habla hispaña procedentes de Europa y muy especialmente de Sudamérica; en este saludo quiero incluír también a los botánicos de habla portuguesa.

¡Bienvenidos a Berlin!

Многоуважаемые дамы и господа!

Я рад, что имею возможность приветствовать и многочисленных биологов - ботаников из славянских стран. Ваше присутствие здесь является для меня подтверждением тому, что сотрудничество, с давних пор развивавшееся среди ученых, будет не только продолжаться, но и все более углубляться. В связи с этим мне хотелось бы выразить надежду на то, что это наше сотрудничество будет содействовать дальнейшему более интенсивному обмену мнениями.

御来場の皆様！

遠路はるはる、ようこそベルリンにお出で下さいました。東アジアの植物学者代表として本会議に御出席の皆様と学術上の意見と情報を交換し、友情をあたためる機会が与えられましたことを大変嬉しく思っております　なれない日本語でお聞き苦しいとは存じますが、私共の心からの歓迎の気持ちをお伝えし、次回は東京でお目に掛かれることを希望して、御挨拶に代えさせて頂きます。

Sehr verehrte Damen, sehr geehrte Herren,

Es ist mir eine ganz besondere Freude, die Botaniker aus den deutschsprachigen Ländern hier in Berlin begrüßen zu dürfen. Ich darf diese Gelegenheit benutzen, einen Dank an diejenigen unter Ihnen auszusprechen, die uns bei der Organisation von Symposien und anderen Veranstaltungen sehr uneigennützig geholfen haben. Fühlen Sie sich mit uns im

Organisationskomitee als Gastgeber und helfen Sie uns, allen Gästen den Aufenthalt in Berlin so angenehm wie möglich zu machen, damit für Sie und damit auch für uns dieser Kongreß zu einem Erfolgserlebnis wird.

Auch Ihnen ein herzliches Willkommen in Berlin!

Purposely, we have kept the official opening rather short in order to be able to enjoy some German music and, afterwards, to take part in the reception, for which we are indebted to the Governing Mayor and to the Senate of Berlin.

I hope that this being together and the other social events will stress our emphasis that we may proceed on the way that all botanists feel as a great international family, sharing scientific work as well as enjoyment afterwards.

Again, welcome to Berlin, the old German capital, which celebrates its 750th anniversary this year, welcome to Germany and welcome to the XIV International Botanical Congress, which I herewith declare to be opened. I wish you all a very successful and enjoyable meeting.

Welcome to Berlin* – by G. Turner, Senator for Science and Research

Presidents and Chairmen, Ladies and Gentlemen,

The Berlin Senate, especially the Governing Mayor, cordially welcomes you to our city.

Botany – like other sciences – has been at home in Berlin for a long time. Visible evidence of this are the Botanical Gardens which have existed in this city for more than 300 years.

The Electors of Brandenburg and later the Kings of Prussia began in good time after the ravages of the Thirty Years' War and at the start of the modern age to promote science and up-to-date production methods by farsighted and intelligent government policies.

Many important scientists – among them Alexander von Humboldt – established Berlin at that time as a significant centre of German science.

In this year of celebrating the 750th anniversary of our city we look back with pride at the achievements made.

---

* This is the official English translation of Professor Turner's text, that was presented in German.

However, merely looking back and nurturing tradition is not enough. What is more important is to face up to the challenges of the present and the future. In this, science and research have major roles to play. Berlin with its great number of scientific institutions is one of the leading locations of research, and not only in Germany.

After the end of World War Two the first challenge was to repair the damage it caused. The reconstruction and extension of existing research establishments took priority. It was, for instance, during that time that we were able to complete the reconstruction phase in the Botanical Gardens by re-opening the library and herbarium wings of the Botanical Museum. The damage to the herbarium was almost completely repaired and the library is again among the major German specialised libraries. You will be able to visit the Botanical Gardens yourselves and see the new and fine buildings there.

As part of its efforts to emphasize future-oriented fields the Berlin Senate also promotes botany in the widest sense. Having established the Institute for Gene Biology Research Berlin Ltd. with emphasis on plant genetics, we are now attempting to coordinate many individual approaches in the field of biotechnology in Berlin in order to be able further to develop this field into a dynamic area of research.

It is, for this reason, particularly gratifying to find that your Congress is also dealing with the subject of biotechnology, thus offering Berlin scientists the opportunity to gain instructive suggestions and information and to form new contacts.

The central topic of your meeting, however, "The World's Forests", will attract attention not only from Berlin's scientists but the whole city. One need not invoke the special emotional ties of the Germans to their forests to explain the Berliners' special interest in the wooded areas of their city. I hope that you will want to gain an impression of the special geo-political situation of the city and its political implications in order to understand the importance of the woodlands of this city for the lives of the people. Climate and atmospheric conditions, water supply, leisure and recreation are all influenced by the belts of forests of the city. Any change, any damage, therefore, attracts the attention of large sections of the population. So it is understandable that an interdepartmental Berlin research project is looking at forest ecosystems close to conurbations. The problems of the present – and not only in the environmental context – appear to be focused on conurbations like Berlin.

Aware of the importance of these comparatively small wooded areas for our existence, we see with great concern the recession of the tropical rainforests and the global implications involved.

Your joint efforts to solve problems affecting people all over the world could be a fine example of the strength of scientific work beyond national borders. In a divided city, the symbol of our divided Europe, this could be a hopeful sign of future peaceful cooperation.

I wish your Congress every success.

I hope that you will see a little of our city on this occasion and that you will come back soon.

Greetings from the International Union of Biological Sciences

by W. D. L. Ride, Vice-President of the IUBS

Mr. President, Professor Dr. Karl Esser, and Distinguished Guests,

It gives me great pleasure to bring to the XIV International Botanical Congress the warm greetings and good wishes for the success of the Congress from the President and members of the International Union of Biological Sciences.

The International Botanical Congresses have played a vital role in the development of plant sciences for more than a century. In particular, they have promoted and maintained understanding and communication between botanists irrespective of the many artificial boundaries that are created by humans from time-to-time. The free communication of scientific thought and the free movement of scientists are ideals which all the Unions of the ICSU firmly hold. This International Congress and the congresses of other member bodies of IUBS are an embodiment of that ideal.

It gives me particular pleasure also, to greet and thank on behalf of the International Union, the citizens of the city of Berlin in their role as hosts to the Congress. We in Australia were the hosts to the XIII International Botanical Congress in Sydney in 1981 and I am sure that my Australian colleagues will tell you that in addition to the friends you will make, you will find that the generous and voluntary exchange of ideas with your visitors will be a great benefit to botanists in Germany and especially in your city.

Finally, I am sure that my colleague, the President of IUBS, Professor Otto Solbrig, who is himself a distinguished botanist, and who sends his apologies for not being here in person, would wish me to congratulate Professor Dr. Esser and Professor Dr. Greuter and their colleagues of the Organizing Committee for the successful conclusion of their work in bringing this great Congress together.

If I may speak on behalf of all your visitors, many of us know you already as great contributors to international science in many areas of scientific endeavour. The success of the Congress will increas the debt which international botanical science owes to you already.

Welcome Address – by W. Nultsch, President of the German Botanical Society

Mr. President, Herr Senator, Mr. Honorary President, dear Colleagues, Ladies and Gentlemen,

On behalf of the Deutsche Botanische Gesellschaft, the German Botanical Society, I would like to welcome you to Berlin. As you are probably aware, the German Botanical Society is one of the oldest Botanical Societies in the world. It was founded in 1882 in Berlin, and Berlin was also the administrative seat of our society until quite recently. Therefore we were justly proud when Berlin was elected as the location for the XIV International Botanical Congress.

The German Botanical Society comprises plant biologists in the broadest sense from all German-speaking countries, but has also members in neighbouring European countries. Honorary members include scientists from the USA, USSR, Switzerland, the Netherlands and, of course, Germany. We are happy that a number of them are among us, and I would like to welcome them in particular. The German Botanical Society also represents German botanists in international federations, such as FESPP and the International Phycological Society, and cooperates with the Botanical Societies of many other countries. In this connection I should not fail to mention that the German Botanical Society has cooperated with the German Association for Applied Botany in providing financial assistance to help young scientists participate in this congress.

Although our Society is actually a little more than hundred years old, our traditional Journal, the "Berichte der Deutschen Botanischen Gesellschaft", celebrates its 100th anniversary this year with a special volume. This deals with some historical aspects of our Society and contains some selected topics of the history of German botany.

However, a hundredth anniversary is also an appropriate opportunity to make thoughts about the future, and, therefore, we have decided to revamp our journal, renaming it *Botanica Acta,* beginning with Vol. 101. You may have already seen in the pilot issue, that the format, the publisher and the editing procedure have been changed. We hope that it will enjoy acclaim as an international journal of plant science.

I would like to take this opportunity to express my thanks first to our Organizing Committee for having done such a splendid task of organizing this Congress, and of course secondly to the numerous sponsors whose donations made this undertaking a feasible proposition. Last, but not least, I would like to thank you, ladies and gentlemen, for coming, for without your continued support and interest there would have been no hope that we could get this far.

Finally it remains for me to wish you a stimulating and rewarding Congress. In so doing, we hope that you will take the opportunity to participate not only scientifically, but also culturally by partaking in some of the numerous excursions in and around Berlin.

The History of Botany in Germany – by F. A. Stafleu, Honorary President of the Congress

Mr. President, Herr Senator, Ladies and Gentlemen,

I should like to take this opportunity offered to me by the Congress to pay a tribute to German botany. I want to do this by telling you something about the 1905 forerunner in Vienna of the present Congress and by giving just a few examples of the impact made by Germans on our branch of science, as well as a few statistics on German botany through the ages.

The International Botanical Congresses originated in the nineteenth century. The first Congress was held in Brussels in 1864. This Congress, however, just as all others in the 19th century, was both botanical and horticultural and each coincided with a large horticultural exhibition. Before 1864 there had already been numerous regional meetings of groups of scientists, such as, for instance, those of the Deutsche Naturforscher. These German meetings of natural scientists actually had an international character almost from the beginning, because scientists from all over Europe used to attend. We have to realize that in this period, between 1830 and 1864, the greater part of the work in the natural sciences was done in Central and Western Europe.

These German meetings of natural scientists soon became unwieldy with the growth of science as well as of the railroads, steamboats and comfortable hotels. As the need for more specialized congresses arose, the geographers and the geophycists took the lead. The first International Zoological Congress was held in 1889, that of the chemists in 1892.

The modern botanical congresses started with Paris, 1900. This was also the Congress with which the present numbering started. This first Congress of the 20th century in Paris was still held in conjunction with a great

"Exposition universelle", a world fair. The first independent botanical congress, however, fully comparable with the present Congress, was that of Vienna in 1905.

I think that we should use th term "German botany" in the sense of botany in the German language area. The enormous body of German botany in the previous centuries arose not just from the territory now covered by the Federal Republic, the city of Berlin and the DDR. German language and culture, and consequently German botany, belongs also to German Switzerland and Austria in the sense of the pre-1918 Austro-Hungarian empire. The Vienna Congress of 1905, the first modern botanical congress, was therefore also the first German botanical congress, and until today it has been the only one. It is therefore very appropriate for us now to meet once again on German soil. The Botanical Congresses, from 1930 onward have fallen under the auspices of the International Union of Biological Sciences, in particular under those of its Division of Botany. The site of most congresses was by invitation, but sometimes IUBS took the initiative, as was the case with Edinborough (1964) and, now, Berlin. Some twenty years ago an IUBS council member came to Berlin, and visited – he was a plant systematist – the Botanical Museum and Herbarium, one of the greater German botanical institutions, not least because one of its former directors was the grandiose organizer and scientist Adolf Engler. The visitor then suggested to the staff that they should start thinking about an International Botanical Congress in Berlin. The staff was surprised and doubtful; but here we are, in Berlin, a classical metropolis of botany; and the Botanical Museum has been the nerve centre of the organization.

Our president has just referred to the Paris Botanical Congress of 1954, which for both him and myself was our first. He pointed out the different size and scope of that Congress. The Vienna Congress of 1905 was of course even smaller: it had some 500 participants against the 4000 today. Even so 500 was a good show for that period. It was a good time for travel and international meetings. One could travel freely all over Europe without visa or exit permits. Vienna, then the metropolis of German culture *sensu lato* was a fantastic host, and the Congress was a fascinating event at which German botany presented itself to its best advantage.

In all there were 12 general, all-Congress sessions in addition to the more specialized events. The two Austrian botanists in charge were Richard von Wettstein, a great systematist, and the plant physiologist Julius Wiesner.

The opening all-Congress lecture was given by the German botanist from Kiel, Johannes Reinke, at that time a foremost experimental morphologist and algologist who spoke about "Hypotheses, suppositions and problems in biology".

A major theme, treated in various sessions, was the Development of the Flora of Europe from the Tertiary era onward, introduced and organized by Adolf Engler. Here, in Berlin, the city in which he brought his institute to world-wide eminence, he stands as a shining example of German contributors to descriptive botany.

A second important general theme at Vienna was the current state of knowledge of carbondioxide assimilation, treated in a symposium organized by Strasburger and Wiesner, who paid a tribute to the two great founders of modern plant physiology, Julius Sachs and Wilhelm Pfefer. The personal influence of Sachs on experimental botany has probably never been surpassed. Botanists of my generation on the continent of Europe have all been brought up with Sachs, Pfeffer and Strasburger: another tribute to German botany.

The third major symposium at Vienna was on regeneration, presided over by another giant of those days, this time from Munich, Karl von Goebel. Goebel and von Wettstein, together with the Dutch geneticist Lotsy and the Pole Raciborski, stood at the cradle of the then brand-new International Association of Botanists, the forerunner of the botanical division of IUBS.

Our president has also referred to the important place taken at this Congress by environmental botany, especially with respect to the devastation of the large tropical forests. At Vienna a resolution was passed in 1905 to stop the devastation of the forests in Bosnia (now part of Yugoslavia), then in progress.

I could give more examples of the pioneer efforts of German botany. After all, modern botany as it evolved after the development of book printing (and the shift to a heliocentric rather than a geocentric concept of the universe) was founded by three German scientists: Brunfels, Bock and Fuchs, the German fathers of botany. Botanical exploration on a large scale and especially on scientific principles began with Alexander von Humboldt; studies of sexuality, pollination and fertilization go back to Camerarius, Kölreuter and Christian Konrad Sprengel. In genetics the Moravian Mendel may be mentioned; in morphology a great series of eminent men from Goethe to Goebel. I could elaborate this theme indefinitely, but it is not necessary. Rather than mentioning more names I should like to give you a few rather remarkable statistics illustrating the contributions of German scientists on the world's botanical literature and therefore knowledge. Such statistics are rare, but I can give you some data with respect, especially, to descriptive botany between 1750 and 1950, based upon a recently published encyclopedic work dealing with some 7000 botanical authors from all over the world who published their work in those two centuries. The number of publications treated in this selective and critical bibliography of these 7,000 authors is c. 16,000.

I shall give you only five figures: of those 7,000 authors 14% were French, 12.5% came from the United Kingdom, 11% from the USA, and, now it comes: 25% came from Germany in a restricted sense. However, as I said before, the German cultural area was equivalent to the German language area, the percentage for which is 33.6% (and this against 23.5% of the English language area). This is an imposing illustration of the historical impact of German culture on botany during these two centuries.

I have no similar figures for, let us say, experimental botany or genetics, but a survey of lichenological literature, covering all disciplines applied to lichens and aiming at complete coverage between 1750 and 1950 based on 3,865 authors, provides almost identical figures.

Ladies and gentlemen, the purpose of this short address was in the first place to pay tribute to German botany and botanists. Here in Berlin we find ourselves, as I said before, in one of the cultural capitals of the world, and, for us more especially a capital of 19th and 20th century botany. I will mention only one further aspect, often overlooked: the enlightened policy in Prussia and, after 1871, in the whole of Germany, of strong governmental science policy. During the hectic industrial, economic and financial expansion of this country after 1850 the German states played an important role in setting up research institutions, such as experiment stations, in Europe as well as in the tropics, and in introducing a rigorous selection of the scientists appointed to be the leaders of pure science, both in the institutions and the universities. For my German colleagues I need to mention only the name of the great Berlin administrator Friedrich Althoff, who had a preeminently beneficial influence on shaping the natural sciences. Traces of Althoff still exist and flourish today; I mention only the Deutsche Forschungsgemeinschaft, the Max-Planck-Gesellschaft and, to come back to our Congress, the Berlin Botanical Garden and Museum.

In conclusion I congratulate our German colleagues on the organization of this Congress and thank them for having brought us all to Berlin to participate in what will be remembered as the most comprehensive manifestation of botany so far in history.

The Musical Programme* of the Opening Ceremony – by K. Esser, President of the Congress

With this musical programme, we would like to give our foreign guests a short glance at the tradition of symphonic music in Germany – from classic to romantic. The Academic Festival Overture serves as a greeting. May the gaily and easily flowing music of Johann Strauss, which ends the musical event, help you to a pleasant stay in our country.

Akademische Festouvertüre, op. 80 Johannes Brahms (1833-1897)

As an expression of gratefulness to the University of Breslau for lending him the title of "Doctor honoris causa", and in some measure as a graduation speech, Brahms thanked the faculty of philosophy by means of the Akademische Festouvertüre (Academic Festival Overture), premiered in 1881. It is a sort of "potpourri overture", based on popular student songs of that time. The most famous "Gaudeamus igitur", which is still today played or sung at academic festivities in Central European countries, forms the mighty finale in this symphonic composition, which was somewhat unusual for its time.

Prelude to "Die Meistersinger von Nürnberg" Richard Wagner (1813-1884)

With "Die Meistersinger" (The Master Singers), Wagner set a great monument to the powerfull guilds of the Middle Ages, which were the carrying strength of the democratically ruled free cities independent of the German principalities. In the prelude, Wagner gives an overview of the scenes of the opera, from the ceremonial entrance of the master singers in the festival hall, to the "Singers' Congress", the songs of the popular figure of the shoemaker Hans Sachs, and the tender love story of the Knight von

* The musical programme was performed by the Westfälisches Sinfonieorchester (Conductor: Walter Gillessen).

Stolzing and a burgess girl; von Stolzing is finally accepted by the township after a singing contest. It becomes obvious that the guilds had not just political influence but also cultural impact on the towns of the Middle Ages.

Overture to "Leonore" 3, op. 72 a — Ludwig van Beethoven (1770-1827)

Beethoven wrote a total of four overtures to his single opera "Fidelio": the first three became known as the "Leonore Overtures". Leonore 3 (premiered in 1806) is a good example of the unremitting compositional style which Beethoven used while trying to capture the essence of the Fidelio material, that is, not the actual scenes from the opera itself, but rather the conception of the Fidelio drama, namely the faithfulness of a wife to her husband, who was unjustly imprisoned by a tyrant.

Symphony No. 3 in E Major (Die Rheinische), op. 97, first movement — Robert Schumann (1810-1856)

Schumann, who, during his work in Düsseldorf, was inspired by the easy way of life in the Rhineland, dedicated his third symphony (premiered in 1851) to his home of choice. Above all, Schumann consciously sought to pay homage to the Rhinelanders' cheerfulness, which is expressed in the liveliness of the first movement, whereby he, at the same time, sought to characterize the sentimental side of the Rhinelanders.

Don Juan, op. 20 — Richard Strauss (1864-1949)

Richard Strauss achieved his great breakthrough at the age of 25 with his composition based on the just-finished poem "Don Juan". The supple, vivacious sound which, from then on, became a typical characteristic of his instrumental style was displayed for the first time in the premiere of "Don Juan" in 1889. Lenau's poem served as the leitmotif for Strauss, who freely composed this work in the form of a sonata. The contents arose in a manner totally different from Mozart's "Don Giovanni", that is, Strauss tried to vividly describe the ups and downs and, finally, the tragic end of the title figure Don Juan.

Figure 9. The front table at the Closing Ceremony. First row (from the left): M. Furuya, W. Greuter, K. Esser, F. A. Stafleu, R. Robertson. Second row (from the left): D. von Wettstein, F.-H. Wang, G. Melchers, A. Lang, J. Kornaś, C. C. Heyn, J. Heslop-Harrison, N. Grobbelaar (half-hidden), E. I. Gabrielian (hidden), K. Faegri, J. McNeill, J. Cram, A. Takhtajan, W. Franke, W. Nultsch, O. Kandler, W. Haupt, H.-D. Behnke, H. Sukopp, B. Zimmer.

Figure 10. The president, K. Esser, congratulates Brigitte Zimmer, scientific secretary, while awarding her the Congress Medal.

PART III

CLOSING CEREMONY OF THE CONGRESS

Opening Address – by K. Esser, President of the Congress

I hereby open the Final Plenary Session of the Congress, which is in the same time the General Assembly of the International Association of Botanical and Mycological Societies, abbreviated IABMS. This organization is part of the International Union of Biological Sciences (IUBS) and has replaced the former Division of Botany, which was abolished when the IUBS was reorganized some years ago. The IABMS has sections and commissions which embrace altogether the broad spectrum of botanical disciplines.

I think it is necessary before considering the business before us to explain in just a few words our organization, especially since many of our members may be attending an International Botanical Congress for the first time.
According to the by-laws of the IABMS, the President of the Congress is ex-officio also President of this organization. The Chairman at present is Professor Greuter and the Secretary at present is Professor McNeill. They were elected by the member organizations.

There is one section, namely General Botany, which is responsible for the organization of the botanical congresses. This section is chaired by Professor Frans Stafleu, to whom we are indebted for the continuity of botanical congresses during the last 40 years and we all hope that this continuity will go on and on and on.

[The opening address was followed by a report on the Congress, by the Secretary General, W. Greuter; this is not repeated here since this would duplicate information given elsewhere (see Part I).]

Invitation for the XV International Botanical Congress

A Committee to receive invitations for the XV IBC had been appointed by the President of the Congress. As announced in the first Daily Bulletin it comprised F. A. Stafleu (Chairman), K. Esser, W. Greuter, M. Furuya, E. Gabrielian, and P. H. Raven.

The Chairman of the Committee reported that the Committee had received one invitation, namely to hold the XV International Botanical Congress in 1993 in Japan. The Committee, having satisfied itself that this was a *bona fide* invitation and that, in particular, the International Council of Scientific Unions' (ICSU) rules on the free circulation of scientists would be followed, recommended that this invitation be accepted.

The audience accepted the invitation by acclamation and without dissent.

Address of Welcome to the XV IBC – By M. Furuya, Chairman of the Committe for the XV International Botanical Congress, speaking on behalf of Japanese plant biologists.

President Professor Esser, Ladies and Gentlemen,

On behalf of Japanese plant biologists we gratefully appreciate your decision to hold the XV International Botanical Congress in Japan. We are especially honored as this will be the first time for the Congress to be held in Asia. We do hope, that all of you, those who were unable to be here, and many who have not yet attended an International Botanical Congress will join us in Tokyo. We look forward to seeing you in Japan in August 1993.

Congress Resolutions

The President of the Congress had appointed a Resolutions Committee, consisting of himself as chairman and by F. A. Stafleu, W. Greuter, K. Faegri, N. Grobbelaar, J. Heslop-Harrison, C. C. Heyn, J. Kornaś, A. Lang, and A. Takhtajan. The committee had been announced in the first issue of the Daily Bulletin, together with the rules to be followed to submit resolutions to the Congress.

On behalf of the Committee, F. A. Stafleu reported that all resolutions that had been submitted had been considered in great detail. The Committee had endeavoured to make their text as concise as possible, to avoid duplication and conflicting statements. The text so revised had been printed in No. 7 of the Daily Bulletin. He moved that the audience accept the resolutions, as printed, as Resolutions of the Congress.

The motion having been seconded, and after some discussion and amendments, the Resolutions were accepted without dissent, as reprinted below.

Resolutions of the XIV International Botanical Congress as adopted at its Final Plenary Session, August 1, 1987

Resolution 1

Whereas the whole world has a real stake in the survival of biological diversity;

whereas local expertise, involvement and dedication will be the safeguard for the preservation of the world's biological resources, especially in the tropics, which the population of the countries concerned will inherit;

the XIV International Botanical Congress urges botanical institutions of the world to render assistance to institutions in other countries, where appropriate, to train local experts and so to enable local biologists to take leadership roles in efforts to maintain their biological patrimony, not only for their own benefit, but for that of the whole world.

Resolution 2

Recognizing the severe threat to life on earth through the global degradation and destruction of forests and other ecosystems;

the XIV International Botanical Congress urges the governments of all countries and the appropriate non-governmental organizations to work toward the conservation of natural resources and the diversity of life through the preservation of natural habitats, and to support the maintenance and development of systems of human land use on a sustainable basis.

Resolution 3

Mindful of the world-wide unabated decline in natural vegetation, particularly forests, and of the consequent threat to global resources;

recognizing the need for factual information for the development of meaningful conservation strategies;

the XIV International Botanical Congress urges that universities and other appropriate centres maintain and develop strong teaching and research programmes in systematic and ecological biology.

Resolution 4

Whereas there is an increasing concern among the scientific community and the general public regarding world-wide forest decline;

whereas the symptoms of this decline suggest that in some instances the process is temporary and/or natural, but in other cases is certainly due to the complex interaction of a number of anthropogenic factors, including air pollution;

whereas the problems of air pollution and the symptoms of forest decline are international in scope;

the XIV International Botanical Congress urges the International Union of Biological Sciences and other appropriate organizations to facilitate international standardization in describing symptoms of the decline and in gathering pertinent biotic and abiotic data, and to promote the exchange of information between research groups dealing with these problems.

Resolution 5

Noting the great importance of botanical gardens as cultural and scientific centres with great educational value;

recognizing their importance for the conservation of plants through living collections;

the XIV International Botanical Congress urges authorities in all countries to maintain, develop, and, where appropriate, create such gardens, and to ensure their adequate finanical support.

Resolution 6

Noting the resolution adopted by the International Union of Biological Sciences General Assembly in support of the International Programme for the Study of Global Change, and the resolution adopted by the XIII International Botanical Congress on global mapping of the vegetation of the earth;

recognizing that advances in space technology and computer sciences open new horizons in the observation of the earth from space;

the XIV International Botanical Congress requests the International Union of Biological Sciences to institute, with the International Council of Scientific Unions, the preparation of a vegetation map of the earth.

Resolution 7

Considering the great importance of a stable system of scientific names of plants for all users in the pure and applied sciences and in many other domains of public life and economy;

recognizing the frequent difficulties arising in the choice of the correct names under the International Code of Botanical Nomenclature;

the XIV International Botanical Congress urges the International Union of Biological Sciences to promote the study of the development of a system for the registration of plant names.

Resolution 8

The XIV International Botanical Congress resolves that the decisions of its Nomenclature Section with respect to the International Code of Botanical Nomenclature, as well as the appointment of officers and members of the nomenclature committees, made by that Section during its meetings, July 20 to 24, be accepted.

Presentation of the Hedwig Medal

H. Inoue, President of the International Association of Bryologists (IAB), consigned the medal to Barbara M. Thiers to hand it over to the awardee, W. C. Steere of the New York Botanical Garden.

Presentation of the Engler Medal

S. W. Greene, Past President of the IAPT, presented the first Engler Medal in Gold, as follows.

Some years ago the members of the International Association for Plant Taxonomy decided to initiate an award system as a means of honouring taxonomic botanists who are making outstanding contributions to the development of plant taxonomy. Clearly such an award had to bear an appropriate name and when we considered the legacy of former plant taxonomists and the awards and distinctions with which many of their names are already associated we came to the conclusion that the medal I am called upon to present today should bear the name of Adolf Engler, the renowned Director of the world famous Botanical Garden and Museum here in Berlin. Engler's botanical philosophy and vision together with his achievement of creating a series of unifying taxonomic works, of which "Die natürlichen Pflanzenfamilien" is surely the greatest, seemed to us to epitomize perfectly the sort of achievement we wished to recognize.

People with such abilities are few in number and for this reason this gold medal is intended to be awarded to only one person at a time at succes sive botanical congresses. I give this background information since this is the first occasion on which the award is being made, by a happy coincidence here in the city with which Engler's name will for ever be associated.

As you can imagine it was not easy to select a recipient for this award. Yet as we tried to stand apart from the current scene and identify a taxonomic botanist with a comparable breadth of vision to Engler's and who in his or her work was creating something that for long will be acclaimed as an achievement of outstanding scholarship and a major contribution to the literature of our subject one name stood out.

Although that person's own research publications are not great in number by some standards, they are rich in content, perceptive in interpretation, broad in scope yet so skilfully organized and presented as to give a masterly synthesis of a complex subject. Through the clarity of the printed record, they provide an abundance of knowledge and stimulation to phanerogamists and cryptogamists alike. By virtue of the recipient's generosity with ideas and a willingness and ability to communicate these to others, this person has additionally brought into being, amongst other things, an impressive series of publications many of which are already widely used as essential reference works in the field of the history and practices of taxonomic botany.

His legacy is "Regnum vegetabile", a series already numbering well over 100 volumes of which his own 7 volume "Taxonomic literature" is to many the most influential work in this outstanding series.

I refer to Prof. Frans Antonie Stafleu of the University of Utrecht, to whom, on behalf of the officers and members of the International Association for Plant Taxonomy, and with great personal pleasure, I present this medal.

Presentation of the Eriksson Medal

On behalf of V. Umaerus, President of the Eriksson Prize Fund Commission, the President of the Congress, K. Esser, presented the Eriksson Medal to Dr. Paul S. Teng. He spoke as follows:

In 1923 a fund was created in honour of the renowned Swedish mycologist and plant pathologist, Jakob Eriksson. The fund is administered by the Swedish Academy of Science. The International Botanical Congress at Stockholm in 1950 adopted a resolution that the Section of General Botany of IUBS (International Union of Biological Sciences), through a committee of experts elected by the Section, before each Botanical Congress nominates a candidate for the Jakob Eriksson prize which consists of a golden medal. The winner shall receive the medal on the plenary session of the Congress.

The mandate of the committee is to nominate "a candidate of distinction, belonging to the younger generation, in recognition of his research in mycology, in plant pathology or in virus diseases, or of a particular publication dealing with such subjects, with the understanding that the work being so recognized is of a distinct international value and merit".

The awardee of 1987 is: Dr. Paul S. Teng

He is native of Malaysia and went to university there. He received his Ph.D. in New Zealand and spent one half year in Wageningen (Netherlands). After this he made a sky-rocketing carreer in the USA, where he holds a professorship at the Department of Plant Pathology, University of Minnesota. At present he keeps a two-year assignment with IRRI (Philippines), as a rice pathologist.

His contributions to phytopathology are especially in the areas of epidemiology and crop loss assessment. He contributed to the penetration of new concepts into these areas, making them more amenable to application in practice. At an early stage he saw the value of systems analysis and computer technology for the science of phytopathology and he was one of the leaders to implement the new technology in crop protection. He has given proof of a vast knowledge of agriculture and plant protection in the temperate zone as well as in the tropics.

Final Address and Presentation of the Congress Medals – by K. Esser, President of the Congress

Upon the completion of a series of experiments, a scientist must ask himself what are the results and whether or not the desired end of the experiments was reached. Further, he or she must also accept questions from outsiders who will ask if there was a good reason for carrying out the experiment in the first place and if the funds spent for it were justified. A Botanical Congress is also an experiment. Its organizers are faced with a pre-set goal and have to make experimental plans in an attempt to achieve this goal. That is why we Organizers should apply the same parameters to ourselves with regard to an evaluation of our Congress. If I should attempt to make a short assessment of this Congress, coming from the point of view of the President, it would certainly be very subjective. It goes without saying that external criticism is essential. Optimal results and further progress can be attained only through permanent and mutual constructive criticism, be it in science or with regard to organizing a congress.

To what did we aspire? Our goal was to present not merely the subject areas of botany, but also and above all, to feature developments in plant science as they relate to the current world situation. For this purpose, we chose to focus on two problem areas that have impact in completely different areas of botany, namely, "Forests of the World", which falls under ecology, and "Biotechnology", which falls under microbiology and genetics.

In the former case, we were prompted by global interest in and, above all, world-wide concern about the maintenance of tropical rain forests and forests of the temperate zones, especially in Europe. We wanted to hear first-hand information on the condition of these forests from scientists who had performed tests there. Especially here in this country, there has been a lot of talk about the forests over the past few years, which has not always originated from competent sources. We believe that, with these presentations, we have made a contribution as scientists towards factual reporting of this indisputably serious problem – which has in the recent past received much attention in press.

We are convinced that we have attained the same goal with the broad treatment of biotechnology, because, here too, there is an on-going public discussion of "pro and con", especially with regard to gene technology. We believe that we have done our part here in providing the public with competent informational sources.

We made several modifications to the standard formula for organizing a Scientific Congress Programme, namely, by dividing the scientific contributions essentially into two large groups: symposia and posters. We ourselves do not wish to judge whether and to what extent this concept found success. We will leave that up to you and to the organizers of the next Botanical Congress.

It is my belief that it is useful and perhaps even necessary that you, the participants, let us know what you think, not only for our benefit, so that we can round off the picture for ourselves and our organization, but also for the organizers of further congresses, so that the future Botanical Congresses can be even more effective. Please recall that in the zoological field, there has not been an international congress for many many years. Let us hope that this never becomes the case in our field.

My last task as President is a very pleasant one, namely, saying "thank you" to all the many institutions and persons that have helped us plan and carry out the Congress. I will attempt to do this. However, it is difficult to mention everyone, and therefore I ask for your consideration if I should happen not to call everybody by name.

First I would like to thank those institutions that have made it possible to organize the Congress at all by giving their financial support. Foremost is the City of Berlin, which, upon invitation, allowed the Congress to be located here and which, through its generous support, has allowed us to meet in the rooms of the ICC, which in contrast to my expectations has turned out to be a marvellous location.

Next is the Federal Minister for Research and Technology, who has also supported us generously. Further encouragement and support were received from the Deutsche Forschungsgemeinschaft (German Research Community) and the Deutsche Botanische Gesellschaft (German Botanical Society), the Vereinigung für Angewandte Botanik (Union for Applied Botany), the Alfred Krupp von Bohlen and Halbach Foundation and, last not least, from many industrial companies which we have listed individually in the Programme Book. At this time, I would like to express my hearty thanks to them all once again.

Next I would like to express my gratitude to our Honorary President, Professor Stafleu, the father of the Botanical Congresses since 1954, who, in his inconspicuous and very efficient way, has seen to it, that our congresses have not died out. I personally am very indebted to him, above all, for the many hours of tutoring he gave me before the XIII IBC in Sydney, when I had the honour of representing the IUBS there, and also for his invaluable support during this Congress. Therefore, please permit me to present him the first Congress Medal.

Next, I wish to thank the members of the Organizing Committee, headed by our Secretary General Werner Greuter, for their long and infatigable assistance. Without Werner Greuter, this Congress would not have been at all possible. His exemplary personal engagement, which ran parallel to his extensive official obligations as the Director of the Botanical Garden and Botanical Museum, was a full-time job. As we all know, if you called Berlin on Saturdays, Sundays, evenings and mornings, he was always on the job. I believe that no words could adequately characterize his dedication to this Congress. I think that in this case, two small words are much more precious – "Thank you", or in his mother tongue "Merci vielmals".

What would have become of our Berlin Secretariat if it had not been for our Scientific Secretary, Brigitte Zimmer. Her dedication to the Congress was in no way inferior to that of Werner Greuter. Anyone who was in Berlin often and observed the expansion of the Congress tasks can attest to the fact that the amount of work increased in inverse proportion to time remaining until the Congress. Here, too, are words not enough to match this meritous achievement. Again, "Thank you", "Danke schön".

My next acknowledgement goes to our Treasurer Herbert Sukopp, who, in his quiet way, has made sure that our accounts have always been correct. In the long years of planning, Professors Behnke, Haupt and Kandler have, as members of the Organizing Committee, not only directed the Programme Committee of their Sections and organized the Symposia, but have also assisted us by advising in all congress matters.

Now I have to think back and say a few words about the Organizing Committee. It is almost a pity that we have to part now, when we have finally learned how it should be done. We would not have to invest nearly as much work in organizing another congress, if we could put to use what we have learned while organizing this one.

I would like to thank the members of the various Programme Committees also, who have done their homework and have moulded the success to the Congress. It was namely left up to them to organize the symposia, to maintain contact with the conveners of the symposia, to arrange for the invitations for the guest speakers, and, above all, to make it clear to the "stars" within their group that a botanical congress is not a chemical congress, that is, that not every invited speaker could be fully subsidized. To them and to the conveners, my special thanks.

Almost all of the people in the last group to be named are members of the Deutsche Botanische Gesellschaft (German Botanical Society) or the Vereinigung für Angewandte Botanik (Union for Applied Botany). One should not overlook the fact that the total organizational weight of the Congress could only be born because of the many members of these two scientific organizations who helped us with the Scientific Programme here. I would therefore like to express my gratitude to both Chairmen of these two scientific societies.

I would further like to present the Congress medal to the Secretary General of the IABMS, Professor McNeill, and to our Honorary Vice-Presidents, who have supported us with their advice and active participation in the Scientific Programme and their work on the Congress Committees.

Not least, I would like to thank all the many people who helped in preparing the individual phases of the Congress or in winding them up. To name a few, the ladies in our Berlin Secretariat, headed by Mrs. R. Ziegler, and the helpers from Berlin and Bochum, who have supported us unselfishly. They are young scientists or doctoral students who put aside their interest in participating in the scientific programme to help us.

I hope you will forgive me if I will say something very personal. I am deeply indepted to Professor Stahl, who has supported me in organizing the section of Genetics and Plant Breeding in relieving me from my duties in order to give me more time for the general organization.

Vote of thanks – presented by Sir Rutherford Robertson, President of the XIII International Botanical Congress

I am honoured to move the Vote of Thanks on behalf of all those who have come as visitors from other countries. I know it will be warmly supported. As the XIV International Botanical Congress comes to a close, we all realize what a success it has been. The wide variety of interests in plants, ranging from the molecular to the community, from pure learning to practical applications for the welfare of humans and of the biosphere, has been catered for in the well-arranged symposia, general lectures, poster sessions, special interest groups and informal but often profound discussions. The success of the whole series of meetings, the excursions and the satellite activities, refutes those who say that Botany no longer lends itself to this kind of Congress. Let those who hold that view stay away, for we have had about 4300 from 81 different countries gathered to broaden each other's knowledge and stimulate each other's enthusiasms. Science is richer in consequence.

None of this would have been possible without the range of expertise and the devoted work of the many people concerned in the organization of this colossal task. I doubt whether anyone who has not been involved in arranging such a congress has any idea of the magnitude of the task; I certainly did not before the Sydney Congress. In the six years since the last Congress the work has been increasing in intensity, culminating in the splendid management of the many activities in this remarkable building, so suitable for a gathering of this type. Add to this the elaborate programme for excursions to many parts of Europe and we have witnessed a magnificant accomplishment. So much has been done by so many people but before I move the formal vote of thanks, special tribute must be paid to the guidance, enthusiasm and drive of our untiring President. Thank you Professor Esser.

I now move that the thanks of the members of the Congress be conveyed to the Governing Mayor of Berlin and our great gratitude by conveyed to Karl Esser, President of the Congress, to Frans Stafleu, Honorary President of the Congress, to Werner Greuter, Secretary General, to Brigitte Zimmer, Scientific Secretary, to the other Congress Officers and to all those who have been involved in helping them to achieve this great success, including the many sponsors from industry and commerce. We hereby resolve to congratulate them and to express our profound thanks.

The vote of thanks was approved by acclamation.

Closing of the Congress – by K. Esser, President of the Congress

Now we are at the very end of our Congress. But we should not leave until I give my very special thanks to all of you. All of you who followed our invitation to come to Berlin, join us with our Congress full of science but also, as I hope, full of happyness.

I now pronounce the XIV International Botanical Congress closed and say good bye to all of you, Auf Wiedersehen in Tokyo. Sayonara.

PART IV

GENERAL AND PUBLIC LECTURES

Introductory remarks

The full texts of the general lectures presented at the XIV International Botanical Congress, in so far as they were submitted by their authors, are included in this final portion of the Proceedings volume. They are followed by the English translation of one of the two public lectures.

When comparing the title list with the overview of the general lectures as printed on page 24, the reader will note a major title change: N. Myers submitted a free version of his Congress lecture under the new heading "Tropical forests and the botanists' community". G. Schatz and P. Quail agreed to having an extended summary of their lectures included since the full contents had already been published elsewhere. Unfortunately, three of the main lecturers (R. Goldberg, L. Bogorad and J. White) failed to hand in their texts. Nevertheless we feel that the 22 papers here published provide a representative, informative and comprehensive overall picture of present-day botanical sciences.

It is our wish to thank all contributors to this section for their helpfulness and friendly cooperation.

Berlin, November 1988 The editors

Bioenergetics and plant productivity

J. Coombs

Abstract

Coombs, J. 1988: Bioenergetics and plant productivity. — In: Greuter, W. & Zimmer, B. (eds.): Proceedings of the XIV International Botanical Congress: 99–116. – Koeltz, Königstein/Taunus.

The relationship between intercepted radiation and long-term net biomass production reflects the interaction of physical, chemical, physiological and genetic components. The optimum short term rates of photosynthesis can be described in terms of the partial processes of "light" and "dark" reactions of photosynthesis in order to define maximum rates of dry matter accumulation on the basis of unit leaf area. Actual yields reflect both the extent to which environmental and genetic factors prevent such optimum rates being reached and the development of effective leaf surface capable of light interception as well as carbon losses resulting from both photorespiration and dark respiration used both for growth related processes such as nitrogen fixation, protein synthesis and maintenance. The extent to which an understanding of these processes has increased due to development of techniques for genetic manipulation and the extent to which they will permit increased plant yields to be realized is discussed.

Introduction

The growth of all oxygen evolving green plants depends on the trapping of light energy of visible wavelength into a stable form as organic matter through the process of photosynthesis. Estimation of maximal rates of photosynthesis based on a theoretical analysis of suggested bioenergetic mechanisms can be compared with actual productivities obtained from experimental systems, natural ecosystems, agricultural crops or managed forests. Such comparisons indicate that there are significant differences between plant productivities, expressed in terms of the net gain in dry weight of biomass per unit of land area *per annum,* and those which might be expected by extrapolation from the theory, or from short term observations of maximum rates of photosynthesis. The purpose of this paper is to give an overview of present understanding of photosynthesis, and in particular what is now known about the genetics and molecular biology of key reactions, in order to assess to what extent plant productivities are determined by the primary reactions and to indicate where opportunities may exist to manipulate the process as a means of increasing plant yield.

Biomass yields

Many of the factors which limit plant productivity can easily be identified and shown to have little, or only an indirect, effect on photosynthesis. Deficiencies in mineral nutrition, extremes of soil pH, water availability or temperature as well as inherent genetic determination of plant size, leaf area, growth habit and reproductive cycle will all affect plant growth which will also be affected by a multitude of different pests and diseases (Beadle et al. 1985). Furthermore, every plant is in competition with its neighbours, whether of the same species in a monoculture stand or other taxa in a more complex ecosystem.

In many instances the effects of such environmental factors can be directly or indirectly linked to changes in photosynthetic capacity, whether through depletion of leaf area by pests, competition for light through shading, protein deficiency linked to low nitrogen availability or chlorosis associated with lack of minerals such as magnesium. Water stress, temperature extremes and low carbon dioxide concentrations can lead to stomatal closure restricting carbon assimilation. On the other hand it is equally possible to recognize highly productive systems which in general are not restricted by the factors listed above and have few genetic limitations to size ranging from unicellular algae (Barclay & McIntosh 1986), through aquatic and emergent aquatic plants such as water hyacinth (Gopal 1987) and rushes, and crops which have been bred for high levels of seed production or carbohydrate accumulation in perennating organs, to coppiced broad leaf trees such as poplar and willow. The highest productivities are recorded in monoculture stands of grasses, such as maize, sorghum, sugar cane and *Pennisetum* (Alexander 1985) which exhibit the C_4 pathway of carbon fixation (Coombs 1985).

All such systems are characterised by having a large leaf area, or in the case of the algae high cell density, which affords complete ground cover for much of the year resulting in a maximum annual interception of light, and continued growth or development of storage organs ensuring provision of a suitable sink for deposition of accumulated carbon, reducing effects of feedback inhibition on the primary processes of photosynthesis. High rates of net biomass accumulation are also associated with those plants which show lower rates of carbon loss through photorespiration and dark respiration and hence by those plants which accumulate carbohydrates rather than proteins or lipids. The reason for this is that significant amounts of respiratory energy are used in converting the sugars formed in the chloroplasts to other metabolites with conversion coefficients (grams glucose to grams product) of 0.84 for carbohydrates, 0.38 for proteins and 0.31 for lipids. These figures assume that nitrogen is being supplied in a fixed form. For plants which are capable of dinitrogen fixation additional glucose will be consumed in the process of nitrogen reduction by symbiotic or associative organisms.

Physiological basis of carbon assimilation

When measured as either rates of oxygen evolution or carbon dioxide assimilation and expressed as a function of light intensity the photosynthetic

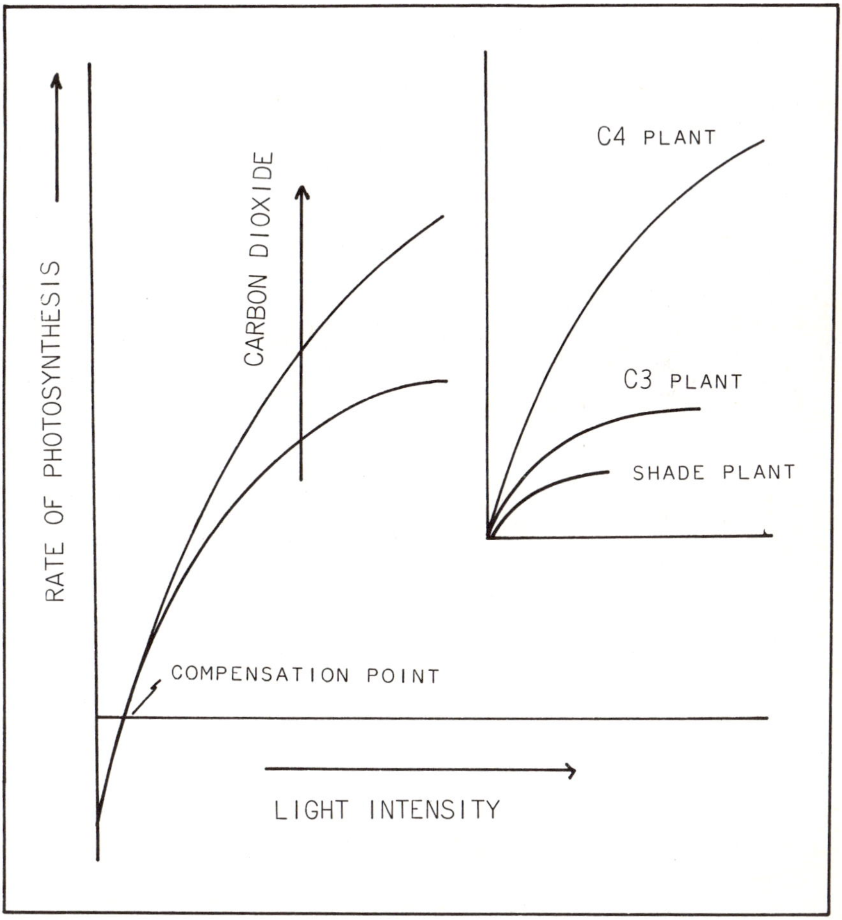

Fig. 1. Photosynthetic response curves for higher plants. The inset shows the variation found between plants adapted to low light (shade plants), C_3 plants and C_4 plants.

response curve is similar to that shown in fig. 1, with the process limited by light at low light intensities and by carbon dioxide at high intensities, although the actual response curve obtained with different species may differ significantly (inset, fig. 1). The slope at low light intensities indicates the maximum quantum efficiency (quanta per atom of carbon fixed) which may be measured in terms of μmoles of CO_2 assimilated per μE of light intercepted. Experimental values can be divided into three groups: <8, 8–10 and >10. Although low values are claimed periodically these are rejected here on the basis of evidence as will be discussed further on. Values of 8–10 are accepted as an indication of the upper limits of the quantum efficiency of the photosynthetic mechanisms now generally accepted, whilst higher values reflect actual observations in the field (Coombs et al. 1983).

As indicated in fig. 1 there are significant differences in the responses, with plants adapted to shade at one extreme and C_4 plants at the other. These differences include variations both in the intensity at which photosynthesis becomes light saturated and in the extent of carbon dioxide efflux at low light intensities, which in turn results in characteristic differences in compensation point.

If the measurements plotted in such curves are expressed in terms of unit leaf area the rate of carbon assimilation at saturating light (net assimilation rate or NAR) reflects the maximum capacity of the leaf to carry out photosynthesis on a short term basis. The capacity of a particular plant to assimilate CO_2 then reflects the product of NAR and the total leaf area (LA) and actual fixation rates will reflect the product (NAR $\times$ LA) multiplied by a factor to compensate for mutual shading of leaves within the canopy. Due to such canopy effects, rates of carbon assimilation related to unit area of the earth surface are more relevant to actual productivities than short term maximum rates of photosynthesis expressed on a unit leaf basis. Table 1 indicates some typical values for NAR, short term rates of photosynthesis and annual net productivities (Beadle et al. 1985, Coombs et al. 1985).

Carbon dioxide flux into the leaf can be expressed in terms of resistance analogue models of the following general form:

$$P = ([\,CO_2 \text{ gradient}] \times \text{diffusion coefficient}) / r_s + r_m \ldots + r_n$$

where r_s is the main variable resistance to gas diffusion represented by the stomata (see table 1), r_m the resistance to internal gaseous diffusion, and further hypothetical resistances reflect the activities of the physical and chemical reactions occurring within the chloroplasts as discussed below.

Table 1. Typical short term rates of photosynthesis, annual productivities and photosynthetic efficiencies of C_3 and C_4 plants or crop stands (derived from Beadle et al. 1985 and Coombs et al. 1983).

	C_3	C_4
Maximum rate of carbon assimilation (mg carbon dioxide per square metre leaf per second)	0.4–1.1	1.1–2.9
Maximum efficiency	4.3%	5.8%
Minimum mesophyll restistance (seconds per metre)	300–800	50–150
Minimum stomatal resistance (seconds per metre)	50–200	200–400
High short term productivities (grams carbon fixed per day per square metre land area)	18–40	39–54
High short term efficiencies recorded for field crops	1.4–4.3%	2.9–5.3%
High annual productivities (tonnes per hectare harvested dry matter)	14–40	60–85
High annual efficiencies	0.5–1.4%	0.7–2.4%

Light and dark reactions of photosynthesis

The light response curves as shown in fig. 1 may be interpreted on the basis that the overall process of photosynthesis reflects the interaction between two separate sets of reactions or partial processes (Hall & Rao 1987):

Light reactions: $2H_2X \rightarrow 4H^+ + 4e^- + 2X$
Dark reactions: $CO_2 + 4H^+ + 4e^- \rightarrow (CH_2O) + H_2O$

The light reactions involved in the capture of light energy by pigments, followed by charge separation and electron transfer, are associated with the internal (thylakoid) membranes of the chloroplasts (see fig. 2) resulting in production of oxygen, a reduced hydrogen carrier (NADPH) and an energy source (ATP). The thylakoids form closed vesicular structures in which an inner lumen can be distinguished from the outer (stromal) compartment where the dark (enzyme catalyzed) reactions of carbon assimilation as well as other anabolic processes such as amino acid, starch, lipid and protein synthesis occur (Coombs & Greenwood 1976).

The absorption spectra of the photosynthetic pigments of various plants cover the entire visible spectrum from around 360 nm to over 700 nm, with chlorophyll a/b and a range of xanthophylls and carotenoids absorbing at the lower wavelengths and the long wavelength peaks of the chlorophylls accounting for absorption in the red region (620–700 nm). In addition the blue green algae absorb light at the blue end of the spectrum through phycobiliproteins, which like many of the other pigments act as "antenna" systems whereby an exciton produced by any of several hundred light harvesting molecules is rapidly (in the picosecond range) transferred to a reaction centre (consisting of a specific form or aggregate of chlorophyll a). Charge separation results in formation of quasi-stable oxidants and reductants. These subsequent processes are dealt with in more detail in the following section which describes some of the thylakoid proteins which have now been identified using techniques of molecular biology. Such discoveries confirm and strengthen concepts which have evolved mainly from the use of sophisticated spectrophotometric techniques to identify oxidation/reduction couples linked through the so called "Z" scheme. The combined process resulting in extraction of electrons and protons from water at +1000 mV and the reduction of NADP at around −1600 mV requires, on a theoretical basis, more energy than is available in a single quantum of light at the wavelengths used in photosynthesis. To achieve the required energy input two photosystems have evolved, linked in series through an intermediary electron transport chain which includes cytochromes, iron sulphur centres, quinones and copper/protein electron carriers of various redox potentials. The two reaction centres, which are known as PSII and PSI, have characteristic light harvesting systems and active centres (specific forms of chlorophyll a described on the basis of their absorption maxima in the red region as P680 for PSII and P700 for PSI). As a consequence of absorption of light PSII generates the strong oxidant necessary to split water and PSI generates a strong reductant capable of reducing NADP. The weak reductant of PSII and the weak oxidant of PSI interact via the series of electron carriers with the free energy made available, partially conserved by the synthesis of ATP from ADP.

Fig. 2. Structure of chloroplasts. – **a:** Section of chloroplast from *Spinacia oleracea* showing typical grana (Gr), stroma (St) and envelope (En); **b:** enlarged portion of membranes showing appressed thylakoids (Th) and intergranal membranes (Ig); **c:** section of the chloroplast of *Nostoc punctiforme* showing the presence of additional pigmented bodies, the phycobilisomes (Ph); **d:** mesophyll chloroplast from a C_4 plant (*Sporobolus airoides*) showing the characteristic outer peripheral reticulum (Pr); **e:** membrane structure lacking grana as found in the bundle sheath cells of an extreme C_4 species, *Zea mays*; **f:** starch grain (St) accumulation in the bundle sheath chloroplast of the C_4 plant *Sporobolus airoides*. Based on electron micrographs previously published by Coombs & Greenwood (1976) from where further details on fixation techniques etc. may be obtained. ▷

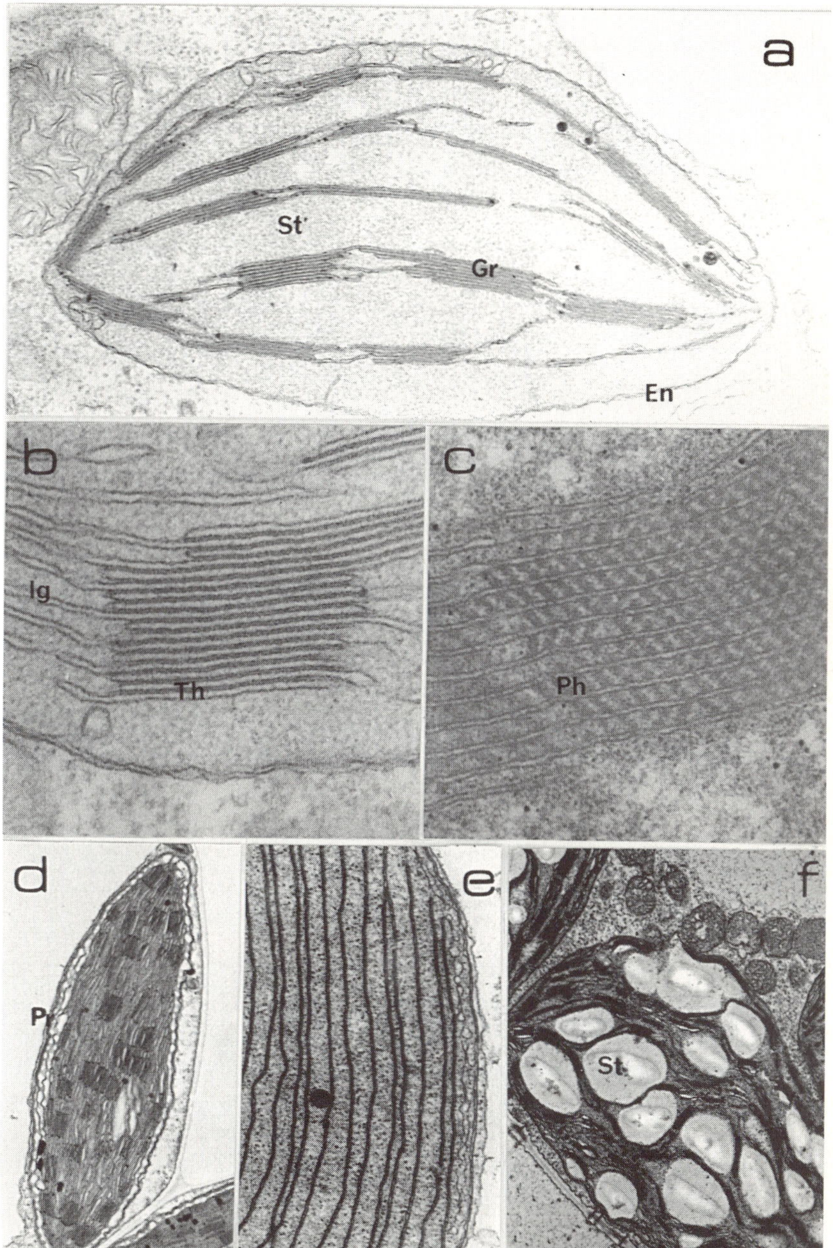
a
St
Gr
En
b
Ig
Th
c
Ph
d
Pr
e
f
St

Experimental evidence exists that other (cyclic) electron flows may occur, resulting in additional proton pumping which may be associated with the production of additional ATP or direct reduction of enzymes in light activation processes. These can be accommodated within the basic scheme as described, without needing to depart from the fundamental two-light reaction concept, but could account for quantum efficiency being around 10–12 rather than 8–10.

The ATP and NADPH are used to drive the photosynthetic carbon reduction cycle in the direction of net carbon assimilation. Although the cycle is complex in detail the main reactions are straightforward and may be summarized in four steps as follows:

- Carbon dioxide is assimilated into phosphoglyceric acid (PGA) in a reaction catalyzed by ribulose bisphosphate carboxylase/oxygenase (Rubisco).
- PGA is reduced to triose-P in a reaction where additional energy is provided by ATP and reducing power by NADPH.
- Triose-P may be exported from the chloroplast to form sucrose or be converted to starch within the chloroplast.
- Triose-P may be used to regenerate the initial substrate used in carbon fixation (ribulose bisphosphate – RBP).

The main factor affecting the efficiency of the PCR cycle is associated with a deficiency in the catalytic activity of Rubisco which results in competition for RBP between carboxylation and an alternative oxygenase reaction which proceeds as follows:

$$\text{RBP} \xrightarrow[O_2]{\text{Rubisco}} \text{PGA} + \text{P-glycollate}$$

The two-carbon fragment (p-glycollate) may be converted back to PGA via the C_2 or photorespiratory cycle during which part of the previously assimilated and reduced carbon is reoxidized to CO_2 which is then released into the substomatal cavity. Net carbon fixation is decreased due to either loss of the carboxylation substrate (RBP) and other intermediates from the cycle; or loss of carbon as CO_2; or decrease in size of the CO_2 diffusion gradient into the leaf. The combined effects of such oxygenase and photorespiratory activity may be to decrease net photosynthesis by between 30 and 50% under extreme conditions.

A number of plant genera, mainly grasses and a few herbaceous Dicotyledons, have evolved mechanisms which effectively counter the deleterious effects of the oxygenase reaction. The term "C_4 species" has been used to describe these plants where the photosynthetic machinery is divided between two cell layers (mesophyll and bundle sheath) on the basis that the first stable compound observed in ^{14}C isotopic labelling experiments is a 4-carbon organic acid, rather than the 3-carbon product PGA found in conventional C_3 temperate species.

The detailed biochemistry of C_4 species differs quite markedly from genus to genus (Coombs 1976). However, the underlining mechanism which accounts for higher productivity can be summarized in terms of the C_4 pathway, which acts as a pump transferring carbon dioxide from the atmosphere (fixed into an organic acid in the cyptoplasm of the mesophyll cells in a reaction catalyzed by the enzyme phosphoenol pyruvate [PEP] carboxylase) to the site of reductive assimilation by Rubisco and subsequent metabolism through the conventional PCR cycle located in these bundle sheath cells. In extreme C_4 plants the chloroplasts of these cells lack grana (fig. 2e) and in general have a much reduced PSII activity, resulting in much lower levels of oxygen evolution in the vicinity of Rubisco reducing photorespiration. Any carbon dioxide released by dark respiration or photorespiration is reassimilated by the cytoplasmic PEP carboxylase, resulting in low or zero compensation points. The benefits of this mechanism as far as plant productivity is concerned are indicated by the comparative figures given in table 1. These modifications have evolved over millions of years. Suggestions have been made that application of genetic engineering to higher plants might lead to increased plant productivity, or decreased cost of nitrogen fixation, by actions such as changing the characteristics of Rubisco, increasing levels of PEP carboxylase, changing the ratio of light harvesting pigments to reaction centres or incorporating nitrogen fixation (NIF) genes into non-legumes. Any such action will depend on a full understanding of the genetics and molecular biology of the relevant processes.

The genetic and molecular basis of photosynthesis

Both nuclear genes born on chromosomes showing normal Mendelian inheritance and chloroplast genes showing maternal inheritance are found to code for intermediates of both the light and the dark steps of photosynthesis. In most higher plants the chloroplast genome consists of a circular portion of supercoiled DNA which varies in size from species to species but is typically about 150 kilobase pairs (kbp). Each chloroplast contains several hundred copies of the genome, which is large enough to code for around 100 different proteins or polypeptides, located in discrete regions termed nucleoids. The chloroplast genome shows a high degree of conservation between species in both composition (with a guanine/cytosine content of 37–44%) and organization. Most species show large (80–100 kbp) and small (13–30 kbp) single copy regions of the genome separated by a pair of inverted repeats bearing the genes associated with chloroplast ribosomal RNA genes. A high degree of homology is also found in actual gene sequences as illustrated by interspecific chloroplast DNA hybridization.

Hybridization is only one of the many techniques of molecular biology which have been used to study both the nuclear and chromosomal genes associated with the production of membrane proteins and soluble enzymes of photosynthesis. Other techniques include: gene mapping by identification of the polypeptide product of cell free translation of mRNA; use of specific antibodies to such polypeptides; cloning of nucleotide sequences for specific proteins into alternative hosts; production of DNA probes; base sequencing of the genes; amino acid analysis of the products; and in particular the use of mutants with deletions in genes coding for photosynthetic proteins.

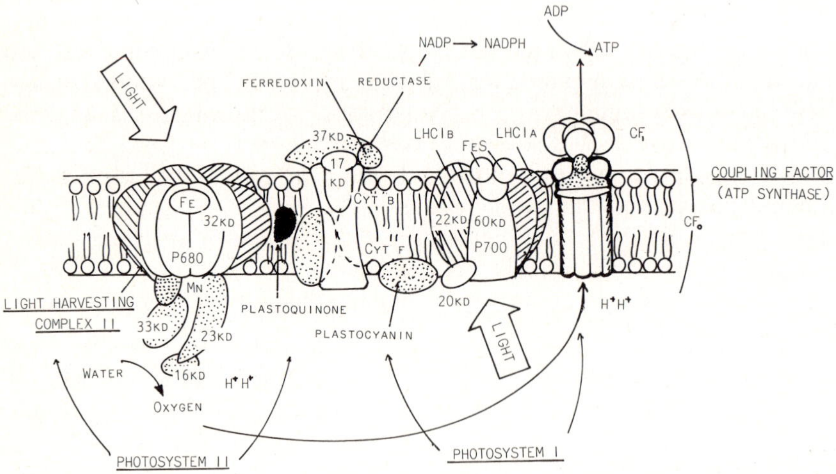

Fig. 3. Hypothetical arrangement of components of light harvesting complex of photosystem II, photosystem II, cytochrome b/f complex, photosystem I and ATP synthase in the chloroplast thylakoid, based on evidence from investigations of the molecular biology of chloroplast proteins. Shaded (light harvesting proteins) and dotted components are encoded on nuclear genes; other components are encoded on the chloroplast genome. Diagram based on information presented by Hall & Rao (1987) and Barber & Marder (1986).

The use of such mutants has been of special value where they are deficient in a single protein complex that contains proteins synthesized on either organelle or cytosolic ribosomes. Examples are known for many of the intermediates of PSI, PSII, the intermediate electron transport chain (cytochrome b/f complex), the ATP synthase complex and Rubisco in particular. As a result of such studies the genes of more than 40 of the 50 or so proteins associated with thylakoids have been identified and mapped. The structural and functional characteristics of the major components are summarized in fig. 3 and briefly described below in terms which relate to parallel observations on membrane ultrastructure using freeze fracture and related techniques to reveal surface and subsurface details.

The thylakoid membrane proteins which so far have been identified and shown to actively participate in photosynthesis may be described in terms of five functionally discrete peptide clusters as follows:

- light harvesting complex of PSII,
- reaction centre PSII,
- cytochrome b/f complex,
- reaction centre/light harvesting complex PSI, and
- ATP synthase (Cf_0/Cf_I) complex.

These peptide complexes are embedded in a matrix of polar lipids dominated by digalactosyldiacylglycerol and monogalactosyldiacylglycerol, possessing acyl chains with a high level of unsaturation associated with the presence of C18:3 linolenic acid. As shown in figure 2b these membranes are characterized by having polarity (an inside and an outside) as well as appressed and/or stacked and unstacked regions.

In more detail the light harvesting complex of PSII consists of two types of peptides of around 27 and 25 kd which bridge the membrane (Barber & Marder 1986). These nucleus encoded peptides contain both chlorophyll a and chlorophyll b in approximately equal amounts as well as significant quantities of xanthophyll pigments. They may account for up to 50% of the total membrane protein. In freeze etch images they may be identified as 7nm diameter particles contributing to 15–18nm particles formed by clusters of 4 to 6 peptides associated with the PSII reaction centre complex to which they pass absorbed quanta in the form of excitons by resonance. Since these light harvesting complexes contain no reaction centres their only function is to capture light energy, most of which is transferred to PSII, although under some conditions they may also transfer energy to PSI.

The PSII complex contains two sets of proteins. Some within the membrane are associated with the active centre whilst other surface bound peptides are associated with oxygen evolution. The major peptides are of 43 and 47 kd and contain chlorophyll a and bound carotene which may have a protective function preventing photodamage. The reaction centre chlorophyll (P680) is bound to 32 kd herbicide binding proteins which are associated with further 10 kd and 4 kd proteins which bind cytochrome b-559; all of these components are encoded by chloroplast genes. The other proteins, including 33, 23 and 16 kd polypeptides associated with splitting of water, are encoded in nuclear genes. Charge separation at the active centre of the PSII complex results in production of oxygen and protons on the inner membrane surface and reduction of a pool of plastoquinone situated in the membrane.

The cytochrome b/f complex which acts as an intermediate between PSII and PSI having plastoquinol/plastocyanin oxidoreductase activity consists of at least five polypeptides of 34, 33, 23, 20 and 17.5 kd which contain cytochrome f, two cytochromes b-563, an Fe-S protein, two non-haem irons and some bound plastoquinones. The 20 kd component as well as a 37 kd reductase which may be associated with the complex are coded on nuclear genes. The other components (23, 34 and 17.5 kd) are chloroplast coded. The main function of this complex is to reduce plastocyanin, although it may also accept electrons from ferredoxin in cyclic electron transport.

Microscopic examination of freeze fractured thylakoids suggests that the PSI complex has a diameter of around 10nm. This complex contains about 30% of the total chlorophyll, with an estimated 200 molecules per complex. Each complex contains a reaction centre (P700) which probably consists of a dimeric form of chlorophyll a associated with a polypeptide of 60 kd coded for in the chloroplast. This is surrounded by nucleus-coded light harvesting complexes containing about 50 chlorophylls each. The PSI complex also contains smaller (20 kd) proteins bearing the Fe-S centres. The PSI complex receives electrons from the cytochrome b/f complex through plastocyanin and acts as a reductant for ferredoxin.

The process of ATP formation depends on the establishment of an pH gradient across the membrane through the activity of the ATP synthase (CF_0/CF_I) complex which consists of two parts; the CF_0 portion embedded in the membrane and the CF_I portion which lies on the outer thylakoid surface. The inner portion consists of three proteins known as subunits I, II and III, of apparent molecular weights of 15, 13 and 8 kd respectively, with I and III coded for by chloroplast genes and II probably coded for by a nuclear gene. Six peptides of type I form a hexagonal proton-conducting channel across the membrane, held in place by the other peptides with I acting as an attachment for CF_I. CF_I is a spherical complex lying on the outer surface of the membrane where it appears as 10nm particles of MW around 400 kd, which can be resolved into five subunits (alpha, beta, gamma, delta and epsilon) of approximate MW of 60, 56, 39, 19 and 14 kd respectively, with a stoichiometric relationship of 3:3:1:1:1. The gamma and delta subunits are probably coded for by nuclear genes whilst the others are products of chloroplast genes. The sequences for these proteins have been derived for a number of plants, and similarity has been shown to proteins of similar function from mitochondrial and bacterial ATPase.

The stromal protein which has attracted most attention is Rubisco. This multimeric enzyme, which is the major soluble protein in chloroplasts, consists of 8 large subunits of around 52 kd and 8 small subunits of around 10–15 kd (Siegelman & Hind 1978). The catalytic activity is associated with the large chloroplast encoded subunits which are synthesized on 70s ribosomes, whereas the small subunits (believed to have a control function) are coded for by nuclear genes and synthesized on 80s ribosomes.

The amino acid sequence of the large catalytic subunit varies from species to species. However, there are regions of high homology which are thought to correspond to the active sites associated with divalent metal binding and carbon assimilation. Carbon dioxide is not only a substrate for this enzyme, but also functions as an activator or regulator through formation of a complex with magnesium binding at the site of a specific amino group of a lysine residue associated with the catalytic site. Attempts have been made to alter the catalytic activity and in particular the ratio of oxygenase to carboxylase by using site specific mutagenesis to alter the nature of some of the amino acids in this region. However, so far no successful alterations have been achieved in terms of reducing oxygenase activity, either using this approach or more general mutagenesis and/or screening for low photorespiration in C_3 plants.

Light/dark control and assimilate partitioning

Regulation of photosynthesis in the short term requires interaction between the light and dark reactions. In the longer term regulatory mechanisms must exist to determine whether newly fixed carbon is used in the generation of further substrate for carboxylation, is retained in the chloroplast as stored starch or passes from the chloroplast to be translocated as sucrose. A number of such mechanisms have been suggested which include phosphorylation or reduction of specific allosteric sites on particular membrane proteins or stromal enzymes; the effects of light induced changes in pH or stromal magnesium concentration; and effects of competition for inorganic phosphate.

Of particular interest as far as assimilate partitioning is concerned is the role of a phosphate translocator located on the inner membrane of the chloroplast envelope. If triose-P is exported from the chloroplast, inorganic phosphate released during sucrose synthesis must be returned to the chloroplast through the activity of this translocator which becomes a limiting factor since depletion of phosphate within the chloroplast will result in decreased capacity for photophosphorylation. Alternatively, if the assimilated carbon is used to form starch within the chloroplast, as occurs in C_4 plants under conditions favourable to high rates of carbon assimilation (see fig. 2f), inorganic phosphate is released within the plastid and does not become limiting.

Flux of other ions such as the light induced pumping of magnesium into the stroma can also play a regulatory role since this will facilitate the formation of the magnesium/carbon dioxide/lysine activation complex of Rubisco as discussed above. Other enzymes of the PCR cycle, such as fructose bisphosphate phosphatase and pyruvate/Pi dikinase of C_4 plants may be activated through reduction by ferredoxin immediately following illumination. Other complex short term control mechanisms exist, many of which have yet to be elucidated. These include not only those associated with the actual physical and chemical reactions but also mechanisms such as stomatal movement with indirect effects. The main need, if basic processes are to be modified in order to increase productivity, is to identify those reactions or control mechanisms which are limiting under present conditions.

Factors limiting short term photosynthesis

On the basis of the information presented so far a wide variety of factors can be identified as limiting photosynthesis. These include light, carbon dioxide concentration, light harvesting capacity (amount of light harvesting protein complexes), number of reaction centres, concentration of electron transport intermediates, the capacity to carry out photophosphorylation, the level of Rubisco in the leaf, oxygenase effects and photorespiration, activity of the Pi-translocator, feedback inhibition from accumulation of end products, and so on. However, it is not possible at present to isolate any particular factor as the primary limitation, each coming into importance under certain specific conditions.

An alternative approach is to consider the maximum potential for photosynthesis under conditions where all the above factors are optimized. Then the photosynthetic efficiency can be expressed in terms of:

$$P = I_o \cdot \varepsilon_p \cdot \varepsilon_b$$

where I_o is the proportion of incident radiation utilized (around 80% of the visible or photosynthetically active radiation [PAR] which in turn is equal to about 50% of the total electromagnetic radiation penetrating the atmosphere); ε_p is the efficiency of the light conversion steps (about 80%); and ε_b is the efficiency of the biochemical reactions (about 29% if it is assumed that 10 quanta of 575 nm light are used per carbon fixed). On this basis the overall efficiency, expressed in terms of energy trapped in the first reduced stable organic product of photosynthesis (triose-P) as a percentage of the light energy of the total incident spectrum, would be around 9%. However, energy is lost during dark respiration for growth, and in C_3 plants due to photorespiration, reducing the actual maximum to around 5 to 6%. Under field conditions this value is further reduced by environmental factors, pests and diseases, so that actual efficiencies are often around 1 to 2% at best (table 1).

Improving productivity

In temperate agricultural systems a major factor in determining what this conversion efficiency will mean on an annual basis is the rate at which the new leaves develop in spring to produce a closed canopy so that maximum light interception occurs, rather than part of the radiation reaching bare ground. Other important factors are those which affect the subsequent duration of the canopy as well as partitioning of fixed carbon into useful biomass.

So far genetic modification of such factors as early leaf development and increased harvest index (*i.e.* modification of growth patterns as shown in fig. 4) has accounted for increased productivity to a much greater extent than any modification of fundamental bioenergetic reactions. However, once an optimal growth pattern has been developed for a given crop and problems of water stress, nutrient limitations, pests and diseases have been overcome by management techniques and chemical inputs then further increases will require that the inherent capacity for photosynthesis, defined as the net assimilation rate per unit leaf area, will have to be improved in order to obtain still higher productivities.

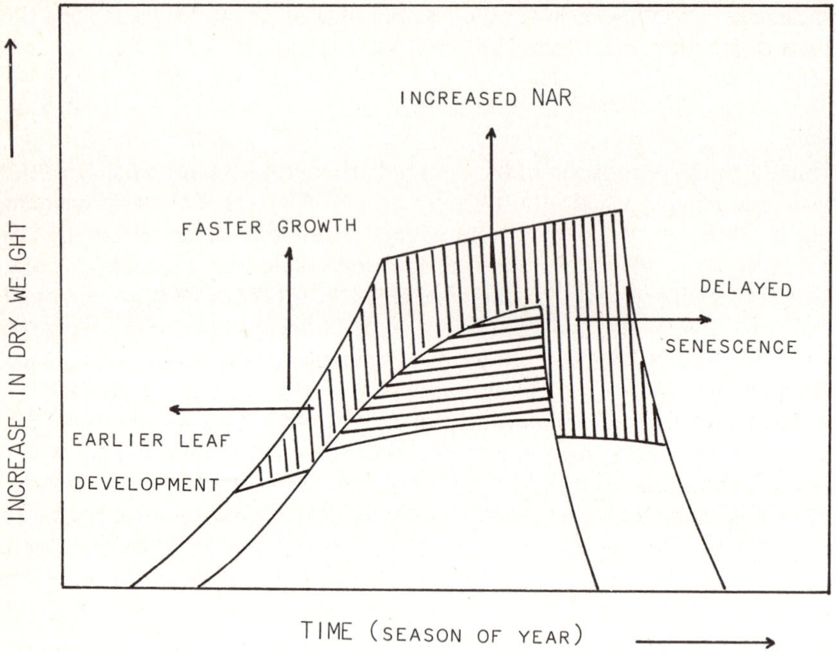

Fig. 4. Diagram showing changes in plant growth pattern which can increase productivity, from Coombs et al. (1983).

At present, advances in molecular biology of photosynthesis serve mostly as an elegant method of confirming various hypothetical reaction schemes derived on the basis of very diverse techniques. It is encouraging that in general this new information does support the theories. However, so far attempts to modify these processes have met with little success and it is clear that much more information concerning metabolic regulation is required before such advances will be possible.

Commercial needs

A major effect of improved plant productivity resulting from changes in agricultural management as well as breeding for higher harvest index has been a marked worldwide increase in cereal (see table 2) and sugar production resulting in massive surpluses, disruption of prices on unsupported world markets and an ever increasing cost burden to areas, such as the European Community, where prices have been kept high by farm support programmes.

In view of such problems it could be suggested that investigations aimed at further increases in plant productivity are unjustified. However, this is not the case since production costs remain high for intensive agriculture and productivities remain low for those farmers who are not able to afford such inputs. Hence, it is hoped that further studies of relationships between basic processes of photosynthesis and plant productivity will enable high productivities to be maintained with lower inputs. This will be beneficial if it reduces environmental problems associated with nitrates and pesticide residues in soil and ground water as well as improving the economics of using agricultural crops and other forms of plant biomass as raw materials for the production of fuels and chemicals.

Table 2. Production of cereals (yields as tonnes per hectare) and total production in millions of tonnes for 1984. The table also shows the percentage change which has occurred since 1974. (Based on figures produced by the Food and Agricultural Organisation of the United Nations – Annual Yearbook for Statistics.)

	Ammounts		% Change 1974–84	
Country	Yield	Production	Yield	Production
China	3.9	365	57	50
India	1.2	169	31	42
USSR	1.5	162	−2	−9
USA	4.4	314	31	32
France	5.9	58	58	61
UK	6.6	26	67	86
World	2.5	1801	26	29

References

Alexander, A. G. 1985: The energy cane alternative. – Elsevier, Amsterdam.

Barber, J. & Marder, J. B. 1986: Photosynthesis and the application of molecular genetics. – Biotechnology & Genet. Engineering Ref. **4:** 355–405.

Barclay, W. R. & McIntosh, R. P. (eds.) 1986: Algal biomass technologies. – Nova Hedwigia Beih. **83.**

Beadle, C. L. et al. 1985: Photosynthesis in relation to plant production in terrestrial environments. – UNEP, Nairobi.

Coombs, J. 1976: Interaction between chloroplast and cytoplasm in C_4 plants. – In Barber, J. (ed.), The intact chloroplast: 279–314. – Elsevier, London.

– 1985: Carbon metabolism. – In: Coombs, J. et al. (1985): 139–157.
– & Greenwood, A. D. 1976: Compartmentation of the photosynthetic apparatus. – Barber, J. (ed.), The intact chloroplast: 1–51. – Elsevier, London.
– et al. 1983: Plants as solar collectors – Reidel, Dordrecht.
– et al. (eds.) 1985: Techniques in bioproductivity and photosynthesis, ed. 2. – Pergamon, Oxford.
Gopal, B. 1987: Water hyacinth. – Elsevier, Amsterdam.
Hall, D. O. & Rao, K. K. 1987: Photosynthesis (fourth edition). – Edward Arnold, London.
Siegelman, H. W. & Hind, G. (eds.) 1978: Photosynthetic carbon assimilation. – Plenum Press, New York.

Address of the author: Dr. J. Coombs, Bio-Services, King's College London, Campden Hill Road, London W8 7AH; or: CPL Scientific Ltd, PO Box 8, Checkendon, Reading RG8 OBP, UK.

Biochemical evolution in plants

D. Boulter

Abstract

Boulter, D. 1988: Biochemical evolution in plants. – In: Greuter, W. & Zimmer, B. (eds.): Proceedings of the XIV International Botanical Congress: 117–131. – Koeltz, Königstein/Taunus.

The concepts and problems associated with using chemical characters to establish phylogenies are described and discussed; the usefulness of semantides is compared with that of epi-semantides, the latter being separated into primary metabolites and secondary metabolites. After a brief summary of the extent of the present data, it is concluded that, due to the very large gaps in the data-base and some conceptual difficulties as to how the data should be assessed, most phylogenetic deductions, using the existing data, must carry a large element of uncertainty; this situation could, with sufficient effort, be rectified in the future. Some of the mechanisms operating during plant biochemical evolution are deduced from the molecular structures of the fifteen enzymes of the glycolytic pathway and from a comparison of the structures of seed storage protein genes and their products.

Introduction

A complete account of biochemical evolution in higher plants would give the changes that have taken place in the biochemistry of the major plant groups since the origin of the land plants and relate these to the evolution of the plants themselves. Such an account would be woefully incomplete because, on the one hand, we have only a patchy knowledge of the comparative biochemistry of different plants and, on the other, the fossil record, the major way of establishing evolutionary relationships, is inadequate in higher plants. It is precisely because of the lack of suitable fossil evidence and the weakness of using the comparative morphology of present-day plants in its stead, that it has been suggested that comparative biochemical studies of extant plants might be a powerful approach to the problem of establishing evolutionary relationship. The other main use of such biochemical data would be to determine the mechanisms which have operated during the course of biochemical evolution.

In using chemical characters, two problems have to be addressed: the first is the establishment of the direction of chemical advance, *i.e.* which variants are the most primitive, and the second, how to combine information from different sources, *e.g.* that of two different protein data sets or of two different biogenetic chemical groups.

The chemical compounds that occur in plants have been classified by Zuckerkandl & Pauling (1965), according to their potential usefulness as indicators of evolutionary history; this classification is based on the well-etablished findings of molecular biology of as hierarchical relationship between different types of molecules as set out in fig. 1.

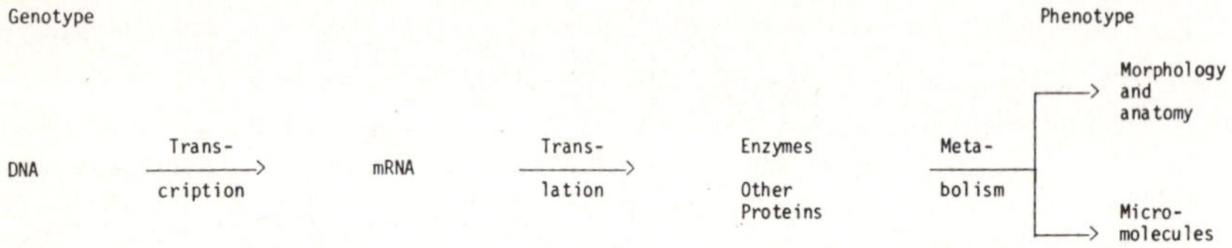

Fig. 1. Informational levels.

The information-carrying molecules are called semantides as they comprise either the primary genetic information itself (DNA), or secondary (mRNA), or tertiary (protein) derivatives of it. Semantides, together with some other large molecules, *e.g.* polysaccharides, are also known as macromolecules, as distinct from smaller micromolecules which are the products of the action of enzymic proteins and which carry much less information; these are classified by Zuckerkandl & Pauling (1965) as epi-semantides.

An evolutionary classification should consider all the hierarchical levels in an integrated system, but that ideal is not yet possible. Existing classifications use data from principally one level, usually morphological characters, and taxonomists try to relate other information, such as chemical data, to these.

Proteins (tertiary semantides)

Initially, partly for technical reasons and partly because of their great resolution, proteins (tertiary semantides) rather than primary or secondary semantides were used most frequently in evolutionary studies. Thus, at present the most extensive data sets of plant chemical characters are those of proteins and phenolics (micromolecules). However, the pace of data acquisition of primary and secondary semantides is accelerating quickly and also the advantages of using DNA-DNA hybridization as a means of overcoming inaccuracies, due to "erratic" protein clocks (see later), may well change the emphasis in the future (Sibley & Ahlquist 1984). Whilst various measures of structure differences in proteins have been used in evolutionary studies, *e.g.* microcomplement fixation, electrophoretic data, this paper reports on amino acid sequence data. In 1958, Crick stated: "It can be argued that these sequences are the most delicate expression of the phenotype of an organism and that vast amounts of evolutionary information may be hidden away within them." Put in another way, evolutionary relationships are directly related to genetic differences, *e.g.* changes in amino acid sequence, whereas morphological change does not necessarily directly relate to genetic change.

If amino acid sequence changes have been primarily divergent, the bigger the difference between sequences, the greater the evolutionary distance between the organisms from which they were derived, and by using fossil datings as fixed time points in evolution, evolutionary relationships can be established from a comparison of proteins of extant organisms. In practice, amino acid sequences are often converted to inferred nucleotide sequences using the genetic code (Penny et al. 1980) and positions at which silent substitutions are possible are standardised; computer programs normally permit ambiguities which arise occasionally from variation or uncertainty at a site.

The first comprehensive data set using this approach with a single protein, cytochrome c from vertebrates, gave a result which was compatible with the fossil evidence and with the classical ideas of the phylogeny of vertebrates, confirming that in this instance, amino acid sequence differences and evolutionary distance are directly related. Since that time there has been a large increase in the amino acid sequence data available, not only from cytochrome c, but also from many other proteins, both from vertebrates and from other groups. Data from different proteins which support one another and the accepted phylogeny are commonplace but in some instances, so-called anomalous positionings are found (Schwabe 1986), either because the data disagree with the majority of other data available or because the results from different proteins disagree with one another.

Reasons for the anomalies may be due to one or more of the following:

- Incorrect data acquisition, since errors at one or two residue positions can often change the predicted phylogeny.
- The use of paralogous rather than orthologous data, since gene duplication is commonplace (*e.g.* see later comments on alcohol dehydrogenase).
- Distortion due to unequal rates of evolution not taken into account by the data-handling method. Detailed molecular structures of over one hundred proteins have now been established and it is clear from the crystallographic data that it is the 3D structure and the requirement for flexibility of the protein to fulfil its function that are conserved in evolution and not the

amino acid sequence *per se.* Thus, the functional requirements of a particular protein may be compatible with considerable changes in the amino acid sequence, so that many of the changes in amino acid residue positions in a data set of a protein are neutral. This leads to a direct relationship between the number of amino acid differences between two sequences of the same protein, and the elapsed time since the divergence of the source plant from a common ancestor, *i.e.* there is a protein clock. This ticks at different rates with different proteins and moreover individual protein clocks keep a somewhat erratic time (Fitch 1976).

– Distortions due to convergence. Examples where the functional parts of two non-homologous proteins have a similar amino acid sequence due to

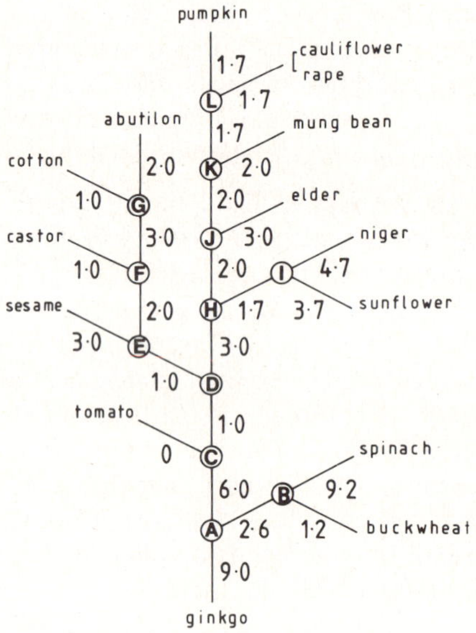

Fig. 2. A phylogenetic tree relating fifteen plant species constructed from cytochrome c using the “ancestral sequence“ method. – The sequences used in constructing the tree are: *Phaseolus aureus* L. (= *Vigna radiata,* mung bean): Thompson et al. (1970 a); *Helianthus annuus* L. (sunflower): Ramshaw et al. (1970); *Ricinus communis* L. (castor) and *Sesamum indicum* L. (sesame): Thompson et al. (1970b); *Cucurbita maxima* L. (pumpkin): Thompson et al. (1971a); *Fagopyrum esculentum* Moench (buckwheat) and *Brassica oleracea* L. (cauliflower): Thompson et al. (1971b); *Abutilon theophrasti* Medic. (abutilon) and *Gossypium barbadense* L. (cotton): Thompson et al. (1971c); *Ginkgo biloba* L. (ginkgo): Ramshaw et al. (1971); *Brassica napus* L. (rape): Richardson et al. (1971); *Guizotia abyssinica* Cass. (niger): J. A. M. Ramshaw, unpublished; *Lycopersicon esculentum* Mill. (tomato): Scogin et al. (1972); *Spinacea oleracea* L. (spinach): Brown et al. (1973); *Sambucus nigra* L. (elder): Brown et al. (1974). Reproduced with permission from The Royal Society, London.

the requirement of a common function are not uncommon (see glycolytic pathway enzymes, later). Furthermore, parallel, neutral substitutions at particular residues are commonplace (see Peacock 1981).

– Data-handling methods are not wholly rigorous since various assumptions have to be made in phylogenetic tree constructions.

The first plant phylogenetic tree, *i.e.* that using cytochrome c sequence data, was published by Boulter et al. in 1972 (fig. 2), and subsequently a 25 species cytochrome c tree was presented (Boulter 1973). From the results the authors did "not consider phylogenetic speculation to be profitable", but hoped that further data would confirm and develop the ideas suggested by the data, *e.g.* the early divergence of buckwheat (*Polygonaceae*) and spinach (*Chenopodiaceae*). With this object in mind Boulter et al. (1979) assembled a plastocyanin data set but due to the faster rate of change of plastocyanin, it was concluded that this protein could not be used effectively to confirm evolutionary events over the whole time-scale of higher plant evolution. More recently, Martin et al. (1985) have derived a phylogenetic tree for 11 Angiosperm families using a strategy and methodology which take into account, so far as it is possible, the various distorting factors given above and also compensate for the lack of resolution of single protein plant trees. Thus, a phylogeny based on the data of one protein uses the information in only *c.* 500–1000 nucleotides, depending on the protein in question (*i.e., c.* 1/50,000th of that in the genome) and it is understandable, therefore, that the degree of resolution should not be great, bearing in mind also the crypticity of

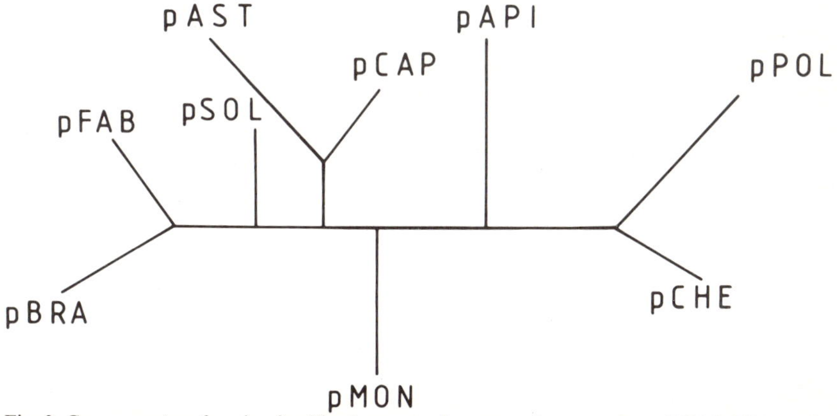

Fig. 3. Consensus tree for nine families from cytochrome c, plastocyanin and RBC-SSU. – The prefix p (for "pro") recognises the small size of samples. APIaceae, ASTeraceae, BRAssicaceae, CAPrifoliaceae, CHEnopodiaceae, FABaceae, MONocotyledons, POLygonaceae, SOlanaceae. Reproduced with permission from "Taxon".

the genetic code, the occurrence of convergent or chance similarities and the "sloppy" time-keeping of single protein clocks. Phylogenetic trees for 11 Angiosperm families were constructed using data from 106 macromolecular sequences for 3 proteins and one ribosomal RNA (Martin et al. 1985). It was shown that the phylogenetic tree of these plant families obtained by using one protein was not the same as that derived by using a different protein. However, in this instance, relatively few sequences were analysed and in some cases, only one sequence for a whole family, so that chance may have badly affected the results.

For nine families a phylogenetic tree derived from three macromolecules remained the same after the addition of the data-set for a fourth and it is thought to be reliable (fig. 3), whereas the positions of the other two families investigated, *Malvaceae* and *Ranunculaceae,* were not fixed. Although detailed comparisons with established phylogenies are not justified, nevertheless, the usefulness of the method for plant phylogenetic studies appears secure. Thus, the close relationship found between *Chenopodiaceae* and *Polygonaceae* confirms other taxonomic work, but the equally close relationship between the *Fabaceae* and *Brassicaceae,* previously not recorded, suggests a taxonomic reinvestigation.

Whilst the method is time-consuming, requiring extensive data acquisition and some expensive automatic sequencing equipment, the uncertainty which still surrounds the evolutionary relationship of plants means that further work in this field is fully justified and the outlook for such, optimistic. However, we can expect, in the future, amino acid sequence data to be supplemented increasingly by DNA-DNA hybridization and gene sequence data; by their relationship via the genetic code data from both macromolecules are fully compatible.

Micromolecules (epi-semantides)

Zuckerkandl & Pauling (1965) classified the micromolecules, which carry much less evolutionary history in their structure, as epi-semantides. In considering the latter it is customary to distinguish primary metabolites, such as the protein amino acids, nucleotides, cholesterol, glucose, some fatty acids, etc., which comprise the building blocks of the proteins, nucleic acids, membranes and other structural components, from the secondary metabolites formed from primary metabolites by enzymic reactions. The secondary metabolites exhibit a very large structural variability which cannot be explained solely by their function as storage compounds or hormones and many of these compounds are increasingly being shown to be allelochemics, *i.e.* chemicals produced by plants and animals as signals in their interaction with other organisms, as well as chemicals affording protection against environmental stress, *i.e.* secondary metabolites play an ecological role. For our present purpose, micromolecules, in contrast to macromolecules, do not have the following advantages: complexity of structure carrying evolutionary history; advancement pattern discernable; character unaffected by environmental influence; and being non-adaptive.

Present evidence suggests that the biochemical pathways give rise to the primary metabolites, *e.g.* glycolysis, oxidative phosphorylation and photosynthesis, have evolved prior to the origin of the higher plants. Consequently, in discussing the evolution of biochemical pathways in higher plants, one is concerned with the evolution of secondary metabolism, rather than that of the "major" pathways, which give rise to the primary metabolites; in general, the primary metabolites are the same in all organisms.

A large amount of data has been assembled over several decades on individual secondary metabolites and their natural distribution, but then it became apparent that more wide-ranging studies of the whole secondary chemistry of a plant group might be more profitably interpreted against information from similar studies of other groups for use in chemotaxonomy. In the context of this paper, therefore, can these data be used to discern the evolution of plant secondary metabolism? The major problem in using these data for such a purpose is the difficulty in establishing the direction of advancement of chemical characters. Also, there is the problem of reconciling data from the twenty or so different biogenetic groups of secondary metabolites which have been investigated. In attempting to overcome these shortcomings, Gottlieb (1982) set out to systematize the data in the hope that they would be more accessible and useful to taxonomists, and to put forward several ideas and principles. Thus, he has suggested that since one biogenetic group of secondary metabolites may functionally replace another in morphologically related taxa, *i.e.* they are analogous, performing the same function, *e.g.* protectant or attractant, it may be possible to find a measure of evolutionary advancement, irrespective of which biogenetic group the compound belongs to. In general, the precursors of all biogenetic groups, the primary metabolites, are present in all plants, and after surveying a large body of chemotaxonomic data, Gottlieb (1982) postulated that evolution proceeds by blocking reaction steps. So that if A, B and C represent primary metabolites, a taxon accumulating derivatives of B evolved from a taxon characterised by derivatives of C. Within each line, evolution of the metabolites of a biogenetic group proceeds by diversification and deoxygenation. If A, B, C, represent biogenetic groups of secondary metabolites with progressively higher oxidation states, a taxon accumulating C derived from a taxon characterised by B.

These postulates shift the emphasis of chemotaxonomic investigations from the structures of the secondary metabolites themselves, to alterations in biosynthetic pathways, as was suggested earlier by Birch (1973). Gottlieb's more dynamic approach to chemotaxonomy also considers the function a metabolite plays, as an important consideration in assessing chemical advancement. As an example of this we can consider the shikimic acid pathway. Metabolism via the shikimic acid pathway produces a large number of aromatic compounds related to the aromatic amino acids, phenylalanine and tyrosine; the starting materials for the biosynthesis of shikimic acid itself, erythrose 4-phosphate and phosphoenol-pyruvate, are both involved in the primary metabolism of sugars and photosynthesis, and shikimic acid is present in both algae and Cyanophyta. The shikimic acid pathway is very important in higher plants for the production of lignins, and the evolution of the structure strengthening lignins was essential for the proper colonization of the land by plants which required the development of upright structures which could eventually carry leaves and compete successfully. Thus, lignins were not present in the earliest land plants or in present-day Bryophyta. The shikimate pathway gives rise to lignins via cinnamic acids (see fig. 4), and this development was possibly aided by the NH_3 partial pressure being lower on land than in the sea (Gottlieb 1982).

Micromolecules formed by the condensation of intermediates of the shikimic acid and polyketide pathways (fig. 4) also exist in all classes of the algae and in

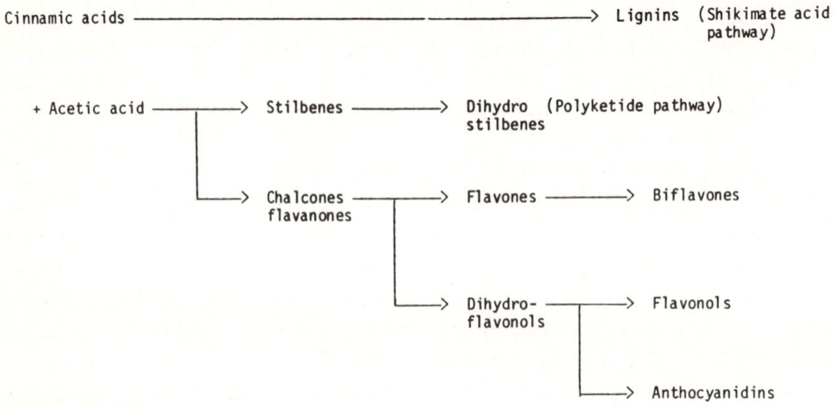

Fig. 4. Biosynthetic relationships of cinnamate derived natural products.

the Cyanophyta. The switch to flavonoid synthesis of the shikimic acid-polyketide pathway from production of stilbenes (see fig. 4) also occurred very early since flavonoids screen UV, so allowing the development of the land plants, and flavonoids are present in all vascular plants. Later on, with the origin of the flowering plants, there appears to have been a switch from flavonoids to anthocyanidins for flower colour, accompanied by an increased production of flavones as UV absorbers. Lastly, a trend from woody to herbaceous as seen in the flowering plants, which are derived from the woody Gymnosperms, has been accompanied by the replacement of lignin and condensed tannins as general protectants by compounds in other biogenetic groups of more specific action (Feeny 1976).

There now exists a large amount of phytochemical data which have been incorporated, to a greater or lesser extent, into the main Angiosperm classifications of Takhtajan (1969), Cronquist (1981), Thorne (1976) and Dahlgren (1980). Whilst it is clear that micromolecules have been useful at all levels of taxonomic investigation (Stace 1980), the extent to which the distribution of micromolecules has allowed the establishment of a phylogenetic classification of the flowering plants is still seriously questioned (Crawford 1978; Cronquist 1980). In fact, Cronquist (1983) has cautioned against constructing phylogenetic classifications based on what is called concealed evidence, *e.g.* chemical characters and microstructures, lest practical taxonomists ignore more useful phenetic classifications.

Glycolytic pathway enzymes

Support for Gottlieb's claim to have established criteria to follow chemical advancement would be forthcoming if corroborated by amino acid sequence data on the enzymes involved, although this is not possible at present since little data exist. Similarly, there are few detailed studies of molecular structures of different enzymes of the primary biosynthetic pathways in plants. However, detailed crystallographic and amino acid sequence data are available for the fifteen enzymes involved in glycolysis, from animals and yeast (Fothergill-Gilmore 1986). These data show that, in the main, the glycolytic pathway is the result of a chance assembly of independently evolving enzymes. Nevertheless, there exists a striking structural similarity among many of these enzymes which would, therefore, appear to have arisen from convergent evolution towards stable structures that bind the appropriate ligands. For example, of the seven enzymes which bind the related nucleotides, ATP and NAD^+, only glyceraldehyde-3-P dehydrogenase and phosphoglycerate kinase would appear to have diverged from a common distant ancestor; all the others are convergent. Again, the four kinases provide no evidence for a common ancestor, although diphosphoglycerate mutase and monophosphoglycerate mutase, which are consecutive enzymes in the glycolytic pathway, have diverged from a common ancestor, apparently by relatively recent gene duplication. Gene duplication with subsequent divergence has also given rise to the multi-isoenzymes which are tissue-specific and species-specific for the glycolytic enzymes, although an enzyme catalysing the same reaction in different species may not necessarily be homologous. Thus, *Drosophila* alcohol dehydrogenase isoenzymes are not homologous with those from mammals or yeast (Fothergill-Gilmore 1986).

Seed storage proteins

Apart from the few examples given above where amino acid sequence data have been determined specifically for phylogenetic studies, the other major data sets available are those of the seed storage proteins which have been determined as part of investigations into the molecular biology of seed development; in many cases, amino acid sequences have often been inferred from gene or cDNA nucleotide sequences. As cloned genes or copy genes become available, it is now possible to correlate nucleotide and amino acid sequences. This has demonstrated that many, but not all, of the plant genes investigated contain intervening sequences which do not form part of the coding messenger RNA, in contrast to the coding sequences which are called exons. Thus, amino acid sequence data can predict the presence of exons and trace evolutionary events at the gene level whilst reciprocally, nucleotide sequences, from which amino acid sequences may be inferred, can aid in studying protein evolution. Furthermore, by comparing the inferred amino acid sequence with the actual amino acid sequence of the protein itself, it is possible to detect the extensive post-translational modifications that are commonplace in plant proteins.

In Dicotyledons the major storage proteins are the globulins, legumin and vicilin (best described from peas), whilst in cereals, apart from some exceptions, prolamins fill this role. Although these data sets do not involve enough taxa to come to any general phylogenetic conclusions about the plants from which they have been obtained, they do allow some insight into the ways in which genes and hence their products, proteins, have changed over relatively short evolutionary periods, and there is no reason to suppose that the changes seen with the storage proteins have not been of general importance during evolution. Vicilin and legumin have sufficient similarities in their amino acid sequences, as first noted by Jackson et al. (1969), to have arisen from a common ancestor, but since both proteins are found in most Dicotyledons that have been investigated, their divergence dates from at least a very early stage of Dicotyledon evolution. In fact, similar proteins are also found in Monocotyledons (Derbyshire et al. 1976), so it is possible these proteins predate even the Angiosperms.

In what follows I shall not present all the data available, but use them to give examples of the types of molecular changes that have occurred during the evolution of these genes. Both vicilin and legumin are encoded by small multigene families, *e.g.*, fig. 5 gives the relationship between the 10 legumin genes of pea. There are three subfamilies and at least the main two are present in all the legumes so far examined, so it would appear probable that the duplication event that gave rise to the two gene families occurred either before or early on

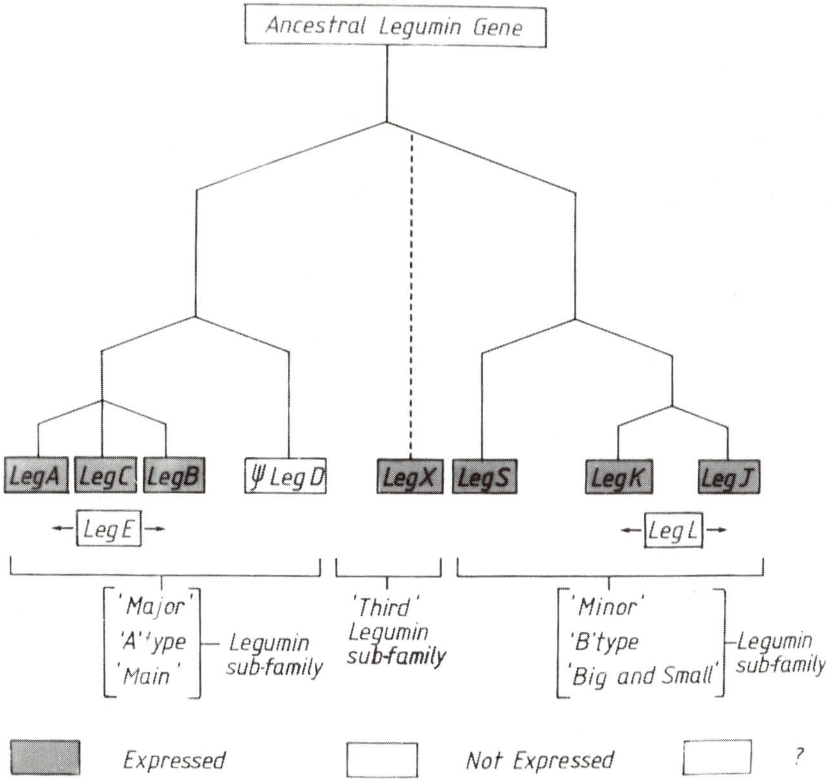

Fig. 5. Pea legumin gene family. – Reproduced with permission from R. R. D. Croy.

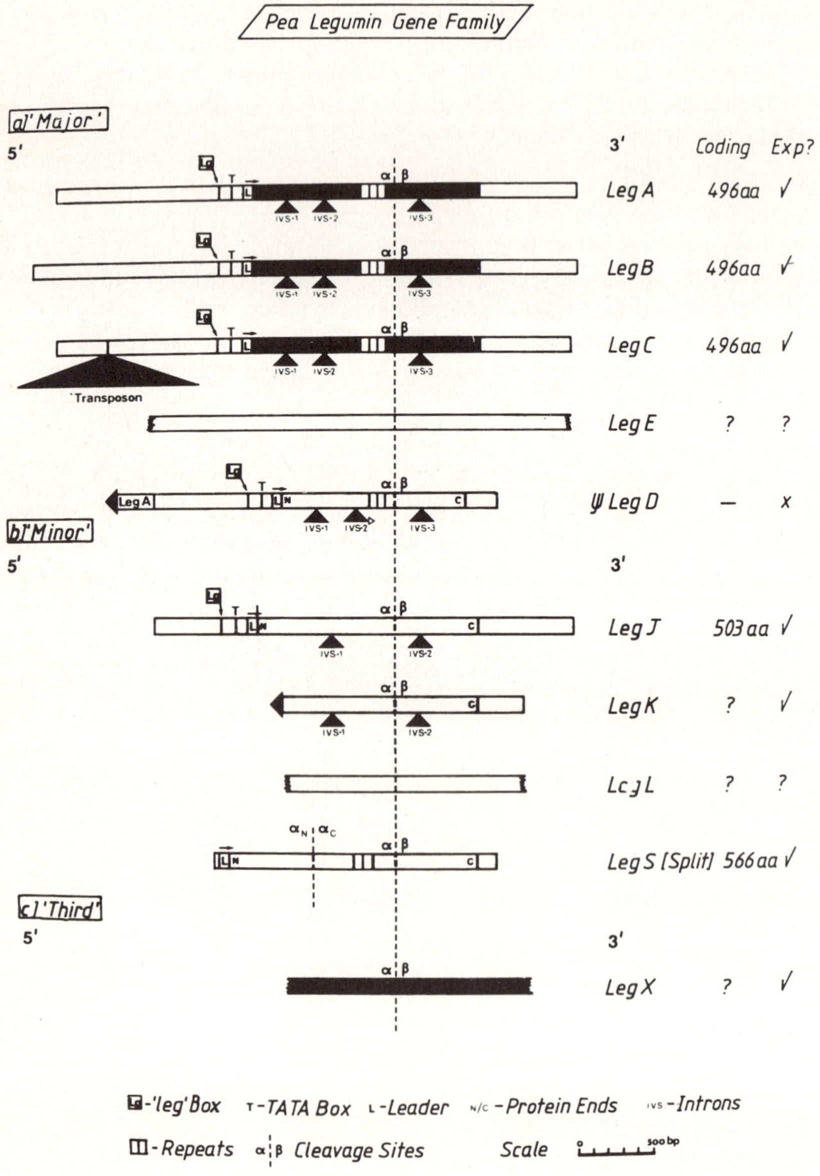

Fig. 6. Main structural features of pea legumin gene family. – Reproduced with permission from R. R. D. Croy.

in the evolution of the *Papilionaceae.* Most of these genes are expressed, since cDNA and/or polypeptides corresponding to them have been isolated and sequenced; fig. 6 gives a diagrammatical representation of their structures. Differences between the different genes include not only point mutations, but also additions, deletions which are common, as are repeats. Nucleotide sequences corresponding to a transposon have been detected in leg C (A. Shirsat, unpublished data), although the importance of the role that transposons may have played in plant evolution is uncertain. One pseudogene is present but most, if not all, of the other genes are expressed. Since both coding and flanking sequences of genes leg A, B and C are very similar, either the duplication events which gave rise to these genes are relatively recent, or gene conversion has occurred.

Prolamins, for example, the main storage proteins of barley, rye and wheat, are of three types, high molecular weight (HMW), S-rich and S-poor prolamins. The S-poor C-hordeins of barley may be derived from the repetitive domain (see later) of the S-rich B-hordeins. The γ-secalin of rye consists of 2 domains, one repetitive the other not (Kreiss et al. 1985). The sequence of the C-terminal part of this S-rich γ-secalin prolamin of rye shows homology with the non-repetitive domain of the S-rich prolamins of other cereals, the N and C termini of HMW prolamin, the trypsin and amylase inhibitors from cereals and with the 2S storage protein from rape and castor bean.

These results suggest that the genes for the storage proteins of cereals have a dual origin, having been derived partly from an ancient gene family encoding enzyme inhibitors present in both monocotyledonous and dicotyledonous plants and from more recently evolved repeat sequences only found in grasses. Kreiss et al. (1985) also suggest that it is possible that the ancient gene for a protein inhibitor with a single domain has duplicated to give the double specificity protease inhibitors of the Bowman Birk type and that subsequently, a further duplication gave rise to the A, B, C, non-repetitive domain of present-day prolamins.

In general, whilst both divergent and convergent changes have taken place during plant evolution, there is no evidence, so far, of lateral foreign higher plant gene transfer having occurred.

In conclusion, we need not abandon the hope that the use of semantide data might provide an independant method to establish a higher plant phylogeny, but this will require a considerable effort of data collection. Whilst many more plant genes and protein sequences will be established in the future for other purposes, that in itself would not suffice, and additional studies specifically directed to this taxonomic objective are required.

References

Birch, A. J. 1973: Biosynthetic pathways in chemical phylogeny. – In: Bendz, G. & Santesson, J. (eds.), Chemistry in botanical classification: 261–270. – Nobel Foundation, Stockholm; Academic Press, New York & London.

Boulter, D. 1973: The use of amino acid sequence data in the classification of higher plants. – In: Bendz, G. & Santesson, J. (eds.), Chemistry in botanical classification: 211–216. – Nobel Foundation, Stockholm; Academic Press, New York & London.

– et. al. 1972: A phylogeny of higher plants based on the amino acid sequences of cytochrome c and its biological implications. – Proc. Roy Soc. London, Ser. B, Biol. Sci. **181:** 441–455.

– et. al. 1979: Relationships between the partial amino acid sequences of plastocyanin from members of ten families of flowering plants. – Phytochemistry **18:** 603–608.

Brown, R. et al. 1973: The amino acid sequence of cytochrome c from *Spinacea oleracea* L. (spinach). – Biochem. J. **131:** 253–256.

– et. al. 1974: The amino acid sequences of cytochrome c from four plant sources. – Biochem. J. **137:** 93–100.

Crawford, D. J. 1978: Flavonoid chemistry and Angiosperm evolution. – Bot. Rev. (Lancaster) **44:** 431–456.

Crick, F. H. C. 1958: On protein synthesis in biological replication of macromolecules. – Symp. Soc. Exp. Biol. **12:**138–163.

Cronquist, A. 1980: Chemistry in plant taxonomy: an assessment of where we stand. – In: Bisby, F. A., Vaughan, J. G. & Wright, C. A. (eds.), Chemosystematics: principles and practice: 1–27. – Academic Press, London & New York.

– 1981: An integrated system of classification of flowering plants. – Columbia University Press, New York.

– 1983: Some realignments in the Dicotyledons. – Nordic J. Bot. **3:** 75–83.

Dahlgren, R. 1980: A revised system of classification of the Angiosperms. – Bot. J. Linn. Soc. **80:** 91–124.

Derbyshire, E. et al. 1976: Legumin and vicilin, storage proteins of legume seeds. – Phytochemistry **15:** 3–24.

Feeny, P. 1976: Plant apparency and chemical defense. – Recent Advances Phytochem. **10:** 1–40.

Fitch, W. M. 1976: Molecular evolutionary clocks. – In: Ayala, F. J. (ed.), Molecular evolution: 160–178. – Sinauer, Sunderland Mass.

Fothergill-Gilmore, L. A. 1986: The evolution of the glycolytic pathway. – Trends Biochem. Sci. **11:** 47–51.

Gottlieb, O. R. 1982: Micromolecular evolution, systematics and ecology. – Springer, Berlin, Heidelberg & New York.

Jackson, P. et al. 1969: A comparison of some properties of vicilin and legumin isolated from seeds of *Pisum sativum, Vicia faba* and *Cicer arietinum.* – New Phytol. **68:** 25–33.

Kreiss, M. et al. 1985: Molecular evolution of the seed storage proteins of barley, rye and wheat. – J. Molec Biol. **183:** 499–502.

Martin, P. G. et al. 1985: Angiosperm phylogeny studied using sequences of five macromolecules. – Taxon **34:** 393–400.

Peacock, D. 1981: Data handling for phylogenetic trees. – In: Gutfreund, H. H. (ed.), Biochemical evolution: 88–115. – Cambridge University Press, London.

Penny, D. et al. 1980: Techniques for the verification of minimal phylogenetic trees illustrated with ten mammalian haemoglobin sequences. – Biochem. J. **187:** 65–74.

Ramshaw, J. A. M. et al. 1970: The amino acid sequence of *Helianthus annuus* L. (sunflower) cytochrome c deduced from chymotryptic peptides. – Biochem. J. **119:** 535–539.

– et al. 1971: The amino acid sequence of the cytochrome c of *Ginkgo biloba* L. – Eur. J. Biochem. **23:** 475–483.
Richardson, M. et al. 1971: The amino acid sequence of rape (*Brassica napus* L.) cytochrome c. – Biochim. Biophys. Acta **251:** 331–333.
Schwabe, C. 1986: On the validity of molecular evolution. – Trends Biochem. Sci. **11:** 280–283.
Scogin, R. 1972: The amino acid sequence of cytochrome c from tomato (*Lycopersicon esculentum* Mill.). – Arch. Biochem. Biophys. **150:** 489–492.
Sibley, C. G. & Ahlquist, J. E. 1984: The phylogeny of the hominoid primates, as indicated by DNA-DNA hybridization. – J. Molec. Evol. **20:** 2–15.
Stace, C. A. 1980: Plant taxonomy and biosystematics. – Edward Arnold, London.
Takhtajan, A. 1969: Flowering plants. Origin and dispersal. – Oliver & Boyd, Edinburgh.
Thompson, E. W. et al. 1970a: The amino acid sequence of *Phaseolus aureus* L. (mung bean) cytochrome c. – Biochem. J. **117:** 183–192.
– et. al. 1970b: The amino acid sequence of sesame (*sesamum indicum* L.) and castor (*Ricinus communis* L.) cytochrome c. – Biochem. J. **121:** 439–446.
– et al. 1971a: The amino acid sequence of cytochrome c from *Cucurbita maxima* L. (pumpkin). – Biochem. J. **124:** 779–781.
– et. al. 1971b: The amino acid sequence of cytochrome c of *Fagopyrum esculentum* Moench (buckwheat) and *Brassica oleracea* L. (cauliflower). – Biochem. J. **124:** 783–785.
– et. al. 1971c: The amino acid sequence of cytochrome c from *Abutilon theophrasti* Medic. and *Gossypium barbadense* L. (cotton). – Biochem. J. **124:** 787–791.
Thorne, R. F. 1976: A phylogenetic classification of the Angiospermae. – Evol. Biol. **9:** 35–106.
Zuckerkandl, E. & Pauling, L. 1965: Molecules as documents of evolutionary history. – J. Theor. Biol. **8:** 357–366.

Address of the author: Professor D. Boulter, Department of Botany, University of Durham, South Road, Durham DH1 3LE, UK.

Formation of organelles and the organization of the eukaryotic cytoplasm

G. Schatz

Abstract

Schatz, G. 1988: Formation of organelles and the organization of the eukaryotic cytoplasm. – In: Greuter, W. & Zimmer, B. (eds.): Proceedings of the XIV International Botanical Congress: 133–135. – Koeltz, Königstein/Taunus.

Recent advances have greatly heightened our awareness of the complexity of the eukaryotic cytoplasm. This complexity now reveals itself in a multitude of cytoskeletal elements, and in specific interactions between these elements and different organelles and domains within the cytoplasm. Experiments on the origin of mitochondrial addressing sequences for nuclear gene-encoded mitochondrial proteins suggest molecular mechanisms by which mitochondria may have become integrated within the eukaryotic cytoplasm.

Our view on the organization of the eukaryotic cytoplasm is undergoing significant change (Yaffe & Schatz 1984). Before the electron microscope was applied to living systems it appeared to many that the cytoplasm was populated by several different organelles endowed with considerable independence. When the resolving power of the electron microscope began to uncover a multitude of membrane-bounded intracellular compartments, these were soon widely viewed as mere specializations of a "unit membrane" originating from the rough endoplasmic reticulum. The discovery of DNA in chloroplasts and mitochondria in the early sixties reemphasized the partial genetic autonomy of these important organelles and led to renewed speculations about their evolutionary origin from free-living bacteria. Since even the DNA-less peroxisomes (or their variants, the glyoxysomes) proved to be made independently of membranes derived from the endoplasmic reticulum (Borst 1986) it appears possible that they, too, are the product of an ancient endosymbiotic event. During the past few years, the application of video-enhanced contrast optical microscopy has taught us that many of the eukaryotic organelles do not diffuse "freely" in the cytoplasm, but are driven along cytoskeletal tracks by molecular "motors" that are powered by the hydrolysis of high-energy phosphate bonds (Allen 1985). Even the apparently "structure-less" cytosol may be organized into domains (Bridgeman et al. 1986) or supermolecular assemblies which only exist at the very high protein concentrations prevailing within a living cell. It ist not unreasonable to suspect that many of the key structural features of the eukaryotic cytoplasm are yet to be discovered. It is already clear, however, that the major membrane-bounded organelles are structurally tightly integrated with the rest of the cell.

How did this integration come about? Our laboratory has approached this question by studying possible mechanisms for the evolution of peptide sequences that direct nuclear gene-encoded proteins specifically into mitochondria. During the evolution of mitochondria from endosymbionts, the genes for most of these proteins were probably transferred from the endosymbiont to the host nucleus. However, in order for the protein products to reach the "protomitochondria", the newly transferred genes had to acquire DNA sequences encoding mitochondrial targeting sequences. It is very unlikely that such sequences could have arisen by sequential point mutations as no single mutation would have conferred a selective advantage upon the evolving system.

Mitochondrial targeting sequences are degenerate sequences whose common denominator is their potential to form positively charged amphiphilic secondary structures of neutral overall hydrophobicity (Roise et al. 1986). No specific amino acid sequence is required: functional mitochondrial targeting sequences can be constructed from only three types of amino acids: arginine, serine and leucine (Allison & Schatz 1986). Such positively charged amphiphilic helices are encoded by many sequences in the *Escherichia coli* genome (Baker & Schatz 1987), and by the coding strand (Hurt & Schatz 1987) or adventitious open reading frames in either of the two strands (Baker & Schatz 1987) of a gene encoding a cytosolic protein such as the mouse enzyme, dihydrofolate reductase. As many as 5–10% of the random sequences from the *E. coli* genome encode functional mitochondrial targeting sequences. Recent experiments in our laboratory have provided direct evidence that mitochondrial targeting sequences can arise by DNA rearrangements within adjacent regions of the genome (Bibus et al. 1988). These rather surprising results suggest that mitochondrial targeting sequences could have evolved quite easily.

References

Allen, R. D. 1985: New observations on cell architecture and dynamics by video-enhanced contrast optical microscopy. – Annual Rev. Biophys. & Biophys. Chem. **14**: 265–269.

Allison, D. S. & Schatz, G. 1986: Artifical mitonchondrial presequences. – Proc. Natl. Acad. USA **83**: 9011–9015.

Baker, A. & Schatz, G. 1987: Sequences from a prokaryotic genome or the mouse dihydrofolate reductase gene can restore the import of a truncated precursor protein into yeast mitochondria. – Proc. Natl. Acad. USA **84**: 3117–3121.

Bibus, C. R., Lemire, B. D., Suda, K. & Schatz, G. 1988: Mutations restoring import of a yeast mitochondrial protein with an non-functional presequences. – J. Biol. Chem. (in press).

Borst, P. 1986: How proteins get into microbodies (peroxisomes, glyoxysomes, glycosomes). – Biochim. Biophys. Acta **866**: 179–203.

Bridgeman, P. C., Kachar, B. & Reese, T. S. 1986: The structure of the cytoplasm in directly frozen cultured cells II. Cytoplasmic domains associated with organelle movements. – J. Cell Biol. **102**: 1510–1521.

Hurt, E. C. & Schatz, G. 1987: A cytosolic protein contains a cryptic mitochondrial targeting sequence. – Nature **325**: 499–503.

Roise, D., Horvath, S. J., Tomich, J. M., Richards, J. H. & Schatz, G. 1986: A chemically-synthesized pre-sequence of an imported mitochondrial protein can form an amphiphilic helix and perturb natural and artificial phospholipid bilayers. – EMBO J. **5**: 1327–1334.

Yaffe, M. P. & Schatz, G. 1984: The future of mitochondrial research. – Trends Biochem. Sci. **9**: 179–181.

Address of the author: Prof. Dr. G. Schatz, Biocenter, University of Basel, CH-4056 Basel, Switzerland

Mechanisms that maintain the genetic integrity of plants[1]

E. J. Klekowski

Abstract

Klekowski, E. J. 1988: Mechanisms that maintain the genetic integrity of plants. – In: Greuter, W. & Zimmer, B. (eds.): Proceedings of the XIV International Botanical Congress: 137–152. – Koeltz, Königstein/Taunus.

Vascular plants have many characteristics which promote the accumulation of mutations. These include the lack of a germline and immune system, long life spans, open systems of growth, flexible meristem organizations, and the fact that most somatic mutations are not immediately life threatening. With continued growth the accumulation of mutations in the apical initials is similar in many respects to the accumulation of mutations in a bacterial chemostat. Many characteristics can influence the rate of mutation in such an organismal chemostat in addition to the biochemistry of DNA metabolism, since the final mutation rate is an expression of the frequency of mutations retained in the apical meristems. In addition the high mutational loads often documented in long-lived species are compensated to some extent by soft selection during sexual reproduction and possibly even recombination itself.

Origins of genetic instability

Before considering the mechanisms that promote genetic integrity in plants, one must first review those characteristics of plants that may contribute to genetic instability. At least six such characteristics distinguish vascular plants from higher animals.

Plants lack a germline

Weismann's doctrine (Weismann 1892) of the separation of soma and germ is invalid in plants. Although plant germ cells ordinarily are produced from undifferentiated cell lineages, these cells are not set aside as in the sex gonads of many mammals (Buss 1983; Walbot & Cullis 1983). Mutation rates are generally related to units of biological time (*e.g.* cell cycles, generations) rather than real time. Thus, every time the genetic material is replicated and segregated at mitosis there is a probability of error. The overall mutation rate

[1] Freely excerpted from Klekowski (1988). The reader is referred to this work for more inclusive discussions and citations.

(per generation for example) is, therefore, a function of the number of mitotic divisions between zygote formation and meiosis in those cell lineages giving rise to meiocytes. One would predict, then, that the mutation rate (per generation) would be higher in organisms with many mitotic divisions between zygote and meiocyte than in organisms with few such mitotic divisions (Slatkin 1984). In humans this seems to be the case. The female oocyte passes through approximately 22 divisions at the time of birth. From birth to sexual maturity and fertilization, this cell will undergo only 2 mitotic divisions regardless of the female's age at which fertilization occurs. In contrast, the number of cell divisions in spermatogenesis is a function of the male's age at which sperm is formed (30 divisions at puberty, 380 divisions at age 28, 540 divisions at age 35). In accordance with expectation, mutation rate is a function of male age whereas the female mutation rate is constant throughout life (Vogel & Motulsky 1986). Similar studies in plants are critically needed. If mutation rate (and frequency) are a function of the individual plant's age, then mutational load will be, in part, a function of the age distribution of the population of plants being studied. The age of plants in a population (with special regard to the relative numbers of mitotic divisions between zygote and meiocyte) is a neglected statistic in plant population biology.

Many plants are potentially immortal

Excluding some specialized forms (annuals, biennials, etc.), the majority of plant species do not have defined life spans. Thus, in many instances the number of mitotic divisions between zygote and meiocyte may be very large.

Open systems of growth are common

In many instances apical meristems have great growth potential, either the meristem itself can grow almost indefinitely (as long as other life support systems are operative) or sub-samples of cells from the apical meristem in the form of axillary buds can perpetuate the original meristem-derived cell lines for very long time periods. In addition apical meristems may be replaced or displaced without critically affecting the viability of the individual genet.

Plant cells are diploid

In vascular plants the sporophyte is the dominant generation; consequently, the majority of mutations occur in diploid somatic cells (polyploid cells are not uncommon in plant tissues, but the majority of cells in meristems are diploid, Jones 1935). The selection against mutant cells in meristems is a function of how physiologically handicapped the mutant cells are in relation to neighbouring wildtype cells (Klekowski & Kazarinova-Fukshansky 1984 a, b). The recessive nature of many mutations confers selective neutrality to most mutant cells with meristems. Cells heterozygous for mutant alleles may become homozygous through somatic crossing-over or mitotic gene conversion. Such cells still may be relatively neutral regarding selection within meristems because of cross-feeding from neighbouring non-mutant cells (Langridge 1958).

Somatic mutations generally are not life threatening

This is one of the major differences between plants and higher animals; plants are relatively immune from cancer-like phenomena (some of which presumably arise from somatic mutations). This cancer resistance is due to the impossibility of metastasis in vascular plants. Plant somatic cells have permanent walls which strongly adhere to the cell walls of adjacent cells. At the organismal level, vascular plants lack a circulatory system through which cells can move to other locations within the organism. These two plant characteristics (cell walls and lack of a circulatory system) confer almost total immobility to plant cells within the organism; consequently, malignancy is not possible (Jones 1935).

Although malignant tumors are lacking in plants, localized aberrant growths (presumably the result of somatic mutation) occur. Such abnormal growths may be classified into two general phenotypic classes, tumors and fasciations. Tumors occur spontaneously in natural plant populations and can be induced by ionizing radiation; many are thought to represent somatic mutations (Kehr 1965), although tumors may also result from viral, bacterial and environmental infection, insect predation and environmental treatments (Bloch 1965). Fasciation is a change from the normal or polygonal stem to one that is flat, banded or ribbon shaped; in some cases the apical meristem may develop numerous growing points forming a witch's broom. Fasciations may have either environmental or genetic causes, the latter often a consequence of somatic mutation (White 1948, Waxman 1975). The tumors or fasciations arising through somatic mutation are seldom life threatening to the individual plant.

Plants lack an immune system

Burnet (1970) noted that "an important and possibly primary function of immunological mechanisms is to eliminate cells which are a result of somatic mutation . . ." In plants such a surveillance system is not present (probably because somatic mutations are not life threatening).

Given the above higher plant characteristics, a suitable model for considering the long-term accumulation of mutations in long-lived plants is based upon the dynamics of mutation accumulation in a microbial chemostat. A chemostat is a device whereby a microbial population may be maintained in a continuous growth phase for an indefinite period of time. This device is based upon the principle that a cell population growing within a defined and enclosed environment will undergo a period of logarithmic growth followed by a prolonged stationary phase where cell numbers remain relatively constant. The classic growth curve response is manipulated in a chemostat so that the cell population is in continuous logarithmic growth. This is accomplished by a continuous flow of fresh nutrient medium into the culture vessel and the simultaneous removal of old medium and cells from the culture vessel.

In many respects the meristems of a vascular plant are analogous to a chemostat. If we consider an apical meristem as an example, the population of apical initials is in a phase of continuous growth (at least during the growing season), the excess cells formed are "washed out" to the soma, and finally there is a continuous flow of nutrient medium (water, minerals, photosynthate) into the meristem. Novick & Szilard (1950, 1951) studied the dynamics of spontaneous mutation in bacterial chemostats. The general principles they developed can be applied to multicellular plants to give a rough appreciation of the impact of continued somatic mutation.

The growth of a newly established cell culture is characterized by a phase of logarithmic growth. The derivative of this part of the growth curve is

$$dn/dt = \alpha n$$

where t is time, n is cell number per unit volume and α is the growth rate. The reciprocal value, $\gamma = 1/\alpha$, is designated as the generation time. In a chemostat the cell population is maintained in growth phase indefinitely. The consequences of spontaneous mutations in such a system are interesting. If the mutations are neutral (*i.e.* no selection for or against the mutant with respect to the wildtype cell), then the mutations will increase linearly with time. If n^* is the cell number per unit volume of the mutant and if we disregard back mutation then

$$dn^*/dt = (\lambda/\gamma)n$$

where n^* is the mutant, n the wildtype, and λ the number of mutations produced per generation per cell. Separating variables and integrating both sides we get

$$dn^*/n = (\lambda/\gamma)\,dt$$

$$\int dn^*/n = \int (\lambda/\gamma)\,dt$$

$$n^*/n = (\lambda/\gamma)\,t + \text{constant}$$

It is clear from the above that the relative abundance of the mutants increases linearly with time.

Estimates of mutation rates

The significance of mutation as a destabilizing force in plants is totally dependent upon the magnitudes of mutation rates in different species and life forms. The following will review the scanty information available for this important metric.

Eriksson et al. (1966) reported that 2×10^{-6} pollen grains of the conifer *Larix leptolepis* exhibited the waxy phenotype, 16% of the mature pollen grains were sterile and 1.9% of anaphase II stages of meiosis exhibited chromosome aberrations. Since cloned grafts (branches) growing in pots were sampled rather than large trees, these values are probably fair estimates of the mutation rates for these characteristics per meiosis in this species. Stadler (1942) determined the spontaneous mutation frequency for genes not selected for high mutability in *Zea mays* (an annual). Measurable mutation frequencies for individual genes varies from *c.* 5×10^{-4} to 1×10^{-6} mutations per gamete.

The evolutionary and developmental consequences of somatic mutation are not based upon the mutation rates of single genes but rather on the summed mutation rates for sets of genes in the genotype. Fitness in any organism is multigenic; thus if the negative effects of mutation are considered, mutation rates must be summed across all loci that can mutate to decrease fitness. In *Drosophila melanogaster* ($2n = 8$) such estimates yield values that are surprisingly high. Wallace (1968) estimated 0.005 to 0.006 lethals per chromosome per generation for the second and third chromosomes, or an overall mutation rate per generation in excess of 1% for lethals. Dobzhansky et al. (1977) calculated that the mean rate of all viability mutations is no less than 0.7055 mutations per individual per generation. In *Zea mays* haploids occur in low frequency and may be identified in the seedling stage if suitable genetic markers are present. A small percentage of such haploids may form diploid homozygous sectors. If such sectors include all or portions of the inflorescences (ear and tassel) some seeds may result from self-fertilization. The resulting sib progeny should be genetically identical since they are derived from completely homozygous parental tissues. Sprague et al. (1960) studied such progeny from individual haploids, inbreeding these plants and their offspring through six generations of selfing. These data formed the basis for estimating the mutation frequencies for the set of genes that may mutate to affect agronomically important quantitative attributes. These attributes include plant height, leaf width, number of tassel branches, number of kernel rows, ear length and diameter, weight per 100 kernels, weight of shelled grains per plant and date of silking. For any given attribute, the conservative estimate of mutability was 4.5 mutations per attribute per 100 gametes tested. As these authors noted, this value is remarkably high. Also in *Z. mays* the mutation rate for that set of genes that can mutate to form the chlorophyll-deficient condition was estimated as 0.002 mutations per gamete per generation (Crumpacker 1967). In barley (*Hordeum vulgare*), the spontaneous mutation rate for chlorophyll deficiencies was measured as 0.0006 mutations per diploid genome per generation (Jørgensen & Jensen 1986). In ferns the mutation rate for gametophyte mutations is approximately 0.02 gametophyte mutations per apical initial per ramet doubling generation (Klekowski 1984, Klekowski & Masuyama unpublished). The spontaneous chromosome aberration frequency at mitosis has been variously estimated as approximately 1×10^{-3} in plants (Giles 1940, Rutishauser & La Cour 1956, Davidson 1960). In addition to the above "conventional" estimates of genetic instability, the genomic instabilities caused by the activities of mobile elements (transposons; Döring & Starlinger 1986) and the relatively rapid and often concerted changes associated with repetitive components of the genome (Dover 1982, Flavell 1986) have led many authors to stress the significance of genomic instability as an important aspect of plant biology (Whitham & Slobodchikoff 1981, Buss 1983, Walbot & Cullis 1983).

Johns et al. (1983) studied restriction site polymorphism in a 20 kb segment of DNA that included the *Adh* 1 locus in maize. Seven lines of maize carrying different *Adh* 1 alleles were mapped using seven restriction enzymes. Two patterns of DNA variability were observed; within the *Adh* 1 gene (transcriptional unit) no polymorphism was detected whereas the DNA in the flanking regions was surprisingly variable. These authors noted that there is as much variability for restriction site polymorphisms among different lines of corn as had been found among *Drosophila* species that have been separate species for millions of years. The domestication of maize probably occurred 10,000 years ago.

Recent research in plant tissue culture has demonstrated dramatically the frequency of somatic mutations in plants. Cell and protoplast culture and regeneration have documented high frequencies of variant phenotypes in regenerated plants of various plant species (Meins 1983). This genetic variation originates in both the cultural and the regenerative process as well as being present within the intact plant (*in planta,* Meins 1983). The *in planta* genetic variability suggests that many plant organs may consist of cells heterozygous for postzygotic mutations.

Mechanisms of genetic stability

The question of genetic integrity (or stability) in plants concerns the mechanism available to reduce the impact of mutational load. An important point to note is that although the accumulation of mutations in a long-lived plant in some ways resembles a chemostat, there are important differences. Plant cells are not randomly interacting as in a cell culture but, rather, are organized into tissues and organs with specific growth dynamics and interactions. Thus, genetic stability does not simply involve biochemical precision during DNA metabolism and mitotic chromatid separation. Errors at this molecular level fix the base line mutation rate per cell cycle, but other plant characteristics determine the long-term significance of mutations. The critical point is whether mutant cells are maintained (or fixed) in apical (or even cambial) meristems or whether the mutant cells are lost to tissues and organs that soon become metabolically moribund.

The majority of cells in a plant are genetically dead so to speak. They are either in tissues and organs that will never give rise to meiotic cells (Dyer 1976) or parts of meristem systems that undergo programmed senescence (Hardwick 1986). Many such tissues and organs are populated by cells having ploidy levels in excess of diploid (which reduces the phenotypic expression of subsequent mutations; D'Amato 1977) and have developed from tunica-corpus-type apical meristems (in almost all Angiosperms and some Gymnosperms). Such development insures that the tissue or organ consists of a sandwich of cells derived from different meristems and, consequently, if somatic mutations have become fixed in these meristems the organ is a chimera of mutant and wildtype cells. Through cross-feeding (Langridge 1958) and other mechanisms (unknown), such chimeric organs often develop and function normally (see Pohlheim 1981, 1983 for examples of growth compensations in chimeras). The development of many plant tissues and organs often appears so autocatalytic that some authors have suggested that once a pattern is established it proceeds in an almost "agenic" manner (see Lintilhac 1984 for discussion).

All of the above characteristics lessen the immediate developmental importance of somatic mutations; in other words, the development of many plant tissues appears well buffered against disruption from somatic mutations. This is not to say that somatic mutations which alter form do not occur (Gottlieb 1984) but, rather, that the majority of somatic mutant cells with metabolic deficiencies of various kinds probably have very little immediate phenotypic impact. With regard to loss or fixation of somatic mutations from meristematic cell pools, apical meristem characteristics play an important role.

The topology of an apical meristem strongly determines the morphology of the primary plant body, yet often similar morphologies may develop from a variety of apical topologies. Thus, apical meristem topology and growth may be viewed as important determinants of form which can tolerate and compensate for a wide variety of variations and still achieve similar end points. Although many of these variations may not have great morphological significance, interestingly they often may have considerable genetic significance regarding the loss or fixation of somatic mutations from the apical meristem cell pools.

In vascular plants, apical meristems may have permanent or impermanent apical initials; the number of apical initials may vary from one to nine or more; if apical initials are impermanent then the size of the cell pool from which initials are selected can vary considerably; the number of cell divisions that the apical initials undergo per node of growth can vary; whether a *méristème d'attente* occurs, whether the apical meristem is stratified into component meristems (tunica-corpus organization) or unstratified, the frequency of periclinal divisions resulting in the movement of cells between tunica layers or tunica and corpus, and the changes in these parameters during different stages of growth: all this can affect the loss or fixation of somatic mutations within the apical meristem (Klekowski et al. 1986). In table 1 the apical meristem characteristics that promote the loss of mutant cells in unstratified meristems with impermanent apical initials are tabulated (these characteristics also apply individually to a tunica or corpus which has impermanent initials).

Apical meristems also have characteristics that can modify the effective mutation rate. Mutation rate per biological time unit is in part a function of the number of times a genome has been replicated and chromatids divided during the biological time unit. For a generation this is the number of cell

Table 1. Characteristics which promote the loss of mutant cells in stochastic meristems. Mutants may be either less viable (mutant CF 〈 wildtype CF) or more viable (mutant CF 〉 wildtype CF) than wildtype cells.

Mutant viability is less than wildtype	*Mutant viability is greater than wildtype*
Mutant loss is promoted by:	
1. Increasing the number of apical initials, high α	1. Decreasing the number of apical initials, low α
2. Increasing the cell pool from which initials are selected, high r	2. Decreasing the cell pool from which initials are selected, low r

divisions from zygote to meiocyte for the cell lines giving rise to reproductive cells. Consequently, the maintenance of cell pools within the meristem which seldom divide but which give rise to meiocytes (*méristème d'attente*) may reduce mutation rates. Also, since the newly synthesized DNA strand is more error prone than the parental strand, meristems with permanent apical initials and non-random chromatid segregation at mitosis will also have lower mutation rates (Cairns 1975, Klekowski et al. 1986). In addition to the characteristics of apical meristems, mutation loss is strongly influenced by branching patterns; highly branched forms would be expected to be less mutation loaded than less branched forms (Klekowski & Fukshansky, unpublished).

Probably one of the most significant losses of mutation occurs during sexual reproduction. Many reproductive traits seem to have been "designed" to filter defective from effective genotypes at the least cost to the plant. Plant species often have life cycle characteristics that allow the operation of soft selection through various and sequential phases of the reproductive cycle. This internal soft selection may eliminate lethals as well as mutations with less harmful consequences without seriously reducing the reproductive capacity of the plant. Buchholz (1922) was the first to note that this type of selection could be operative in plants. He coined the term developmental selection to describe these selective processes within the reproductive cycle of the plant. Developmental selection is possible whenever competition occurs between spores, gametophytes, gametes, zygotes, embryos, ovules, fruits. Such competitive interactions are characteristic of sexual reproduction in vascular plants and probably represent an important family of mechanisms for maintaining genetic integrity.

Consequences of the loss of genetic integrity

The primary consequence of the loss of genetic integrity in long-lived plants is an increase in genetic load (specifically mutational load). Mutational load represents the mutations inherited from a plant's ancestors plus the mutations accumulated during the life of the individual genet. Thus, two factors will strongly influence load levels, the reproductive system with regard to selfing and crossing and the various developmental and organographic parameters influencing the accumulation of mutations during a plant's growth. In general, one might expect that load levels would be higher in long-lived plants than in animals with germlines.

In *Homo sapiens,* Vogel & Motulsky (1986) summarized a number of studies and concluded that the load of lethal equivalents per gamete varied from 0 to 4 with a median of slightly above 1. Thus, perhaps each of us carries more than two lethal equivalents in our genotype, which, of course, is an underestimate of the actual number of recessive detrimental genes. In vascular plants different species and life forms vary enormously in genetic loads. In the homosporous ferns some species are essentially loadless (*e.g. Ceratopteris,* an annual, Lloyd & Warne 1978) and others have in excess of 4 lethal equivalents per gamete. In conifers loads vary from loadless (*Pinus resinosa*) to as many as 13 lethal equivalents per gamete in some trees of *Pseudotsuga* (high loads are typical of many conifers; Park & Fowler 1984, Sorensen 1982). In Angiosperms, based upon very few and often inconclusive studies, 2 to 3 lethal equivalents per gamete have been reported. But these estimates are very conservative and may be much too low. It should be remembered that load is estimated on the projected viability of a gamete genotype converted into a homozygous zygote. The numerous possibilities of soft selection in plant life cycles (especially in Angiosperms) may result in large underestimates of load because of non-random gamete and/or zygote selections in the progeny resulting from selfing.

Another problem with comparisons of human and plant genetic loads relates to the mathematical formulations used in calculating lethal equivalents. If S is survivorship and $\mathcal{L}$ is load,

$$S = 1 - \text{Ł}$$
$$S = e^{-(A+BF)}$$

where A is a measure of expressed genetic damage (F= 0) plus environmental damage, B is a measure of hidden genetic damage that could manifest itself only with complete homozygosity (F = 1), and F is the inbreeding coefficient (Crow & Kimura 1970, Jacquard 1974). The load (due to environment and genetics) in a randomly mating population is

$$1 - S_1 = 1 - e^{-A}$$

Lethal equivalents per gamete (B) are calculated from the ratio of the viable progeny resulting from inbreeding and random mating (Sorensen 1969)

$$S/S_1 = (e^{-(A+BF)})/e^{-A}$$
$$S/S_1 = e^{-BF}$$

In many plants the survivorship of outcrossed progeny is very often less than 0.9 (see table 2). In general as life span increases, the value of S_1 generally decreases. In addition, in plants S_1 values show considerable intraspecific variation. The total genetic damage is the sum of B and the genetic component of A and, therefore, lies between B and B+A (Vogel & Motulsky 1986). In organisms where S_1 values are high, A values are low and lethal equivalents (B) may be a useful metric. If S_1 values are low, then A values are high and, consequently, lethal equivalents (B) may seriously underestimate total genetic damage since it lies between B and B+A. The crucial question is, of course, what proportion of the low S_1 values in many plants is due to genetics (the segregation of lethals expressed in the haploid gametophytes as well as dominant lethals expressed in the embryos, or endosperm in Angiosperms) and what proportion is due to environmental influences. Two points are suggestive that genetic influences may be important. The high levels of intraspecific variation for S_1 values within plant populations may evidence genetic rather than environmental causes, as one would have expected environmental effects to show greater variability between populations than between individuals in the same population. The second point concerns the results of reciprocal crosses. One would expect that environmental effects may be more important for the female rather than male parent (since zygote and embryo development occur on the female). If S_1 values from reciprocal crosses are similar, this is strong evidence for genetic rather than environmental causes. Where reciprocal crosses differ in S_1 values, little can be concluded since either environmental variables or genetics (the segregation of lethals

Table 2. Survivorship of progeny resulting from outcrossing (S_1) and phenotypic criteria used to determine viability.

Species	S_1	Phenotype	References
Homo sapiens	0.9	various criteria	see Vogel & Motulsky (1986) for review
Drosophila willistoni	0.843	egg to adult viability	Malogolowkin-Cohen et al. (1964)
Tribolium castaneum	0.756	egg to adult viability	Levene et al. (1965)
Tribolium confusum	0.763	egg to adult viability	Levene et al. (1965)
Phlox drummondii (annual)	0.876	seed abortion	Levin (1984)
Ulmus americana	0.55	seed set	Lester (1971)
	0.68	seed germination	
	(0.55) (0.68) = 0.374	general viability	
Acer saccharum	0.349	seed set	Gabriel (1967)
Larix laricina	0.389	sound seeds	Park & Fowler (1982)
	0.750	seedling survival	
	(0.389) (0.750) = 0.291	general viability	
Picea glauca	0.322, 0.507	sound seeds	Coles & Fowler (1976)
Picea mariana	0.63, 0.70	seed germination	Park & Fowler (1984)
	0.71, 0.80	seedling survival	
	(0.70) (0.80) = 0.56	general viability	
Picea omorika	0.334	full seeds	Langner (1959)
	0.497	seed germination	
	(0.334) (0.497) = 0.166	general viability	
Pinus resinosa	0.72	full seeds	Fowler (1965)
Pseudotsuga menziesii	0.685	sound seeds	Sorensen (1969)

expressed only in either the mega- or microgametophyte) can be invoked. In sugar maple (*Acer saccharum*) Gabriel (1967) reported that three out of four reciprocal crosses gave similar S_1 values and, thus, support the hypothesis that the low S_1 values have genetic causes at least for those six trees.

Because of the variable and often low S_1 values found in plants, in addition to reporting lethal equivalents, S_1 values should also be reported. The data in table 2 show an inverse relationship between life span und S_1 values. If the S_1 depression has a significant genetic component, then it may represent the somatic accumulation of recessive mutants expressed in the gametophytes and/or dominant mutants expressed during embryogenesis or endosperm development.

Mutational load in plants is manifested both in low S_1 values and inbreeding depression. It is generally conceded that inbreeding depression is the only general factor in large outcrossing populations that can prevent the evolution of selfing in most plant species (Charlesworth & Charlesworth 1979). It has also been hypothesized that the relationship between inbreeding depression and selfing determines, in large part, the direction of mating system evolution in plants (Lande & Schemske 1985, Schemske & Lande 1985). Mutational load may, therefore, be the major determinant in plant mating system evolution. *i.e.* loadless species may evolve systems of selfing whereas species with high loads must be outcrossed. Kondrashov (1982, 1984) concluded that selection against deleterious mutations is one of the main factors maintaining sexual reproduction and recombination. Thus, the loss of genetic integrity due to molecular, cellular, histological and organographic constraints may be ultimately compensated for by the genetics of sexual reproduction, *i.e.* heterozygosity covering genetic damage and recombination generating offspring with fewer mutations than the parents.

Conclusions

Although the loss of genetic integrity is primarily a consequence of biochemical events occurring at the cellular level, the maintenance of genetic integrity involves all levels of plant biology. The molecular organization of many aspects of plant genomes seems adapted to reduce the immediate consequences of mutation. The loss of mutant cells may occur within the plant by the displacement of mutant cell lineages to non-meristematic cell populations, mutant genotypes may be lost through the various mechanisms of developmental selection associated with sexual reproduction, non-mutant genotypes may be reconstituted from mutant genotypes during meiosis and syngamy, and, finally, whether a plant species manifests adaptations for selfing or outcrossing may be a consequence of mutational load levels. Thus, the promotion of genetic stasis against the forces of mutation may be an important underlying principle in plant biology.

Acknowledgements

Supported by grants from Alexander von Humboldt Stiftung (Bonn) and US National Science Foundation (DCB 8519018).

References

Bloch, R. 1965: Spontaneous and induced abnormal growth in plants. – In: Ruhland, W. (ed.): Handbuch der Pflanzenphysiologie, **15:** 156–183. – Berlin.

Buchholz, J. T. 1922: Developmental selection in vascular plants. – Bot. Gaz. (Crawfordsville) **73**: 249–286.

Burnet, F. M. 1970: Immunological surveillance. – Sydney.

Buss, L. W. 1983: Evolution, development, and the units of selection. – Proc. Natl. Acad. USA **80**: 1387–1391.

Cairns, J. 1975: Mutation, selection and the natural history of cancer. – Nature **255**: 197–200.

Charlesworth, B. & Charlesworth, D. 1979: The evolutionary genetics of sexual systems in flowering plants. – Proc. Roy. Soc. London, Ser. B, Biol. Sci. **205**: 513–530.

Coles, J. F. & Fowler, D. P. 1976: Inbreeding in neighbouring trees in two white spruce populations. – Silvae Genet. **25**: 29–34.

Crow, J. F. & Kimura, M. 1970: An introduction to population genetics theory. – New York.

Crumpacker, D. W. 1967: Genetic loads in maize (*Zea mays* L.) and other cross-fertilized plants and animals. – Evol. Biol. **1**: 306–424.

D'Amato, F. 1977: Nuclear cytology in relation to development. – Cambridge.

Davidson, D. 1960: Meristem initial cells in irradiated roots of *Vicia faba.* Ann. Bot. (London) **24**: 287–295.

Dobzhansky, T. et al. 1977: Evolution. – San Francisco.

Döring, H.-P. & Starlinger, P. 1986: Molecular genetics of transposable elements in plants. – Ann. Rev. Genet. **20**: 175–200.

Dover, G. 1982: Molecular drive: a cohesive mode of species evolution. – Nature **299**: 111–117.

Dyer, A. F. 1976: Modification and errors of mitotic cell division in relation to differentiation. – In: Yeoman, M. M. (ed.): Cell division in higher plants: 199–249. – London.

Eriksson, G. et al. 1966: Genetic changes induced by semiacute irradiation of pollen mother cells in *Larix leptolepis* (Sieb. et Zucc.) Gord. – Hereditas **55**: 213–226.

Flavell, R. B. 1986: Repetitive DNA and chromosome evolution in plants. – Philos. Trans., Ser. B **312**: 227–242.

Fowler, D. P. 1965: Effects of inbreeding in red pine, *Pinus resinosa* Ait. IV. Comparison with other northeastern *Pinus* species. – Silvae Genet. **14**: 76–81.

Gabriel, W. J. 1967: Reproductive behavior in sugar maple: self-compatibility, cross-compatibility, agamospermy, and agamocarpy. – Silvae Genet. **16**: 165–168.

Giles, N. 1940: Spontaneous chromosome aberrations in *Tradescantia.* – Genetics **25**: 69–87.

Gottlieb, L. D. 1984: Genetics and morphological evolution in plants. – Amer. Naturalist **123**: 681–709.

Hardwick, R. C. 1986: Physiological consequences of modular growth in plants. – Philos. Trans., Ser. B **313:** 161–173.

Jacquard, A. 1974: The genetic structure of populations. – Berlin.

Johns, M. A. et al. 1983: Exceptionally high levels of restriction site polymorphism in DNA near the maize *Adhl* gene. – Genetics **105**: 733–743.

Jones, D. F. 1935: The similarity between fasciations in plants and tumors in animals and their genetic basis. – Science **81**: 75–76.

Jørgensen, J. H. & Jensen, H. P. 1986: The spontaneous chlorophyll mutation frequency in barley. – Hereditas **105**: 71 – 72.

Kehr, A. E. 1965: The growth and development of spontaneous plant tumors. – In: Ruhland, W. (ed.), Handbuch der Pflanzenphysiologie, 15: 184–196. – Berlin.

Klekowski, E. J., jr. 1984: Mutational load in clonal plants: a study of two fern species. – Evolution **38**: 417–426.

– 1988: Mutation, developmental selection and plant evolution. – Columbia University Press, New York.

– & Kazarinova-Fukshansky, N. 1984a: Shoot apical meristems and mutation: fixation of selectively neutral cell genotypes. – Amer. J. Bot. **71**: 22–27.

– & – 1984b: Shoot apical meristems and mutation: selective loss of disadvantageous cell genotypes. – Amer. J. Bot. **71**: 28–34.

– et al. 1986: Mutation, apical meristems and developmental selection in plants. – In: Gustafson, J. P. & Stebbins, G. L. & Ayala, F. J. (eds.), Genetics, development and evolution: 79–113. – New York.

Kondrashov, A. S. 1982: Selection against harmful mutations in large sexual and asexual populations. – Genet. Res. **40**: 325–332.

– 1984: Deleterious mutations as an evolutionary factor. 1. The advantage of recombination. – Genet. Res. **44**: 199–217.

Lande, R. & Schemske, D. W. 1985: The evolution of self-fertilization and inbreeding depression in plants. I. Genetic models. – Evolution **39**: 24–40.

Langner, W. 1959: Selbstfertilität und Inzucht bei *Picea omorika* (Pančić) Purkyne. – Silvae Genet. **8**: 84–93.

Langridge, J. 1958: A hypothesis of developmental selection exemplified by lethal and semi-lethal mutants of *Arabidopsis*. – Austral. J. Biol. Sci. **11**: 58–68.

Lester, D. T. 1971: Self-compatibility and inbreeding depression in American elm. – Forest Sci. **17**: 321–322.

Levene, H. et al. 1965: Genetic load in *Tribolium*. – Proc. Natl. Acad. USA **53**: 1042–1050.

Levin, D. A. 1984: Inbreeding depression and proximity-dependent crossing success in *Phlox drummondii*. – Evolution **38**: 116–127.

Lintilhac, P. M. 1984: Positional controls in meristem development: a caveat and an alternative. – In: Barlow, P. W. & Carr, D. J. (eds.): Positional controls in plant development: 83–105. – Cambridge.

Lloyd, R. M. & Warne. T. R. 1978: The absence of genetic load in a morphologically variable sexual species, *Ceratopteris thalictroides* (Parkeriaceae). – Syst. Bot. **3**: 20–36.

Malogolowkin-Cohen, C. et al. 1964: Inbreeding and the mutational and balanced loads in natural populations of *Drosophila willistoni*. – Genetics **50**: 1299–1311.

Meins, F. jr. 1983: Heritable variation in plant cell culture. – Annual Rev. Pl. Physiol. **34**: 327–346.

Novick, A. & Szilard, L. 1950: Experiments with the chemostat on spontaneous mutations of bacteria. – Proc. Natl. Acad. USA **36**: 708–719.

– & – 1951: Genetic mechanisms in bacteria and bacterial viruses I. Experiments on spontaneous and chemically induced mutations of bacteria growing in the chemostat. – Cold Spring Harbor Symp. Quant. Biol. **16**: 337–343.

Park, Y. S. & Fowler, D. P. 1982: Effects of inbreeding and genetic variances in a natural population of tamarack (*Larix laricina* [Du Roi] K. Koch) in eastern Canada. – Silvae Genet. **31**: 21–26.

– & – 1984: Inbreeding in black spruce (*Picea mariana* [Mill.] B.S.P.): self-fertility, genetic load, and performance. – Canad. J. Forest Res. **14**: 17–21.

Pohlheim, F. 1981: Induced mutations for investigation of histogenetic processes as the basis for optimal mutant selection. – In: Proceedings of the international symposium on induced mutations as a tool for crop plant improvement, International Atomic Energy Agency Wien: 489–495.

– 1983: Vergleichende Untersuchungen zur Änderung der Richtung von Zellteilungen in Blattepidermen. – Biol. Zentralbl. **102**: 323–336.

Rutishauser, A. & La Cour, L. F. 1956: Spontaneous chromosome breakage in hybrid endosperms. – Chromosoma **8**: 317–340.

Schemske, D. W. & Lande, R. 1985: The evolution of self-fertilization and inbreeding depression in plants. II. Empirical observations. – Evolution **39**: 41–52.

Slatkin, M. 1984: Somatic mutations as an evolutionary force. – In: Greenwood, P. J., Harvey, P. H. & Slatkin, M. (eds.), Evolution: 19–30. – Cambridge.

Sorensen, F. 1969: Embryonic genetic load in coastal Douglas-fir, *Pseudotsuga menziesii* var. *menziesii*. – Amer. Naturalist **103**: 389–398.

– 1982: The roles of polyembryony and embryo viability in the genetic system of conifers. – Evolution **36**: 725–733.

Sprague, G. R. et al. 1960: Mutations affecting quantitative traits in the selfed progeny of doubled monoploid maize stocks. – Genetics **45**: 855–866.

Stadler, L. J. 1942: Some observations on gene variability and spontaneous mutation. – In: "The Spragg memorial lectures on plant breeding (third series)". – East Lansing, Michigan.

Vogel, F. & Motulsky, A. G. 1986: Human genetics (ed. 2). – Berlin.

Walbot, V. & Cullis, C. A. 1983: The plasticity of the plant genome – is it a requirement for success? – Pl. Molec. Biol. Rep. **1**: 3–11.

Wallace, B. 1968: Topics in population genetics. – New York.

Waxman, S. 1975: Witches'-brooms, sources of new and interesting dwarf forms of *Picea, Pinus* and *Tsuga* species. – Acta Hort. **54**: 25–32.

Weismann, A. 1892: Das Keimplasma. Eine Theorie der Vererbung. – Jena.

White, O. E. 1948: Fasciation. – Bot. Rev. **14**: 319–358.

Whitham, T. G. & Slobodchikoff, D. N. 1981: Evolution by individuals, plant-herbivore interactions, and mosaics of genetic variability: the adaptive significance of somatic mutations in plants. – Oecologia (Berlin) **49**: 287–292.

Address of the author: Professor Edward J. Klekowski, Botany Department, University of Massachusetts, Amherst, Mass. 01003, USA

Molecular biology of phytochrome

P. H. Quail

Abstract

Quail, P. H. 1988: Molecular biology of phytochrome. – In: Greuter, W. & Zimmer, B. (eds.): Proceedings of the XIV International Botanical Congress: 153–156. – Koeltz, Königstein/Taunus.

The mechanism of phytochrome action can be investigated by examining the molecular properties of the photorecptor and the expression of genes under its control. Biochemical and molecular biological studies have provided valuable insights into the structure of phytochrome, including features conserved between monocot and dicot phytochromes; regions of the molecule involved in photoconversion–induced conformational changes and dimerization; and sequences involved in protein-chromophore interactions. The expression of the phytochrome genes is regulated rapidly by the photoreceptor itself in *Avena* in negative feedback fashion, with evidence for regulation at both transcriptional and posttranscriptional levels. However, the degree of down-regulation is less in most other species examined than in *Avena.* Moreover, *Avena* tissue also contains a distinct molecular species of phytochrome that appears to be constitutively expressed, unaffected by light treatments. The molecular basis for the phytochrome-deficient *aurea* mutant of tomato and the consequences of phytochrome deficiency to the plant are being studied.

Molecular Biology of Phytochrome

We are approaching the question of the mechanism of phytochrome action by: (a) assembling basic information on the structural properties of the photoreceptor molecule, in the hope of obtaining clues as to its functional properties; and (b) examining genes whose expression is changed rapidly by phytochrome, in the hope that this will aid identification of the molecular events between Pfr formation and altered transcription (Quail et al. 1986, 1987 a-c).

Data thus far obtained have provided valuable insights into the primary, secondary, tertiary and quaternary structure of the chromoprotein. From cDNA sequence analysis, we have delineated sequences in the polypeptide that are highly conserved between phytochromes from the evolutionarily divergent monocot and dicot genera, *Avena* and *Cucurbita* (Hershey, et al. 1985, Sharrock et al. 1986). These sequences predominate in the chromophore-bearing, NH_2-terminal domain. We have also identified segments of the polypeptide that undergo conformational changes upon photoconversion from the inactive Pr to the active Pfr form and are therefore candidates for involvement in the regulatory action of the photoreceptor (Jones et al. 1985, Quail et al. 1986, Vierstra et al. 1984). One such segment includes Arg-52 at the NH_2-terminus, as indicated by the preferential cleavage of the polypeptide that is caused by the protease subtilisin at this residue when the molecule is in the Pr form (A. M. Jones, unpublished). By analysis of chromopeptides generated by proteolytic digestion of purified *Avena* phytochrome, we have localized a conserved, hydrophobic sequence between residues 190 and 210 that may play a major role in providing the noncovalent interactions that stabilize the chromophore in the Pr configuration (A. M. Jones, unpublished; Quail et al. 1987 b). Finally, we have provided evidence that the native phytochrome molecule is an elongated dimer (Jones & Quail 1986), with the polypeptide chain of each monomer being folded into a globular, 74−kDa NH_2-terminal domain and a more extended COOH-terminal domain (Jones & Quail, 1986, Jones et al. 1985, Quail et al. 1987 b). The data indicate further that the dimerization site is in the COOH-terminal domain (Jones & Quail 1986, Quail et al. 1987 b, Vierstra et al. 1984).

Our investigations of phytochrome-regulated gene expression have focused primarily on the genes for the phytochrome polypeptide. The expression of these genes is regulated rapidly by the photoreceptor itself in *Avena* in negative feedback fashion, with evidence for regulation at both transcriptional and post-transcriptional levels (Colbert et al. 1983, 1985, Quail et al. 1987 a, b). To investigate the molecular basis for this regulation, we have isolated and sequenced phytochrome genes from *Avena* (Hershey et al. 1987) and have begun to explore both transient expression and stable transformation approaches to identifying sequences and cellular factors involved in the regulation. Electroporation of protoplasts has proven to be of limited usefulness as a transient expression system because of the apparent loss of light-regulation of expression in the osmotically stressed cells. Moreover, we have found considerable variability between plant species in the degree of negative feedback control that phytochrome exerts over its own mRNA levels. In particular, the degree of down-regulation in some species such as tomato, a potentially useful species for transgenic plant studies, is considerably less than in *Avena* (Sharrock et al. 1988). This observation indicates the need for caution in selecting a transformable and regenerable host plant suitable for studying regulation of phytochrome gene expression by stable transformation procedures.

An additional level of complexity that has become apparent in recent times, is the existence of a molecular species of phytochrome with an array of characteristics distinct from those established for the more extensively studied molecule isolated from etiolated tissue (Tokuhisa et al. 1985, Tokuhisa & Quail 1987). In *Avena*, this newly recognized molecule appears to be constitutively expressed, unaffected by light treatments. Whether this molecular species represents a distinct gene product or some form of post-translational modification of the more familiar phytochrome molecule is yet to be definitively determined.

Experiments designed to understand the molecular basis for the phytochrome-deficient *aurea* mutant of tomato show that translationally active phytochrome mRNA is produced at normal levels but that the polypeptide is deficient in the mutant tissue (Parks et al. 1987, Sharrock et al. 1988). These data indicate that although normal levels of the phytochrome polypeptide are likely synthesized, the molecule is unstable in the mutant cells. Possible reasons for this observation include lesions in the synthesis or attachment of the chromophore and an alteration in the process responsible for intracellular degradation of phytochrome. One of the consequences of phytochrome deficiency in *aurea* is greatly reduced expression of *cab* genes in response to light (Sharrock et al. 1988).

References

Colbert, J. T., Hershey, H. P. & Quail, P. H. 1983: Autoregulatory control of translatable phytochrome mRNA levels. – Proc. Natl. Acad. Sci. USA **80:** 2248–2252.

–, – & – 1985: Phytochrome regulation of phytochrome mRNA abundance. – Pl. Molec. Biol. **5:** 91–102.

Hershey, H. P., Barker, R. F., Idler, K. B., Lissemore, J. L. & Quail, P. H. 1985: Analysis of cloned cDNA and genomic sequences for phytochrome: Complete amino acid sequence for two gene products expressed in etiolated *Avena.* – Nucleic Acids Res. **13:** 8543–8559.

–, –, –, Murray, M. G. & Quail, P. H. 1987: Nucleotide sequence and characterization of a gene encoding the phytochrome polypeptide from *Avena.* – Gene **61:** 339–348.

Jones, A. M. & Quail, P. H. 1986: Quaternary structure of 124–kilodalton phytochrome from *Avena sativa* L. – Biochemistry **25:** 2987–2995.

– Vierstra, R. D., Daniels, S. M. & Quail, P. H. 1985: The role of separate molecular domains in the structure of phytochrome from *Avena sativa.* – Planta **164:** 501–506.

Parks, B. M., Jones, A. M., Adamse, P., Koorneef, M., Kendrick, R. E. & Quail, P. H. 1987: The *aurea* mutant of tomato is deficient in spectrophotometrically and immunochemically detectable phytochrome. – Pl. Molec. Biol. **9:** 97–107.

Quail, P. H., Barker, R. F., Colbert, J. T., Daniels S. M., Hershey, H. P., Idler, K. B., Jones, A. M. & Lissemore, J. L. 1986: Structural features of the phytochrome molecule and feedback regulation of the expression of its genes in *Avena.* – In: Fox J. E. & Jacobs, M. (eds.), "Molecular Biology of Plant Growth Control". – UCLA Symposia on Molecular and Cellular Biology, New Series, **44:** 425–439. – Alan R. Liss, Inc., New York.

– Christensen, A. H., Jones, A. M., Lissemore, J. L., Parks, B. M. & Sharrock, R. A. 1987a: The phytochrome molecule and the regulation of its genes. In: Kon, O. L., Chung, M. C.-M., Hwang, P., L., Leong, S.-F., Loke, K. H., Thiyagarajah, P., Wong, P. T.-H. (eds.), Integration and Control of Metabolic Processes: Pure and Applied Aspects: 41–54. Proceedings 4th FOAB Congress, Singapore, ICSU Press, Cambridge.

– Colbert, J. T., Peters, K. P., Christensen, A., Sharrock, R. A. & Lissmore, J. L. 1987b: Phytochrome and the regulation of the expression of its genes. – Philos. Trans., Ser. B, **314:** 469–480.

– Gatz, C., Hershey, H. P., Jones, A. M., Lissemore, J. L., Parks, B. M., Sharrock, R. A., Barker, R. F., Idler, K., Murray, M. G., Koornneef, M. & Kendrick, R. E. 1988: Molecular Biology of phytochrome. In: Furuya, M. (ed.), Phytochrome and Photomorphogenesis in Plants. – Proceedings of XVI Yamada Conference. Academic Press, New York (in press.)

Sharrock, R. A., Lissemore, J. L. & Quail, P. H. 1986: Nucleotide and derived amino acid sequence of a *Cucurbita* phytochrome cDNA clone: Identification of conserved features by comparison with *Avena* phytochrome. – Gene **47:** 287–295.

–, Parks, B. M., Koornneef, M. & Quail, P. H. 1988: Molecular analysis of the phytochrome deficiency in an *aurea* mutant of tomato. – Molec. Gen. Genetics (in press).

Tokuhisa, J. G., Daniels, S. M. & Quail, P. H. 1985: Phytochrome in green tissue: Spectral and immunochemical evidence for two distinct molecular species of phytochrome in light-grown *Avena sativa.* – Planta **164:** 321–322.

– & Quail, P. H. 1987: The levels of two distinct species of phytochrome are regulated differently during germination in *Avena.* – Planta **172:** 371–377.

Vierstra, R. D., Cordonnier, M.-M., Pratt, L. H. & Quail, P. H. 1984: Immunoblot analysis of apparent molecular weight and proteolytic degradation of native phytochrome from several plant species. – Planta **160:** 521–528.

Address of the author: Professor Peter H. Quail, Plant Gene Expression Center, University of California, Berkeley, 800 Buchenau St., Albany, CA 94710, USA.

Botany and mycology

J. Webster

Abstract

Webster, J. 1988: Botany and mycology. – In: Greuter, W. & Zimmer, B. (eds.): Proceedings of the XIV International Botanical Congress: 157–180. – Koeltz, Königstein/Taunus.

Are fungi plants? Despite the difficulties of defining fungi, it is concluded that there is little evidence for close relationship between fungi and extant plants. In place of the traditional 2-kingdom classification in which fungi are classified with plants, many biologists now prefer to classify living things into about five kingdoms, fungi being accorded the status of a separate kingdom. Following the theme of the Congress, "Forests of the world", a mycologist looks at forests and discusses saprotrophic activities of Basidiomycetes in relation to litter and wood decay and individualism. The succession of Basidiomycetes causing sheathing mycorrhizas is illustrated in relation to forest age, and the role of sheathing mycorrhizas in protecting root systems from attack by fungal pathogens is discussed. The possibility of using saprotrophic wood-decaying Basidiomycetes in the biological control of tree diseases caused by fungi is evaluated. In view of the closely intertwined activities of fungi and plants, it is important that the training of botanists should include mycology, and vice-versa, and that a flourishing mutualistic symbiosis should be encouraged between the two disciplines.

Introduction

I shall begin by asking the question "Are fungi plants?" This question was addressed by Martin (1955) who stated then that "it is probable that an overwhelming majority of botanists and a substantial number of those whose particular interest is with fungi would answer this with an unqualified affirmative". Some thirty years later, it is extremely doubtful if the same conclusion would be drawn. The reasons underlying this change of opinion have to do, in part, with a questioning of whether two-kingdom classification of living things is satisfactory, and of the assumption that heterotrophic organisms must have had autotrophic ancestors. There has also been a better appreciation of relationships between groups of organisms stemming from a comparison of fine structure, molecular architecture and metabolism of different groups of living things; new knowledge obtained from modern techniques of electron microscopy, biochemistry and physiology.

It is difficult to define the fungi concisely but I will accept the definition provided by Ainsworth & Bisby's "Dictionary of the Fungi" (Hawksworth et al. 1983).

> "The main characteristics are: *Nutrition:* heterotrophic (never photosynthetic) and absorptive (ingestion rare). *Thallus:* typically non-plasmodial (Eumycota) but sometimes plasmodial (Myxomycota), the former unicellular of filamentous (mycelial) and septate or non-septate, typically non-motile (with protoplasmic flow through the mycelium), but motile stages (*e.g.* zoospores) may occur. *Cell wall:* in non-plasmodial forms well-defined, typically chitinized (cellulose in Oomycetes). *Nuclear status:* (*q.v.*): eukaryotic, multinucleate, the mycelium being homo- or heterokaryotic, haploid, dikaryotic, or diploid, the last being usually of limited duration. *Life-cycle:* simple to complex. *Sexuality:* asexual or sexual and homo- or heterothallic. *Sporocarps:* microscopic or macroscopic and showing limited tissue differentiation."

What to include in the fungi depends on the definition chosen, and some would exclude the slime moulds (Olive 1975) or forms with zoospores (*e.g.* Margulis & Schwartz 1982, Barnes 1984).

If one is required to place fungi, whether defined in a broad or a restricted sense, into one of the two kingdoms, plants or animals, one would probably choose the former on the grounds that fungi more closely resemble plants in their habit (sedentary rather than motile), nutrition (absorptive, not usually ingestive) and reproduction (by spores) than animals. However, opinion has been growing, possibly first expressed in writing by Wiggers (1780), that fungi are not plants. The difficulty in separating morphologically simpler "plants" from "animals" led Haeckel (1866, 1894) to the concept of Protista.

Many mycologists now believe that the 5-kingdom classification proposed by Whittaker (1969) provides a more satisfactory conceptual framework within which to classify fungi. The five kingdoms proposed by Whittaker are based on levels of organization (prokaryotic *vs.* eukaryotic, unicellular *vs.* multicellular) in relation to three principal modes of nutrition (photosynthesis, absorption and ingestion). They are:

Monera: prokaryotes.
Protista: unicellular eukaryotes.
Plantae: multicellular green plants and higher algae.
Fungi: multinucleate higher fungi.
Animalia: multicellular animals.

Whittaker has advanced several reasons why fungi are judged not to be plants.

- Despite earlier speculations that fungi were derived from algae, it is now believed that the lower fungal groups had a polyphyletic origin from different colourless flagellate ancestors. The details of fine structure of zoospores of Mastigomycotina (with the exception of the Oomycetes) support this idea. It is also believed that the higher fungi (Ascomycetes and Basidiomycetes) were derived from one of the groups of lower fungi.
- Their organization is very different from that of plants. Their non-motile condition and mycelial organization reflect their distinctive absorptive nutrition.
- Reproductive structures and the dikaryotic condition are distinctive from the reproductive structures and diploid condition of higher plants.

Whittaker's five-kingdom classification has been accepted by Margulis (1971) and by Margulis & Schwartz (1982), although in the latter work the kingdom Protoctista has been substituted for the Protista to include not only unicellular eukaryotes, but their multinucleate and multicellular relatives. The Protoctista are not a natural assemblage of phyletically related organisms, but a group "defined by exclusion: its members are neither animals (which develop from a blastula), plants (which develop from an embryo), fungi (which lack undulipodia and develop from spores), nor prokaryotes."

There are, of course, alternative ways of classifying the organisms included in the Protoctista *sensu* Margulis & Schwartz, and some of the groups have been the traditional field of study of mycologists. However, their scheme indicates no or few clear affinities between heterotrophic undulipodiate protoctists and photosynthetic forms at a parallel level of organization (with the exception of the Oomycota-Xanthophyta-Phaeophyta), nor are the 'higher fungi' (Zygomycota, Ascomycota, Basidiomycota) shown as having any obvious link with extant algal groups.

Classification is essentially a subjective matter, and any scheme or outline classification must rely on the judgement of the biologist as he assesses the evidence of relatedness between organisms, and seeks to present it in a form reflecting his opinion. It is clear, however, that there is a growing body of opinion, not only amongst mycologists (*e.g.* Kendrick 1985) but amongst algologists (*e.g.* Leedale 1974), that green plants and fungi are not closely related.

If one accepts the view that fungi are not plants, one might then ask "Should botanists study fungi?" or the converse question "Should mycologists study botany?" The answer to both questions must surely be emphatically "yes" for several reasons.

- The saprotrophic role of fungi, along with bacteria and the animal population of the soil, is vital in decomposition which is linked to soil fertility.
- Most vascular plants have symbiotic relationships with fungi in the form of mycorrhiza, the development of which is often essential to thrifty growth and nutrition (Harley 1969, Harley & Smith 1983). A modified population of soil micro-organisms including fungi is also stimulated to develop around plant roots (Curl & Truelove 1985).
- Many plant diseases are caused by fungi.
- Fungi are often the most suitable material for studying problems of general biological interest, and knowledge gained in studying fungi may have much wider significance.

Since the overall theme of the XIVth International Botanical Congress was "Forests of the world", I propose to illustrate the symbiosis between mycology and botany by reference to fungi associated with trees.

A mycologist looks at a forest: Saprotrophs

The annual litter production in forests ranges from about 1–20 t ha^{-1} (Bray & Gorham 1964). A diverse population of bacteria, fungi and animals is engaged in the decomposition of this litter, but within the decomposer community, Basidiomycetes play an important and central role (Swift 1982). They are well-equipped with enzymes to enable them to degrade the ligno-cellulose complex of which plant litter, and especially woody debris, is largely composed. It has been estimated that Basidiomycetes may account for 60% of the *living* biomass of microbes in a mull soil profile (Frankland 1982), whilst in mor profiles the proportion is probably higher.

Basidiocarps found on the forest floor belong to several ecologically distinct kinds of fungi, *e.g.* leaf litter decomposers, lignicolous fungi (some saprotrophs and some necrotrophic parasites), and mycorrhizal symbionts (especially of sheathing mycorrhizas). The distinction between leaf litter and lignicolous fungi is not absolute, because it has been observed that fungi such as *Phallus impudicus* and *Tricholomopsis platyphylla* which grow on tree stumps can also decompose leaf litter in the vicinity of bulky woody substrata (Thompson 1984), whilst *Mycena galopus* can fruit on leaf litter and small twigs, but not on tree trunks. Hintikka (1982) has shown that there are physiological differences between leaf litter fungi and lignicoles. Lignicolous fungi have a higher tolerance for CO_2 than litter decomposers, and their growth may even be stimulated by 10% CO_2. Some wood-decomposers are able to grow at low concentrations of oxygen, and appear to be more tolerant of certain products of anaerobic respiration such as acetate, formate and proprionate than leaf litter fungi.

In the course of decay of both leaf litter and woody debris, fungi concentrate minerals, so that the mineral content of mycelium and fruit-bodies, on a dry weight basis, usually exceeds that of the litter on which the fungi are growing (Frankland 1982, Swift 1977, Witkamp 1969). It has been suggested that the immobilisation of mineral nutrients by fungal mycelium may be advantageous to the nutrition of the higher plant by a reduction in leaching, but as Satchell (1974) has pointed out in connection with the concentration of minerals by soil animals, this is difficult to prove. One important consequence of the concentration of minerals is that the nitrogen content of the litter is increased and the C/N ratio is greatly reduced (Hering 1982). This renders the leaf litter more palatable and nutritious to soil invertebrates, which comminute the litter and hasten the process of decay.

Autecological studies have been made on several decomposer Basidiomycetes. The most detailed picture has been obtained for *Mycena galopus* (Frankland 1984). It occurs on a wide range of host materials including coniferous and deciduous leaf litter, as a secondary colonist, causing a typical white rot by degrading lignin and cellulose. The mycelium has been identified in leaf litter by a variety of techniques, of which the fluorescent antibody method is fairly specific (Frankland et al. 1981, Chard et al. 1983). It has also been possible to trace the distribution of mycelium in the litter from fruit-bodies formed at the surface. In litter of *Picea sitchensis, Mycena galopus* often fruits in association with *Marasmius androsaceus* and *Cystoderma amianthinum,* apparently in competition with them. The distribution of mycelium from which fruit-bodies of the three fungi arose (fruiting depths in fig. 1) was not identical, and was influenced by whether fruiting of some of them occurred in single- or mixed-species clumps. Newell (1984 a, b) has suggested that preferential grazing of the mycelium of *Marasmius androsaceus* by the collembolan *Onychiurus latus* altered the outcome of competition between it and *Mycena galopus.* Without grazing, *Marasmius androsaceus* had a higher capacity to colonise spruce litter in the laboratory, but when grazing was permitted, the situation was reversed. This conclusion was supported by field experiments in which the collembolan was excluded from spruce litter, resulting in a reduction of the activity of

Marasmius androsaceus and an increase in that of *Mycena galopus.* In the surface litter layers, the density of the collembolan was not sufficiently high to affect the outcome of competition between the two Basidiomycetes, which may explain the dominance of *Marasmius androsaceus* mycelium at the surface.

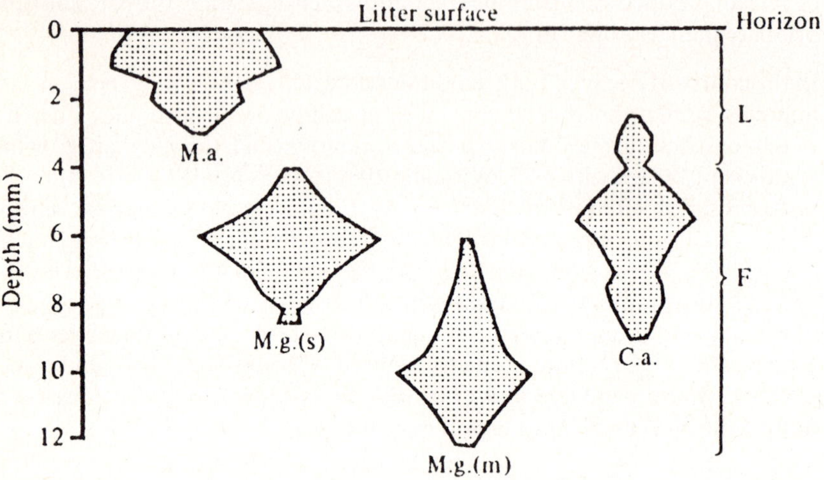

Fig. 1. The fruiting depths of *Marasmius androsaceus* **(M.a.)**, *Mycena galopus* **(M.g.)** and *Cystoderma amianthinum* **(C.a.)** in a *Picea sitchensis* plantation. The width of each "kite" at any depth is proportional to the percentage number of basidiocarps originating at that depth. – **s**, single-species clumps of *M. galopus*; **m**, mixed-species clumps of *M. galopus* with *M. androsaceus.* (From Newell 1980, in Frankland 1984.)

The mycelium of many litter-decomposing Basidiomycetes is perennial. Examples of perennial mycelia among pasture-inhabiting fungi are those agarics associated with fairy-rings. The best-known is *Marasmius oreades.* The dikaryotic mycelium developing from a central point grows outwards as an annulus, extending at the margin, and dying off behind. Measurements of the annual rate of advance of the mycelial front and of the diameter of fairy rings show that some are probably several centuries old. Analysis of the mating types from spores collected from basidiocarps on a single ring show that they are identical, a striking confirmation of the stability of the original dikaryon (Burnett & Evans 1966).

A characteristic feature of several basidiomycetous wood-rotting saprotrophs is the possession of mycelial cords. Mycelial cords are formed by the aggregation of individual hyphae, resulting from a change from divergent to coherent growth (Thompson 1984). Typical examples of wood-rotting cord-formers are

Phanerochaete velutina, Phallus impudicus and *Tricholomopsis platyphylla.* The precise stimulus to cord-formation is not always clear, although it is known that cords fail to develop in sterile soil. Volatile exudates, *e.g.* from decaying wood, can stimulate directional growth. The cords are often differentiated into wider central vessel hyphae which may facilitate economical movement of water and nutrients from a colonised resource unit (food base) towards the advancing front (Jennings 1984). The surface of the cords is heavily encrusted with calcium oxalate, and Thompson (1984) has claimed that the reduction of pH in the litter surrounding the cords, which is a consequence of calcium oxalate production, may suppress the growth of bacteria. Other advantages in the possession of mycelial cords are the protection of hyphae from environmental fluctuations, and an increase the inoculum potential of the fungus, enabling it to colonise woody substrata.

The presence of mycelial cords permits autecological studies of the distribution of the mycelium of Basidiomycetes in woodland litter. Moreover, cultures can be obtained from mycelial cords, from pieces of decaying wood and from basidiocarps (flesh and spores). When genetically identical dikaryotic mycelia are inoculated side by side, anastomosis of the mycelia takes place and healthy growth is continued, but when genetically distinct mycelia are opposed to each other, either on agar, or as mycelia or cords developing from wood blocks, contact between the two strains results in death of the hyphae or cords (fig. 2; Thompson & Rayner 1983). Death of the dissimilar hyphae is a result of vegetative (somatic) incompatibility. Pairing of isolates in this way permits recognition of identical genotypes or individuals (Rayner & Todd 1979, 1982 a, b). Using these techniques, Thompson & Rayner (1982) have analysed the spatial structure of a population of *Tricholomopsis platyphylla* in an oak stand some 70 years old in the Forest of Dean, Gloucestershire, UK.

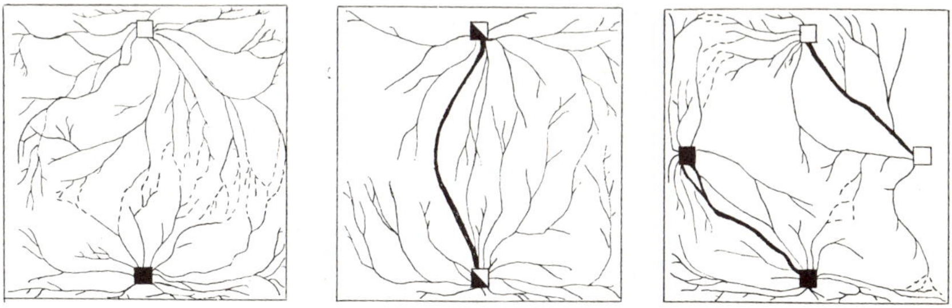

□ Mycelial type 1 ◪ Mycelial type 3 ——— Living mycelial cords
■ Mycelial type 2 - - - - - Dead and dying mycelial cords

Fig. 2. Pairings in soil between like and unlike strains of *Phanerochaete velutina.* (Reproduced, by permission, from Thompson & Rayner 1983.)

Cord-forming mycelia of *T. platyphylla* were mapped. Mycelial isolates, including 113 from cords, 12 from basidiocarp tissue and 13 from wood were obtained from 113 separate locations as shown in fig. 3. Pairings indicated that they were of 22 different mycelial types. Some of these, *e.g.* 16, 17, 21, were sampled only rarely, but others, *e.g.* type 10, were much more extensive. In most cases, isolate 10 was obtained from closely neighbouring locations,

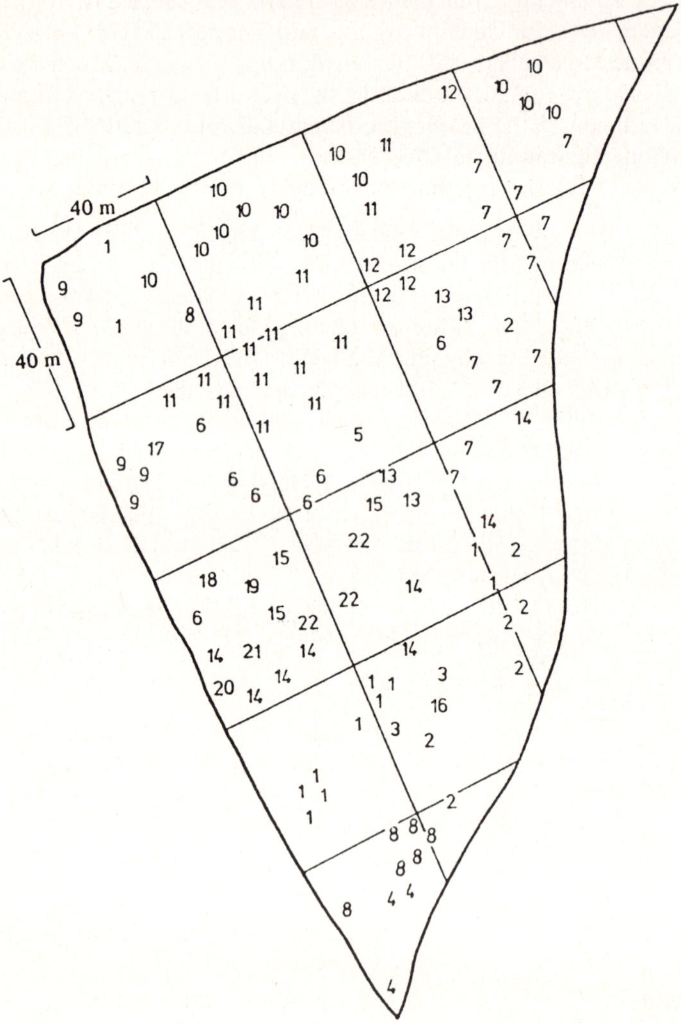

Fig. 3. Distribution of different mycelial types of *Tricholomopsis platyphylla,* indicated by different numbers (Thompson & Rayner 1982).

often in places where other isolates were not found. In other cases, *e.g.* isolate 1, the same isolate was obtained from points as much as 150m apart, separated by areas occupied by other mycelial types.

In interpreting the distribution of the different mycelial isolates, Thompson & Rayner (1982) think that it is highly likely that the isolates of the same mycelial type originate from the same mycelial individual, associated with the fact that *Tricholomopsis platyphylla* is capable of propagation by vegetative mycelium growing between food bases in the soil. The alternative possibility, that dikaryons sufficiently similar to intermingle in culture should develop from separate colonization by basidiospores, is extremely unlikely.

A number of different distribution patterns can be distinguished:

- Extensive mycelial types, *e.g.* type 11 in fig. 3. These may represent individuals which were etablished early on in the development of the population and have persisted. An alternative is that the mycelium may be extensive only temporarily, *i.e.* at the time of study.
- Individuals of more limited extent *e.g.* types 5, 6 and 12, may represent remnants of a formerly more extensive mycelium, or possibly the initial stages of development. It could be that they have arisen from adjacent, more extensive mycelia such as type 11, possibly as new dikaryons originating from basidiospores.
- Certain mycelial types occupied defined areas from which other types were excluded, *e.g.* type 11, whereas others had a discontinuous distribution, having isolates of the same type located considerable distances apart, with intervening spaces occupied by different types. Thompson & Rayner have suggested that this discontinuous distribution may result from the break-up of an originally continuous mycelium. How this might be achieved is not known, but one possibility is that of direct spatial competition between different mycelial types.

It is clear that there are many unresolved questions raised by this study and, as the authors have recognised, repeated samplings over a longer period would be necessary to answer them. However, there emerges a picture of dynamic interactions within the population of *Tricholomopsis platyphylla,* and evidence of the existence of individual mycelia which retain their identity over prolonged periods.

The comparatively small number (22) of mycelial tyes of *T. platyphylla* in an area of over 20 ha, *i.e.* an average of about one per ha, contrasts sharply with the populations of some other wood-rotting fungi such as *Coriolus versicolor.* This is a polypore which causes white-rot decay on a wide range of hard-wood hosts. Its basidiocarps are very variable in morphology and colour, ranging from pale fawn to almost black. On cut stumps of certain trees such as birch, several different morphs may be found fruiting side by side. The studies of Rayner & Todd (1979, 1982 a, b), Todd & Rayner (1978, 1980), Williams et al. (1981 a, b) and Williams & Todd (1985) have provided interesting information on the behaviour of individual dikaryons and on intraspecific antagonism in this fungus. Serial transverse sections of a birch stump taken at 2 cm intervals by means of a band-saw showed that the stump was occupied by a series of vertical columns of dikaryotic mycelia. The number of columns in a single stump is variable, but in some cases 10 distinct columns have been detected. The continuity of the columns could be traced throughout the serial sections, and each decay column could be related to a basidiocarp (or group of basidiocarps) on the exposed cut surface of the stump. The individual columns of decayed wood were separated by narrow dark zones (black lines). Isolations onto malt extract agar were made from the dikaryotic mycelia within the wood and from the dikaryotic flesh of the basidiocarps. Monokaryotic isolates were also prepared from single basidiospores or by dedikaryotization of dikaryotic mycelia from cultures. *Coriolus versicolor* shows tetrapolar (bifactorial) multiple allele heterothallism. Mating between monokaryons derived from the different dikaryotic decay colums was successful, indicating that the individual mycelia were conspecific. However, when different dikaryotic isolates were paired together by placing separate inocula about 2 cm apart on 3% malt agar plates, although mycelial anastomosis occurred at the interface between the two dikaryons, this was quickly followed by a series of antagonistic reactions. The fusion cell formed by plasmogamy often became enlarged and spindle-shaped. The hyphal tips died, and the dead zone of hyphae separating the two colonies was often strongly pigmented by dark-brown coloration. The intensity of reaction is dependent on the degree of relatedness between the opposed dikaryons. Those which have a nucleus in common show less intense pigmentation. These reactions are interpreted as being due to heterogenic somatic incompatibility. In contrast, pairings between identical dikaryons, whether derived from decay columns or basidiocarps, showed complete intermingling, with no adverse reactions following contact. Examination of the interface between adjacent dikaryons within the wood (especially after incubation under moist conditions) showed that there are usually two dark zones. The space in-between, which contains no living *Coriolus* hyphae, is often occupied by unrelated dematiaceous fungi (Rayner 1976).

These observations have been interpreted in relation to the colonization of stumps by basidiospores. Germination of the basidiospores results in the establishment of monokaryotic mycelia. These are probably only short-lived, and in the decaying wood they would become quickly converted into dikaryons, either by fusion with compatible monokaryons or by nuclear transfer from already-established dikaryons (di-mon mating) to which, paradoxically, they may eventually become antagonistic. The dikaryons show stability over several seasons, and apart from the di-mon matings referred to, show no capacity for nuclear exchange with adjacent, different, dikaryons. The decay of the wood is thus brought about by a population of distinct individual mycelia, rather than by a single mycelium with a heterogeneous nuclear content. These findings seem to be of general application, and virtually identical results have been obtained by analysing populations of *Bjerkandera adusta, Hypholoma fasciculare, Piptoporus betulinus* and *Stereum hirsutum* (Rayner & Todd 1982a).

Fungal parasites in the forest

There are numerous fungal pathogens of forest trees. Plant pathologists distinguish between *biotrophic* pathogens which are ecologically obligate parasites of living hosts, and *necrotrophic* pathogens which bring about the death of host tissue and usually survive as saprotrophs on the dead tissue. The rust fungi provide good examples of biotophs, whilst *Heterobasidion annosum* and *Armillaria mellea* are necrotrophs. The Ascomycete *Rhytisma acerinum* is regarded by some as a hemi-biotroph. It causes the well-known tar-spot disease of living leaves of sycamore, *Acer pseudoplatanus,* but continues its development during winter and early spring on fallen leaves. It is ready to eject ascospores in May to June as the new sycamore leaves expand. Interest has been shown in this fungus as an indicator of aerial pollution, and close to the centre of industrial cities the disease is absent, but it increases with increasing distance from the source of pollution. The studies of Bevan & Greenhalgh (1976) and Greenhalgh & Bevan (1978) have shown that the incidence of tar-spots on sycamore leaves can be correlated with the SO_2 concentration in the atmosphere, whether measured directly or by comparison with SO_2 levels estimated from the tolerance levels of lichens such as *Xanthoria* or *Parmelia.* Tar-spot indices (TSI) obtained by counting the number of spots per 100 cm^2 of leaf surface have been correlated with SO_2 concentrations estimated from volumetric sampling or from lichen data. The tolerance limit for the fungus is approximately 90 $\mu g\ m^{-3}$ SO_2, but below a level of about 25–30 $\mu g\ m^{-3}$ there is a very rapid increase in the incidence of tar-spot lesions. The effect of SO_2 on the fungus is probably due to the inhibition of ascospore germination, and it has been shown that rainwater with a sulphate level of 22.5 ppm inhibits ascospore germination on the surface of sycamore leaves.

Although fungal pathogens bring about disease and death of trees in 'natural' forests, their incidence is undoubtedly much greater in single species plantations, and as a result of disturbance by man, *e.g.* through felling, logging, road-making or fire. Some diseases associated with man's activities will now be considered.

Rhizina undulata is a discomycete which forms conspicuous large convex apothecia on burnt ground in conifer forests. It causes a serious disease of conifers termed "group dying", affecting *Picea* and *Pinus* spp. (Murray & Young 1961). Research by Jalaluddin (1967a, b) has shown that the disease is only prevalent on acid sites. Ascospore germination is stimulated by heat, *e.g.* by exposure to temperatures around 37° C for 3 days, and is also stimulated by exudates from heat-treated living *Pinus* roots. Ascospores which are liberated form June to November remain viable in soil for up to 2 years. Jalaluddin has postulated that, on acidic soils when fires are made adjacent to living trees or to recently-felled stumps, there is a saucer-shaped zone in the soil beneath the fire where soil temperatures would remain at a suitable level, and where root exudation induced by heat would provide a stimulus to ascospore germination. This might be followed by mycelial strand growth on the surface of the coniferous host root. In the light of this knowledge, mechanical chopping of tree debris instead of burning on clear-felled sites, and delayed planting of young conifers, is recommended, especially where spruce is grown on acid soils.

The most serious disease of plantation-grown conifers in Europe is buttrot, which is especially common on pine and spruce. The causal agent is *Heterobasidion annosum (Fomes annosus)* (Peace 1962). The orange-brown, corky, perennial basidiocarps occur at the base of infected trees, and the fungus also reproduces asexually by means of *Oedocephalum*-like conidia which form on the surface of moist infected wood and bark. Basidiospores are shed throughout the year and can be dispersed over large distances. Rishbeth (1959) used freshly-cut slices of *Pinus* stems as a selective bait for basidiospores of *H. annosum.* After exposing them to air outdoors for periods ranging from 10–90 min, the slices were incubated for 10–15 days, and after this time the presence of *H. annosum* colonies could be detected from their characteristic conidia. Depending on their proximity to sources, viable basidiospores were deposited on the slices. In one exceptional case, 280 spores/100 cm^2/hr were recorded. The selectivity of freshly-cut pine wood surfaces is significant because *H. annosum* is one of the first fungi to colonise the stumps exposed by thinning operations in recently-planted forests. The mycelium rapidly penetrates from the stumps into the roots. Root-to-root contact or even root grafting between the roots of the thinning stump and those of adjacent healthy tree rows provides a subterranean pathway of infection which can lead to very high incidence of infection. Once the importance of exposed thinning stumps as infection courts had been understood, control measures were aimed at treatment of the stumps with fungicides. It was later discovered that biological control of stump infection was more effective. *Peniophora gigantea,* which forms greyish flat crust-like basidiocarps on coniferous stumps, is also a very effective primary coloniser of freshly-cut furfaces. It is, however, saprotrophic, *i.e.* although it brings about wood decay, it does not cause disease. It also reproduces asexually by the formation of chains of cylindrical arthroconidia. Rishbeth (1963) has shown that the application of conidia of *P. gigantea* to pine stumps at the time of felling is a very effective method of controlling butt-rot disease. This biological control measure has the advantage that it is relatively cheap, and does not lead to the accumulation of toxic residues as is the case with chemical fungicidal stump treatments. Moreover, a single application of *P. gigantea* conidia provides good control, whilst it may be necessary to make several applications of chemical fungicides to obtain worthwhile reductions in the level of stump infection. Webster et al. (1970) and Ikediugwu (1976) have studied the interactions between hyphae of the two fungi in culture. There was no evidence that the inhibition of growth of the *Heterobasidion* was mediated by diffusible toxins or antibiotics produced by *Peniophora.* When hyphae of the two fungi come into direct contact, the hyphae of *Heterobasidion* become disorganised, the cytoplasm becomes vacuolate and the cells lose turgor. They also show increased permeability, shown by the uptake of dilute neutral red dye by the affected cells, whilst unaffected cells remain impermeable. This phenomenon of contact inhibition has been termed *hyphal interference.* Although the cause of the phenomenon is not understood at the cellular or chemical level, it is a widespread method of antagonism (combat) which is especially common amongst Basidiomycetes (Rayner & Webber 1984).

Phytophthora cinnamomi causes a root-rot of woody plants, especially in the moist subtropics. Zentmyer (1980) has written a comprehensive monograph on this fungus and the diseases which it causes. There is controversy over its centre of origin, but good evidence has been presented for a centre "extending from the New-Guinea–Celebes–Malaysian region into northeastern Australia and possibly in other parts of eastern Asia". However, the possibility of a Latin American, Mexican or Central American centre of origin cannot be ruled out. Although there is still debate as to whether the fungus was introduced or is indigenous to Australia, there is strong circumstantial evidence that the fungus was introduced there and into many other parts of the world along with the soil and roots of imported plants. Once present in a country, it can be spread by soil, e.g. soil and gravel used to make surface forest roads, and by water channels which convey zoospores. It is particularly common in badly-drained soils. This is probably related to the fact that infection can be caused by zoospores which swim in soil water and are attracted chemotactically to young roots. In many cases, the results of introducing the pathogen have been catastrophic. To quote from Zentmyer: "This disease will undoubtedly become one of the outstanding and classic examples of the disastrous epidemic proportions attained by a root disease in a forest. The story of the dissemination of the pathogen, the rapid build-up of infection, and the susceptibility of an amazing number of other native plants in the Western Australian forest . . . constitutes an awesome example of the devastation possible from invasion of a native forest by one plant pathogen." Features of *P. cinnamomi* include the large range of hosts attacked and its wide distribution: over 950 varieties and species of plant, reported from 68 countries. Amongst the serious disease it causes are little-leaf disease of pine, and dieback of jarrah *(Eucalyptus marginata)* in Australia.

The control of diseases caused by *P. cinnamomi* is difficult because the fungus can persist in the soil for long periods in the absence of a host. It is known to survive in the form of chlamydospores and oospores. Given suitable conditions of soil moisture and temperature and the presence of a susceptible host, there is a rapid build-up of infectious propagules. The wide host range, the fact that the hosts are generally woody, and the general problems connected with the control of soil-borne pathogens are further complications. The main control measures which have been attempted include prevention, chemical, physical and biological control, and genetical control involving resistance. However, on the forest scale, biological control appears the most promising, and this will be discussed in relation to the effects of sheathing mycorrhizas.

In the 100 years which have passed since Frank (1885) coined the term mycorrhiza, enormous advances have been made in understanding the mutualistic symbiotic relationships between fungi and higher plants (Harley 1969, Harley & Smith 1983). Forest trees in regions of high altitude and latitude have roots with sheathing (ectotrophic) mycorrhizas (Read 1984). The fungal partners in these associations are generally Basidiomycetes belonging to genera such as *Thelephora, Suillus, Amanita, Hebeloma, Laccaria, Russula, Pisolithus, Scleroderma* and *Rhizopogon.* Some of the associations are highly specific, *e.g.* that between *Suillus grevillei* and *Larix,* whilst others are not, *e.g. Amanita muscaria* can form mycorrhizas with *Betula* and *Pinus.* A single tree may have a root system infected with several different Basidiomycetes, and experienced workers have learnt to recognize from external morphology the characteristic features of mycorrhizal root tips formed from different partners. Recognition and certain identification of mycorrhizal types are, however, complicated by the fact that a single ectomycorrhizal root may yield isolates of at least three different fungi (Zak & Marx 1964). An interesting discovery is that there is a succession of mycorrhizal associates as a tree develops: the characteristic fungi associated with seedling roots may be displaced by others as the tree matures, and as the environment of the forest floor changes with closure of the canopy and accumulation of litter. Two different kinds of evidence point to this conclusion. Observations on the fruiting of Basidiomycetes known to form mycorrhizal associations with *Betula pendula* and *B. pubescens* grown in plantations of known age have shown that, in the third year after planting, there were four species, and this number increased to about thirty after ten years (Last et al. 1983, Mason et al. 1983). Within two years of planting, fruit bodies of *Hebeloma crustuliniforme* and *Laccaria tortilis* were observed. In year 4, *Inocybe lanuginella* and *Lactarius pubescens* appeared. In year 6, *Cortinarius* and *Leccinum* were recorded, and *Russula* spp. fruited in year 10 (Dighton & Mason 1985). More direct evidence of a succession of mycorrhizal fungi has been obtained by dissecting cores containing mycorrhizal root tips taken at intervals of 25 cm from the base an 8-year-old *B. pubescens* (Deacon et al. 1983). Mycorrhizal root tips of *Hebeloma*-type were most abundant in the outer part of the root systems, *i.e.* on the newest roots, whilst tips of the *Leccinum*-type were infrequent in this zone. Conversely, near the centre of the root system, *Leccinum*-type tips were more frequent, whilst *Hebeloma*-type tips were less frequent (fig. 4).

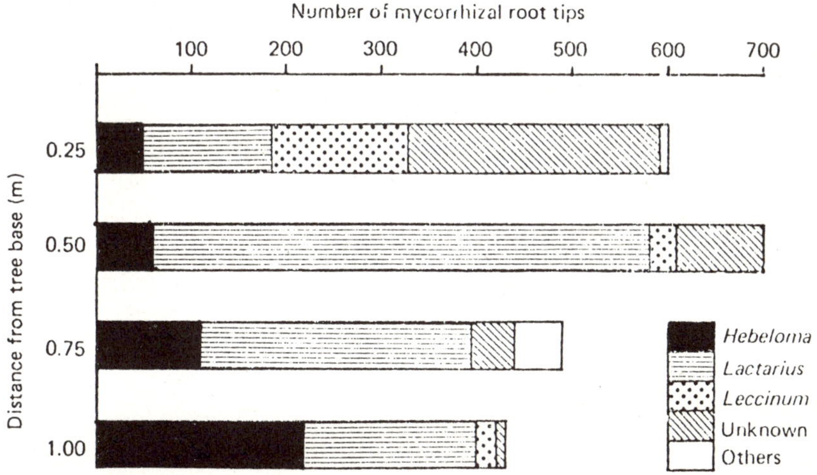

Fig. 4. Number of mycorrhizas of different types in 15 soil cores taken at increasing distances from the base of a *Betula pubescens* tree. (After Deacon et al. 1983, in Dighton & Mason 1985.)

These findings are supported by similar studies on other hosts, *e.g.* on *Pinus radiata* in Australia (Marks & Foster 1967) and on *Pseudotsuga menziesii* in New Zealand (Chu-Chou & Grace 1981, 1983).

Other experiments have provided information on the ability of "early stage" and "late stage" mycorrhizal fungi to form mycorrhizal associations with seedling trees. Deacon et al. (1983) grew birch seedlings in soil cores collected beneath fruit bodies of the "early stage" fungi *Laccaria* and *Inocybe.* The seedlings invariably developed mycorrhizas belonging to these fungi. The seedlings almost without exception failed to develop mycorrhizas of the "late stage" *Lactarius* or *Leccinum* type, although both types of mycorrhizas were abundant in the cores. Fox (1983) planted birch seedlings into non-sterile soil to which basidiospores had been added. The seedlings developed mycorrhizas with *Hebeloma, Inocybe* and *Laccaria* ("early stage" fungi) but failed to do so with *Cortinarius, Lactarius, Leccinum* and *Russula.*

In attemping to explain the differences between "early stage" and "late stage" fungi in their ability to establish succesful mycorrhizal associations with root-tips from trees of different age, attention has been focussed on the carbohydrate requirements of the fungi, and their enzymic capabilities in relation to the degradation of litter. Dighton et al. (1981) have suggested that the carbohydrate demand, as supplied by the host tree, is low for "early stage" fungi, whilst that of "late stage" fungi is higher. Some evidence for this has been obtained from studies of the growth of representatives of the different groups in agar cultures containing graded levels of glucose. It has also been shown that, in general, the fruit bodies of "early stage" fungi are smaller and lighter than those of "late stage" associates.

A distinction should be made between the colonization of seedling roots at a distance from a stand of mature trees (or physically isolated from their roots) and those which develop in close association with mature trees. Beneath mature trees, seedling roots may be colonized by "late stage" fungi, and it has been suggested that this is due to the extension of the mycelium of a mycorrhizal fungus from the roots of the mature tree to the seedling roots, and that this might be associated with translocation of carbohydrate synthesised by the older tree to the seedling, *i.e.* the mycorrhizal fungus on the seedling roots might be obtaining part of its carbohydrate supply from a "mother" tree (Dighton & Mason 1985).

As the trees in a plantation age, there are changes not only in the physiology of the trees, but in the properties of the soil, and especially the litter which accumulates over it. The quality of the litter as a resource supplying nutrients to the trees by recycling declines as the proportion of recalcitrant, slowly-decomposing components such as lignin and polyphenols increases. More evidence is needed on the enzymic capabilities of "late stage" mycorrhizal symbionts, but it is possible that some are not entirely dependant on the living host for their carbohydrate supply, and are active in decomposition, during the course of which the mineral nutrition of the host tree would be enhanced.

The demonstration of a succession of mycorrhizal fungi on tree roots as they age has a bearing on forestry practice. Seedlings are often reared in forest nurseries where the soil has been sterilized, and under these conditions may become naturally infected by "early stage" symbionts. When such seedlings are transplanted to their permanent sites, the "early stage" partners may be replaced quite quickly (in a few months) by native mycorrhizal fungi (Bledsloe et al. 1982).

The artificial inoculation of tree seedlings with mycorrhizal fungi is now a commercial practice. Marx (1980), Marx & Kenney (1982) and Marx et al. (1982, 1984) have reviewed the techniques available. A number of different fungi have been used, *e.g. Thelephora terrestris, Suillus plorans, Cenococcum graniforme* and *Pisolithus tinctorius.* It is important to choose a fungus which is able to infect the host tree to be inoculated, and which can also establish itself in the site at which the trees are to be grown (Trappe 1977). For example, *Suillus plorans* is adapted to infect *Pinus cembra* in the low temperatures prevailing in alpine pastures in Austria, where the establishment of trees helps to minimise the incidence of avalanches (Moser 1965). *Cenococcum graniforme* is drought-tolerant and forms mycorrhizas in natural soils with a pH range of 3.4–7.5 (Trappe 1964). In regions with higher temperatures, *e.g.* in the southern part of the United States of America, one of the most successful fungi for artifical mycorrhiza inoculation is the Gasteromycete *Pisolithus tinctorius.* The epithet *tinctorius* refers to the bright yellow pigment present in the basidiocarps, mycelium and mycelial strands. There are several reasons why this fungus has proved so successful. The puff-ball-like sporocarps develop within a few months of planting of infected seedlings, and are easily collected, yielding as much as 1 kg of basidiospores per man-hour. The spores are prolific: 1 gram contains about 1.1×10^9 spores. The spores can be stored dry, and can survive for long periods, *e.g.* 5 years at 5° C. The host range is wide (over 50 species) and the fungus is widely distributed in a variety of habitat types. It has wide temperature range, 17–40° C with an optimum of about 28° C. It is adapted to growth in soils of low fertility, *e.g.* coal spoil, and is tolerant of toxic metals, *e.g.* Fe and S. The characteristic yellow mycelium with conspicuous mycelical cords makes the fungus easy to recognise in soil and on roots. It can be grown in solid and in liquid media, and infection can be brought about by the application of spores or as a mycelial inoculum. When the mycelial inoculum is prepared in vermiculite, it is capable of survival during transport and in soil. This is not to say that there are no problems with the commercial application of this fungus – *e.g.* survival during storage (Hung & Molina 1986) – but it is probable that the difficulties can be overcome and that the fungus will be used on a large scale in the USA (Cordell 1985).

The better growth of mycorrhizal as compared with non-mycorrhizal seedlings when they are transplanted into the field, especially on infertile soils, has generally been ascribed to their improved mineral nutrition and water supply. However, it is also apparent that the survival rate of mycorrhizally-infected transplants is often significantly better than for non-mycorrhizal plants. Seedlings may succumb to infection by pathogens such as *Pythium, Phytophthora, Fusarium* and *Rhizoctonia.* These generally have a wide host range, and often attack immature root tissues such as the unsuberised tips of long and short roots. Zak (1964) has suggested that mycorrhizal infection may help to protect root systems from attack by pathogens. He considered that the mycorrhizal root might protect the root tissues in a number of ways, *e.g.*

(a) by utilizing carbohydrates, thus reducing the attractiveness of the root to pathogens;
(b) by serving as a physical barrier to infection;
(c) by secreting antibiotics;
(d) by favouring, along with the root, protective rhizosphere organisms.

A further possibility is that mycorrhizal infection may stimulate host root cells to elaborate anti-fungal inhibitors which may inhibit infection by pathogens (Marx 1969a).

There is good experimental evidence to support (b) and (c), and this will now be reviewed. Marx (1969a) has reported on *in vitro* tests on agar in which isolates of ectomycorrhizal fungi were tested for their ability to inhibit the growth of a range of pathogenic fungi, and in which the effects of filtrates from liquid cultures were tested for their ability to affect the motility of zoospores of *Phytophthora cinnamomi.* There was great variation in the capacity of the mycorrhizal fungi to inhibit the growth of the pathogens, and this is indicated in table 1.

Analysis of the range of pathogens inhibited by the different mycorrhizal fungi indicated that more than one antibiotic was being produced. The most active antibiotic producer was *Leucopaxillus cerealis* var. *piceinus.* When grown in liquid culture, the culture filtrate produced antifungal and antibacterial antibiotics. Of particular interest was the fact that culture filtrates

Table 1. Inhibition of growth of pathogens by ectomycorrhizal fungi (data from Marx 1969a)

Mycorrhizal fungus	% of pathogens inhibited
Pisolithus tinctorius	0
Lactarius deliciosus	16
Laccaria laccata	35
Suillus luteus	76
Leucopaxillus cerealis var. *piceinus*	92

from this fungus could inhibit motility and germination of zoospores of *Phytophthora cinnamomi,* and this proved to be a very sensitive bioassay. The antibiotic from *Leucopaxillus* was subsequently identified (Marx 1969b) as diatretyne nitrile, a polyacetylene antibiotic previously isolated from *Clitocybe diatreta* and other members of the *Tricholomataceae* (Anchel et al. 1962). At concentrations as low as 2 ppm, the pure antibiotic was fungicidal to zoospores of *P. cinnamomi,* and zoospore germination was inhibited at concentrations of 50–70 ppb. Vegetative mycelium was not killed by concentrations as high as 9 ppm. It was shown that the antibiotic could be produced by *Leucopaxillus* grown in sterile soil supplemented by carbohydrate, and in extracts of roots of *Pinus echinata.* Later work by Marx & Davey (1969a) showed that mycorrhizal roots formed in association with *Leucopaxillus* contained both diatretyne nitrile and diatretyne 3. The experiments to demonstrate this were done by inoculating sterilized seeds of *Pinus* spp. with a number of mycorrhizal symbionts. Not all root tips formed fully-developed mycorrhizas with a complete mantle and Hartig net. When mycorrhizal and non-mycorrhizal root tips formed under these conditions were exposed to dense suspensions of zoospores of *Phytophthora cinnamomi,* it was found that roots with a complete mycorrhizal mantle did not become infected, whilst in those in which the mantle was incomplete, a proportion (17–21%) were infected. When the fungus forming the mycorrhiza was *Leucopaxillus,* it was found that non-mycorrhizal short roots had only a low percentage of infection (25%), compared with 100% in the non-mycorrhizal controls, implying that there might have been some protection form adjacent mycelium even though the mantle was not present. This may have been due to diatretyne production, and the possibility exists that there may have been translocation of antibiotic from mycorrhizal roots to non-mycorrhizal.

Despite the apparent inhability of *Pisolithus tinctorius* or *Thelephora terrestris* to produce a diffusible antibiotic inhibitory to the motility and germination of zoospores of *Phytophthora cinnamomi,* or inhibitory to its mycelial growth, there is experimental support for the conclusion that mycorrhizas formed on *Pinus* by both these fungi form effective barriers to cortical penetration by the pathogen (Marx & Davey 1969b, Marx 1970). As in the experiments with *Leucopaxillus,* provided that the mantle was complete, and the Hartig net well-developed, the proportion of mycorrhizal roots infected was very low, and negligible compared to the 100% infection of non-mycorrhizal roots.

There is thus clear support for the contention that ectomycorrhizas protect growing root-systems from invasion by serious soil-borne pathogen, and that in some cases the protection may be mediated *via* antibiotic production and in others by the physical barrier provided by the mantle and Hartig net.

Conclusion

I have argued that fungi are better classified as belonging to a different kingdom than the green plants. Nevertheless, as the examples which I have given illustrate, there is an inextricable interrelationship between members of the two kingdoms, because they can have a profound influence on eath other. There is thus every reason for botanists to study mycology, and strong reasons for most mycologists to study botany. For the two subjects to flourish, a mutualistic symbiotic relationship as close as that between mycorrhizal fungi and their host plants is essential.

Acknowledgements

It is a pleasure to thank many colleagues who have generously loaned me illustrations, and have helped to form my ideas through their publications or by discussion. In particular, I would like to thank Dr. Juliet Frankland, Dr. David Lonsdale, Dr. Donald Marx, Dr. Philip Mason, Dr. Kathryn Newell, Dr. Alan Rayner, Dr. Roy Watling, Dr. Eirene Williams and Professor George Zentmyer.

References

Anchel, M., Silverman, W. B., Valanjo, N. & Rogerson, C. T. 1962: Patterns of polyacetylene production. I. The diatretynes. – Mycologia **54**: 249–257.

Barnes, R. S. K. (ed.) 1984: A Synoptic classification of living organism. – Blackwell, Oxford etc.

Bevan, R. J. & Greenhalgh, G. N. 1976: *Rhytisma acerinum* as a biological indicator of pollution. – Environmental Poll. **10**: 271–285.

Bledsloe, C. S., Tennyson, K. & Lopushinsky, W. 1982: Survial and growth of outplanted Douglas fir seedlings inoculated with mycorrhizal fungi. – Canad. J. Forest Res. **12**: 720–723.

Bray, J. R. & Gorham, G. 1964: Litter production in forests of the world. – Advances Ecol. Res. **2**: 101–157.

Burnett, J. H. & Evans, E. J.. 1966: Genetical homogeneity and the stability of the mating-type factors of fairy rings of *Marasmius oreades*. – Nature **210**: 1368–1369.

Chard, J. M., Gray, T. R. G. & Frankland, J. C. 1983: Antigenicity of *Mycena galopus*. – Trans. Brit. Mycol. Soc. **81**: 503–511.

Chu-Chou, M. & Grace, L. J. 1981: Mycorrhizal fungi of *Pseudotsuga menziesii* in the north island of New Zealand. – Soil Biol. Biochem. **13**: 247–249.

– & – 1983: Characterization and identification of mycorrhizas of Douglas fir in New Zealand. – Eur. J. Forest Pathol. **13**: 251–260.

Cordell, C. E. 1985: The application of *Pisolithus tinctorius* ectomycorrhizae in forest land management. In: Molina, R. (ed.), Proceedings of the 6th North American conference on mycorrhizae: 69–71. – Oregon State University, Forest Research Laboratory, Corvallis, Oregon.

Curl, E. A. & Truelove, B. 1985: The rhizosphere. – Springer, Berlin etc.

Deacon, J. W., Donaldson, S. J. & Last, F. T. 1983: Sequence and interactions of mycorrhizal fungi on birch. – Pl. & Soil **71**: 257–262.

Dighton, J., Harrison, A. F. & Mason. P. A. 1981: Is the mycorrhizal succession on trees related to nutrient uptake? – J. Sci. Food Agric. **32**: 629–630.

– & Mason, P. A. 1985: Mycorrhizal dynamics during forest tree development. In: Moore, D., Casselton, L. A., Wood D. A. & Frankland, J. C. (eds.), Developmental biology of higher fungi (Brit. Mycol. Soc. Symp. **10**): 117–139. – University Press, Cambridge.

Fox, F. M. 1983: Role of basidiospores as inocula of mycorrhizal fungi of birch. – Pl. & Soil **71**: 269–273.

Frank, A. B. 1885: Über die auf Wurzelsymbiose beruhende Ernährung gewisser Bäume durch unterirdische Pilze. – Ber. Deutsch. Bot. Ges. **3**: 128–145.

Frankland, J. C. 1982: Biomass and nutrient cycling by decomposer Basidiomycetes. In: Frankland, J. C., Hedger, J. N. & Swift, M. J. (eds.), Decomposer Basidiomycetes: their biology and ecology (Brit. Mycol. Soc. Symp. **4**): 241–261. – University Press, Cambridge.

– 1984: Autecology and the mycelium of a woodland decomposer. In: Jennings, D. H. & Rayner, A. D. M. (eds.), The ecology and physiology of the fungal mycelium (Brit. Mycol. Soc. Symp. **8**): 241–260. – University Press, Cambridge.

–, Bailey, A. D., Gray, T. R. G. & Holland, A. A. 1981: Development of an immunological technique for estimating mycelial biomass of *Mycena galopus* in leaf litter. – Soil Biol. Biochem. **13**: 87–92.

Greenhalgh, G. N. & Bevan, R. J. 1978: Response of *Rhytisma acerinum* to air pollution. – Trans. Brit. Mycol. Soc. **71**: 491–494.

Haeckel, E. 1866: Generelle Morphologie der Organismen. – Reimer, Berlin.

– 1894: Systematische Phylogenie I. Systematische Phylogenie der Protisten und Pflanzen. – Reimer, Berlin.

Harley, J. L. 1969: The Biology of Mycorrhiza. – Leonard Hill, London.
– & Smith, S. E. 1983: Mycorrhizal symbiosis. – Academic Press, New York, etc.
Hawksworth, D. L., Sutton, B. C. & Ainsworth, G. C. (eds.) 1983: Ainsworth and Bisby's Dictionary of the fungi (including the Lichens). – Commonwealth Mycological Institute, Kew.
Hering, T. F. 1982: Decomposing activity of basidiomycetes in forest litter. In: Frankland, J. C., Hedger, J. N. & Swift, M. J. (eds.), Decomposer Basidiomycetes: their biology and ecology (Brit. Mycol. Soc. Symp. **4**): 213–225. – University Press, Cambridge.
Hintikka, V. 1982: The colonisation of litter and wood by basidiomycetes in Finnish forests. In: Frankland, J. C., Hedger, J. N. & Swift, M. J. (eds.), Decomposer Basidiomycetes: their biology and ecology (Brit. Mycol. Soc. Symp. **4**): 227–239. – University Press, Cambridge.
Hung, L. L. & Molina, R. 1986: Temperature and time in storage influence the efficacy of selected isolates of fungi in commercially produced ectomycorrhizal inoculum. – Forest Sci. **32**: 534–545.
Ikediugwu, F. E. O. 1976: The interface in hyphal interference by *Peniophora gigantea* against *Heterobasidion annosum*. – Trans. Brit. Mycol. Soc. **66**: 281–290.
Jalaluddin, M. 1967 a: Studies on *Rhizina undulata*. I. Mycelial growth and ascospore germination. – Trans. Brit. Mycol. Soc. **50**: 449–459.
– 1967 b: Studies on *Rhizina undulata*. II. Observations and experiments in East Anglian plantations. – Trans. Brit. Mycol. Soc. **50**: 461–472.
Jennings, D. H. 1984: Water flow through mycelia. In: Jennings, D. H. & Rayner, A. D. M. (eds.), The ecology and physiology of the fungal mycelium (Brit. Mycol. Soc. Symp. **8**): 143–184. – University Press, Cambridge.
Kendrick, B. 1985: The fifth kingdom. – Mycologue Publications, Waterloo, Ontario.
Last, F. T., Mason, P. A., Wilson, J. & Deacon, J. W. 1983: Fine roots and sheathing mycorrhizas: their formation, function and dynamics. – Pl. & Soil **71**: 9–21.
Leedale, G. F. 1974: How many are the kingdoms of organisms? – Taxon **23**: 261–270.
Margulis, L. 1971: Whittaker's five kingdoms of organisms: minor revisions suggested by consideration of the origin of mitosis. – Evolution **25**: 242–245.
– & Schwartz, K. V. 1982: Five kingdoms. An illustrated guide to the pyhla of life on earth. – Freeman, San Francisco.
Marks, G. C. & Foster, R. C. 1967: Succession of mycorrhizal associations on individual roots of radiata pine. – Austral. Forest. **31**: 193–201.
Martin, G. W. 1955: Are fungi plants? – Mycologia **47**: 779–792.
Marx, D. H. 1969 a: The influence of ectotrophic mycorrhizal fungi on the resistance of pine roots to pathogenic infections. I. Antagonism of mycorrhizal fungi to root pathogenic fungi and soil bacteria. Phytopathology **59**: 153–163.
– 1969 b: The influence of ectotrophic mycorrhizae on the resistance of pine roots to pathogenic infections. II. Production, identification and biological activity of antibiotics produced by *Leucopaxillus cerealis* var. *piceina*. – Phytopathology **59**: 411–417.
– 1970: The influence of ectotrophic mycorrhizal fungi on the resistance of pine roots to pathogenic infections. V. Resistance of mycorrhizae to infection by vegetative mycelium of *Phytophthora cinnamomi*. – Phytopathology **60**: 1472–1473.
– 1980: Ectomycorrhizal fungus inoculations: a tool for improving forestation practics. In: Mikola, P. (ed.), Tropical mycorrhiza research: 13–71. – Clarendon Press, Oxford.
– Cordell, C. E., Kenney, D. S., Mexal, J. G., Artman, J. D., Riffle, J. W. & Molina, J. R. 1984: Commercial vegetative inoculum of *Pisolithus tinctorius* and inoculation techniques for development of etomycorrhizae on bare-root seedlings. Forest Sci. Monogr. **25**: 1–101.
– & Davey, C. B. 1969a: The influence of ectotrophic mycorrhizal fungi on the resistance of pine

roots to pathogenic infections. III. Resistance of aseptically formed mycorrhizae to infection by *Phytophthora cinnamomi.* – Phytopathologie **59**: 549–558.

– & – 1969b: The influence of ectotrophic mycorrhizal fungi on the resistance of pine roots to pathogenic infections. IV. Resistance of naturally-occurring mycorrhizae to infection by *Phytophthora cinnamomi.* – Phytopathology **59**: 559–565.

– & Kenney, D. S. 1982: Production of ectomycorrhizal fungus inoculum. In: Schenk, N. C. (ed.), Methods and principles of mycorrhizal research: 131–146. – American Phytopathological Society, St. Paul, Minnesota.

– Rühle, J. L., Kenney, D. S., Cordell, C. E., Riffle, J. W., Molina, R. J., Pawuk, W. H., Navratil, S., Tinus, R. W. & Goodwin, O. C. 1982: Commercial vegetative inoculum for *Pisolithus tinctorius* and inoculation techniques for development of ectomycorrhizae on container-grown tree seedlings. – Forest Sci. **28**: 373–400.

Mason, P. A., Wilson, J., Last, F. T. & Walker, C. 1983: The concept of succession in relation to the spread of sheathing mycorrhizal fungi on inoculated tree seedlings growing in unsterile soils. – Pl. & Soil **71**: 247–256.

Moser, M. 1965: Künstliche Mykorrhiza – Impfung und Forstwirtschaft. – Allg. Forstz. **20**: 6–7.

Murray, J. S. & Young, C. W. T. 1961: Group dying of conifers. (Forest. Commiss., Forest Rec. **46**). – Her Majesty's Stationery Office, London.

Newell, K. 1984a: Interaction between two decomposer Basidiomycetes and a collembolan under Sitka spruce: distribution, abundance and selective grazing. – Soil Biol. & Biochem. **16**: 227–233.

– 1984b: Interaction between two decomposer Basidiomycetes and a collembolan under Sitka spruce: grazing and its potential effects on fungal distribution and litter decomposition. – Soil Biol. & Biochem. **16**: 235–239.

Olive, L. S. 1975: The mycetozoans. – Academic Press, New York, etc.

Peace, W. R. 1962: Pathology of trees and shrubs with special reference to Britain. – University Press, Oxford.

Rayner, A. D. M. 1976: Dematiaceous hyphomycetes and narrow dark zones in decaying wood. – Trans. Brit. Mycol. Soc. **67**: 546–549.

– & Todd, N. K. 1979: Population and community structure and dynamics of fungi in decaying wood. – Advances Bot. Res. **6**: 333–420.

– & – 1982a: Population structure in wood-decaying basidiomycetes. In: Frankland, J. C. Hedger, J. N. & Swift, M. J. (eds.), Decomposer, Basidiomycetes: their biology and ecology (Brit. Mycol. Soc. Symp. **4**): 109–128. – University Press, Cambridge.

– & – 1982b: Ecological genetics of Basidiomycete populations in decaying wood. In: Frankland, J. C., Hedger, J. N. & Swift, M. J. (eds.), Decomposer Basidiomycetes: their biology and ecology (Brit. Mycol. Soc. Symp. **4**): 129–142. – University Press, Cambridge.

– & Webber, J. F. 1984: Interspecific mycelial interactions – an overview. In: Jennings, D. H. & Rayner, A. D. M. (eds.), The ecology and physiology of the fungal mycelium (Brit. Mycol. Soc. Symp. **8**): 382–417. – University Press, Cambridge.

Read, D. J. 1984: The structure and function of the vegetative mycelium of mycorrhizal roots. In: Jennings, D. H. & Rayner, A. D. M. (eds.), The ecology and physiology of the fungal mycelium (Brit. Mycol. Soc. Symp. **8**): 215–240. – University Press, Cambridge.

Rishbeth, B. 1959: Dispersal of *Fomes annosus* Fr. and *Peniophora gigantea* (Fr.) Massee. – Trans. Brit. Mycol. Soc. **42**: 243–260.

Rishbeth, J. 1963: Stump protection against *Fomes annosus.* III. Inoculation with *Peniophora gigantea.* – Ann. Appl. Biol. **52**: 63–77.

Satchell, J. E. 1974: Litter-interface of animate/inanimate matter. In: Dickinson, C. H. & Pugh, G. J. F. (eds.), Biology of plant litter decomposition **1**: XIII–XLIV. – Academic Press.

Swift, M. J. 1977: The roles of fungi and animals in the immobilisation and release of nutrient elements from decomposing branch wood. In: Lohn, U. & Persson, T. (eds.), Soil organisms as components of ecosystems: (Ecol. Bull., Stockholm, **25**): 193–202. – Swedish Natural Science Research Council, Stockholm.

– 1982: Basidiomycetes as components of forest ecosystems. In: Frankland, J. C., Hedger, J. N. & Swift, M. J. (eds.), Decomposer Basidiomycetes: their biology and ecology (Brit. Mycol. Soc. Symp. **4**): 309–337. – University Press, Cambridge.

Thompson, W. 1984: Distribution, development and functioning of mycelial cord systems of decomposer Basidiomycetes of the deciduous woodland floor. In: Jennings, D. H. & Rayner, A. D. M. (eds.), The ecology and physiology of the fungal mycelium (Brit. Mycol. Soc. Symp. **8**): 185–214. – University Press, Cambridge.

– & Rayner, A. D. M. 1982: Spatial structure of a population of *Tricholomopsis platyphylla* in a woodland site. – New Phytol. **92**: 103–114.

– & – 1983: Extent, development and functioning of mycelial cord systems in soil. – Trans. Brit. Mycol. Soc. **81**: 333–345.

Todd, N. K. & Rayner, A. D. M. 1978: Genetic structure of a natural population of *Coriolus versicolor* (L. ex Fr.) Quel. – Genet. Res. **32**: 55–65.

– & – 1980: Fungal individualism. – Sci. Progr. (Oxford) **66**: 331–354.

Trappe, J. M. 1964: Mycorrhizal hosts and distribution of *Cenococcum graniforme.* – Lloydia **27**: 100–106.

– 1977: Selection of fungi for ectomycorrhizal inoculation in nurseries. – Annual Rev. Phytopathol. **15**: 203–222.

Webster, J., Ikediugwu, F. E. O. & Dennis, C. 1970: Hyphal interference by *Peniophora gigantea* against *Heterobasidion annosum.* – Trans. Brit. Mycol. Soc. **54**: 307–309.

Whittaker, R. H. 1969: New concepts of kingdoms of organisms. – Science, **163**: 150–160.

Wiggers, F. H. 1780: Primitiae Flora Holsaticae. – Kiliae.

Williams, E. N. D. & Todd, N. K. 1985: Numbers and distribution of individuals and mating-type alleles in populations of *Coriolus versicolor.* – Genet. Res. (Cambridge) **46**: 251–263.

– Todd, N. K. & Rayner, A. D. M. 1981a: Spatial development of populations of *Coriolus versicolor.* – New Phytol. **89**: 307–319.

– – & – 1981b: Propagation and development of fruit bodies of *Coriolus versicolor.* – Trans. Brit. Mycol. Soc. **77**: 409–414.

Witkamp, M. 1969: Environmental effects on microbial turnover of some mineral elements. II. Biotic factors. – Soil Biol. Biochem. **1**: 177–184.

Zak, B. 1964: Role of mycorrhizae in root disease. – Annual Rev. Phytopathol. **2**: 377–392.

– & Marx, D. H. 1964: Isolations of mycorrhizal fungi from roots of individual slash pines. – Forest Sci. **10**: 214–222.

Zentmyer, G. A. 1980: *Phytophthora cinnamomi* and the diseases it causes. – Amer. Phytopathol. Soc. Monogr., **10**.

Address of the author: Professor J. Webster, Department of Biological Sciences, University of Exeter, EX4 4PS, U.K.

Botany and biotechnology

D. von Wettstein

Abstract

Wettstein, D. von 1988: Botany and biotechnology. – In: Greuter, W. & Zimmer, B. (eds.): Proceedings of the XIV International Botanical Congress: 181–201. – Koeltz, Königstein/Taunus.

Various new achievements in the field of botanical biotechnology are presented. The barley cDNA gene encoding β-glucanase, cloned in *Escherichia coli,* was used to transform yeast strains including a lager brewing strain, whereby residual β-glucans can be reduced during beer fermentation and maturation. Insect resistent tobacco plants were engineered using a gene encoding a toxin found in *Bacillus thuringensis.* Similarly, tobacco, potato and tomato plants resistent to a herbicide were engineered using an gene from *Streptomyces hygroscopicus* encoding a detoxifying enzyme. A DNA-mediated transformation system was established in the genus *Mucor,* using *M. circinelloides,* whereby it may become possible to produce calf chymosin or improved *Mucor* acid proteases – enzymes used for the clotting of milk in cheese manufacturing – in *Mucor* species. New *Saccharomyces* transformation systems enable the transformation of brewers' yeasts and the obtention of improved strains, *e.g.* with reduced off-flavour compounds, obviating the need for a long secondary fermentation (lagering). Perhaps the title of this presentation should rather be biotechnology and botany, since many of the biotechnological tools evolved for medical research and its applications in diagnosis or treatments of diseases are helpful in defining and solving interesting problems in botany. In the first part of my paper I will give an example for this. Biotechnological approaches will be treated in the second part, and concern plant breeding for resistance to insects and herbicides. Finally I will address myself to improvements of cheese production and beer brewing.

Introduction

The genetic information is written in the 64 codons of DNA (fig. 1). It is replicated in connection with each cell division by a machinery of more than 15 proteins, comprising over 30 polypeptides (Kornberg 1982). The DNA is transcribed into messenger RNA in a tissue and cell specific manner with the aid of RNA polymerase and proteins binding to the 5' region of the gene. Identification of some of these proteins is in progress and expectations are high that it will be possible to determine the regulatory molecules, such as hormones, which associate with the DNA binding proteins in order to modulate transcription.

The messenger RNA is translated with the aid of transfer RNA molecules into polypeptide chains composed of 20 amino acids, the chains growing with a speed of 15 residue additions per second. Retroviruses such as cauliflower mosaic virus encode in their genetic information a reverse transcriptase, which transcribes messenger RNA molecules into DNA molecules, and this enzyme is an important biotechnical tool. The DNA can be cut at specific base sequences with restriction endonucleases produced by bacterial species, each one being designed to cut a defined sequence of 3 to 8 bases. These enzymes and their mode of action were discovered roughly 100 years after the isolation and first characterization of DNA in 1878. The fragments with their sticky, single stranded ends can be ligated together in the same or a new way. Small circular DNA molecules, plasmids, into which the restriction endonuclease fragments have been ligated are introduced into bacteria, fungi, higher plant, insect or mammalian cells, where the genetic information of the plasmid is multiplied and expressed.

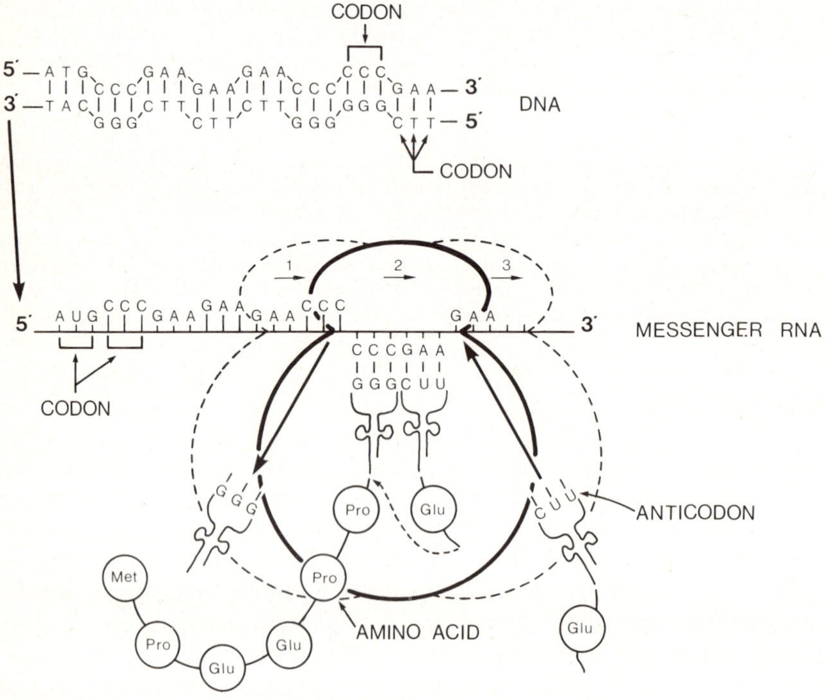

Fig. 1. The base pairing rule of nucleic acids governs protein synthesis.

Obtaining an interesting gene

We can illustrate the importance of availability of products and equipment from biotechnological companies by describing how one has obtained a gene of interest for studying germination in barley. This gene is also of prime importance in the controlled germination of hundreds of tons of barley in the production of malt for whisky and beer. Germination is initiated by the synthesis in the aleurone and scutellum of the enzyme 1→3, 1→4-β-D-glucan 4 glucanohydrolase which is secreted into the endosperm to degrade (1→3,1→4)-β-D-glucans (fig. 2), the major cell wall material of the barley

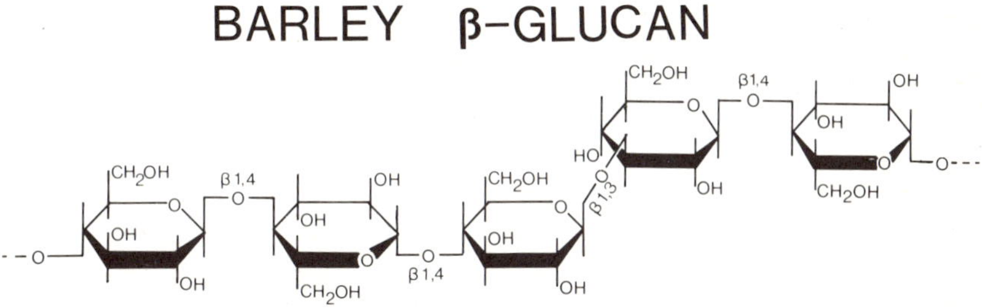

Fig. 2. The chemical structure of barley β-glucan.

storage tissue. The enzyme specifically breaks a β 1,4 bond next to a β 1,3 bond and its activity is required to remove the endosperm walls so that amylases and proteases can get access to the starch and protein storage material.

A cDNA gene for this enzyme was obtained in the following way: Two forms of the enzyme were purified (Woodward & Fincher 1982) from 10 kg of barley, steeped and germinated over 5 days, after homogenization and an initial ammonium sulphate precipitation at 40–80% saturation. For this purpose a series of chromatography, ion exchange and gel filtration steps was carried outh with DEAE cellulose, carboxymethyl cellulose, carboxymethyl Sepharose and Sephadex material purchased from a biotechnology company. Amino acid sequence data of the isoenzyme II (Fincher et al. 1986) were obtained by isolating tryptic peptides by reversed phased separation on a high performance liquid chromatograph with a Bakerbond C_8 column. Eluted peptides (fig. 3) are placed on polybrene coated glass fiber sheets and the amino acid sequence is determined in a solid phase sequencer which can perform Edman degradation with as little as nanomole quantities of protein, the sequencer again being an acquisition from a biotechnology company. The next step is to purchase a solid-phase oligonucleotide synthesizer. Using

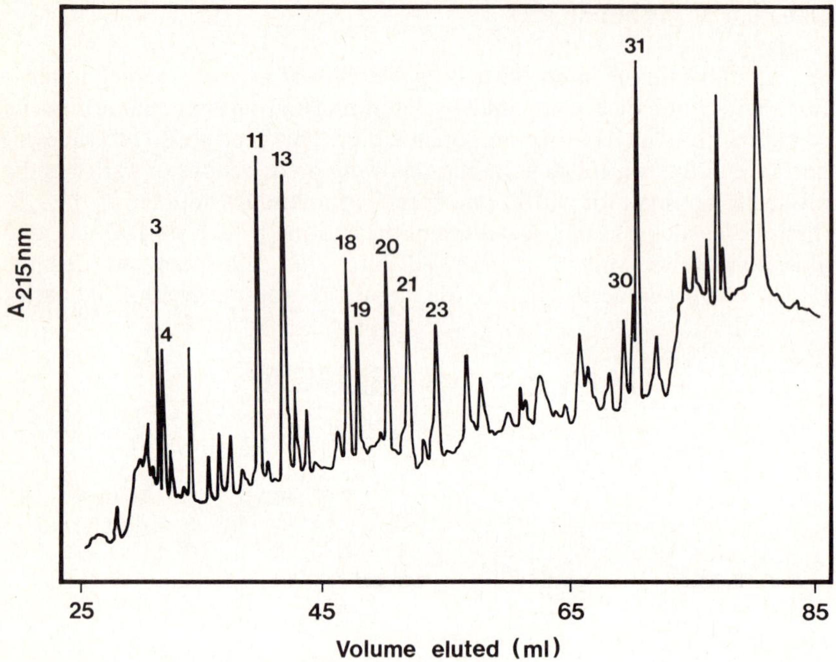

Fig. 3. Elution profile of polypeptides separated by high performance liquid chromatography.

phosphoramidite chemistry a 15-base mixed oligonucleotide probe of 32-fold redundancy (to cover all codon possibilities) encoding the sequence Phe-Tyr-Asn-Glu-His of tryptic peptide 11 was prepared to screen for appropriate cDNA clones. The cDNA library is prepared by isolating aleuron messenger RNA and reverse transcribing it with the aid of a kit containing the appropriate enzymes, linkers and vectors. After insertion of the cDNA into a plasmid like pUC9, bacteria are transformed and the plasmid multiplied. The nucleotide sequence of the cDNA insert is determined and the amino acid sequence deduced from the nucleotide sequence. Often the primary structure of a protein with known function is established by comparing the cDNA deduced amino acid sequence with an number of sequences from peptides of the protein. The amino acid sequence of the (1→3, 1→4)-β-glucanase from barley can be compared with that of a (1→3)-β-glucanase from tobacco cells (fig. 4). Strongly conserved homologous domains are recognized and these include acidic amino acid residues (marked with asterisks) that may play a role in the catalytic function. This then provides a basis to characterize the functional domains of the enzyme. If one studies the codon usage of the gene for the

```
                                                                1
(1→3)                                                           L G
(1→3,1→4)                                         I G V C Y G M S A

                                                                                   32
(1→3)       N N L P N H W E V I Q L Y K S R N I G R L R L Y D P N H G A
(1→3,1→4)   N N L P A A S T V V S M F K S N G I K S M R L Y A P N Q A A

                                                                                   62
(1→3)       L Q A L K G E N I E V M L G L P N S D V - K H - I A S G M E
(1→3,1→4)   L Q A V G G T G I N V V V G A P N - D V L S N L A A S P A A
                                                *

                                                                                   92
(1→3)       H A R W W V Q K N V K D F W P D V K I K Y I A V G N E I S P
(1→3,1→4)   A A S W - - - - - V K - S N - I Q A - - Y P K V S F R Y V C

                                                                                  122
(1→3)       V T G T - S Y L - T S F L T P A M V N I Y K A I G E A G L G
(1→3,1→4)   V G N E V A G G A T R N L V P A M K N V H G A L V A A G L G

                                                                                  152
(1→3)       N N I K V S T S V D M T L I G N S Y P P S Q G S F R N D A R
(1→3,1→4)   H - I K V T T S V S Q A I L G V F S P P S A G S F T G E A A

                                                                                  182
(1→3)       W F T D F I V G F L R D T R A P L L V N I Y P Y F S Y S G N
(1→3,1→4)   A F M G P V V Q F L A R T N A P L M A N I Y P Y L A W A Y N

                                                                                  212
(1→3)       P G Q I S L F Y S L F T A P N V V V Q D G S R Q Y P N L F D
(1→3,1→4)   P S A M D M G Y A L F N A S G T V V R D G A Y G Y Q N L F D
                                                  *                         *

                                                                                  242
(1→3)       A M L D S V Y A A L E R S G G A S V G I V V S E S G W P S A
(1→3,1→4)   T T V D A F Y T A M G K H G G S S V K L V V S E S G W P S G
                  *                                     *

                                                                                  272
(1→3)       G A F G A T Y D N A A T Y L R N L I Q H A K E G S P R R P G
(1→3,1→4)   G G T A A T P A N A R F Y N Q H L I N H V G R G T P R H P G

                                                                                  302
(1→3)       P I E T Y I F A M F D E N N K N P E L E K H F G L F S P N K
(1→3,1→4)   A I E T Y I F A M F N E N Q K D S G V E Q N W G L F Y P N M
                *                 *                 *

(1→3)       Q P K Y N L N F G V S G G V W D S S V E T N A T
(1→3,1→4)   Q H V Y P I N F
```

Fig. 4. Amino acid sequences of barley 1→3,1→4- and tobacco 1→3-β-glucanase.

Table 1: Codon usage for each of the listed amino acids in the genes for (1→3, 1→4)-β-glucanase of barley aleurone, α-amylase 2 of barley aleurone and B_1 hordein of barley endosperm.

		β-glucanase (1→3, 1→4)	β-glucanase (1→3)	α-amylase	B_1-hordein
Leu	n	17	26	34	25
TTA	%	0	31	0	0
CTT		6	15	6	16
CTA		0	4	3	16
TTG		0	15	3	24
CTC		53	23	47	20
CTG		41	12	41	24
Ser	n	23	28	22	12
TCT	%	0	36	5	17
TCA		0	29	0	0
AGT		0	11	0	25
AGC		44	7	32	17
TCC		28	14	45	17
TCG		28	3	18	25
Arg	n	8	13	17	6
CGT	%	0	8	0	33
CGA		0	8	0	17
AGA		0	31	0	17
AGG		24	46	24	17
CGC		38	8	52	17
CGG		38	0	24	0

(1→3, 1→4)-β-glucanase of barley (table 1) a marked preference towards C and G in the wobble position is observed as is also the case for the α-amylase isoenzyme 1 (Rogers & Milliman 1983, Chandler et al. 1984) from aleurone cells, while this avoidance of the A and T nucleotide in the third position is not observed in the codon usage of the (1→3)-β-glucanase from tobacco tissue culture cells or the B-hordein storage protein synthesized in the barley endosperm. We do not know why genes expressed in aleuron cells of barley show this extreme biased codon usage.

As to the practical use of this gene the following is relevant. Insufficient β-glucanase activity during malting gives rise to high levels of residual β-glucans in the wort, which can create problems during filtration of wort and beer. High β-glucan content may lead to precipitates in the finished beer. The

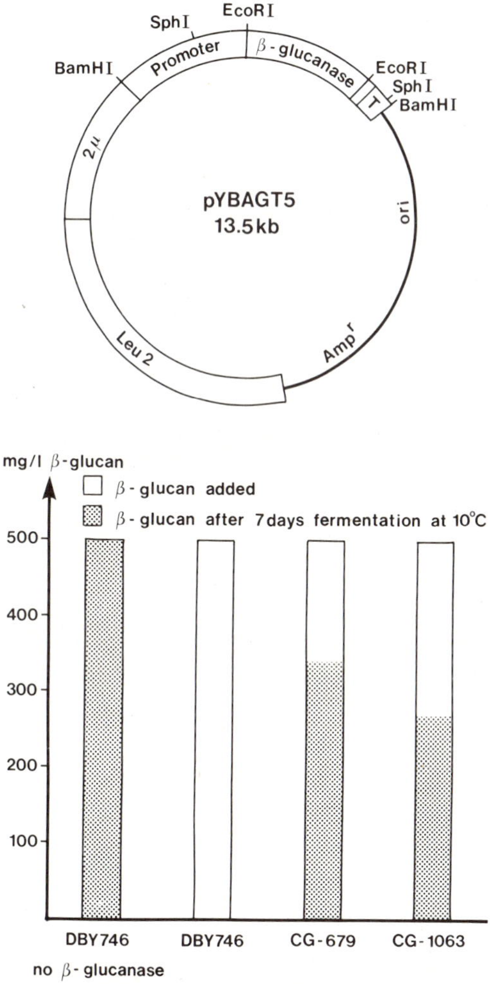

Fig. 5. Yeast-*Escherichia coli* shuttle vector containing yeast promoter, barley β-glucanase gene and terminator in addition to genes for replication in yeast (2μ) and *E. coli* (ori) as well as a *LEU2* gene for selection in yeast and an ampicillin resistance gene for selection in *E. coli*. Yeast strains containing such plasmids degrade β-glucans.

barley β-glucanase is almost completely inactivated during kilning of the malt and mashing. However, (1→3, 1→4)-β-glucanases from microbial sources (*Bacillus subtilis, Aspergillus niger* or *Trichoderma reesii*) are less heat labile than barley β-glucanase. Such preparations are added during mashing or at a later stage during fermentation to reduce the content of high molecular weight β-glucan. With the isolation of a cDNA gene for barley β-glucanase a number of new approaches to alleviate these problems have become feasible.

The cDNA gene was supplied with the nucleotide sequence for the signal peptide of mouse α-amylase to permit transport of the barley β-glucanase through the secretory pathway of an eukaryotic cell. This construction was inserted into a yeast-*Escherichia coli* shuttle plasmid behind the gene promoter of yeast alcohol dehydrogenase I. A laboratory yeast strain (DBY 746) was transformed with the plasmid and the transformed strain secreted barley β-glucanase (Jackson et al. 1986, Thomsen et al. 1987) which permitted during fermentation for one week at 10°C to degrade concentrations of β-glucans (500 mg/l) which are found troublesome in wort and beer filtration (fig. 5). The β-glucanase gene was additionally provided with a terminator sequence and inserted into a plasmid containing a yeast *LEU2* gene (fig. 5). Transformants of a leucine requiring lager brewing strain also secreted the barley β-glucanase and lowered β-glucan contents during fermentation (fig. 5).

While a lager yeast which synthesizes and secretes β-glucanase during fermentation and maturation of beer alleviates the problems of beer filtration and formation of precipitates in the finished beer, this will not help to solve problems of wort filtration. Here yeast strains which synthesize and secrete barley β-glucanase will provide a source of the natural enzyme which can be added during mashing. If the brewery has available a lager yeast strain secreting barley β-glucanase, it can produce the enzyme with its normal fermentation equipment, possibly even during the production of green beer. Such β-glucanase might be added during mashing instead of microbial enzymes. While the pH optimum of the latter enzymes is about 6.7, the barley enzyme operates optimally at pH 4.7. The barley enzyme is therefore expected to be more suitable for the brewing industry since wort fermentations typically are carried out at a pH of 4.5.

In the long run modification of the barley enzyme by site directed mutagenesis of its gene towards increased heat stability, so that it survives the kilning process, might be feasible. Availability of genetic transformation of barley is required to insert such a heat stable β-glucanase gene into future malting barley varieties.

Plant breeding

Transfer of individual genes or groups of genes by genetic transformation is a promising strategy for breeding useful plants, if we know of genes that encode enzymes or other proteins supplying a valuable property for a crop plant. Two recent examples can serve as models for such an approach: Vaeck et al. (1986) have engineered insect resistant tobacco plants using a gene encoding a toxin in *Bacillus thuringensis* and Block et al. (1987) have engineered tobacco, potato and tomato plants resistant to a herbicide using a gene from *Streptomyces hygroscopicus* encoding a detoxifying enzyme.

Bacillus thuringensis strain berliner contains a gene on a plasmid encoding a protein of 1155 amino acids (130, 533 mol. weight) which forms crystals in the spores of the bacterium. Proteases in the midgut of larvae of *Pieris brassicae* (cabbage butterfly) and *Manduca sexta* (tobacco hornworm) as well as other insects will cleave the protein into a fragment of 60,000 mol. weight which is toxic to the cells and kills the larvae. The gene has been cloned and expressed in *Escherichia coli* (Schnepf & Whiteley 1981, Klier et al. 1982, Shibano et al. 1985, Adang et al. 1985). By Bal 31 exonuclease deletion mapping the minimal protein fragment killing cell lines of lepidoptera and diptera was determined to comprise 579 amino acids. The corresponding DNA fragment was inserted into a disarmed Ti-plasmid vector and the gene transferred into tobacco plants using a leaf disc transformation system. The transformed plants expressed the gene as shown by immunochemistry of plant extracts and toxicity tests. Freshly hatched larvae of *Manduca sexta* were placed on leaves of 40 cm high control and transgenic tobacco plants. After 10–15 days the leaves of the control plants were completely eaten by the larvae. In contrast, the larvae on transgenic plants died within 4 days and leaf damage was restricted to a few small holes.

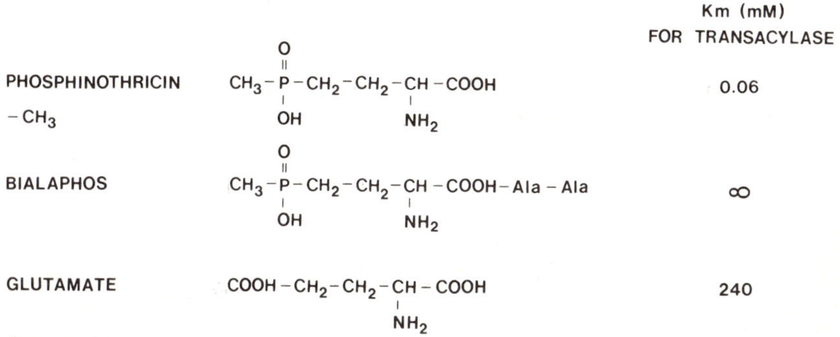

Fig. 6. Affinity of acetyltransferase to the herbicide phosphinothricin, the tripeptide bialaphos and glutamate.

Commercial formulations of the crystal protein have been used to destroy in African rivers the diptera which transmit the parasites causing river blindness (Onchocerciasis). While successful the toxic protein rapidly sinks and does not stay in the surface regions of the water, where the insect larvae feed. Perhaps incorporation and expression of the gene encoding the toxic fragment in cyanobacteria which occupy the same ecological niche as the diptera larvae may be a more durable system to control the larval populations.

The tripeptide bialaphos (fig. 6) consisting of two alanine residues and an analogue of glutamate called phosphinothricin is produced by *Streptomyces hygroscopicus.* In bacteria and plants peptidases remove the alanine residues and release phosphinothricin which then blocks glutamine synthetase, the only enzyme which can detoxify ammonia released by nitrate reduction, amino acid catabolism and photorespiration. Phosphinothricin is thus a very potent herbicide (Block et al. 1987) which is used commercially either in the form of the chemically synthesized compound or in the form of bialaphos isolated from *Streptomyces.* The mycobacterium synthesizes bialaphos in 13 steps from pyruvate. One of the enzymes in the pathway is phosphinothricine acetyltransferase which simultaneously protects the organism against the poisonous intermediate by acetylation of the free amino group. The enzyme is encoded by the *bar* gene (Thompson et al. 1987) which has been cloned, sequenced and placed under the control of the 35S promoter of the cauliflower mosaic virus and transferred to plant cells using a Ti plasmid derived vector

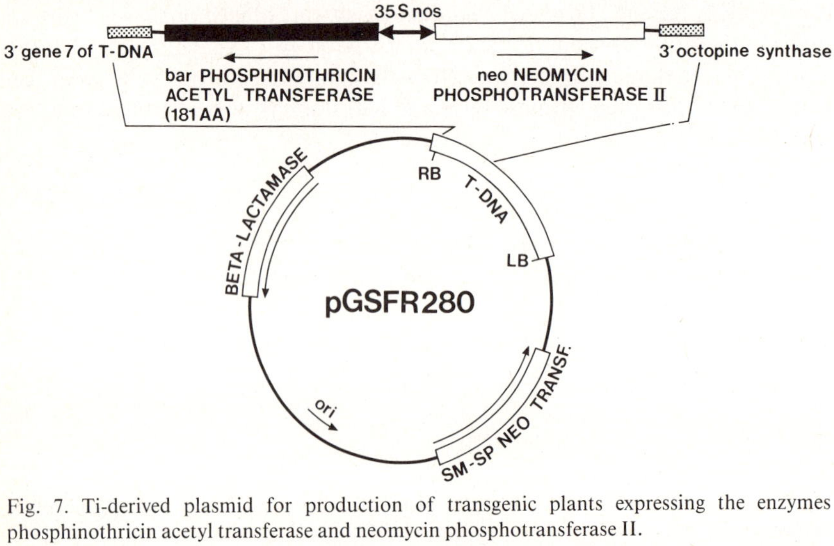

Fig. 7. Ti-derived plasmid for production of transgenic plants expressing the enzymes phosphinothricin acetyl transferase and neomycin phosphotransferase II.

(fig. 7). The GTG initiator codon of the *bar* gene was substituted with the ATG codon used by higher plants with the aid of two complementary oligonucleotides and appropriate restriction enzymes. As can be seen from fig. 7 the gene was inserted between the 35S promoter and the 3' termination and polyadenylation signals of gene 7 of the T-DNA. Adjacent to this cassette and with the ooposite direction of transcription a gene encoding neomycin

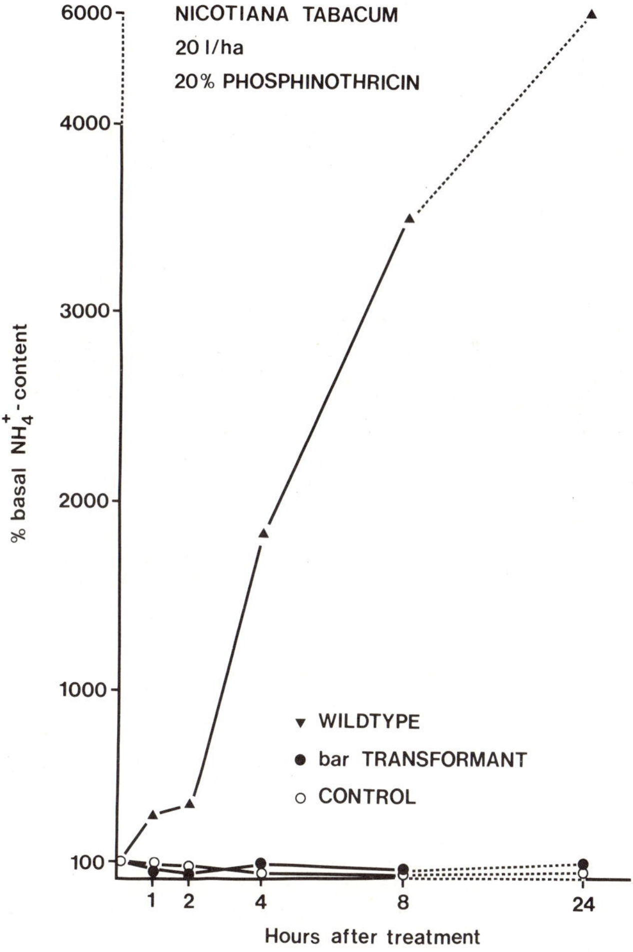

Fig. 8. Accumulation of ammonia in plants sprayed with phosphinothricin and its prevention in transformants expressing phosphinothricin transferase.

phosphotransferase was placed under the nopalin synthase promoter and using the 3' termination and polyadenylation signals of the octopine synthase gene of the T-DNA. This provides selection opportunity for neomycin and kanamycin resistance of transformants in addition to the phosphinothricine resistance. The right (RB) and left (LB) border repeats of the T-DNA insure the mobilization of this DNA from the *Agrobacterium* strain used for transformation and the integration of the T-DNA into the plant chromosomes. Transgenic calli of tobacco, potato and tomato resistant to phosphinothricin or kanamycin were isolated after cocultivation of protoplasts with *Agrobacterium* of after leaf disc infection with the *Agrobacterium* strain. The regenerated plants of the three species revealed resistance towards commercially used doses of phosphinothricin or bialaphos. The expressed acetyltransferase (transacylase) with its high affinity (low Km, fig. 6) to phosphinothricin of demethylphosphinothricin but low affinity to the normal substrate glutamate was able to detoxify the herbicide and protects the plants from ammonia poisoning. This is illustrated in fig. 8 (Block et al. 1987). Within 24 hours phosphinothricin causes an enormous accumulation of ammonia in the leaves of the treated tobacco plants as a result of blocking the glutaminsynthase enzyme. The plants transformed with the *bar* gene revealed normal levels of ammonia in their leaves.

Alternative options for engineering plants resistant to the phosphinothricin comprise to produce plants with a target enzyme resistant to the herbicide or plants that overproduce the target enzyme. It turns out that the gene for glutaminsynthetase from alfalfa (*Medicago sativa*) can complement a mutation in the glutaminsynthetase gene (glnA) of *Escherichia coli* and it has therefore been possible to select in *E. coli* mutant forms of the plant enzyme resistant to phosphinothricin (Das Sarma et al. 1986) and then incorporate the mutant gene by transformation into alfalfa. Phosphinothricin tolerant alfalfa cell suspension cultures have been shown to be due to amplification of the gene encoding the glutaminsynthetase (Donn et al. 1984, Tischer et al. 1986) and tolerant plants could be obtained due to a high expression of the glutaminsynthetase enzyme.

Cheese production

The two thermophilic species *Mucor miehei* and *Mucor pusillus* of the fungal class Zygomycetes secrete into the culture medium aspartic proteases which are used in a crucial step in the manufacturing of cheese: the clotting of milk as a result of κ-casein proteolysis. Approximately 40% of the milk clotting enzymes employed in cheese making today is *Mucor miehei* acid protease. This enzyme hast first been applied for cheese production in 1966 instead of calf chymosin, the classical milk coagulating enzyme which is prepared from the 4th ventricle stomach of suckling newborn calves. Other species also produce milk clotting enzymes, *e.g. Mucor circinelloides*, a species which turned out to be more amenable to genetic manipulation. Since the industrial technology of manufacturing enzymes from *Mucor* species is well developed, one can contemplate to produce calf chymosin or improved *Mucor miehei* acid protease in *Mucor* species. This requires the establishment of a DNA-mediated transformation system in this genus and we have therefore explored this possibility over the last few years. Almost no genetics of molecular biology was available at the outset.

Mucor circinelloides can grow yeast-like on 2% glucose under anaerobic conditions. The first task was to produce protoplasts and regenerate them. It turned out that stable protoplasts with regeneration capacity could be obtained by dissolving the germ tube wall (9,4% chitin and 32,7% chitosan) of sporangiospore germlings using a mixture of snail gut enzyme (Novozym 234) and a chitosanase isolated from *Streptomyces* species 6, if certain temperature

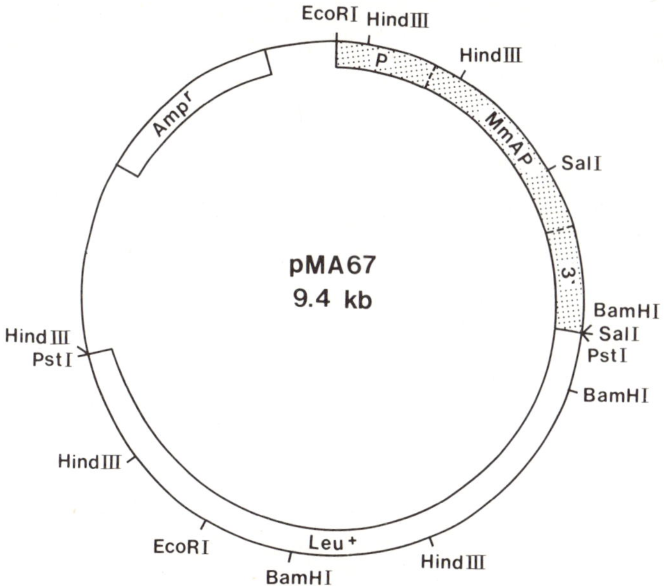

Fig. 9. *Mucor*-yeast-*Escherichia coli* shuttle plasmid for expression of *Mucor miehei* acid protease in *Mucor circinelloides.*

regimes are observed (Heeswijck 1984). *Mucor circinelloides* is resistant to the standard antibodies used for selection of transformants such as G418, neomycin, hygromycin, oligomycin, benomyl and methotrexate. Therefore UV induced auxotrophic mutants were isolated using N-glycosylpolifungin as a selective agent for killing prototrophs prior to germination of the auxotrophic mutants (Roncero 1984). A gene library of *Mucor* DNA was constructed by insertion of fragments obtained by partial digestion with the restriction endonuclease MboI into the BamHI site of the *Escherichia coli*-yeast shuttle vector YRp17, which contains origins of replication for *E. coli* and yeast as well as selection markers for both organisms. High frequency transformation (7000 transformants per μg DNA) was obtainable for a DNA fragment which could complement a leucin auxotroph recipient (Heeswijck & Roncero 1984). Interestingly enough *Mucor* genes do not readily complement mutant genes of *E. coli* or yeast. The molecular analysis of the transformants uncovered in the *Mucor* DNA fragment an autonomous replication

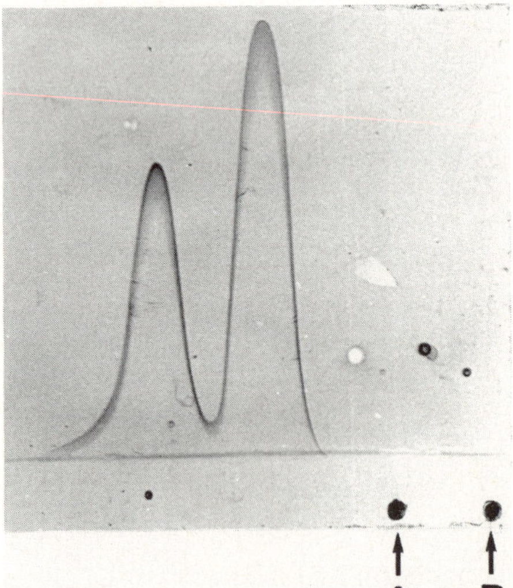

Fig. 10. Tandem crossed immunoelectrophoresis of culture medium of *Mucor circinelloides* transformant expressing *M. miehei* acid protease (left rocket) and authentic *M. miehei* protease (right rocket).

sequence allowing the plasmid to replicate in *Mucor* as well as in *E. coli* and yeast (Heeswijck 1986). The nucleotide sequence of the *Mucor* Leu^+ gene revealed it to encode a β-isopropylmalate dehydrogenase, the enzyme catalyzing the second step in leucine biosynthesis. With the gene encoding the *Mucor miehei* aspartic prepro-protease (Gray et al. 1986) the vector pM67 (fig. 9) was constructed which additionally contains the gene providing ampicillin resistance (Amp^r) for selection in *E. coli* as well as the Leu^+ gene and ARS sequence from *Mucor circinelloides.* A leucine requiring strain of this species was transformed and the transformants shown to excrete *Mucor miehei* aspartic protease (Dickinson et al. 1987). The heterologous gene expression was proven in several ways:

- As shown in fig. 10 tandem crossed immunoelectrophoresis revealed immunological identity between the acid protease produced by *Mucor miehei* and the transformants of *Mucor circinelloides.* In A 0.62 μg *Mucor miehei* acid protease was applied and in B 1 μg of acid protease produced by a transformant. After electrophoresis in the first dimension, electrophoresis was carried out at right angles using 80 μl antiserum against *Mucor miehei* acid protease in the gel. Identity of the two enzymes is demonstrated by fusion of the two precipitates and by identical precipitate rocket morphology.
- The acid protease of *Mucor miehei* and *Mucor circinelloides* secreted by the transformants was separated by ion-exchange chromatography using Whatman DE-52 cellulose. The former protease binds to the column, while the latter runs through.

The *Mucor miehei* aspartyl protease produced in *Mucor circinelloides* shows the same milk clotting activity as authentic enzyme and reveals the same apparent molecular weight in western blots after denaturing polyacrylamide gel electrophoresis. The monospecific antibodies against the *Mucor miehei* aspartyl protease inhibited the milk clotting activity of the heterologously expressed enzyme but did not inhibit the simultaneously produced *Mucor circinelloides* protease. Establishment of a genetic transformation system for *Mucor miehei* is in progress.

Active sites of calf chymosin and *Mucor* aspartyl protease are conserved: Aspartyl residue 32 is neighboured by the same amino acids in both enzymes (Phe·Asp·Thr·Gly·Ser·Ser·) as is aspartyl residue 215 (Asp·Thr·Gly·Thr). Otherwise amino acid homology is quite low (25%). The *Mucor* enzyme is too stable above 60° C and under alkaline conditions. Therefore the enzyme continues to work after the cheese is finished and decreases its quality.

Heterologous expression and genetic transformation opens the way to tailor the *Mucor* acid protease by site directed mutagenesis to become more similar in its properties to calf chymosin.

Beer brewing

A successful breeding effort aiming at the reduction of fermentation time or improved flocculation or diminished off-flavour compounds requires the demonstration that hybrids, mutants, recombinants or transformants of brewers yeast can produce beer of the same quality as that obtained conventionally from pure strains. It has been demonstrated that yeast after prolonged growth on a synthetic medium reveals excellent brewing performance (Christensen et al. 1978) and that mutants resistant to isoleucine analogues can change the content of flavour compounds (Mikkelsen et al. 1979). Likewise hybrid brewing strains obtained by mating of meiotic segregants were shown in pilot and full scale brewing tests to yield beer of equal quality to that of the parent strain (Gjermansen & Sigsgaard 1981, Wettstein 1983). A special automated equipment has been developed to screen 60 strains a week for their brewing performance at a 2 l scale (Sigsgaard & Rasmussen 1985).

For genetic transformation an auxotrophic *leu2* mutant induced in a meiotic segregant of the Carlsberg lager strain is available and can be transformed to leucine prototrophy with a plasmid carrying the *LEU2* gene from *Saccharomyces cerevisiae* (Gjermansen 1983). Yeast transformation systems without the use of foreign DNA are established (Kielland-Brandt et al. 1979, 1981, 1984).

A superior transformation system has been developed by R. R. Yocum, Biotechnica International, Cambridge, Mass. (Yocum 1984). It allows the selection of transformants with the aid of a vector containing a gene providing resistance to genticin (G418 sulphate). This gene has been placed under the promoter of the yeast *CYC1* gene. The vector contains sequences for selection and replication in *Escherichia coli,* the *HO* gene from yeast chromosome IV and a LacZ fusion gene with the yeast *GAL1* promoter. After insertion of a desirable gene in the *HO* region and transformation, the G418 resistant transformants contain the desired gene as well as the rest of the vector integrated in chromosome IV. Introduction of the vector results in the production of β-galactosidase giving rise to blue colonies. Looping out of the vector sequences as a result of crossing over between homologous chromosome and vector regions results in white colonies which can be screened for the sole presence of the desired gene.

Several aroma compounds of beer flavour are produced by the yeast during fermentation of the wort. Excessive amounts of the vicinal diketones, diacetyl and pentanedione are formed from acetolactate as side products of the isoleucine-valine pathways during the main fermentation and therefore usually a long secondary fermentation, lagering, is required to reduce their levels. That this is the main function of lagering has been substantiated by the observation that maturation of beer can be effected enzymatically within 24 hours by addition of α-acetolactate decarboxylase which catalyzes the formation of acetoin, a compound rapidly reduced to butandiol (Godtfredsen & Ottesen 1982). It would be interesting to achieve a reduction in the production of acetolactate and α-aceto-α-hydroxybutyrate from fermenting lager strains by curtailing the biosynthesis of the responsible enzymes or by enhancing the flow of the intermediates into valine and isoleucine.

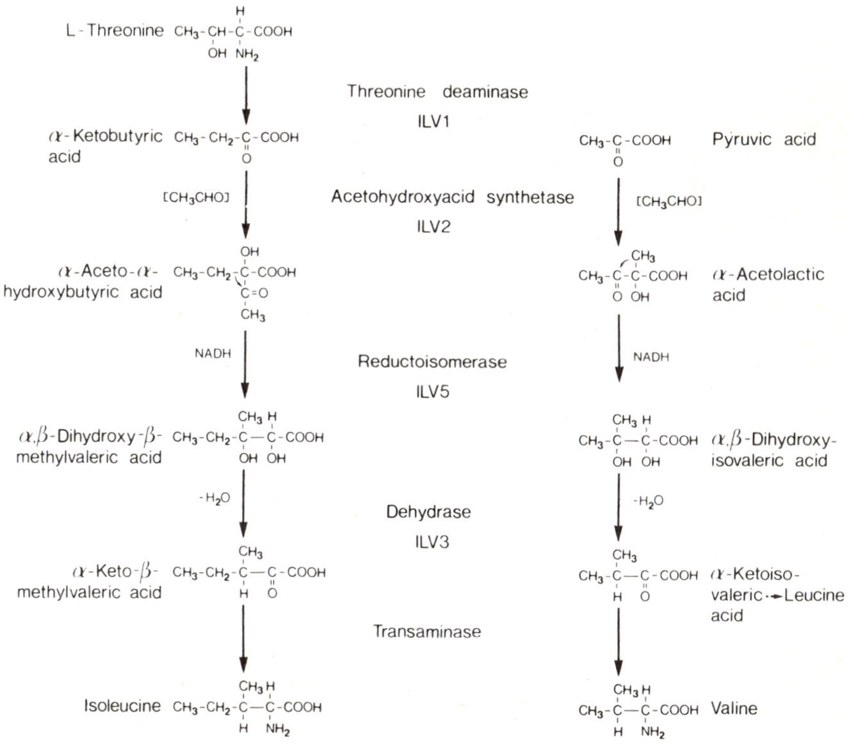

Fig. 11. Biosynthetic pathway of isoleucine and valine in yeast.

Five enzymes catalyze the biosynthesis of isoleucine and valine (fig. 11). The first enzyme for the synthesis of isoleucine is threonine deaminase (encoded in *ILV1*), an allosteric enzyme converting threonine to α-keto-butyrate. The other four enzymes are shared by the biosynthetic pathway for valine, converting α-ketobutyrate to isoleucine and pyruvate to valine. The acetohydroxy acid synthetase is coded for by *ILV2*, acetohydroxy acid reductoisomerase by *ILV5* and dihydroxy acid dehydrase by *ILV3* (Petersen et al. 1983). The genes *ILV1* (Holmberg et al. 1985, 1987, Kielland-Brandt et al. 1984, Petersen et al. 1983a, Polaina 1984), *ILV2* (Falco & Dumas 1985, Falco et al. 1985, Polaina 1984) and *ILV5* (Petersen 1986, Polaina 1984) have been cloned, sequenced and mapped for transcripts. Information has thereby been obtained on the coding regions, regulatory sequences in front of the genes and putative signal sequences for import of the enzymes into the mitochondria. The enzymes in the pathways are regulated in such a way that repression is observed by addition of isoleucine, valine and leucine.

At least 28 different genes coding for enzymes in amino acid biosynthetic pathways are under general control of amino acid biosynthesis: Starvation for a single amino acid elicits the transcription of all genes under general control and this is mediated by the sequence TGACT in front of the genes thus to be activated. Several TGACT general control motives are present in front of the *ILV1, ILV2* and *ILV5* genes in *Saccharomyces cerevisiae* as well as *S. carlsbergensis* chromosomes. The three enzymes threonine deaminase, acetohydroxy acid synthetase and reductoisomerase are subject to regulation at the transcriptional level by the general control system. Multiple 5' transcript ends without recognizable physiological functions are characteristic for *ILV1,* while mRNA of *ILV2* and *ILV5* have determinate single 5' initiation points. The *ILV5* gene is highly expressed yielding large quantities of acetohydroxy acid reducto-isomerase in yeast amounting to a few percent of the soluble cell protein. Apparently the movement of the methyl group from the α to the β position in the conversion of α-acetolactate to a-β-dihydroxy isolvaleric acid is a difficult and inefficient step. Therefore, an increase of the amount of this enzyme is hardly a useful breeding aim. Strategies with the aim of reducing the levels of threonine deaminase and acetohydroxy acid synthase in brewers' yeast strains are therefore favoured. The herbicide sulfometuron methyl is kwown to inhibit the acetohydroxy acid synthase and mutants with increased resistance to this herbicide have been shown to map in the *ILV2* gene (Falco & Dumas 1985). Spontaneous mutants in brewers' yeast with increased resistance to sulfometuron methyl produced fully acceptable beer with an amount of diacetyl released during the fermentation that is close to or below the flavour threshold level.

References

Adang, M. J., Staver, M. J., Rochelean, T. A., Leighton, J., Barker, R. F. & Thompson, D. V. 1985: Characterized full-length and truncated plasmid clones of the crystal protein of *Bacillus thuringiensis* subsp. *kurstaki* HD-73 and their toxicity to *Manduca sexta*. – Gene **36**: 289–300.

Block, M. de, Botterman, J., Vandewiele, M., Dockx, J., Thoen, C., Gosselé, R., Movva, C., Thompson, C., Montagu, M. van & Leemans, J. 1987: Engineering herbicide resistance in plants by expression of a detoxifying enzyme. – EMBO J. **6**: 2513–2518.

Chandler, P. M., Zwar, J. A., Jacobsen, J. V., Higgins, T. J. V. & Inglis, A. S. 1984: The effects of gibberellic acid and abscisic acid on α-amylase mRNA levels in barley aleurone layers studies using an α-amylase cDNA clone. – Pl. Molec. Biol. **3**: 407–418.

Christensen, B. E., Kielland-Brandt, M. C. & Erdal, K. 1978: Brewing performance of a yeast after prolonged growth on a synthetic medium. – Carlsberg Res. Commun. **43**: 1–4.

Das Sarma, S., Tischer, E. & Goodman, H. M. 1986: Plant glutamin synthetase complements of glnA mutation in *Escherichia coli*. – Science **232**: 1242–1244.

Dickinson, L., Harboe, M., Heeswijck, R. van, Strøman, P. & Jepsen, L. P. 1987: Expression of *Mucor miehei* aspartic protease in *Mucor circinelloides*. – Carlsberg Res. Commun. **52** (in print).

Donn. G., Tischer, E., Smith, J. A. & Goodman, H. M. 1984: Herbicide resistant alfalfa cells: an example of gene amplification in plants. – J. Molec. Appl. Genet. **2**: 621–635.

Falco, S. C. & Dumas, K. S. 1985: Genetic analysis of mutants of *Saccharomyces cerevisiae* resistant to the herbicide sulfometuron methyl. – Genetics **109**: 21–35.

–, Dumas, K. S. & Livak, K. J. 1985: Nucleotide sequence of the yeast ILV2 gene which encodes acetolactate synthase. – Nucl. Acids Res. **13**: 4011–4027.

Fincher, G. B., Lock, P. A., Morgan. M. M., Lingelbach, K., Wettenhall, R. E. H., Mercer, J. F. B., Brandt, A. & Thomsen, K. K. 1986: Primary structure of the 1→3,1→4-β-D-glucan 4 glucanohydrolase from barley aleurone. – Proc. Natl. Acad. USA **83**: 2081–2085.

Gjermansen, C. 1983: Mutagenesis and genetic transformation of meiotic segregants of lager yeast. – Carlsberg Res. Commun. **48**: 557–565.

– & Sigsgaard, P. 1981: Construction of a hybrid brewing strain of *Saccharomyces carlsbergensis* by mating of meiotic segregants. – Carlsberg Res. Commun. **46**: 1–11.

Godtfredsen, S. E. & Ottesen, M. 1982: Maturation of beer with α-acetolactate decarboxylase. – Carlsberg Res. Commun. **47**: 93–102.

Gray, G. L., Hayenga, K., Cullen, D., Wilson, J. & Norton S. 1986: Primary structure of *Mucor miehei* aspartyl protease: evidence for a zymogen intermediate. – Gene **48**: 41–53.

Heeswijck, R. van 1984: The formation of protoplasts from *Mucor* species. – Carlsberg Res. Commun. **49**: 597–609.

– 1986: Autonomous replication of plasmids in *Mucor* transformants. – Carlsberg Res. Commun. **51**: 433–443.

– & Roncero, M. I. G. 1984: High frequency transformation of *Mucor* with recombinant plasmid DNA. – Carlsberg Res. Commun. **49**: 691–702.

Holmberg, S., Kielland-Brandt, M., Nilsson-Tillgren, T. & Petersen, J. G. L. 1985: The ILVl gene in *Saccharomyces cerevisiae:* 5' and 3' end mapping of transcripts and their regulation. – Carlsberg Res. Commun. **50**: 163–178.

–, Petersen, J. G. L., Nilsson-Tillgren, T., Gjermansen, C. & Kielland-Brandt, M. C. 1987: Biosynthesis of isoleucine and valine in *Saccharomyces*. Gene structure and regulation. – In: Stewart, G. G., Russel, I., Klein, D., Hiebsch, R. R. (eds.), Biological research on industrial yeasts, **1** – CRC Press, Boca Raton (in print).

Jackson, E. A., Ballance, G. M. & Thomsen, K. K. 1986: Construction of a yeast vector directing the synthesis and release of barley (1→3,1→4)-β-glucanase. – Carlsberg Res. Commun. **51**: 445–458.

Kielland-Brandt, M. C., Holmberg, S., Petersen, J. G. L. & Nilsson-Tillgren, T. 1984: Nucleotide sequence of the gene for threonine deaminase (ILVl) of *Saccharomyces cerevisiae.* – Carlsberg Res. Commun. **49**: 567–575.

–, Nilsson-Tillgren, T., Holmberg, S., Petersen, J. G. L. & Svenningsen, B. 1979: Transformation of yeast without the use of foreign DNA. – Carlsberg Res. Commun. **44**: 77–87.

–, –, Petersen, J. G. L. & Holmberg, S. 1981: Transformation in yeast without the involvement of bacterial plasmids. – In: Wettstein, D. von, Friis, J., Kielland-Brandt, M. & Stenderup, A. (eds.), Molecular genetics in yeast (Alfred Benzon Symposium, **16**): 369–380. – Munksgaard, Copenhagen.

Klier, A., Fargette, F., Ribier, J. & Rapoport, G. 1982: Cloning and expression of the crystal protein genes from *Bacillus thuringiensis* strain berliner 1715. – EMBO J. **1**: 791–799.

Kornberg, A. 1982: Supplement to DNA replication. – Freeman, San Francisco.

Mikkelsen, J. D., Sigsgaard, P., Olsen, A., Erdal, K., Kielland-Brandt, M. C. & Petersen, J. G. L. 1979: Thiaisoleucine resistant mutants in *Saccharomyces carlsbergensis* increase the content of D-amyl alcohol in beer. – Carlsberg Res. Commun. **44**: 219–223.

Petersen, J. G. L. 1986: The ILV5 gene of *Saccharomyces cerevisiae* is highly expressed. – Nucl. Acids Res. **14**: 9631–9651.

–, Holmberg, S., Nilsson-Tillgren, T. & Kielland-Brandt, M. C. 1983a: Molecular cloning and characterization of the threonine deaminase (ILV1) gene of *Saccharomyces cerevisiae.* – Carlsberg Res. Commun. **48**: 149–159.

–, Kielland-Brandt, M. C., Holmberg, S. & Nilsson-Tillgren, T. 1983b: Mutational analysis of isoleucine-valine biosynthesis in *Saccharomyces cerevisiae.* Mapping of ILV2 and ILV5. – Carlsberg Res. Commun. **48**: 21–34.

Polaina, J. 1984: Cloning of the ILV2, ILV3 and ILV5 genes of *Saccharomyces cerevisiae.* – Carlsberg Res. Commun. **49**: 577–584.

Rogers, J. C. & Milliman, C. 1983: Isolation and sequence analysis of a barley α-amylase cDNA clone. – J. Biol. Chem. **258**: 8169–8174.

Roncero, M. I. G. 1984: Enrichment method for the isolation of auxotrophic mutants of *Mucor* using the polyene antibiotic N-glycosyl-polifungin. – Carlsberg Res. Commun. **49**: 685–690.

Schnepf, H. E. & Whiteley, H. R. 1981: Cloning and expression of the *Bacillus thuringiensis* crystal protein gene in *Escherichia coli.* – Proc. Natl. Acad. USA **78**: 2893–2897.

Shibano, Y., Yamagata, A., Nakamura, N., Jizuka, T., Sugisaki, H. & Takanami, M. 1985: Nucleotide sequence coding for the insecticidal fragment of the *Bacillus thuringiensis* crystal protein. – Gene **34**: 243–251.

Sigsgaard, P. & Norager Rasmussen, J. 1985: Screening of the brewing performance of new yeast strains. – ASBC J. **43**: 104–108.

Thompson, C. J., Movva, N. R., Tizard, R., Crameri, R. & Davies, J. E. 1987: Characterization of the herbicide resistance gene bar from *Streptomyces hygroscopicus.* – EMBO J. (in print).

Thomsen, K. K., Jackson. E. A. & Brenner, K. 1987: Genetic engineering of yeast: Construction of strains which degrade β-glucans with the aid of a barley gene. – ASBC J. (in print).

Tischer, E., Das Sarma, S. & Goodman, H. M. 1986: Nucleotide sequence of an alfalfa glutamine synthetase gene. – Mol. Gen. Genet. **203**: 221–229.

Vaeck, M., Reynaerts, A., Höfte, H., Jansens, S., Beuckelaer, M. de, Dean, C., Zabeau, M., Montagu, M. van & Leemans, J. 1987: Transgenic plants protected from insect attack. – Nature **328**: 33–37.

Wettstein, D. von 1983: Emil Christian Hansen Centennial Lecture: From pure yeast culture to genetic engineering of brewers yeast: 97–119. – EBC Congress 1983, London.

Woodward, J. R. & Fincher, G. B. 1982: Purification and chemical properties of two 1,3;1,4-β-glucan endohydrolases from germinating barley. – Eur. J. Biochem. **121**: 663–669.

Yocum, R. R. 1984: Yeast vector. – US patent application 612796 and European patent application 85303625.9.

Address of the author: Professor Diter von Wettstein, Department of Physiology, Carlsberg Laboratory, Gamle Carlsberg Vej 10, DK-2500 Copenhagen Valby, Denmark.

Eukaryote cell evolution

T. Cavalier-Smith

Abstract

Cavalier-Smith, T. 1988: Eukaryote cell evolution. – In: Greuter, W. & Zimmer, B. (eds.): Proceedings of the XIV International Botanical Congress: 203–223. – Koeltz, Königstein/Taunus.

The Eukaryota are divided into 6 kingdoms: Animalia, Plantae, Fungi, Chromista, Protozoa, which all have mitochondria, peroxisomes, and Golgi dictyosomes; and Archezoa, which do not. Archezoa (the phyla Archamoebae, Metamonada, and Microsporidia) are the most primitive eukaryotes. They could have evolved from a wall-less mutant gram-positive eubacterium, which also gave rise to the prokaryote Archaebacteria. Key steps in the origin of the eukaryote cell were the origin of the cytoskeleton, phagocytosis, and other cytotic mechanisms, which together radically transformed the cell's ultrastructure and genetic system. – The distinction between the 5 "higher" eukaryote kingdoms, here grouped together as a superkingdom, Metakaryota, and the Archezoa is far more fundamental than that between plants and animals. Unlike the first eukaryotes, which evolved purely autogenously, the origin of the first metakaryote protozoan was by symbiosis – not serial symbiosis, but the simultaneous conversion of three different bacterial endosymbionts into chloroplasts, mitochondria, and peroxisomes. When this happened in a biciliated [1] phagotrophic metamonad host it stimulated the origin of more complex intracellular targeting of proteins and smooth vesicles and thus the origin of Golgi dictyosomes and 80s ribosomes. – As early metakaryote protozoa diversified many lost chloroplasts, but one became obligately dependent on them to form the ancestor of the kingdom Plantae. A fourth endosymbiosis of a primitive phycobilin-containing dinoflagellate created the "botanical" kingdom Chromista. The third "botanical" kingdom, Fungi, like its sister group Animalia, probably evolved from a choanomonad protozoan.

Life is a symbiosis between chromosomes, ribosomes, enzymes and membranes, which together form the living cell. But the structural basis for this symbiosis is radically different in bacteria and eukaryotes as fig. 1 makes clear. The gulf between bacteria and eukaryotes is the biggest discontinuity in biological history since life began 3500 million years or more ago. To understand the origin of these dramatic differences it is helpful to focus first on features common to both types of cell.

[1] See footnote on p. 221.

Some fundamental similarities between bacterial and eukaryote cells

In both, membrane proteins and secretory proteins are made by ribosomes physically attached to membranes; these ribosomes insert integral membrane proteins directly into the membrane and translocate secretory proteins directly across the membrane. In both cases this is achieved by an N-terminal hydrophobic peptide – the signal peptide, which inserts itself into the membrane with the aid of specific signal-recognition machinery. In vitro, *plant* ribosomes can use the *animal* signal-recognition particle to translocate secretory proteins coded by a *bacterial* messenger into the lumen of *animal* endoplasmic reticulum (ER) vesicles; this shows that the signal mechanism was well conserved during the prokaryote-eukaryote transition. This implies that the eukaryotic ER/nuclear envelope membranes are homologous with, and were probably evolutionarily derived from, the bacterial plasma membrane.

Just as important as membrane-ribosome links is the universal attachment of DNA to membranes. Bacterial chromosomes are circular and replicate bidirectionally from an single replicon origin to a single terminus; they are invariably attached to the plasma membrane by their origins and termini, probably also by their two replication forks. Eukaryote chromosomes are linear, and in interphase invariably attached by their ends (telomeres) to the inside of the inner membrane of the nuclear envelope. Other regions of eukaryote chromosomes, notably centromeres, are also often thus attached to the

Fig. 1. The fundamental differences between bacterial and eukaryote cells. – **a**: Ribosomes making membrane proteins are physically attached to the plasma membrane (P) in bacteria, as is the single chromosome. – **b**: In eukaryotes both are attached to the endoplasmic reticulum (ER)/nuclear envelope (NE) membranes. The ER/NE and the bacterial plasma membrane grow by the direct insertion of newly made lipids and proteins, but the eukaryote plasma membrane grows by the fusion with it of smooth vesicles budded off from the ER. Such vesicle fusion (exocytosis) and budding from the ER (ercytosis: Cavalier-Smith 1987a) are the essential basis, together with phagocytosis and other forms of endocytosis, for the evolutionary origin and permanent existence of two topologically distinct membranes in eukaryotes. The lysosomes (L) with their digestive enzymes (D) constitute a third topologically distinct compartment, which is continually regenerated by budding from the ER; by contrast the plasma membrane and ER/NE each maintains a direct 'genetic' continuity and individuality by the growth and division of preexisting membranes of the same type. To maintain osmotic stability and to mediate cell division and DNA segregation (Cavalier-Smith 1987b) bacteria have an exoskeleton, typically of the peptidoglycan murein (M) as in the gram-positive eubacterium shown here; eukaryotes by contrast have an endoskeleton, the cytoskeleton of microtubules (MT), intermediate (I) and actin (A) filaments, with exactly the same roles. Bacterial motility is by a proton-driven cell surface machinery, either gliding or extracellular rotary flagella (F). Eukaryote motility is by intracellular ATP-driven machinery, *e.g.* dynein-tubulin in cilia (C); actomyosin in cytokinesis; kinesin-tubulin in mitosis.

▷

nuclear envelope. This attachment is indirect and mediated by the nuclear lamina, a proteinaceous sheet lying inside the nuclear envelope, which assembles itself around the condensed chromosomes at mitotic telophase.

To the lamina are also attached the nuclear pore-complex proteins, which prevent the nuclear envelope from fully enclosing the DNA and cutting it off from the cytoplasm, and provide large aqueous channels for the traffic of ribonucleoprotein and protein across the nuclear envelope. The nuclear envelope is not a single membrane, as often incorrectly portrayed (*e.g.*

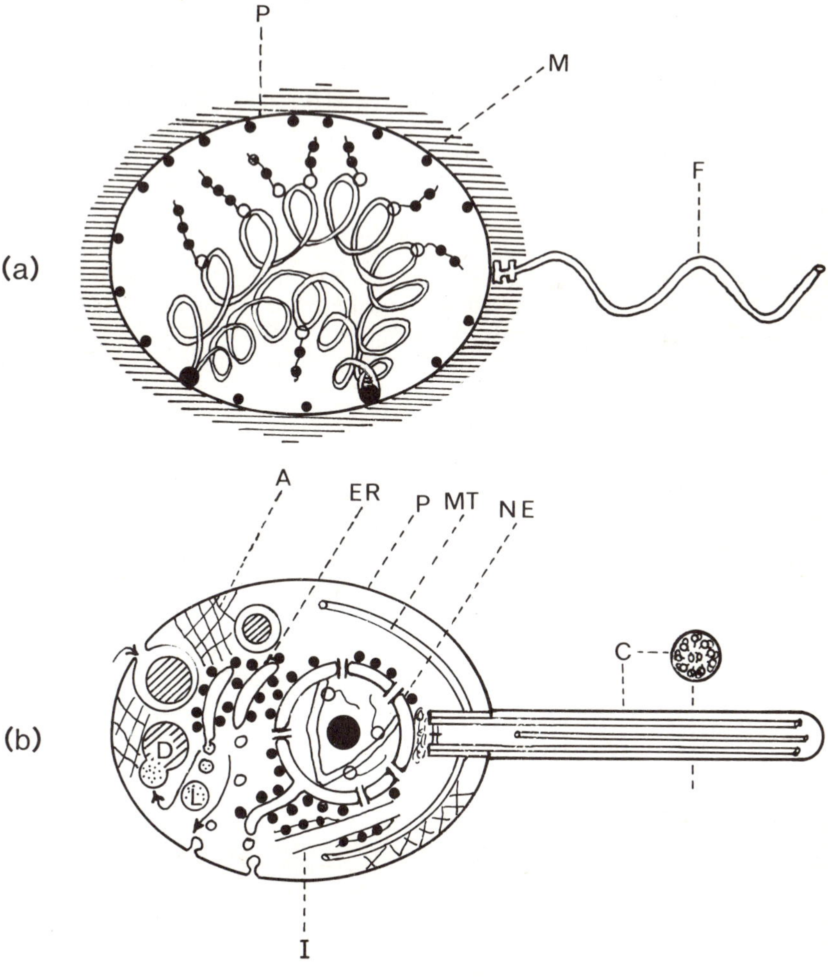

Margulis & Schwartz 1982), but two distinct concentric lipoprotein bilayer membranes, in continuity with each other at the nuclear pores. The outer membrane always bears ribosomes on its cytoplasmic surface, and is the most constant part of the rough ER (RER) which grows and buds from it. Because integral membrane proteins of the RER can diffuse into the inner membrane (by migration within the plane of the fluid bilayer) the nuclear envelope and RER form a single structural and functional unit, so their origins must be considered together.

The cytoskeleton and cytosis: the keys to eukaryogenesis

I have long argued (Cavalier-Smith 1975, 1981, 1987a) that the origin of the nuclear envelope/RER depended on two fundamental molecular innovations:

Fig. 2. The non-symbiotic theory of the origin of the eukaryote cell. – **a**: Following the loss of the murein wall by a bacterium that already secreted extracellular digestive enzymes, the fragile naked cell was stabilised by the evolution of cytoskeletal proteins, and Ca^{++}-stimulated contractility. The basic features of the eukaryote cell cycle evolved before the nucleus *(i)* Replication-origin (O) attachment proteins became primitive centrosomes (CE), nucleating the assembly of microtubules. Centrosome duplication *(ii)*, coupled to the initiation of replication, produced two half-spindles as in protozoan pleuromitosis; *(iii – iv)* the microtubules constrained the contractile actin gel to cleave the cell longitudinally between the centrosomes, and histones compacted the DNA to prevent it breaking during cleavage. (T = chromosome terminus). – **b**: Phagocytosis *(i)* would internalize membranes with attached ribosomes and DNA *(ii)*, initially reversibly *(iii)*. – **c**: Permanent internalization *(i)* was caused by the evolution of ercytosis – budding of *smooth* vesicles (v) from the new endomembrane – and of exocytosis: their fusion with the plasma membrane. Continued phagocytosis denuded the plasma membrane of ribosomes *(ii)* and a new sorting mechanism sequestered digestive enzymes inside smooth vesicles, the early lysosomes (L) which fused with the phagosomes (P) instead of the plasma membrane. Centrosome division continued in the same way *(iii)* even after the origin of the nuclear envelope. Not shown are the non-membrane-bound ribosomes making soluble proteins: 'free' in the bacterial ancestor, but attached via mRNA to the newly-formed intermediate filaments of the early eukaryote. – **d**: Mutant intermediate filaments evolved into the nuclear lamina (L), attached both to the chromatin surface and to endomembranes, thus creating the nuclear envelope. Nuclear pore complexes (NP) evolved, initially to prevent total enclosure of the DNA by endomembrane; partial enclosure protected the DNA from shearing by the new cytoplasmic motility mediated by myosin and kinesin *(i)*. Misdivision of the cell *(ii)* was prevented by attachment *(iii)* of astral microtubules to the cell surface, which evolved into ciliary microtubules (C) and ciliary roots (CR). This constrained the actin-based cleavage furrow to the region between daughter nuclei, centrosomes, and cilia *(iv, v)*. Mutant intermediate filaments were also recruited to make ciliary tektins and thereby create more robust outer doublets for the evolving cilium *(iv, v)* following the origin of dynein arms from mutant kinesins. The resulting uniciliate cell closely resembled modern Archezoa of the class Mastigamoebea. For more details see Cavalier-Smith (1987a). ▷

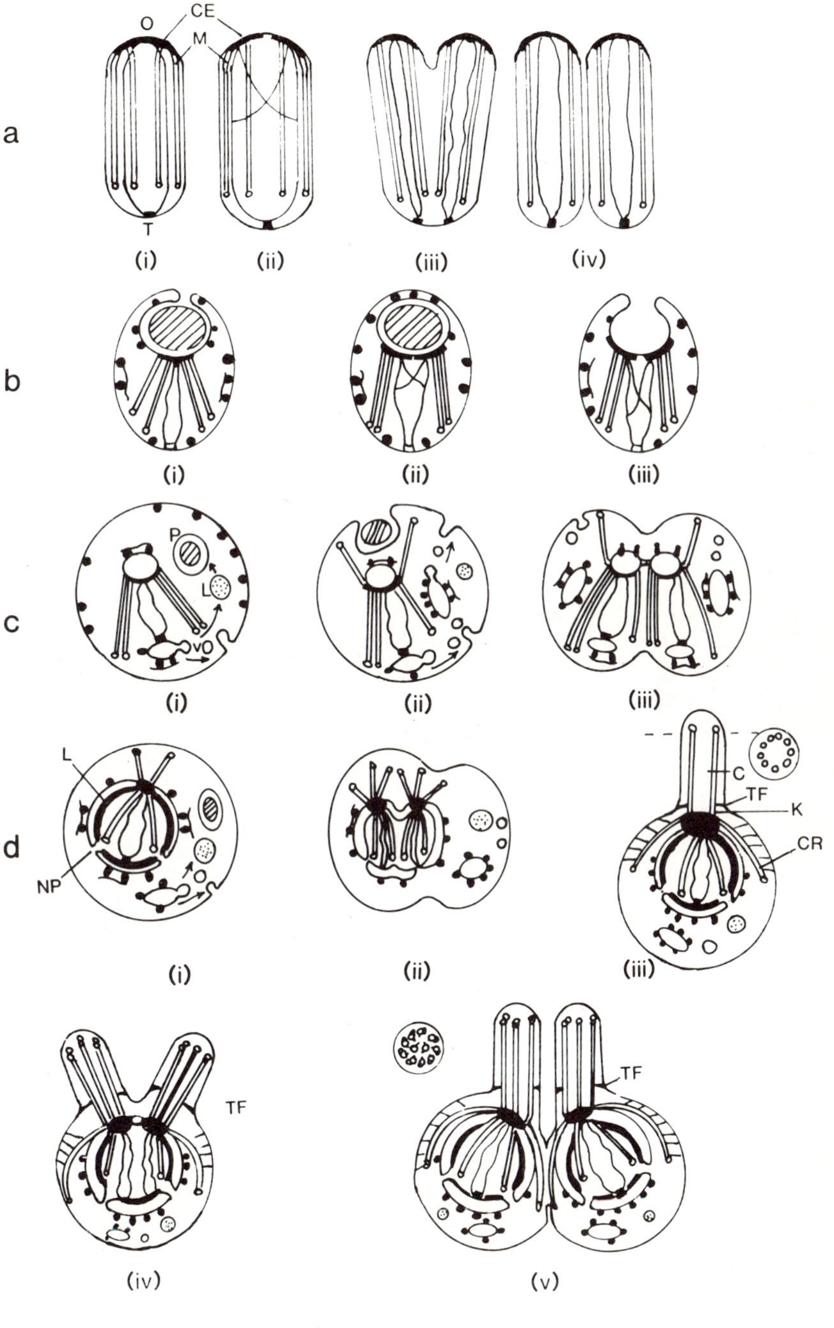
a
O
CE
M
T
(i)
(ii)
(iii)
(iv)
b
(i)
(ii)
(iii)
c
P
L
V
(i)
(ii)
(iii)
d
L
NP
C
TF
K
CR
(i)
(ii)
(iii)
TF
(iv)
TF
(v)

- the origin of the eukaryote cytoskeleton, and
- the origin of membrane budding and fusion processes (exocytosis, endocytosis, including phagocytosis, and ercytosis, which I collectively call either "cytosis" or "cytotic" mechanisms).

These are more fundamental than the existence of the nuclear envelope, because they are what made its evolution possible. As fig. 2 shows, the origin of the nucleus probably occurred in two stages:

- the removal by phagocytosis of pieces of the plasma membrane bearing ribosomes and DNA into the cell interior to form topologically distinct RER cisternae (fig. 2 b, c), and
- aggregation of the cisternae around the DNA to form an nucleus (fig. 2 d).

I have argued that it was the origin of the nuclear lamina that attached early ER cisternae to the outside of the chromatin mass, and thereby created the nuclear envelope. The major structural components of the nuclear lamina and therefore of nuclear architecture, the lamin proteins, belong to the intermediate filament family of cytoskeletal proteins; as the latter also form the major non-motile framework of the cytoplasm they were central to the origin of eukaryote architecture.

Another fundamental reason for thinking that the cytoskeleton evolved before the nuclear envelope it its role in mitosis and eukaryotic cell division. Efficient DNA segregation at cell division is so important for maintaining the life of the cell that any theory of the origin of the eukaryote cell must explain how it was achieved throughout the transition from prokaryote to eukaryote. Since segregation in bacteria depends on DNA attachment to the plasma membrane and rigid cell wall (Cavalier-Smith 1987b), the internalization of the DNA attachment site to form the nuclear envelope would have prevented regular DNA segregation. A cell with a nuclear envelope but with no cytoskeleton, or machinery for intracellular motility, to provide a mitotic apparatus and division mechanism would therefore be inviable. Thus I have argued that the cytoskeleton must have evolved before the nuclear envelope (Cavalier-Smith 1980, 1981, 1987a), and that during the transition from prokaryote to eukaryote a premitotic mechanism based on microtubules evolved at the cell surface while the DNA was still attached to it (fig. 2a): in this way the fundamental problem of the transition between the prokaryotic, exoskeletal DNA segregation mechanism based on the cell wall, and the eukaryotic, endoskeletal DNA segregation mechanism based on a mitotic spindle, is solved without major discontinuity. The origin of the cytoskeleton is therefore a much more fundamental innovation than the origin of the nuclear envelope/ER *per se,* which essentially involved the internalisation of pre-existing structures: DNA-membrane and ribosome-membrane associations (fig. 2b). The latter, I believe, was an inevitable consequence of the evolution of phagocytosis, which was thus pivotal in the origin of eukaryotes. Permanent internalization of the membrane-protein and membrane-lipid biosynthetic machinery to form the ER would not have been possible without the prior evolution of cytotic mechanisms, *i.e.* the budding of membranous vesicles from the ER (ercytosis) and their fusion with the plasma membrane (exocytosis) to allow it to grow. I have proposed that it was the evolution of ercytosis itself that actually caused the fundamental distinction in structure and function between the ER, nuclear envelope and the plasma membrane (fig. 2c). Ercytosis by its very nature acts as a filter during membrane flow from the endomembrane system to the plasma membrane: during the budding of smooth vesicles it screens out ribosome-receptor proteins and the phospholipid-synthesising machinery and thus prevents their transport to the plasma membrane. If ercytosis and exocytosis were the only way to add new membrane to the plasma membrane then the removal of all the ancestral cell's ribosome-receptors and DNA-attachment proteins from the plasma membrane by any mechanism – I argue phagocytosis was the actual one – would automatically be irreversible; they would be located ever after on internal membranes topologically distinct from the plasma membrane, *i.e.* the endomembrane system; this latter differentiated into two topologically distinct

Table 1. Distinctive genomic features of eukaryotes absent in bacteria.

1. Plural replication origins per chromosome.
2. Plural chromosomes per genome.
3. Histones (secondarily reduced in dinoflagellates).
4. Linear chromosomes.
5. Large and variable genome size (0.007–700 pg of DNA, compared with 0.0005–0.014 pg in bacteria).
6. Introns in protein-coding genes.
7. Separate RNA polymerases for mRNA, tRNA and rRNA.
8. Capped and polyadenylated stable mRNAs.
9. Monocistronic mRNA.

subsystems: the nuclear envelope/ER and the lysosomes. I have also shown how, once such internalization occurred and a cytoskeleton-based mitotic mechanism evolved, this would so radically alter the selective forces acting on the genome that the evolution of the nucleus, and all the distinctive characters of the eukaryotic genome (table 1), would automatically follow (Cavalier-Smith 1987a).

The phylogenetic relationship between eukaryotes and bacteria

In my view the idea of an "urkaryotic line of descent" possessing such eukaryotic genomic characters but lacking the cytoskeletal, endomembrane and cytotic properties found in all eukaryotic cells, as suggested by the Woese school of rRNA phylogeny, is fundamentally misconceived, because these genomic features can be understood only as *consequences* of the prior evolution of the eukaryotic cytoskeletal and membrane properties. There is no reason to think that these genomic features evolved, or could have evolved, prior to the cytoskeleton and endomembrane system.

The marked divergence between the rRNA sequences of eukaryotes, archaebacteria, and eubacteria, which is the sole basis for the idea of such an 'urkaryotic line of descent', does not entail such a view. Quite the reverse: the fact that archaebacterial and eubacterial 16s rRNA sequences are almost as different from each other as either is from eukaryotic sequences, clearly shows that marked divergence in rDNA is possible in the total absence of the genomic changes (table 1) that separate prokaryotes and eukaryotes. In cellular and genomic organization, archaebacteria are fundamentally pro-

Fig. 3. The five kinds of cell. – Archaebacteria are unique in having isoprenyl ether membrane lipids rather than acyl ester lipids in all other organisms. Negibacteria *(e.g., Escherichia coli)* are unique in having an outer lipoprotein membrane, which I argue arose during the origin of the first cell (Cavalier-Smith 1987c), and which was lost only once during the whole history of life – during the origin of the Posibacteria *(e.g., Bacillus subtilis)* by murein hypertrophy (Cavalier-Smith 1980). Archezoa differ from higher eukaryotes (Metakaryota) in not having mitochondria (M), chloroplasts (C), or peroxisomes (P), and in lacking permanent Golgi dictyosomes (D), and in at least some cases in having 70s ribosomes like bacteria. Dictyosomes (in mammals at least) consist of 3 topologically distinct compartments interconnected by vesicle budding, shuttling, and fusion. The negibacterial outer membrane was retained in the mitochondrial and chloroplast envelopes during their symbiotic origin (Cavalier-Smith 1982, 1983, 1987d). Most cyanobacteria (uniquely among prokaryotes) have intracellular membranes topologically distinct from the plasma membrane (*i.e.* their thylakoids, which also directly became the thylakoids (T) of chloroplasts). Peroxisomes, archaebacteria, and archezoa probably all evolved from posibacteria, which like them have only a single bounding membrane (Cavalier-Smith 1987a, d). ▷

karyotic, *i.e.* lacking an endomembrane system and cytosis, and segregating their DNA by a rigid exoskeleton rather than a cytoskeleton. They do however share certain molecular features with eukaryotes, notably details of RNA polymerase and ribosomes, and introns in tRNA and rRNA, which suggest a close relationship between their ancestor and that of eukaryotes.

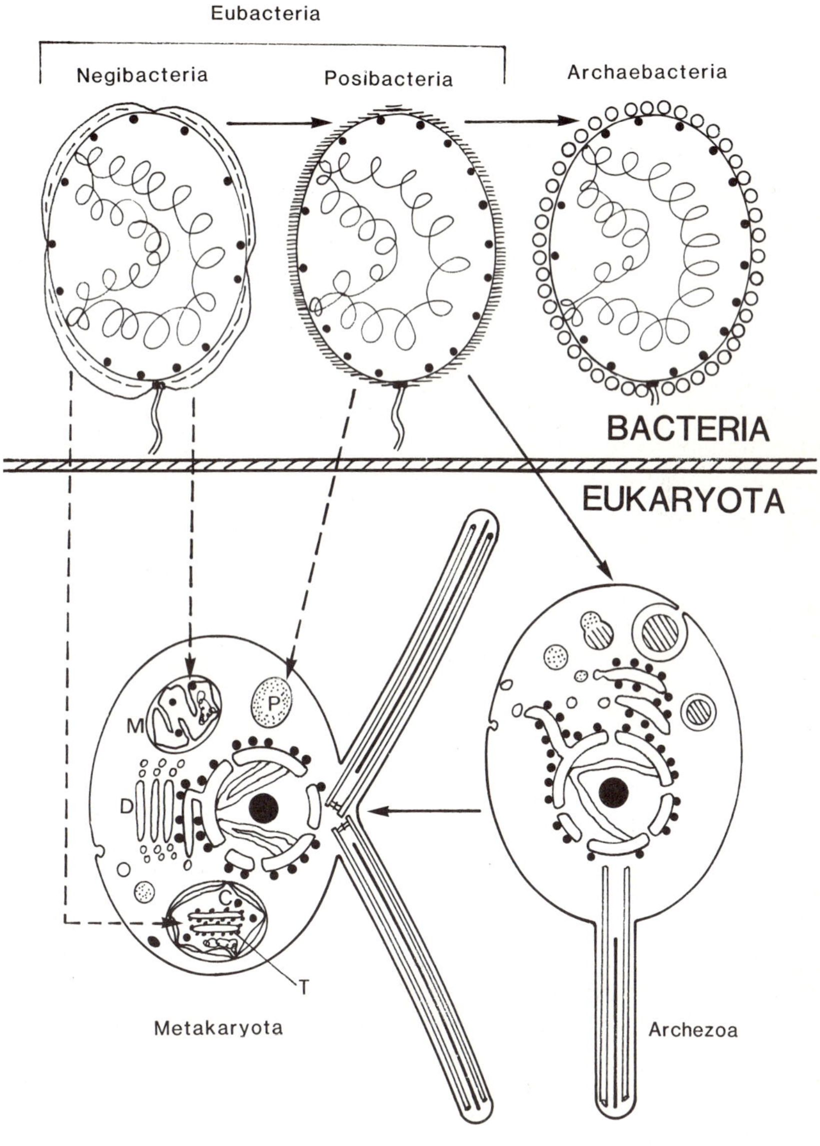

The Woese school has tended to view the great divergence in rRNA sequences between eukaryotes, archaebacteria, and eubacteria as having occurred during a burst of exceptionally rapid differentiation in the earliest stages of cellular (or even precellular) evolution, and therefore as a sign of very early divergence of the three lineages. But this was only an assumption based on the gut-feeling that big changes might have been easier early on. Its only test lies in the fossil record, which does not support an early origin for eukaryotes. There is no evidence for eukaryotes before about 1500 My ago; even the evidence before roughly 900 My ago based on cell size is debatable though probably valid, whereas bacteria go back at least 3500 My. The occurrence then of stromatolites indicates that photosynthetic eubacteria already existed. The antiquity of archaebacteria is less clear, but there is no evidence that they go back as far as eubacteria. Because of the shared features with eukaryotes they must be at least as old as eukaryotes; but possibly no older, as I have argued.

The universal rRNA phylogenetic tree is very useful, but one cannot use rRNA data alone to determine its root or the time of divergence of the three major lineages because no outgroup exists and because rates of ribosomal RNA base replacement vary several-fold in different lineages, *i.e.* by more than the uncertainty in the time of origin of eukaryotes and archaebacteria. But fossils and rRNA trees are not the only tools for inferring relationships. Also vitally important are the biological mechanisms underlying the megaevolutionary transitions between the major groups implicit in any particular phylogeny. If one phylogeny involves transitions which are implausible or incomprehensible, whether from a mutational, epigenetic, or selective viewpoint, whereas another phylogeny also compatible with the fossil record and sequence data involves transitions for which a rational explanantion can be found, then the latter phylogeny is to be preferred. Thus rival phylogenies cannot be properly judged unless one also develops detailed and explicit scenarios as to the mutational, epigenetic, and selective basis for all the major transitions that they imply.

Taking into account ultrastructural as well as molecular divergences, there are not three but five major kinds of cell, three prokaryote and two eukaryote, as fig. 3 makes clear. Within eubacteria, the distinction between negibacteria with two surrounding membranes and posibacteria with but a single membrane is profoundly important. The rRNA phylogeny (Woese 1987) and other considerations (Cavalier-Smith 1987a, b) suggest that the ancestral eubacterium was a double-membraned photosynthetic negibacterium and that the single-membraned posibacterial condition is derived. I have argued that the loss of the outer membrane resulted from a sudden thickening of the peptidoglycan layer, causing the outer membrane to break away from its contact with the inner membrane, thus preventing transfer of lipids and proteins needed for the continued growth of the outer membrane. The difficulty of explaining the origin of the negibacterial outer membrane other than by the folding up of a precellular inside-out cell (obcell) to form the very first cell is a major reason for rooting the universal tree (fig. 4) in the eubacteria.

Eukaryote cell diversity and the 8 kingdom classification

Like bacteria, eukaryotes are divisible into two fundamentally different cell types: Archezoa and Metakaryota. Metakaryotes, or "higher eukaryotes", typically have mitochondria, peroxisomes, and Golgi dictyosomes, and often also chloroplasts in addition to the basic eukaryotic structures shown in fig. 1b. By contrast Archezoa, a neglected group of unicellular eukaryotes, lack all four organelles, and represent the eukaryotic condition at its simplest. My suggestion that the 3 archezoan phyla (Microsporidia, Metamonada and Archamoebae) are descendants of the earliest eukaryotes that never evolved these four organelles is supported by the radical differences of microsporidian and metamonad ribosomes from those of all higher eukaryotes (Cavalier-Smith 1987e). The rRNA of the microsporidian *Vairimorpha* and the metamonad *Giardia* are markedly shorter molecules than those of metakaryotes; no longer than those of bacteria. The *Vairimorpha* 23s rRNA, like bacterial 23s rRNA, contains the sequence homologous to metakaryote 5.8s rRNA, whereas *Giardia* has a 127 nucleotide RNA that may be homologous to the 160 nucleotide 5.8s rRNA of metakaryotes. Like bacteria, microsporidia have 70s rather than 80s ribosomes, and the *Vairimorpha* 16s rRNA sequence diverges more from the 18s rRNA sequences of metakaryotes than any of the latter do from each other. Ultrastructurally and in rRNA sequence the gulf between Archezoa and Metakaryota is much more profound than that between conventional eukaryote kingdoms such as Fungi, Animalia, and Plantae. In order to recognize this gulf at the highest taxonomic level within Eukaryota I separated Archezoa from the kingdom Protozoa, as a distinct kingdom and superkingdom, and grouped the 5 kingdoms Animalia, Plantae, Fungi, Chromista, and Protozoa as a new superkingdom, the Metakaryota (Cavalier-Smith 1987c). The even greater gulf between Eukaryota and Bacteria was recognised by raising both taxa to the rank of Empire (table 2).

Recognising the fundamental distinction between Archezoa and Metakaryota clarifies the problems of eukaryote cell evolution by dividing it into two phases: firstly the origin of the eukaryote cell itself, *i.e.* of the first archezoan, which I argue was a purely autogenous process not involving symbiosis of any kind; and secondly the origin of the metakaryote cell, *i.e.* the first true protozoan with chloroplasts, mitochondria, peroxisomes, dictyosomes, and 80s ribosomes, in which I argue multiple symbioses played the major role.

Wall loss: the key to the origin of both eukaryote and archaebacterial cells

Since phagocytosis could not have evolved in a typical bacterium with a cell wall the key step in the origin of eukaryotes must have been the loss of the cell wall to form a naked cell. When such cells are produced in the laboratory they have great problems because of their osmotic fragility and aberrant divisions and DNA segregation, which may produce cells with no DNA or too much DNA. Such cells could survive in the wild only if additional mutations created new skeletal structures: actin filaments, intermediate filaments, and microtubules probably arose by gene duplication and mutation from yet unidentified prokaryotic proteins, and were selected because they helped reduce the osmotic instability and division and DNA segregation errors of a mutant posibacterium that has lost its murein wall. (Evolutionary loss of murein by a negibacterium, as in *Planctomyces* and chlamydias, does not abolish the outer membrane, so does not produce a cell bounded by a single membrane).

Because the wall-free mutant evolved an endoskeleton rather than a new cell wall it was able to evolve phagocytosis which is what inevitably transformed it into a eukaryote. If instead it had evolved a new external cell wall it could not

Fig. 4. Phylogenetic relationships between the 8 kingdoms of cellular organisms. – The most important megaevolutionary changes in the history of life are shown in boxes. Changes in skeletal structures played the dominant role not only in the origin of the eukaryote cell but also in the origin of the prokaryote cell by the folding of a precellular inside-out-cell known as an obcell (Cavalier-Smith 1987c), and in the origin of the kingdoms Archaebacteria, Fungi (the chitinous wall: Cavalier-Smith 1987f), and Animalia (collagen and glycoprotein cell-adhesion molecules). Changes in internal membranes played the dominant role in the origin of the kingdoms Protozoa, Plantae and Chromista. The time of origin of the Metakaryota is deduced from the date of the first protist fossils having cell walls with complex surface projections, which probably depend on a well-defined Golgi dictyosome for their secretion; that of the origin of Eukaryota is marked by the sudden increase in size of fossil cells, which probably depended on a cytoskeleton and intracellular motility, coincident with an increase in fossil steranes; that of the Eubacteria corresponds with the first fossil cells and stromatolites (laminated fossils probably made by photosynthetic eubacteria). ▷

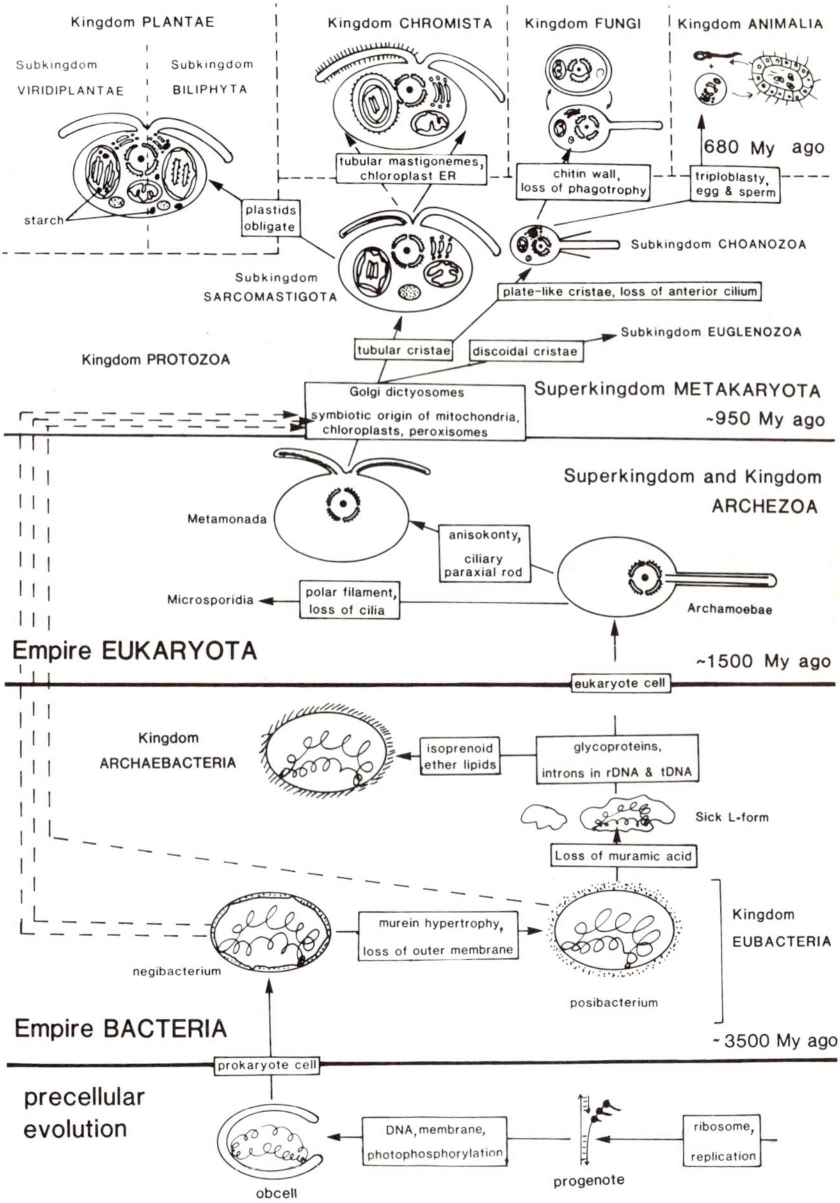
Kingdom PLANTAE
Subkingdom VIRIDIPLANTAE
Subkingdom BILIPHYTA
starch
plastids obligate
Kingdom CHROMISTA
tubular mastigonemes, chloroplast ER
Kingdom FUNGI
chitin wall, loss of phagotrophy
Kingdom ANIMALIA
680 My ago
triploblasty, egg & sperm
Subkingdom CHOANOZOA
Subkingdom SARCOMASTIGOTA
plate-like cristae, loss of anterior cilium
Subkingdom EUGLENOZOA
Kingdom PROTOZOA
tubular cristae
discoidal cristae
Golgi dictyosomes
symbiotic origin of mitochondria, chloroplasts, peroxisomes
Superkingdom METAKARYOTA
~950 My ago
Superkingdom and Kingdom ARCHEZOA
Metamonada
anisokonty, ciliary paraxial rod
Microsporidia
polar filament, loss of cilia
Archamoebae
Empire EUKARYOTA
~1500 My ago
eukaryote cell
Kingdom ARCHAEBACTERIA
isoprenoid ether lipids
glycoproteins, introns in rDNA & tDNA
Sick L-form
Loss of muramic acid
murein hypertrophy, loss of outer membrane
negibacterium
posibacterium
Kingdom EUBACTERIA
Empire BACTERIA
~3500 My ago
prokaryote cell
precellular evolution
DNA, membrane, photophosphorylation
ribosome, replication
obcell
progenote

have evolved phagocytosis and so would necessarily have remained fundamentally prokaryotic. Exactly this, I suggested (Cavalier-Smith 1987a), happened to the ancestor of archaebacteria, which also evolved from a wall-less mutant posibacterium. As archaebacteria and eukaryotes both totally lack muramic acid, the constituent of murein that is covalently cross-linked by peptide bridges, the initial step in the origin of both groups was probably the mutational loss of the ability to make muramic acid (or to incorporate it into peptidoglycan). Because, as already mentioned, eukaryotes and ar-

Table 2. The 8 kingdoms of cellular organisms (*i.e.* excluding viruses, viroids and virusoids).

Empire 1.	**Bacteria** (syn. Prokaryota; DNA not separated from ribosomes by an envelope; endomembrane system and cytoskeleton absent: DNA attached to "envelope skeleton").		
	Kingdom 1.	Eubacteria (acylglycerol membrane lipids; murein).	
		Subkingdom 1.	Negibacteria (plasma membrane + outer membrane).
		Subkingdom 2.	Posibacteria (outer membrane absent).
	Kingdom 2.	Archaebacteria (isoprenoid ether lipids; no murein; outer membrane absent).	
Empire 2.	**Eukaryota** (nucleated cells; with endomembrane systems, cytosis & internal cytoskeleton; acyl glycerol membrane lipids; no peptidoglycan walls).		
	Superkingdom 1.	Archezoa (70s ribosomes; Golgi dictyosomes, mitochondria, chloroplasts, and peroxisomes absent).	
	Kingdom 1.	Archezoa.	
	Superkingdom 2.	Metakaryota (80s ribosomes; Golgi dictyosomes, mitochondria, peroxisomes, and often chloroplasts, typically present).	
	Kingdom 1.	Protozoa (chloroplast envelope of 3 membranes; includes Mycetozoa; predominantly phagotrophic).	
	Kingdom 2.	Chromista (tubular ciliary mastigonemes and/or chloroplast endoplasmic reticulum present; Cavalier-Smith 1986, 1989).	
	Kingdom 3.	Plantae (plastids invariably present; chloroplast envelope of 2 membranes; plastids not in ER).	
	Kingdom 4.	Fungi (typically with chitinous walls and no chloroplasts or phagocytosis; excludes Mycetozoa & Pseudofungi, Cavalier-Smith 1987f).	
	Kingdom 5.	Animalia (triploblastic multicellular phagotrophs).	

chaebacteria share positive characters not found in eubacteria, notably the presence of introns in tRNA and rRNA and the ability to make N-asparagine-linked glycoproteins, I suggested that both groups descended from the very same muramic-acid-deficient wall-less posibacterial mutant, and that these common features all evolved in this common ancestor prior to the divergence of the two groups (fig. 4).

Rudiments of the cytoskeleton may possibly have evolved in that common ancestor, as hinted by the cytochalasin sensitivity of the archaebacterium *Thermoplasma,* the only archaebacterium without a cell wall. But the main way that the archaebacterial ancestor solved its osmotic and segregational problems was to stiffen its membrane by evolving isoprenoid tetraether lipids that in effect converted it into a monolayer rather than a bilayer. This enabled it to colonize acid thermal habitats unsuitable for eubacteria (all having the more fluid and acid-sensitive acyl ester lipids). Even with such lipids, adequate stability in the absence of an efficient cytoskeleton was attainable only by evolving a very small size (*Thermoplasma*) or a new external cell wall (other archaebacteria), both of which would have precluded the origin of phagocytosis.

The early eukaryote ancestor probably also evolved new isoprenoid derivatives that helped rigidify its membranes (*i.e.* sterols), but as these remained a minor constituent of the membrane, and as the cytoskeleton and intracellular contractility was evolving at the same time, phagocytosis was able to evolve, and transform the cell into a eukaryote as sketched in fig. 2. Probably the cytoskeleton was not only a prerequisite for phagocytosis – by making a large naked and osmotically stable cell – but also through its contractility intimately partook in the phagocytis mechanism itself, and in intracellular transport of vesicles destined for exocytosis, without which sustained phagocytosis would have been impossible.

Coevolution of nucleus, mitosis, and cilia

One of the most crucial changes in the origin of eukaryotes was the replacement of the eubacterial DNA segregation and cell division mechanisms based on the rigid murein wall by the eukaryotic one based on a microtubular endoskeleton. This did not occur because the eukaryote mechanism is better for larger genomes with several chromosomes, as often asserted. It should be seen not as an improvement but as a rescue mechanism for the defective wall-less mutant that had just undergone a worsening of its segregation mechanism. As shown in fig. 2 the mitotic mechanism probably evolved in two stages:

firstly, a premitotic system that segregated the chromosomes while they were still attached to the cell surface (fig. 2a) – rigid intracellular microtubules replaced the rigid extracellular murein as the mechanical framework for the attachment of DNA, while a contractile actin gel replaced septation by rigid murein growth as the mechanical force causing cell division. Secondly, following the internalization of the membrane-attachment sites for DNA, there simultaneously evolved the nucleus, the cilium, and a primitive pleuromitotic mechanism (fig. 2d) from which the various forms of orthomitosis subsequently evolved (Cavalier-Smith 1987a).

The key step in the origin of both eukaryotic cell division and of cilia was the orthogonal attachment of nine of the astral microtubules to the plasma membrane by the nine transitional fibres. This caused the growth of the presumptive ciliary microtubules to push the plasma membrane outwards to form the ciliary membrane. By bringing the centrosome – embedded in the nuclear envelope – and the remaining astral microtubules close to the cell surface it allowed the latter to attach to the surface and become ciliary roots, and also to mechanically constrain the cleavage furrow so that it had to lie between the two daughter centrosomes. The involvement not only of microtubules, but also of intermediate-filament-related tektins and of actin, in ciliary outer doublet structure clearly supports the autogenous origin of cilia (which term includes also eukaryotic flagella) from the cytoskeleton; good arguments or evidence for a symbiotic origin for cilia (Margulis 1970) do not exist.

Genetic consequences of the origin of mitosis and the nuclear envelope

In my view (Cavalier-Smith 1987a) all eight of the fundamental genetic properties of eukaryotes (table 1) evolved as inevitable consequences of the origin of mitosis, cytokinesis, and/or the nuclear envelope. The prokaryotic segregation mechanism requires a circular chromosome with a single origin and a single terminus of replication (Cavalier-Smith 1987b). Its replacement by mitosis made possible the evolution of plural replicon origins per chromosome, several chromosomes per genome, and linear chromosomes: mutation pressure would inevitably generate plural replicons, and sex (greatly favoured by nakedness and the eukaryotic propensity for membrane fusion) would favour multiple linear chromosomes. Cytokinesis by active contraction favoured the evolution of histones to fold the DNA more compactly and prevent breakage. Plural replicon origins made indefinitely large genomes possible and a new nucleoskeletal role for DNA arising with the origin of the nuclear envelope made them advantageous (Cavalier-Smith 1985): larger cells need proportionally larger nuclei, and expansion of their genome size it the simplest way of achieving this. Evolution of stabler messengers (by capping and polyadenylation) was a necessary concomitant of the origin of the nuclear enveolope: in its turn this allowed transposable introns to spread for the first time into protein-coding genes (Cavalier-Smith 1987a, 1988), and led to the origin of monocistronic messengers, and of 3 separate RNA polymerases (Cavalier-Smith 1987a).

Some of these genetic changes had important consequences later in evolution (*e.g.* the ability greatly to increase gene numbers without lengthening replication times, and the creation of new genes by exon-shuffling) but it is a fundamental mistake to see these *consequences* as *reasons* for those changes, or any of those changes as *reasons* for the origin of mitosis and the nuclear envelope. The latter were simply concomitants of the evolution of a naked phagotroph: the great success of the first phagotrophs carried them all to fixation, and there was neither a physical mechanism nor a selective advantage for going back to the old prokaryote way of doing things.

Constraints on prokaryote architectural complexity

The cytoskeleton and cytosis are essential prerequisites for making complex cells such as ciliate protozoa, or complex multicells such as blue-whales or blue-gum trees. *Drosophila* apparently has fewer genes than a large cyanobacterial cell. Cyanobacteria can undergo cell differentiation. I suggest that it is neither the small genome size nor the genomic organization of bacteria that makes them unable to grow into trees or whales but the fact that what their genes code for is an evolutionarily and architecturally constricting exoskeleton rather than a liberating endoskeleton. Genetic mutations first produced the cytoskeleton, which by facilitating phagocytosis converted the cell structure and genetic system from prokaryotic to eukaryotic. But the key mutational changes that initiated the transformation took place while the cell was still structurally and genetically prokaryotic. Therefore prokaryotic *genetic* organization is not inimical to major change. Of course plants and fungi have secondarily evolved cell walls, but their ability to use them to make more complex structures than those of bacteria depends on the cytoskeleton and cytosis, evolved by their naked archezoan and protozoan ancestors.

Sex, phagotrophy, and megaevolution

Sex has often been suggested as a cause of eukaryote success. I have argued (Cavalier-Smith 1987a) that it played a key role in the origin by allopolyploidy of the first biciliated metamonad, from which all metakaryotes ultimately descend. But sex was not involved in the origin of the eukaryotes – the most radical megaevolutionary event since the origin of life. This shows that sex is not necessary even for the largest and most important steps in megaevolution (by megaevolution I mean evolution of characters that typify classes, phyla, and yet higher taxa). It was eating, not sex, that as the most powerful force in the early evolution of the eukaryote cell, was in human affairs.

So-called "typical" sex (that of obligately sexual, outbreeding, dioecious species) might even inhibit megaevolution. This is because at the highest levels (phyla, kingdoms, and superkingdoms) megaevolution often, I believe, occurs via hopeful monsters radically different from their parents, which would be unlikely to be able to mate and reproduce sexually with individuals retaining the unaltered ancestral state. But as it is likely that the immediate ancestors of each animal and plant phylum could reproduce clonally and were mostly, if not always, hermaphrodites – possibly even capable of self-fertilization – quasi-saltatory evolution involving hopeful monsters could have been quite general in the origin of phyla. This seems even more probable in unicellular eukaryotes, which can invariably multiply clonally and are often homothallic.

The origin of eukaryotes and of metakaryotes must each have required thousands of mutations, but the key mutations that initiated each process, and each major substituent step, were probably single mutations with a phenotypically major, saltatory, effect that did produce hopeful monsters (*e.g.* the loss of muramic acid, the first step in phagocytosis, the origin of ercytosis, the insertion of the adenine nucleotide carrier into the mitochondrial inner membrane). By creating a radically new kind of organism such mutations would dramatically change the selective forces on many other genes, and favour an explosive rate of change. The origin of metakaryote kingdoms also, can probably be traced to single key macromutations, as I have argued for Chromista (Cavalier-Smith 1986) and Fungi (Cavalier-Smith 1987f) – as well as for their constituent phyla. I suspect that it is not general features of the genome, nor statistically recurrent features of population genetics, but the *specific phenotypic consequences* of rare mutations, and especially *unique* combinations of rare mutations, that are the key determinants of megaevolution.

Symbiosis and the origin and diversification of metakaryotes

Phagocytosis has been profoundly significant for eukaryote cell evolution, not only because it created the eukaryote cell in the first place, but also because, as Stanier (1970) emphasised, it makes it easy for eukaryotes to acquire cellular endosymbionts. Since bacteria are totally unable to acquire cellular endosymbionts I have always opposed Margulis's (1970) symbiotic theory of the origin of eukaryotes, according to which the symbiotic origin of mitochondria in a bacterial host was the *prerequisite* for the origin of the eukaryote cell. Her theory of the symbiotic origin of cilia and the mitotic spindle, equally makes no structural or functional sense.

Recognition of the evolutionary importance of the kingdom Archezoa however makes it possible to dissociate entirely the symbiotic origin of chloroplasts and mitochondria, for which the evidence is now overwhelming, from the purely autogenous origin of eukaryotes. My attempts to reconstruct the common ancestor of metakaryotes and to understand how symbionts could be converted into organelles have led to some radical revisions of the symbiotic theory (Cavalier-Smith 1987d). According to these, mitochondria, chloroplasts, and probably also peroxisomes, all evolved simultaneously from 3 different symbionts in a single biciliated[1] metamonad host; the outer membrane of mitochondria and of plant and chromist chloroplasts and the middle membrane of protozoan chloroplasts, represents the outer membrane of their purple bacterial and cyanobacterial ancestors – the phagosomal membrane remains only as the third outermost membrane of protozoan chloroplasts. Of crucial importance was the origin of organelle-specific post-translational protein-import mechanisms, without which gene transfer from symbiont to nucleus together with loss of transferred genes from the symbiont could not have occurred. The protein-import mechanisms probably evolved by reversing the protein-export mechanisms of the 3 bacterial symbionts: even so the changes involved were so considerable that it is undesirable to assume that this occurred more than once for each organelle unless there is conclusive phylogenetic evidence that this is what happened. Since one can construct a unified metakaryote phylogeny on the assumption of only a single origin for each organelle, I do not favour the traditional view of the multiple symbiotic origin of chloroplasts. The different pigments, and thylakoid and envelope morphologies, of the varied algal phyla can all be explained by divergence from a single cyanobacterial ancestor (Cavalier-Smith 1982, 1987g).

A fourth major symbiosis is necessary to explain the origin of the chloroplast endoplasmic reticulum, which characterizes the metakaryote kingdom Chromista; but as the symbiont was another metakaryote already possessing

[1] Note added in proof: I now think that the first metakaryote and its retortamonad-like metamonad ancestor were both tetraciliated, not biciliated (Cavalier-Smith 1989).

a chloroplast, albeit in an early stage of evolution, this was no independent *origin* of chloroplasts, only their acquisition by a new host.

Therefore, although it is totally wrong to speak of the symbiotic origin of eukaryotes it is perfectly proper to speak of the symbiotic origin of the kingdom Chromista and of the superkingdom Metakaryota. Both events, however, involved more than symbiosis: for the Chromista the origin of tubular ciliary mastigonemes, probably stimulated by the symbiotic acquisition of chloroplasts (Cavalier-Smith 1986); and for the Metakaryota the origin of permanent Golgi dictyosomes and of 80s ribosomes, possibly stimulated by the symbiotic acquisition of the three new types of membranous organelles which might have confused the cotranslational protein insertion and vesicle targetting mechanisms and thus caused them to be modified so as to avoid the confusion.

There has been a tendency for supporters of the symbiotic theory to think that chloroplasts can be gained by symbiosis but not lost. However the phylogenetic evidence for the kingdoms Chromista and Protozoa suggests that chloroplasts have been lost several times in each group. In our own direct ancestry photosynthesis has, I believe, been gained three times and lost three times since the origin of life! The losses were during (1) conversion of a photosynthetic heliobacterium into a saprotrophic posibacterium, (2) conversion of a photosynthetic purple non-S bacterium into a mitochondrion, and (3) loss of plastids by an early protozoan ancestor of our choanomonad ancestor. The gains were (1) the origin of photosynthesis during precellular evolution (Cavalier-Smith 1987c), and (2 and 3) the simultaneous acquisition of cyanobacterial and purple-non-S bacterial symbionts by the first metakaryote.

In Protozoa and Chromista the loss of photosynthesis typically involves the loss of plastids, but in the kingdom Plantae it never does; both in green plants and red algae the plastids remain as leucoplasts. This obligate requirement for non-photosynthetic functions of plastids, together with the absence of a third plastid envelope membrane or any chloroplast endoplasmic reticulum, are the key features of the kingdom Plantae.

Acknowledgements

I thank Chris Perera for help with the figures and Pat Wright for typing.

References

[To save space, references are largely restricted to some of my own writings where extensive references to the work of others will be found.]

Cavalier-Smith, T. 1975: The origin of nuclei and of eukaryote cells. – Nature **256**: 463–468.
– 1980: Cell compartmentation and the origin of eukaryote membranous organelles. – In: Schwemmler, W. & Schenk, H. E. A. (eds.), Endocytobiology: Endosymbiosis and cell biology, a synthesis of recent research: 893–916. – De Gruyter, Berlin.
– 1981: The origin and early evolution of the eukaryote cell. – In: Carlile, M. J., Collins, J. F. & Moseley, B. E. B. (eds.), Molecular and cellular aspects of microbial evolution (Society for General Microbiology Symposium, **32**): 33–84. – University Press, Cambridge.
– 1982: The origins of plastids. – Biol. J. Linn. Soc. **17**: 289–306.
– 1983: Endosymbiotic origin of the mitochondrial envelope. – In: Schwemmler, W. & Schenk, H. E. A. (eds.), Endocytobiology II: 265–279. – De Gruyter, Berlin.
– (ed.) 1985: The evolution of genome size. – Wiley, Chichester.
– 1986: The kingdom Chromista: origin and systematics. – In: Round, F. E. & Chapman, D. J. (eds.), Progress in phycological research, **4**: 309–347. – Biopress, Bristol.
– 1987a: The origin of eukaryote and archaebacterial cells. – Ann. New York Acad. Sci. **503:** 17–54.
– 1987b: Bacterial DNA segregation: its motors and positional control. – J. Theor. Biol. **127:** 361–372.
– 1987c: The origin of cells: a symbiosis between genes, catalysts, and membranes. – Cold Spring Harbor Symp. Quant. Biol. **52:** 805–824.
– 1987d: The simultaneous symbiotic origin of mitochondria, chloroplasts, and microbodies. – Ann. New York Acad. Sci. **503:** 55–71.
– 1987e: Eukaryotes with no mitochondria. – Nature **326**: 332–333.
– 1987f: The origin of Fungi and pseudofungi. – In: Rayner, A. D. M., Brasier, C. M. & Moore, D. (eds.), Evolutionary biology of the fungi. (Symp. Brit. Mycol. Soc. **13**): 339–353. – University Press Cambridge.
– 1987g: Glaucophyceae and the origin of plants. – Evol. Trends Pl. **2:** 75–78.
– 1988: The origin of introns and the eukaryotic genome. – (Submitted).
– 1989: The Kingdom Chromista. – In: Green, J. C. & Leadbeater, B. S. C. (eds.), The chromophyte algae: problems and perspectives. – Oxford University Press (in press).
Margulis, L. 1970: Origin of eukaryotic cells. – Yale University Press, New Haven.
– & Schwartz, K. V. 1982: Five kingdoms: an illustrated guide to the phyla of life on earth. – Freeman, San Francisco.
Stanier, R. Y. 1970: Some aspects of the biology of cells and their possible evolutionary significance. – Symp. Soc. Gen. Microbiol. **20**: 1–38.
Woese, C. R. 1987: Bacterial evolution. – Microbiol. Rev. **51**: 221–271.

Address of the author: Dr. T. Cavalier-Smith, Department of Biophysics, Cell and Molecular Biology, King's College, 26–29 Drury Lane, London WC2B 5RL, UK.

Calcium and development

P. H. Hepler

Abstract

Hepler, P. K. 1988: Calcium and development. – In: Greuter, W. & Zimmer, B. (eds.): Proceedings of the XIV International Botanical Congress: 225–240. – Koeltz, Königstein/Taunus.

The mechanism by which cells convert a signal into a response is fundamental to development. Emerging evidence suggests that calcium (Ca) may act as a second messenger in several different systems. The possibility exists that plant growth regulators (PGRs) cause an influx of Ca into the cytosol, and thereby activate a calmodulin-requiring kinase, which phosphorylates proteins responsible for initiating a new developmental programme. Accumulating evidence shows that developmental processes stimulated by cytokinin, auxin, gibberellin, phytochrome and gravity all appear to utilize Ca. Considerable additional complexity to the second messenger scheme is provided by recent studies showing the involvement of the phosphatidyl inositol pathway in signal transduction. Here the PGR may cause the appearance of inositol 1,4,5-trisphosphate (IP_3) and diacylglycerol (DG). IP_3 stimulates Ca release, while DG activates protein kinase-C. The second messenger system is thus able to provide different pathways that allow separate but coexisting developmental processes to be differentially activated. Elucidation of the details of Ca mobilization, and the phosphatidyl inositol pathway may help explain the mechanism of action of PGRs.

Introduction

How plant cells induce a new developmental programme is a major unsolved problem. Since development usually involves the synthesis of new proteins and other macromolecules is has been attractive to imagine that plant growth regulators (PGRs: hormones, light, gravity) activate new gene expression. However in most instances there appear to be both spatial and temporal gaps between the reception of the PGR and the actual transcription of new mRNA and translation of new protein. Spatial constraints arise from the realization that the PGRs, especially hormones, are received at the plasmalemma, which may be a significant molecular distance from the nucleus, the site of transcription. Temporal limitations become evident from the observations showing that intervals of minutes (Theologis 1986), hours, or even days may separate the reception of a PGR and the appearance of a new gene product. The mechanism by which the cell bridges these spatial and temporal domains may involve the activation of an intermediary "second messenger" system.

Emerging evidence indicates that many different processes utilize calcium ions (Ca) as a second messenger (fig. 1). Mitosis, polarized tip growth, cytoplasmic streaming, phytochrome-induced spore germination, gravitropic root and shoot curvature, cytokinin-induced bud formation, gibberellin-stimulated α-amylase secretion, and polar transport of auxin are but a few of the processes in which a role for Ca has been implicated (Hepler & Wayne 1985, for review). A generally accepted scheme for the sequence of events is one in which the agonist induces an increase in free cytosolic [Ca]. At its elevated level Ca binds to and activates calmodulin, which can then stimulate a response element such as a protein kinase. The subsequent phosphorylation of a target enzyme may enhance its activity and thus initiate a series of developmentally important reactions leading ultimately to the formation of a new gene product (Ranjeva & Boudet 1987) (fig. 2).

A recent and exciting new finding has been the discovery of another messenger system, namely the phosphatidyl inositol (PI) pathway, which may work together with Ca to transmit information between the stimulus and

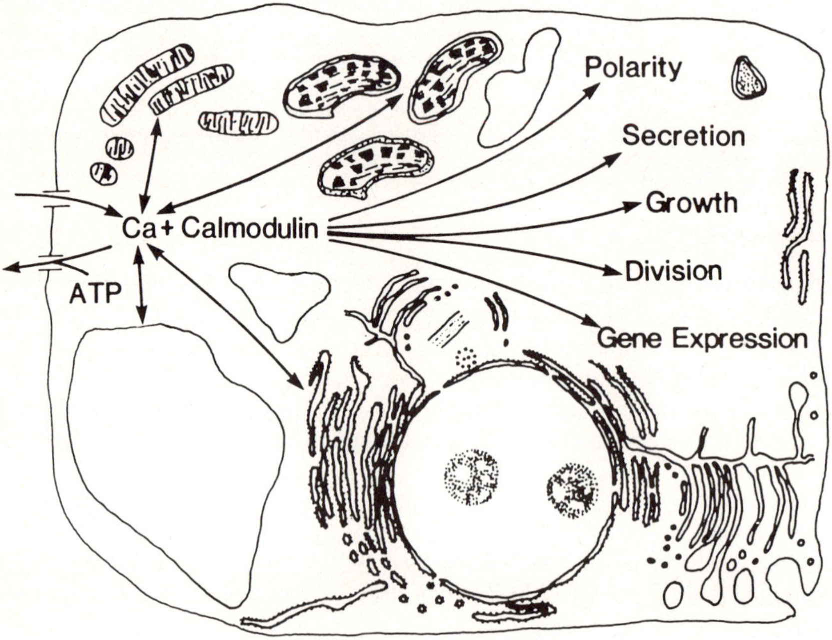

Fig. 1. A schematic drawing of a cell showing involvement of Ca and calmodulin in many of the basic cellular processes. From Trewavas (1986).

response elements of a developmental process (Berridge & Irvine 1984). Research in animal systems shows that agonists of development, working through "G" proteins, activate a phosphodiesterase or phospholipase-C that causes the hydrolysis of plasmalemma-bound phosphatidyl inositol-4,5-bisphosphate into two produces, inositol-1,4,5-trisphosphate (IP_3) and diacylglycerol (DG) (fig. 3). The former moves to the cytosol and stimulates release of Ca from internal stores – ER (Drobak & Ferguson 1985) or vacuoles (Schumaker & Sze 1987) – while the latter remains in the plasmalemma and activates the Ca-requiring enzyme, protein kinase-C or "C-kinase" (Nishizuka 1984). Both products lead to the phosphorylation of certain proteins and thus to the activation or amplification of specific reactions. Preliminary evidence indicates the presence of the PI pathway in plants and thus opens up exciting new opportunities for future research (Heim & Wagner 1987, Morré et al. 1984, Morse et al. 1986).

The contribution of Ca to plant development has been reviewed in recent publications (Hepler & Wayne 1985, Kauss 1987, Marmé & Dieter 1983, Roux 1983, Trewavas 1986), and therefore the current article will be more selective in its coverage. After a brief consideration of the properties and fitness of Ca as a second messenger I will discuss the role of the ion in two systems: cytokinin-induced bud formation in the moss *Funaria*, and phytochrome-stimulated spore germination in the fern *Onoclea*.

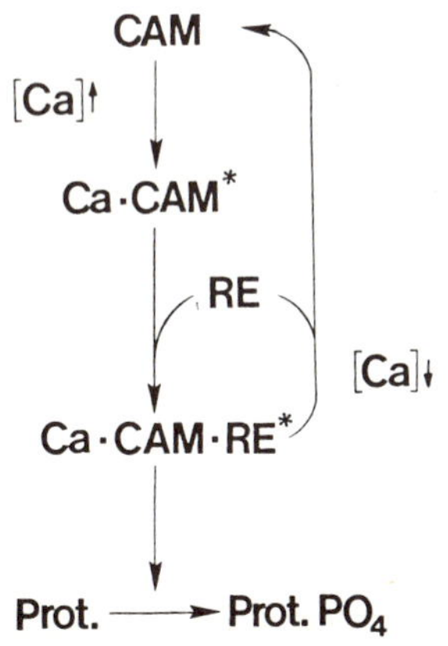

Fig. 2. Amplitude modulation. At elevated [Ca] the ion reacts with calmodulin (CAM) to form a complex that can activate a response element (RE) (kinase). At low [Ca] the ion dissociates from CAM and the system is inactivated.

Why Ca?

Of all the ions that participate in growth and development, Ca is the principal one involved in signal transduction. It is probably true that other ions could participate but owing to certain special conditions that have arisen during evolution, Ca has emerged as the primary ionic transmitter of information. Briefly stated, the special condition is its low intracellular free concentration relative to that outside the cell. Exactly why cells possess a low (0.1 μM) internal free [Ca] is not known but it has been speculated that the existence of

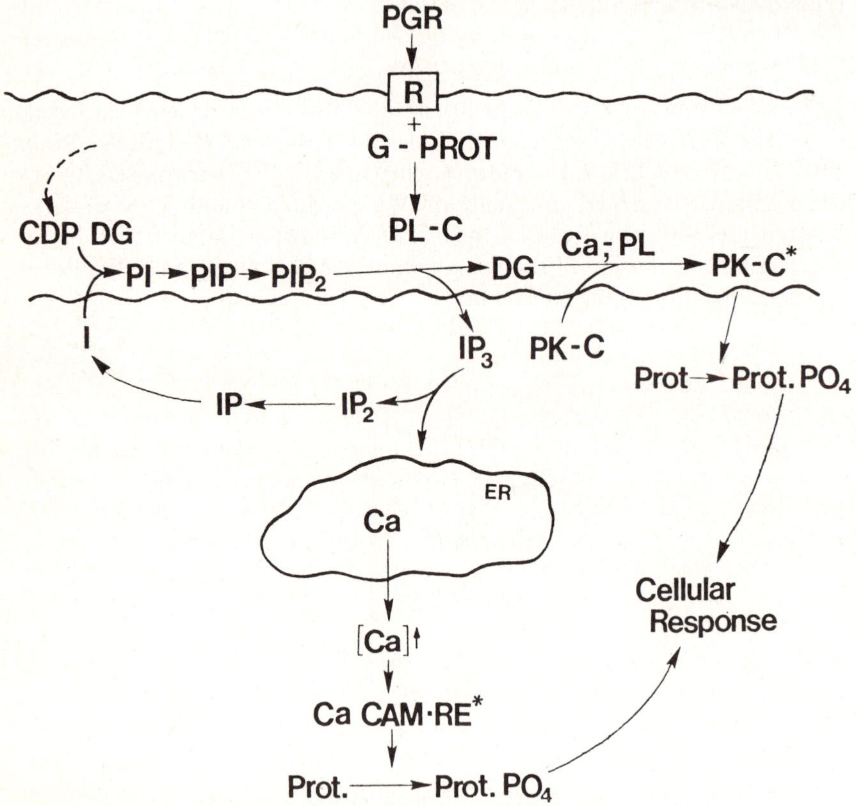

Fig. 3. The phosphatidyl inositol (PI) signal pathway. Adapted from Berridge & Irvine (1984). – DG = diacylglycerol; G-Prot = GTP-binding protein; IP_3 = inositol-1,4,5-trisphosphate; PGR = plant growth regulator; PIP_2 = phosphatidyl inositol-4,5-bisphosphate; PK-C = protein kinase-C; PL = phospholipids; PL-C = phospholipase-C; R = receptor protein.

phosphate-based energy metabolism created a situation that could not survive in the presence of a normal concentration of the ion (1 mM) (Kretsinger 1979). At these relatively high levels Ca would immediately react with phosphate and form a highly insoluble precipitate that would essentially destroy the energy currency of the cell. This potentially lethal condition thus may have provided the selective pressure for the cell to evolve a system that would efflux or sequester Ca to very low levels and render its interaction with phosphate insignificant. Cells do indeed possess a variety of mechanisms for reducing cytosolic [Ca] (Marmé & Dieter 1983, Sze 1985): firstly the plasmalemma is quite impermeable to Ca and secondly there are different energy-driven processes that pump Ca out of the cytoplasm against a steep (10,000–100,000x) concentration gradient.

The presence of the low internal [Ca] held against a steep gradient creates an unique situation that can be exploited by the cell for the purposes of information transfer. To increase the cytosolic [Ca] from 0.1 μM to 1.0 μM requires a relatively small number of ions that would have little impact on cellular metabolism. Yet the ten-fold increase in concentration carries with it an energy change that can be utilized as information. By comparison K, with a resting level at 0.1 M, would have to increase to 1.0 M to achieve an equivalent energy change, a concentration increase that would have enormously detrimental consequences to the metabolism of the cell.

The membranes that separate the cytosol from the enriched stores of Ca, *e.g.*, the plasmalemma, tonoplast, ER, etc., become key regulators of development because they control the influx of Ca. Although these membranes are generally quite impermeable to Ca they contain regulated or gated channels, some of which are able to conduct Ca (Reuter 1983). Largely from work on animal membrane systems we known that Ca channels most commonly open in response to a depolarization of the membrane potential and are thus called "voltage-operated channels". However Ca channels that respond to the binding of an agonist are also present and these are called "receptor-operated channels". Evidence from use of pharmacological agents suggests that the voltage-operated channels exist in plants (Saunders & Hepler 1983). Given the concentration and charge gradient tending to drive Ca into the cell, a substantial increase in intracellular [Ca] can be brought about through the transient opening of relatively few channels. The early steps of signal transmission therefore can be viewed in this context as a problem of understanding how a PGR causes the opening of Ca channels. For example, do PGRs induce a change in membrane potential and thus indirectly cause channel opening, do they bind directly to channel proteins and provoke an opening event through a conformational change, or do they invoke yet another messenger such as IP_3, which causes Ca release?

Even though the cell is poised to generate large, rapid transients in free [Ca] this does not explain why proteins such as calmodulin respond to it and not to magnesium (Mg), which is quite similar to Ca and which is present at much higher free concentration (1 mM). Two properties seem important in constructing binding domains that are far more selective for Ca than Mg: Ca is somewhat larger (99 pm *vs.* 65 pm), and has a much more flexible coordination (6–12) than Mg, which is held at 6 (Levine & Williams 1982, Williams 1975). Within proteins groups of neutral ethers, carbonyl oxygens and hydroxyl oxygens can form cages that will specifically accommodate the large size and flexible coordination of Ca. In a variety of Ca-binding proteins, studies show that these domains are similar and from their structure have been referred to as "E-F" hands (Kretsinger 1979). An additional property that adds to the favourable signalling function of Ca over Mg is its relatively weaker energy of hydration. The significance is that Ca can shed water much more quickly (10^9/sec) than Mg (10^5/sec) and as a consequence can control fast reactions, such as muscle contraction (Levine & Williams 1982).

Not only must calmodulin and other Ca-binding proteins respond to Ca but in the concentration range (0.1–1.0 μM) for which the changes appear to occur. The association constant for calmodulin fits these criteria; at the resting level the protein will lack Ca and be inactive, but at the elevated level its four binding sites become occupied and the molecule undergoes a confirmational shift that enables it to activate a response element (Allan & Hepler 1987). When the pumps return the [Ca] to the resting level the Ca-binding proteins lose their bound ion and also their ability to activate a response element.

Modulation of the signal

The above scheme provides a general mechanism for understanding Ca action. However, the question is often raised, how can the enormous variation of plant development be controlled by a single process? The complete answer is not known but enough information is available to tell us that considerable variation is built into the second messenger system. So far our discussion has focused on one aspect, namely positive amplitude modulation, wherein an elevation of the [Ca] activates a process. The elevation of Ca might, instead, inhibit a process. Cytoplasmic streaming is such an example of negative amplitude modulation wherein a [Ca] increase to 1 μM halts the process (Tazawa et al. 1987). An alternative mechanism referred to as "sensitivity modulation" (Rasmussen 1986) involves the changes in the response of protein to a given level of Ca, an event which may be brought about through phosphorylation and which may be either positive or negative in nature. For example, the phosphorylation of phosphorylase b kinase increases its affinity for Ca from 3 μM to 0.3 μM for half maximal activation, whereas the phosphorylation of myosin light chain kinase decreases its affinity for Ca from 0.8 μM to 8.0 μM (Rasmussen 1983).

The presence of yet another messenger system, namely the PI pathway (Berridge & Irvine 1984), superimposed upon the existing Ca pathways, greatly increases the range of variability and modulation in control mechanisms. On the one hand IP_3 causes Ca release and thereby activates Ca-calmodulin dependent kinases through a conventional amplitude modulation mechanism. On the other hand DG stimulates protein kinase C and its dependent phosphorylations through a sensitivity modulation process that is maximally active at resting levels of Ca (Nishizuka 1984). Experimental evidence from studies on serotonin release in platelets reveals that these two arms of the PI signal pathway act synergistically to support long term developmental events (Nishizuka 1984); IP_3-induced Ca release initiates the event while DG activation of protein kinase C permits the event to be sustained. One important feature of this dual control mechanism, especially for long term developmental processes, is that the free [Ca] need only be elevated for a short time. After the initiation phase, the [Ca] can return to the basal, nontoxic, level without inhibiting the progression of development (Rasmussen 1986).

Despite the variation that can arise through the different modulation pathways it seems unreasonable to imagine that the second messenger system can couple all stimuli and responses in a given cell. We must realize that cells are preprogrammed or destined to respond in a limited number of ways, and this places an upper boundary on the number of individual processes to be regulated. Temporal and spatial factors, for example, play an enormous role in determining the fate of a cell. However there is a subtle and crucial interplay between the genetically defined potential of a cell and its second messenger system, which enables it to respond quickly to signals from the outside.

With these general comments I turn our attention to the role of Ca in two specific systems, as follows: cytokinin-induced bud formation in *Funaria,* and phytochrome-stimulated spore germination in *Onoclea.*

Ca and bud formation in *Funaria*

Application of exogenous cytokinin to protonemata of the moss *Funaria* stimulates the formation of buds at the distal end of the target caulonema cells. The system has been known for years and is well characterized regarding its cytology and physiology (Bopp 1968, Brandes & Kende 1968). To explore the possible role of Ca we have been guided by three rules enunciated by Jaffe (1980) as follows:

- the agonist-stimulated developmental event should be preceded or accompanied by an increase in free [Ca];
- blockade of the [Ca] increase should inhibit agonist-induced development; and
- experimental imposition of a [Ca] increase should stimulate development in the absence of the natural agonist.

Satisfaction of the first rule poses the greatest problem owing to the difficulty of introducing Ca-sensitive reporting molecules and of measuring their Ca-dependent signal. The question, however, is of enormous importance. Indeed the concept of amplitude modulation hinges upon the existence of these transients in [Ca]; if they occur they must be demonstrated. Our approach for the moment has therefore been indirect and has used more accessible methods to follow presumed changes in free [Ca]. In particular we have explored the disposition and change in level of membrane-associated Ca as detected by chlortetracycline (CTC) fluorescence (Saunders & Hepler 1981). Since free Ca is probably derived from this compartment, the changes in fluorescence, which indicate changes in membrane-associated Ca, may be closely related to changes in free [Ca].

When cytokinin-induced caulonemata are stained with CTC results show that the fluorescence increases markedly in the bud site. The fluorescence is apparent well before the first asymmetric division and persists at a high level in the young buds. To ascertain whether the CTC fluorescence was due to a relative increase in Ca or more simply to an increase in total membrane we determined the fluorescence change during budding after staining with N-phenyl naphthalamine, a general membrane marker (Saunders & Hepler 1981). Although the total membrane in the bud site increases 1.5 fold the Ca-dependent CTC fluorescences increase 4 fold, indicating that the existing membranes become greatly enriched with associated Ca.

What do these observations mean and how do they relate to changes in free [Ca] in the bud site? Our first view was that Ca influx, derived from the extracellular wall space and regulated by the plasmalemma, caused the free Ca to become elevated (Saunders & Hepler 1981). Internal membrane organelles (*e.g.*, ER, mitochondria) responded to the increased free Ca by sequestering the ion and thereby causing the observed increase in CTC fluorescence (fig.

4a). An alternative view postulates that the CTC compartment is an intermediary between the plasmalemma and cytosol. In this view the ER forms junctional complexes with the plasmalemma that facilitate transport of extracellular Ca directly into ER cisternae (fig. 4b). Subsequent release of these cisternal stores causes an increase in free [Ca]. The latter idea gains support from our electron micrographs showing that ER-plasmalemma associations exist and increase early in development (Conrad et al. 1986), and also by the pharmacological studies showing that both extracellular (EGTA, D-600, La) and intracellular (TMB-8) inhibitors of Ca release block bud formation (Saunder & Hepler 1983). Evidence from other systems reveals that cortical ER complexes may participate in Ca fluxes during development. For example the ability of *Xenopus* oocytes to respond to activation is correlated with the formation of a cortical system of smooth ER that is spatially close to the plasmalemma and to the cortical granules (Charbonneau & Grey 1984). Upon activation and exocytosis of the granules the distinct ER-plasmalemma relationship is lost. The system thus appears to poise itself for a Ca release but

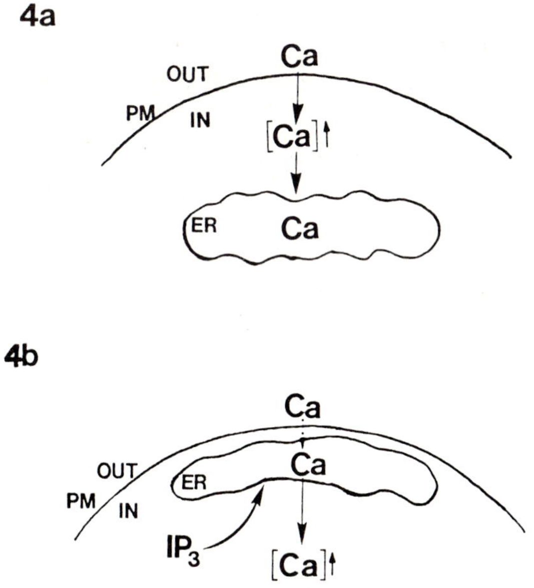

Fig. 4. Two pathways for Ca influx into *Funaria.* In **4a** the suggestion is made that Ca moves into the cytoplasm from the outside and is subsequently sequestered by the ER. **4b** suggests, in contrast, that the ER forms junctional complexes with the plasma membrane (PM) which allow Ca to move directly from the outside to the ER cisternal space. Release by IP_3 accounts for an elevation of free Ca. Adapted from Putney (1986).

after that event the structural elements disorganize. Another example is found in studies of parotid gland tissue in which flux experiments show that the intracellular pools depend almost entirely on extracellular Ca and further that refilling of the pool from the extracellular space occurs rapidly without an apparent increase in free intracellular Ca (Putney 1986). The ion thus appears to be transported directly from the outside of the cell to an internal IP_3-sensitive pool, presumably the ER cisternae. The elucidation of the Ca pathway as well as the kinetics and location of changes in its free concentration will provide valuable information about signal transduction.

Whereas we have not succeeded in directly demonstrating the existence of the presumed transients in free [Ca] we have been able, with the use of pharmacological agents, to block and enhance development in *Funaria* according to Jaffe's guidelines. Thus either removal of Ca from the cell wall and medium or the use of agents that block Ca entry, including EGTA, La, verapamil, and D-600, prevent cytokinin-induced bud formation (Saunders & Hepler 1983). These agents all appear to act at the plasmalemma and thus the results suggest that activator Ca is derived from the extracellular space. However, as previously mentioned, TMB-8, an agent that inhibits Ca release from internal compartments (Chiou & Malagodi 1975), also blocks development.

The experimental imposition of a Ca influx into caulonemata of *Funaria* provides exciting evidence in support of Ca as crucial intermediary in bud development. Using the ionophore A-23187 plus Ca we showed that bud initials formed on every target cell, even in the absence of applied cytokinin (Saunders & Hepler 1982). However these initials seldom formed complete buds; usually they reverted to branches although subsequent application of cytokinin prior to the reversion would induce the initials to complete the budding process. More than Ca entry is needed to give rise to mature buds.

Additional work on this problem has used two different dihydropyridine channel effectors and provides evidence that Ca entry may be controlled by voltage-operated channels on the plasmalemma (Conrad & Hepler 1986a). One of the dihydropyridines, referred to by number, 202–791, exists as stereoisomers in which the (+) form is Ca channel agonist while the (−) form is an antagonist. The results dramatically show that (−)202–791 blocks cytokinin-induced bud formation while (+)202–791 stimulates initiation even in the absence of cytokinin. Here too, as with A-23187, the experimentally induced initials fail to form complete buds. Yet another dihydropyridine Ca channel antagonist, nifedipine, has the useful property that it can be inactivated by irradiation with 365 nm light. Thus cytokinin-induced caulonemata treated with nifedipine fail to form buds. If, however, they are subsequently irradiated, budding will resume (Conrad & Hepler 1986a).

The results support the conclusion that Ca influx is controlled by voltage-operated channels on the plasmalemma. Whereas A-23187 intercalates into the membrane and creates its own avenues of Ca entry, the dihydropyridines preumably act upon existing channels and modify their function. The fact that caulonemata treated with either A-23187 or (+)202-791 form bud initials in the correct position suggests that the polarity of the target cell response is not dependent on the location of Ca entry, rather it seems that the caulonemata are already polarized and that the Ca influx sets in motion a developmental program that contains the crucial spatial information (Conrad & Hepler, in preparation). If voltage-operated channels control Ca entry, by comparison with other systems, one would predict that cytokinin causes a depolarization of the membrane potential, a matter that can be resolved experimentally. The studies continue to emphasize that Ca entry alone is insufficient to support complete bud formation; cytokinin activates other pathways that are essential to development although it should be emphasized that Ca must be available at all stage sof the budding process.

Recent preliminary evidence indicates that the PI pathway operates in *Funaria* (Conrad & Hepler 1986b). Budding can be reduced by culture in lithium (Li), an agent that prevents the dephosphorylation of inositol monophosphate to inositol and thus blocks the cyclic pathway eventually leading to IP_3 production (Berridge & Irvine 1984). The inhibition can be overcome by adding inositol, which in *Funaria* restores bud formation. Evidence for the DG arm of the PI signal pathways has been obtained from studies using tumor promoting agents, the phorbol esters as artificial stimulators of protein kinase C. Although Ca influx alone will not support bud maturation, if A-23187 plus Ca is given together with the phorbol ester, TPA, a marked increase in complete buds is observed (Conrad & Hepler, unpublished observations). The results tentatively support the idea that Ca influx, perhaps partly under the regulation of IP_3, initiates a new developmental programme, while activation of protein kinase C by DG induces reaction cascades that sustain the initiating event and allow development to continue.

Ca and spore germination in *Onoclea*

Dark grown spores of the fern *Onoclea* can be stimulated to germinate by brief irradiation with red light (Wayne & Hepler 1984). Far-red immediately following the red irradiation returns germination to the level of dark controls and provides evidence that the photoreceptor is phytochrome. That Ca participates in the process can be shown by simply preincubating the spores in

EGTA to remove wall-bound Ca, followed by culture in media in which the ion concentration is carefully regluated. Below 1 μM Ca spores fail to germinate but as the extracellular concentration rises so does germination,

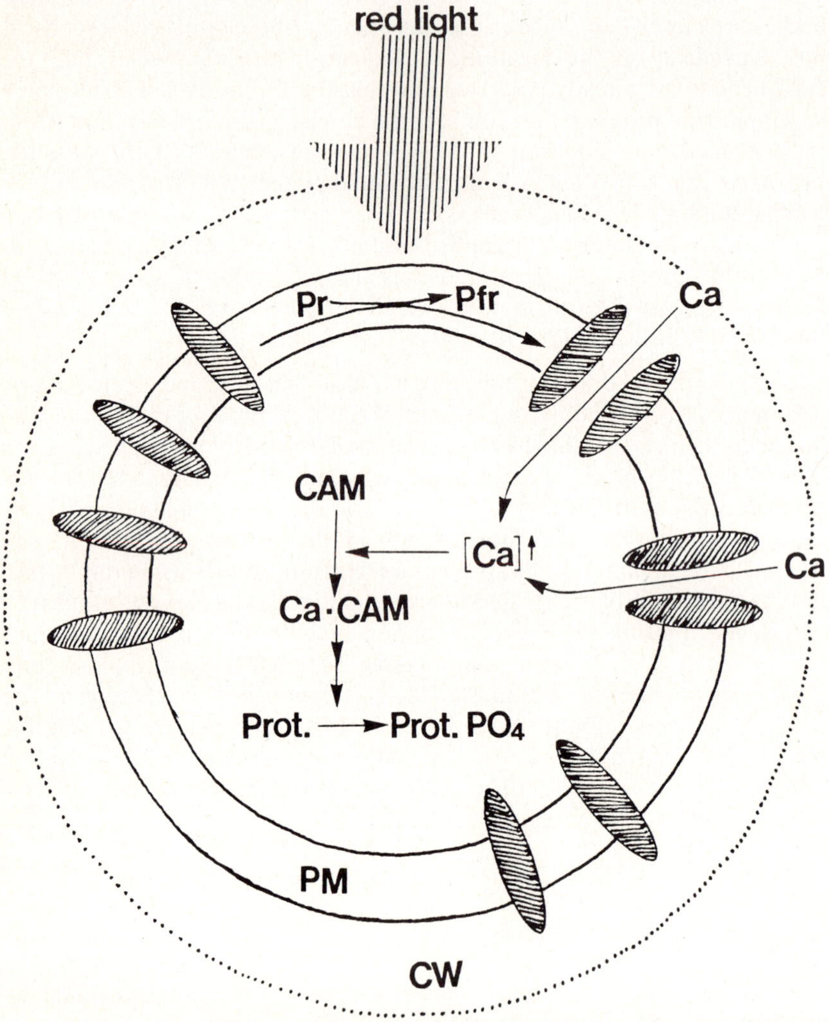

Fig. 5. A scheme to account for the mechanism of phytochrome-induced Ca influx. Red light converts Pr to Pfr. Pfr poises a Ca channel, which can remain in an open state after Pfr has been destroyed or decayed back to Pr. Ca influx, through amplitude modulation (fig. 2), induces many pathways necessary for spore germination.

reaching a maximum at 10μM. If we experimentally block movement of Ca into the cell using La, germination is also inhibited (Wayne & Hepler 1984).

Although we have not been able to examine free Ca due to problems discussed earlier we have measured total Ca by atomic absorption spectroscopy. We have found that the spores grown under conditions which support germination, *e.g.*, red light, show a marked increase in total Ca while those cultured under inhibitory conditions, *e.g.*, complete darkness, red followed by far-red, incubation in La, fail to demonstrate an increase in [Ca] (Wayne & Hepler 1985). The results further show that the rate of Ca increase is greatly accelerated immediately following red irradiation and falls back to steady state within 10 minutes. Additional support for the idea that Ca influx is rapid and immediate upon irradiation is the observation that La, added only five minutes after the beginning of red light, has no effect either on Ca uptake or germination (Wayne & Hepler 1985). Spore germination in *Onoclea* thus contrasts sharply with bud formation in *Funaria* by utilizing a Ca influx system that responds quickly and completely to the developmental stimulus.

An unexpected discovery from these studies has been that the phytochrome-Ca transport system possesses memory. Whereas far-red reversal of Pfr is lost within a few minutes, the Ca transport system, once activated through irradiation, can remain poised in the dark for hours (Wayne & Hepler 1984). To demonstrate this phenomenon we first depleted spores of wall-bound Ca and cultured them in a Ca-free medium. Under these conditions they were irradiated for 5 min with red and returned to the Ca-free medium in the dark. Spores that remain under these conditions will not germinate. However if spores are simply transferred to a Ca-containing medium still in the dark, even up to 8 hours after the initial red irradiation, they will display maximal germination. By way of explanation we (Wayne & Hepler 1984, 1985) suggest the following scenario. Red light transforms Pr to Pfr. Pfr ist able to facilitate the opening of Ca channels on the PM (through phosphorylation?). In the dark Pfr rather quickly degrades or reverts back to Pr, but the Ca channels remain open and only decay to a closed state after several hours (fig. 5). Thus, while Pfr appears to open the Ca uptake system, it becomes uncoupled from that system.

Conclusions

The above discussion of *Funaria* and *Onoclea* provides only a brief glimpse at the role of Ca as a second messenger in plant development. We are still quite uninformed about most systems but our knowledge is increasing rapidly as new Ca-mediated processes are uncovered and described. Among the questions that deserve attention the most pressing concerns the experimental detection and measurement of the hypothetical Ca transients that are required for amplitude modulation. If they exist, then where do they occur, how long to they last, and to what magnitude do they reach? What are the sources of the activator Ca, *e.g.*, the wall-PM, ER, vacuolate-tonoplast? What are the response elements? and what pathway do they activate or inactivate? Finally, what is the role of the PI cycle as a modulator-regulator of Ca? These are but a few the questions that we can now address, and whose answers may greatly elucidate some of the long standing questions concerning the mode of action of PGRs.

Acknowledgements

Special thanks go to three former graduate students, Patricia Conrad, Mary Jane Saunders and Randy Wayne, who carried out the studies on *Funaria* and *Onoclea* summarized herein. I also thank Anna Hepler and Susan Lancelle for help in the preparation of this manuscript. This work has been supported by NSF grants PCM: 84-02414 and DCB: 87-02057.

References

Allan, E. & Hepler, P. K. 1987: Calmodulin and calcium-binding proteins. – In: Marcus, A. (ed.), The biochemistry of plants, 9. – Academic Press, New York (in press).

Berridge, M. J. & Irvine, R. F. 1984: Inositol trisphosphate, a novel second messenger in cellular signal transduction. – Nature **312:** 315–231.

Bopp, M. 1968: Control of differentiation in fern-allies and bryophytes. – Annual Rev. Pl. Physiol. **19:** 361–380.

Brandes, H. & Kende, H. 1968: Studies on cytokinin-controlled bud formation in moss protonemata. – Pl. Physiol. **43:** 827–837.

Charbonneau, M. & Grey, R. D. 1984: The onset of activation responsiveness during maturation coincides with the formation of the cortical endoplasmic reticulum in oocytes of *Xenopus laevis.* – Developm. Biol. **102:** 90–97.

Chiou, C. Y. & Malagodi, C. M. 1975: Studies of the mechanism or action of a new Ca^{2+} antagonist 8-(N,N diethylamino)octyl 3,4,5-trimethoxy-benzoate hydrochloride in smooth and skeletal muscles. – Brit. J. Pharmacol. **53:** 279–285.

Conrad, P. A. & Hepler, P. K. 1986a: The action of dihydropyridines on cytokinin-induced bud formation in the moss *Funaria.* – J. Cell Biol. **103:** 454a.

– 1986b: The PI cycle and cytokinin-induced bud formation in *Funaria.* – Pl. Physiol. **80:** 60.

–, Steucek, G. L. & Hepler, P. K. 1986: Bud formation in *Funaria:* organelle redistribution following cytokinin treatment. – Protoplasma **131:** 211–223.

Drobak, B. K. & Ferguson, I. B. 1985: Release of Ca^{2+} from plant hypocotyl microsomes by inositol-1,4,5-trisphosphate. – Biochem. Biophys. Res. Commun **130:** 1241–1246.

Heim, S. & Wagner, K. G. 1987: Enzyme activation of the phosphatidyl-inositol cycle during growth of suspension cultured plant cells. – Pl. Sci. **49:** 167–173.

Hepler, P. K. & Wayne, R. 1985: Calcium and plant cell development. – Annual Rev. Pl. Physiol. **36:** 397–439.

Jaffe, L. F. 1980: Calcium explosions as triggers of development. – Ann. New York Acad. Sci. **339:** 86–101.

Kauss, H. 1987: Some aspects of calcium-dependent regulation in plant metabolism. – Annual Rev. Pl. Physiol. **38:** 47–72.

Kretsinger, R. H. 1979: The information role of calcium in the cytosol. – Advances Cyclic Nuc. Res. **11:** 1–26.

Levine, B. A. & Williams, R. J. P. 1982: The chemistry of calcium ion and its biological relevance. – In: Anghileri, L. J. & Tuffet-Anghileri, A. M. (eds.), The role of calcium in biological systems, **1:** 3–26. – CRC Press, Boca Raton, Florida.

Marmé, D. & Dieter, P. 1983: Role of Ca^{2+} and calmodulin in plants. – Calcium & Cell Function **4:** 263–311.

Morré, D. J., Gripshover, B., Monroe, A. & Morré, J. T. 1984: Phosphatidyl inositol turnover in isolated soybean membrane stimulated by the synthetic growth hormone, 2,4-dichlorophenoxyacetic acid. – J. Biol. Chem. **259:** 15364–15368.

Morse, M. J., Crain, R. C. & Satter, R. L. 1986: Phosphatidyl inositol turnover in *Samanea* pulvini: a mechanism of phototransduction. – Pl. Physiol. (Lancaster) **80,** Suppl.: 92.

Nishizuka, Y. 1984: The role of protein kinase C in cell surface signal transduction and tumour promotion. – Nature **308:** 693–698.

Putney, J. W. 1986: A model for receptor-regulated calcium entry. – Cell Calcium **7:** 1–12.

Ranjeva, R. & Boudet, A. M. 1987: Phosphorylation of proteins in plants: regulatory effects and potential involvement in stimulus/response coupling. – Annual Rev. Pl. Physiol. **38:** 73–93.

Rasmussen, H. 1983: Pathways of amplitude and sensitivity modulation in the calcium messenger system. – In: Cheung, W. Y. (ed.), Calcium and cell function **4:** 1–61. Academic Press, New York.

– 1986: The calcium messenger system. Parts 1 & 2. – New England J. Med. **314:** 1094–1101, 1164–1170.

Reuter, H. 1983: Calcium channel modulation by neurotransmitters, enzymes and drugs. – Nature **301:** 569–574.

Roux, S. J. 1983: A possible role for Ca^{2+} in mediating phytochrome responses. – Symp. Soc. Exp. Biol. **36:** 561–580.

Saunders, M. J. & Hepler, P. K. 1981: Localization of membrane-associated calcium following cytokinin treatment in *Funaria* using chlorotetracycline. – Planta **152:** 272–281.

– 1982: Calcium ionophore A23187 stimulates cytokinin-like mitosis in *Funaria.* – Science **217:** 943–945.

– 1983: Calcium antagonists and calmodulin inhibitors block cytokinin-induced bud formation in *Funaria.* – Developm. Biol. **99:** 41–49.

Schumaker, K. C. & Sze, H. 1987: Inositol 1,4,5-trisphosphate releases Ca^{2+} from vacuolar membrane vesicles of oat roots. – J. Biol. Chem. **262:** 3944–3946.

Sze, H. 1985: H^+-Translocating ATPases: advances using membrane vesicles. – Annual Rev. Pl. Physiol. **36:** 175–208.

Tazawa, M., Shimmen, T. & Mimura, T. 1987: Membrane control in the *Characeae.* – Annual Rev. Pl. Physiol. **38:** 95–117.

Theologis, A. 1986: Rapide gene regulation by auxin. Annual Rev. Pl. Physiol. **37:** 407–438.

Trewavas, A. J. (ed.) 1986: Molecular and cellular aspects of calcium in plant development. – Plenum Press, New York.

Wayne, R. & Hepler, P. K. 1984: The role of calcium ions in phytochrome- mediated germination of spores of *Onoclea sensibilis* L. – Planta **160:** 12–20.

– 1985: Red light stimulates an increase in intracellular calcium in spores of *Onoclea sensibilis* L. – Pl. Physiol. (Lancaster) **77:** 8–11.

Williams, R. J. P. 1975: The binding of metal ions to membranes and its consequences. – In: Parsons, D. S. (ed.), Biological membranes: 106–121. – Oxford University Press, Clarendon.

Address of the author: Professor Peter K. Hepler, Botany Department, University of Massachusetts, Amherst, MA 01003, USA.

Internal controls of plant morphogenesis

T. Sachs

Abstract

Sachs, T. 1988: Internal controls of plant morphogenesis. – In: Greuter, W. & Zimmer, B. (eds.): Proceedings of the XIV International Botanical Congress: 241–260. – Koeltz, Königstein/Taunus.

Major experimental facts of plant development suggest the following principles which constrain theoretical patterning possibilities. [a] Intracellular determination limits cell competence and serves as a quantitative memory. This memory allows the accumulation of gradual changes. Intracellular programmes in which one event leads to another, however, do not have an important role in the determination of plant form. [b] Developing apices are sources of spatial signals. These orient differentiation and transport, inhibit similar primordial development and enhance the formation of other organs. [c] Known phytohormones, auxins and cytokinins, serve as important spatial signals. [d] Positive feedback relations involving phytohormones account for the disrupted structure of tumors and for the regeneration of form in severely damaged shoots. It is concluded that patterning is governed by quantitative rules of cellular responses to relatively unspecific signals, not by cellular programmes elicited by precise chemical pre-patterns. Ontogeny appears to involve epigenetic selection: at both the organ and cellular levels there is an excess developmental capacity and structures prevail according to their participation in gradual feedback relations.

> *. . . we know sufficiently accurately that all organic formation in which protoplasm is concerned is conditioned by internal factors. In this domain lie the genuine problems of developmental physiology, from any insight into which we are as yet entirely excluded.* *L. Jost (1907).*

> *. . . I am a great believer in saying familiar, well known things backwards and inside out, hoping that from some new vantage point the old facts will take on a deeper significance.* *J. T. Bonner (1958).*

The problem and purposes of this article

Single cells, zygotes, undergo development and become complex, multicellular plants or animals. Though this development can be influenced by environmental cues, the information supplied by the environment is general rather than detailed. The complex cellular patterns could not, therefore, be only a reflection of environmental patterns. It follows that *biological form is specified by information or instructions present in single cells.* The general problem to be considered here concerns this internal specification of what is known as “pattern formation“ (Wolpert 1971, Malacinski 1984): what is the nature of this specification of the time and location of cellular changes and how does it lead to microscopic and macroscopic form.

Genetic differences, such as the ones used by Mendel, show that at least part of this specification of pattern is hereditary and depends on DNA and the structure of proteins. Molecules can undergo considerable self-assembly, but it is not kown how the structure of macromolecules could determine the structure of organisms. The mechanisms of gene regulation, which have been a central topic of developmental biology, could account for differences between cells – but it is not known how or whether gene regulation could account for the patterned location of different cells. Gene regulation, furthermore, occurs even where there is no patterned development: this is seen in crown gall and other tumors, where there is growth and differentiation that are not limited in location, orientation or duration.

The general problem of pattern formation is theoretically challenging because new concepts could be required so as the understand the apparent increase in complexity that occurs during biological development. At present there is no lack of models or “theories“ that could account for pattern formation (Brenner et al. 1981, Malacinski 1984). These various hypotheses generally assume that signals for differentiation move by diffusion, that cells respond to concentrations of substances at critical times, and that there are critical chemical reactions between few molecules. These chemical and physical processes could set up a pre-pattern or a grid of “positional information“ of chemical signals to which the cells later respond by specific differentiation (Wolpert 1971). The information specified by the pre-pattern is often assumed to be in the form of precise chemical concentration. These hypotheses and the assumptions on which they are based are certainly plausible. These assumptions are, however, quite arbitrary: active transport of signals between cells is possible, and cells that receive developmental signals may change these signals rather than passively responding to the conditions they are in.

Advances in the field of pattern formation have been experimental as well as theoretical. In plants, detailed knowledge is available concerning cellular events during normal development. There is also information concerning the regenerative capacity of various tissues. Organs, tissues and isolated cells can be grown in culture on chemically defined media and, in many cases, they can be induced to form new plants. The precise molecular causes for the disruption of organization in crown gall tumors are now known – the unregulated formation of phytohormones of only two groups, auxins and cytokinins (Braun 1956, Hooykaas et al. 1982). The problem to be considered here is, therefore, whether or *how hypotheses of patterning should be modified or changed in view of the facts that are now available.*

Plants have characteristic that make them suitable for the experimental and theoretical study of the general biological problem of pattern formation (Sachs 1978a). Though the forms of plants are undoubtedly complex and magnificent, their structures are relatively simple. This structure, furthermore, develops without any movement of the cells relative to one another (Sinnott 1960). The advantages of plants also include their 'continued embryology': all developmental processes, including organogenesis, continue in meristems of large, vigorous plants. Finally, plant regeneration makes many experimental manipulations possible. These advantages were the basis of important botanical contributions to biological patterning in the last century; Vöchting (1892), for example, was one of the originators of the concepts of cell determination and cell polarity and Jost (1893) demonstrated the induction of cell differentiation by neighbouring tissues.

The roles of intracellular determination

Controls of developmental events – of local gene action – can be ot two types: intercellular and intracellular. The first must depend on signals from sorrounding tissues and may therefore be named spatial controls. Intracellular factors would depend on the developmental history of the cells and may therefore be named temporal controls (Sachs 1978a, Lyndon 1979). A hypothetical organism in which temporal controls have no role can be used to illustrate their importance: in such an organism all differentiation states would be always available and the appropriate one would have to be constantly specified by spatial signals. This entire organism, including its mature tissues, would be one active "morphogenetic field". Considering the size and complexity of plants, this appears unlikely or impossible.

It is thus necessary to seek generalizations concerning the role of temporal controls in determining patterned development. Processes in which temporal controls play a dominant role could be expected to follow an invarible course, one that would continue even when cells or organs that could have spatial effects are removed or grafted at unusual locations. Complete isolation from spatial signals, furthermore, is achieved in tissue and cell cultures. A survey of the many experiments available suggests the following three generalizations.

Solely intracellular programmes do not determine major patterns

Though plants have remarkable abilities of regeneration, when they are cut there is generally no continued development of the tissues close to the wound (fig. 1). This is true even when the isolated regions are large, so it is unlikely to be due only to wound effects. Normal development is autonomous and continues only in meristematic apices of the shoots and roots, the centres of organogenesis. Other tissues do not continue their development, with the exception of some elongation and a "maturation" expressed by an enlargement of cells and their vacuoles. It may be concluded that processes of differentiation are not determined only by intracellular programmes: they do not continue when conditions, and presumably the configuration of spatial signals, are changed.

A limited role of intracellular programmes that would act alone is also indicated by the course ot normal, unperturbed development. In seed plants, at least, there is good evidence that the cell lineages leading to the formation of given mature structures are quite variable (Stewart & Derman 1975, Sachs 1986). For example, even the development of neighbouring stomata is variable in the number of cell divisions and their relative orientations (Sachs 1978b). Processes typical of stomata formation, furthermore, are sometimes reversed in mid-course and regular epidermal cells are formed.

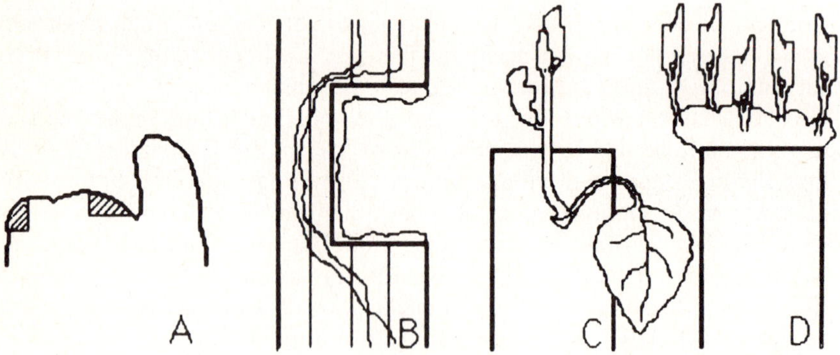

Fig. 1. Major types of regeneration in shoots (similar phenomena are knwon for roots). The facts demonstrate that the direct replacement of removed parts by tissues present near the wound is quite limited. – **A,** Removed tissues are replaced locally only when they are taken from promeristems or very small, and thus young, leaf primordia. **B,** Damaged vascular tissues are also replaced by a redifferentiation of tissues, forming new channels around a wound. **C,** Removed shoots are commonly replaced indirectly, by developmental changes in lateral buds. This is a most common form of regeneration. **D,** When no apices are present, removed shoots may be replaced by new, adventitious apices. These apices appear on callus (as shown) or they may be formed by organized tissues.

The developmental competence of cells at any given time is limited

The regeneration potential of plants has led to the impression that the developmental competence of their cells is virtually unlimited. The remarkable regeneration of plants, however, is indirect – entire organ systems are replaced by changes in the growth of the remaining immature organs or, as a last resort, adventitious apices. The growth of meristematic apices, whether adventitous or not, leads indirectly to the formation of all cell and tissue types – new meristems must be formed before most types of regeneration can take place. The totipotentiality of plant cells is thus real, but it does not contradict the limitation of the differentiation possibilities open to any plant cell.

The direct replacement of missing party by divisions of neighbouring cells occurs within the promeristematic tips, where it involves no knwon changes of cell differentiation. The only other direct regeneration is in the vascular system: new tracheary and sieve elements can be formed by the redifferentiation of parenchyms cells (Jacobs 1952, 1970, Sachs 1981). Related, earlier stages of the same process may be other changes of the orientation of cells, expressed by the direction of growth, cell division and the transport of auxin (Sachs 1984, Gersani & Sachs 1984). A change or loss of dominant orientation is characteristic of the callus that forms on the surface of wounds and whose growth can be unlimited in culture.

Thus at any given time plant cells have a restricted developmental competence. It is limited to continuing normal development at different rates, undergoing reorientation and vasalar loss of orientation with continued division as a callus. There is, finally, the possibility of a large change leading to differentiation and a new, adventitious apex of a root, a shoot or, rarely, to structures resembling normal embryos ("embryoids"). This last change is an extreme differentiation jump, from mature cells to a promeristematic or embryonic state.

There is a cumulative "memory" of past differentiation

The growth of callus has been ascribed to a loss of differentiation, and thus of all temporal controls. However, even in culture the form, rate of growth and requirements from the medium of callus depend on the original differentiation of the cells, when they were part of the intact plant (Wareing & Al-Chalabi 1985, Meins 1986). The traits maintained in prolonged culture even include specific antigen characteristics of tissues (Meins 1986). It may be concluded that the differentiated state of plant cells can be determined: it can be stable or maintained through many if not unlimited divisions, even when the surrounding tissues and the environmental conditions are changed. The states that are maintained are expressed by quantitative traits, such as rate of cytokinin formation (Meins 1986), and not necessarily by all-or-none expressions of differention.

These three generalizations lead to important suggestions concerning the determination of plant form. The spatial signals need not specify a choice between many possibilities, but any small change they induce can be expected to become determined and thus influence further differentiation. This means that temporal controls – intracellular determination – are more than a source of developmental stability: they provide a quantitative "memory" of past development so that small changes can accumulate. It may be suggested, therefore, that overt differentiation depends on neither intracellular programmes nor momentary spatial signals. Instead, development is a response to a continued interplay between the developmental past of the cells and the signals they recieve from their surroundings, a response to the *integrated effect of spatial and temporal signals* during a relatively extended period.

There is physiological and biochemical evidence that would appear to contradict this conclusion. Momentary environmental conditions, such as short exposure to red light, do determine the differentiation of plants. These are cases, however, where plants, conditions and parameters measured have been carefully selected because of their value for research on specific mechanisms. There is also evidence that the application of substances such as auxin results in almost immediate cellular responses, measured as both growth and the synthesis of new gene products (Theologis 1986). In view of the discussion above, it may be suggested that much of the differentiation that plays a role in the organization of the plant can become determined and may respond much more slowly. Distinctions between rapid and gradual responses, between determinate and labile changes, may be difficult and this subject must await additional data.

Fig. 2. Various spatial effects of developing organs. Comparisons of four pairs: in A–B apices were present on the right and removed on the left, in C–F, the reverse is true. – **A, B,** The development of a lateral branch on a cut stem induces the differentation of vascular contacts along new orientations. The same lateral branch may also induce the abscission of the axial tissues that do not connect it with the roots. **C, D,** The presence of an intact organ reduces the growth of similar, though lateral, organs (apical dominance). Intact shoots may also prevent degeneration and, in some plants, even abscission, of axial tissues along their direct contact with the roots. **E, F,** The presence of an intact organ induces the differentation of lateral organs (rather than preventing their development, as in C). In the illustration, the laterals become plagiogeotropic organs. **G, H,** The presence of growing buds on a cutting increases the initiation of roots. The corresponding induction of shoots by roots also occurs, though it is not shown. ▷

Spatial signals of developing apices

The previous section indicated that spatial signals need specify only few developmental possibilities at any given time. It is now necessary to consider what is known concerning the spatial signals, such as where they originate

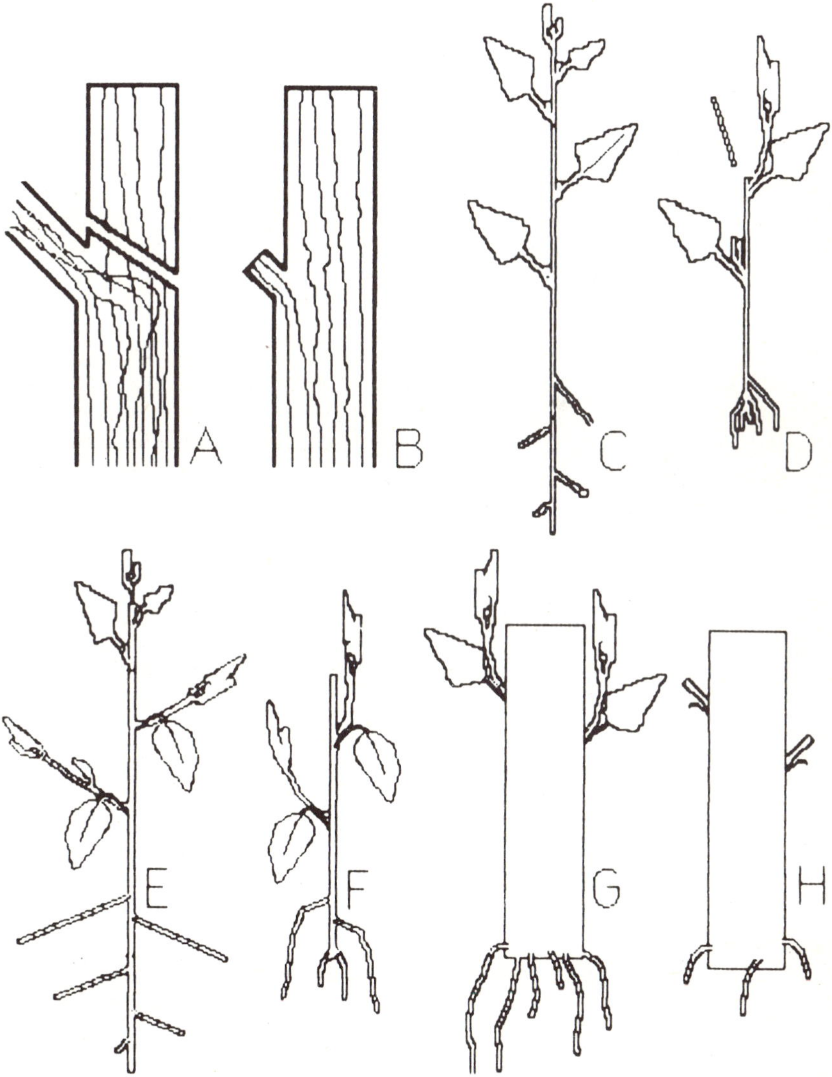

and the distances over which they act. The most pronounced spatial effects on development are a function of the growth and location of developing apices on the plant (Sachs 1986). The influence of a growing apex is evident during normal development: the growth of an apex is associated with the oriented differentiation of vascular tissues that connect it with the rest of the plant and with the correlated development of other apices. The spatial effects of apices are also evident when apices are removed, when they are grafted and, finally, when they develop in unusual locations, either lateral – or adventitious. It thus appears that major, though not necessarily unique, spatial signals originate in developing apices. The available information allows for the following generalizations concerning the effects of these siganals (fig. 2, Sachs 1986).

Induction of vascular differentiation

The presence of an organ is correlated with the differentiation of vascular tissues that connect it with the rest of the plant. Removal and grafting experiments show that this correlation includes an actual induction of differentiation by growing apices (Jost 1893, Jacobs 1952, Sachs 1981). This statement does not contradict the possibility of a reciprocal effect, of vascular differentiation on organ development (Larson 1975).

Maintenance of axial tissues

When an organ is removed the axis of the plant leading to it tends to wither, and, depending on the plant, even to abscind. These processes are inhibited or delayed in axial tissues connecting a growing apex with the rest of the plant (Warren Wilson et al. 1986).

Apical dominance

The presence of an organ reduces the formation of similar organs by the plant. The most common expression of this principles is the demonstration of apical dominance – the increase in bud growth and in the initiation of adventitious shoot apices when growing shoot apices are removed (Thimann & Skoog 1934, Sachs & Thimann 1967).

Induction of apical differentiation

The presence of an apex tends to cause similar apices to differentiate as specialized organs. This is another expression of the inhibition mentioned above, an expression that does not occur in all plants. Examples are the plagiogeotropic growth of lateral shoots and roots (Snow 1945) and the differentiation of lateral buds as storage organs (Woolley & Wareing 1972), thorns and flowers (Umrath 1948).

Promotion of complementary apical initiation

Young, growing shoot tissues promote the initiation and primordial growth of roots and vice versa. This correlative effect is not always obvious when size parameters are measured. The correlations of different organs are apparent, however, in observations of organogenic processes – it is root initiation rather than root growth that is correlated with shoot development (Keeble et al. 1930).

The removal or grafting of apices involve major wounds. Could additional conclusions be suggested by results of more limited structural disruptions? Local wounds often result in the reorientation of differentiation events, expressed most clearly by vascular regeneration. These wounds also result in the growth of callus and in the initiation of news apices. Experiments in which plants were both wounded locally and their growing apices removed, however, indicate that much of the effect of wounds could be due to their interrupting or diverting signals that could come from these apices. There have been virtually no experiments in which specific cells were damaged and the developmental consequences of this damage were observed directly. There is also a technical difficulty: at this scale, the effect of the wounds themselves may be larger than that of removed or damaged cells. Thus only limited conclusions are possible concerning local developmental interactions between cells (Sachs 1986). This must reflect poor knowledge, since the topic has hardly been studied, but if cells had large, local effects on one another, major additional changes of differentation could be expected in the vicinity of large wounds.

The spatial signals of developing apices are not similar to the specific controls that are commonly assumed to determine developmental events. Thus *the origin of the spatial signals is not localized.* Selective removal of various parts of shoot tips shows that the origin of correlative signals is not limited to any one primodium or tissue (Snow 1929). Furthermore, though removing rapidly developing tissues has the largest effect on the rest of the plant, the distinction between the influence of young and mature tissues is only quantitative. *The effects of the signals are not limited to distinct distances* and they are certainly not transported only by diffusion. The growth of a new bud in spring can be associated with the formation of new vessels along the entire trunk of large trees (Tepper & Hollis 1967). This is a dramatic demonstration of the action of spatial signals over distances of many metres. At the same time, the influence of the new bud is expressed at very short distances, between individual cells whose differentation must be coordinated if the vessel is to be functional. All the effects of apices on the plant are *quantitative, rather than all-or-none.* This quantitative response is seen in the size of newly formed vascular systems, the degress of apical inhibition and the number of developing complementary apices – roots in response to shoot tissues and vice versa. *The response to the spatial signals is not limited to tissues of any one type or age.* Vascular differentiation and apical initiation occur most readily in young meristematic tissues. Especially when these most competent tissues are absent, however, the signals may elicit the same differentiation even in mature parenchyma (Sachs 1981). The various spatial responses are also *not specific to any stage or individual developmental process.* Thus there is no clear seperation between the induction of the initiation of new apices and their continued primordial development. The effects of a developing apex act, furthermore, on all aspects of vascular differentiation. However, there is one distinctive effect – the spatial signals *orient the differentiation events* in the cells. This is seen most clearly in the vascular system, but might be expressed by very early transport events (Gersani 1985) and perhaps also by the orientation of cytoplasmic structures (Sachs 1972 a, Kirschner & Sachs 1978).

It may be concluded that there are spatial signals whose effects are remarkably unspecific in terms of their origin, the tissues they act on and the responses they elicit. Alone, the role of such signals in the control of organized development would at best be limited. These signals, however, must be assumed to act together with the quantitative temporal changes of cells, considered above. It has often been assumed that different stages in the formation of an apex and various processes of cell division and differentiation must reflect the action of different spatial signals. However, temporal differences between cells, mentioned above, and possible local interactions, could result in different responses to the very same signals. The thesis to be developed below is that the integrated effect of both spatial and temporal controls could result in specific, localized differentiation. Thus, though there is no doubt that there are various interactions between neighbouring cells, their common occurrence and specificity should not be taken for granted (Sachs 1986).

Known phytohormones as spatial signals

The terms "signals" was used above to indicate any passage of information between parts of the plant. There is no a priori reason to assume that these signals should be substances, electrical oscillations, physical stresses etc. – the possibilities do not exclude one another, and it is even likely that they all play a part in plant organization. There is, however, concrete evidence concerning the chemical nature of some signals. This evidence came from work that led to the isolation of phytohormones – the working hypothesis being that there must be specific controlling substances for each developmental process (Went & Thimann 1937). In now appears that this hypothesis of specific controls is wrong, but it nevertheless led to important results.

Auxin was isolated and defined as a signal that controls elongation of cells (Went & Thimann 1937). As soon as it was characterized, however, it was found to modify plant development in bewildering ways. The list of the various effects of auxin, however, is not random: auxin replaces the cellular and organogenic effects of young shoot tissues on the rest of the plant (Sachs 1975, 1986). Thus auxin inhibits all aspects of the formation of other shoot

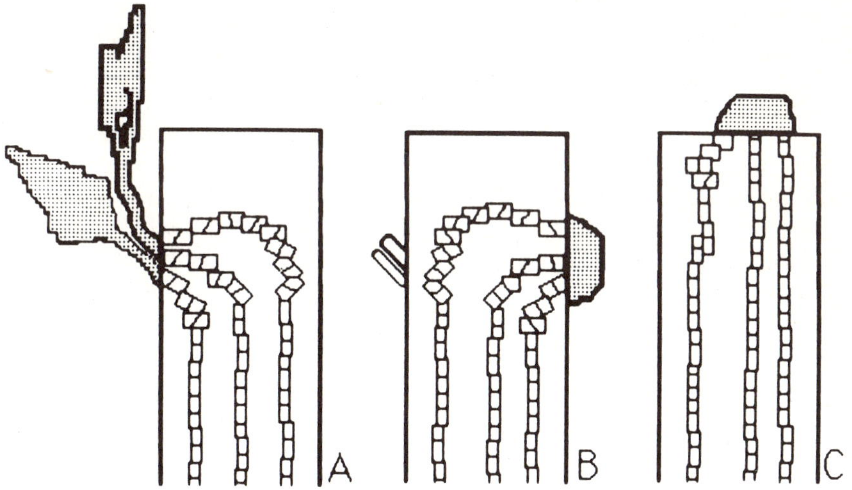

Fig. 3. Schematic representation of the location and orientation of new transport channels, vessels and sieve tubes. For reasons of clarity, these channels are not on the same scale as the cut stems in which they are induced. – **A,** A bud induces the formation of oriented vascular channels. **B,** A localized source of auxin has the same effect, even though it is in an unusual location, opposite the stump of the bud. **C,** Auxin placed in another relation to tissue polarity. New transport channels form along the expected flow of the auxin. No substances other than auxin are known to orient differentiation in relation to a localized source.

tissues, promotes the formation of roots and *orients* (fig. 3) and induces the differentiation of vascular tissues (Sachs 1981). All these effects are repeated in many unrelated plants, though they are certainly not expressed in all conditions. It is hardly likely, to say the least, that this correspondence between the various effects of shoot tissues and of auxin is a matter of chance. It is known, furthermore, that young shoot tissues are major sources of auxin (Thimann & Skoog 1934). Since auxins replace the effects of tissues in which they are known to be formed, it follows as a logical conclusion, not an hypothesis, that *auxins are developmental signals of developing shoot tissues.* This conclusion does not imply, of course, that auxins are the only signals of shoot tissues.

Evidence concerning the formation of auxins in roots is conflicting, but the effects of roots on the plant clearly show that they do not act as sources of auxins. The movement of auxins through the plant, however, suggests that the roots could be sinks for auxin (Bourbouloux & Bonnemain 1979). Such a sink effect on the flow of auxin could account for correlative effects on the plant. This would suggest that the differentiation of vascular tissues is oriented along the auxin flow through the plant (Sachs 1968, 1981). A sink effect of roots on auxin flow could also be expressed by the inhibition of the formation of additional root primordia.

Similar statements can be made concerning a role of cytokinins as correlative signals. Cytokinins were also isolated as specific controls – in this case of cell division. However, cytokinins induce all aspects of the development and maintenance of shoot tissues and inhibit the initiation and development of roots: they thus replace correlative influences of roots (Sachs 1975, 1986). Since it is well established that cytokinins are formed in roots (Kende 1965, Goodwin et al. 1978), they replace the effects of organs in which they are known to be formed. It must be concluded, therefore, that *cytokinins are developmental signals of roots.* This logical conclusion does not depend on the roots being the sole sources of cytokinins in all possible conditions. It is also possible, but by no means proven, that shoot tissues inhibit other shoot development and promote root initiation by acting as sinks for cytokinins. The evidence for roots acting as a sink for auxin is, however, stronger than evidence that shoots influence development by acting as sinks for cytokinins. The difference is due to the unique orientation of differentiation associated with the flow of auxin (Sachs 1981).

There is additional evidence for these far-reaching conclusions concerning the correlative roles of auxins and cytokinins. The unregulated synthesis of only these two factors is sufficient to disrupt organized development, leading to the formation of tumors (Braun 1956, Hookyaas et al. 1982). It is, furthermore, the cause of tumor formation in a number of unrelated systems (Braun 1978). Auxins and cytokinins are also the only general controls of apical initiation on plants and in culture (Skoog & Miller 1957). They are also the two groups of organic molecules that are most often required in small quantities by tissues in culture (Murashige 1974), thus indicating a regulatory role in the plant. Finally, there are mutants that influence the synthesis of various phytohormones – but mutations that influence cytokinin (Horgan 1986) and auxin biosynthesis are missing, presumably because they are lethal.

It may be concluded that auxins, and probably also cytokinins, serve as spatial signals that control development. It is thus significant that the parameters of the action of these two groups of phytohormones – their gradual, multiple effects on differentiation and the distances over which they act – are the ones suggested as characteristic of spatial signals in the previous section. This is not meant to imply, of course, that auxins and cytokinins are the only developmental signals. Other phytohormones are known to coordinate the development of plants with environmental conditions – but these phytohormones are not known to have general roles as spatial signals in the internal control of development. It is reasonable to assume, however, that there are additional internal signals – hormones and other factors – that are at present unknown or whose role is not clearly proven.

There have been numerous objections to the concept of developmental control by phytohormones. Many of these are based on the unjustified expectation that phytohormones should account for all phenomena or that they should specify all different forms of development independently of other factors, such as the competence of developmental history of the responding tissues. There is also a poor correspondence between measured concentrations of phytohormones and development (Goodwin et al. 1978). The conclusions above, however, would not require such a correspondence: plant cells appear to respond to the changes of concentration or to the flow of signals, not necessarily to measurable amounts, averaged for all cell compartments. Finally, it is not clear that the synthesis of auxins and cytokinins is always localized in one type of tissue (Wang & Wareing 1979). Localization would be important for the relative role and specificity of the signals, but possible exceptions would not change the qualitative conclusions reached above.

Pattern formation by epigenetic selection

It is now necessary to turn to the question of how much organized detail could be specified by integrative controls in which phytohormones have an important role. An answer to this question is not available, and the relevant information would require a book rather than a short article. It is, however, possible to indicate general possibilities by considering a few cases in which patterned development is disturbed or even abolished.

An extreme disruption of organization is the formation of crown gall tumors (Braun 1978). As mentioned above, this disruption is known to be the result of the unregulated synthesis of only two phytohormones, auxins and cytokinins. The crown gall tumors have the structural traits of callus tissues that develop on wounds, but the tumor development continues even after the wounded plant has regenerated. This continued growth must mean that the tumor is insensitive to the known inhibitory effects of the other organs of the plant. At the same time the tumor must also be able to dominate the plant sufficiently so as to obtain the resources necessary for continued development. It is possible to account for the dependence of these traits of crown gall and other tumors on the unregulated synthesis of auxin and cytokinins (Sachs 1975). The formation of auxin could be expected to lead to the dominance of the plant by the tumor, though this dominance need not be expressed by the cessation of all other growth. The very same auxin could induce the large vascular contacts of tumors with the rest of the plant, these contacts being an anatomical expression of the ability of a tumor to dominante the plant and presumably divert the metabolites necessary for continued tumor development. The very same tumor tissues that form auxin, furthermore, also produce their own cytokinins. This cytokinin formation could be a reason for the tumors being relatively insensitive to the inhibitory effects of normal shoots.

Thus the unregulated formation of two phytohormones, one (auxin) orienting differentiation and the other (cytokinin) limiting shoot tissue development, is sufficient to account for the relations of tumors with the rest of the plant. The same principles could, furthermore, apply to the lack of organization within tumors. When tumors continue to grow on plants they do not consist of uniform callus tissue. Instead, at least some tumors growing on plants include many meristematic centers that do not inhibit one another. Perhaps this lack of inhibition has the same phytohormonal basis as the disrupted correlations with the rest of the plant. The tumors on the plant also include vascular tissues but the vascular channels have no single consistent axis. This disruption of vascular orientation could be due to the absence of the one dominant source of directional induction present in normal tissues with one defined apical centre. An extreme form of unorganized growth is a callus with no apparent cell differentiation, and such callus, both tumorous and normal, can be readily grown in culture. Culture conditions in which callus is grown are carefully designed so that nothing is limiting, and this may be a reason for the loss of all organization. These facts suggest that *organization is dependent on competition for limited supplies of signals that control development.*

A second case of disturbed development occurs in some plants after an entire shoot system is removed. Form is restored by the formation of adventitious apices and this, too, could be related to the integrative, phytohormonal controls considered above. Thus the change in the shoot/root balance when the entire shoot is removed would cause a relative increase of resources required for shoot development, among them cytokinins. This increase would induce competent cells – if present – to start using cytokinins and forming auxin, and these could be the earliest stages of shoot formation. The auxin formed by the cells that undergo these earliest changes would orient differentiation so the resources the changing cells receive would increase. Additional resources would lead to the increased formation of auxin – a positive feedback would start between the cells that have initiated the change and the rest of the plant. The resources would suffice for the recruitment of additional cells, neighbours of the cells that initiated the change (Sachs 1972b). This recruitment would lead to the formation of a new apex. The size of the apex would increase until it becomes limited by its own use of resources, and demand and supply would be balanced. This could occur when the new apex reached its final size and its consumption of resources increased as a function of its rising developmental rate. Thus the control by limitation of essential morphogenetic signals and *a positive feedback between developmental rate and the supply of limiting resources* could be a basis of the collaboration of cells during the initiation of apices (Sachs 1972b).

These processes of apex formation and development after the entire shoot system is removed would not, however, specify a new plant form. In the early stages of regeneration there is a large excess of apices, whose precise location appears partly random. The large number of apices presumably reflects the initial excess of cytokinins. Once the apices are formed, however, the excess of cytokinins would be consumed. The new apices would then compete for cytokinins, and only few apices would prevail – and inhibit further development of the less successful apices. There would thus be a selection process, but this would not be evolutionary selection since it would occur between apices of the same genotype. This *epigenetic selection could play a major role in determining the localization of continued developmental processes.* Succes in the selection process between shoot apices would depend on the rate of their formation, their location on the plant, the environmental conditions that could influence the formation of auxin and even chance events that could give one bud an advantage over another in the positive feedback relations with the plant.

Finally, a feedback relation and a process of selection between cells could account for the determination of the cellular pattern of vascular channels that connect the adventitious apices with the rest of the plant (Sachs 1969, 1981). The oriented differentiation of vascular tissues can be understood as a gradual response to the flow of auxin through the cells. At the same time, the capacity to transport the very same inductive flow of auxin is increased during the process of differentiation. Thus there is a positive feedback between transport and differentation, and *this feedback leads to the canalization of the flow of auxin to discrete strands* of specialized cells. The precise location of these strands is not determined by the genetic constitution of the plant. Instead, there is a competition for the available auxin between the differentiating channels. Only the channels that are most appropriately placed, or whose cells are most competent to initiate rapid differentiation, are epigenetically selected by the competition for auxin flow. It is these channels that complete their gradual differentiation as functional transport systems.

These examples of the dependence of development on phytohormones suggest the following three generalizations concerning pattern formation:

- Organization can depend on the sustained *competition for limited supplies of signals* critical for continued development.
- Quantitative effects and responses lead to *positive feedback relations which gradually restrict the continuation of processes.*
- The initiation of differentiation events is not the sole determinant of later pattern. The developmental events appropriate for balanced relations between the different parts of the plant are *epigenetically selected* during the processes of development. A mechanism for this selection could be successfull participation in feedback relations with other parts of the plant.

General conclusions

The study of plant and other biological development has been dominated by a search for developmental programmes. These programmes have taken many forms, such as a homunculus, a simplified pre-pattern or, most recently, a stepwise unravelling of gene expression (for example, Stubblefield 1986). The assumption that programmes are essential seems to be based on human activities and lately also on computer operation. A suggestion based on the discussion above is that programmes are not a fruitful concept concerning the way an organism constructs itself. Perhaps it is not the way an architect's plan of a building is followed that is a useful image for morphogenessis. Instead, the epigenetic formation of a new organism resembles the thinking processes that produces the building plan. Thus various alternatives, severely constrained by rules of the possible, are "considered". The one best suited to the conditions or requirements is followed. This selection is then only the basis for the next step in a gradual construction of a new entity.

Regardless of these images, the dependence of cells on long term conditions and on feedback loops could be expected to be more reliable in leading to the final mature structure than any precise specification of all developmental events. Feedbacks loops would mean that chance mutations and other "mistakes" could be compensated for – development plasticity, including the capacity for regeneration, would be part of undisturbed plant development. Furthermore, since the same rules are used in different contexts, their specification is presumably economical – it is cheap in terms of the information that must be coded. This suggests that the basis for orderly development should be sought in differences that build up gradually within apical meristems and not in rapid, all-or-none conversions of cells from one state to another. Relatively simple rules could lead to a rich structure because their expression would differ quantitatively depending on the developmental history of the cells. In other words, the feedback relations between differentation and the responses to differentiation signals would lead to a progression of developmental events. These events could change with time without depending on necessary changes of gene expression. Thus developmental rules may be "integrated over time" (Green & Poethig 1982) or the patterning of differentation may be "differentiation dependent" (Sachs 1978a).

The discussion above is also relevant to the cellular properties essential for organized development. Cell differentiation may not be a response to precise concentrations of critical substances at a critical time. Since cells have a "memory", they could respond to *changes of signal concentration* rather than to the concentrations themselves. Furthermore, since the orientation of cellular events is essential for orderly development, and concentrations alone could not specify orientation, the cells must be sensitive to gradients or to the actual flow of signals (Sachs 1981). An advantage of sensitivity to changes of relative concentrations and to flow through the cells is that these parameters might not be readily perturbed by temperature and other variations. The discussion above also suggests that the gene regulation critical for development involves quantitative changes in the formation of relatively unspecific substances – not qualitative changes of specific states of genes.

Central to evolutionary botany is the evolution of form. This is, necessarily, the evolution of the genes that determine development. The discussion above suggests that specific structures are specified indirectly, through the action of genes that determine quantitative aspects of developmental rules. The assumption that it is structures which evolve could therefore lead to mistakes, especially concerning what are and are not small, gradual steps. The concept of evolving structures has also been used as the basis of considerations of homology and this has led to a complex, quantitative picture in which reality can not be represented by any one set of concepts (Sattler 1974). It may be therefore suggested that homology should be used to refer to processes rather than to their final products, mature structures (Sachs 1982). The discussion above suggests that these homologous processes are gene-dependent and it is they that would undergo Darwinian evolution.

Acknowledgements

Special thanks are due to Dr. P. W. Barlow and Dr. P. B. Green for important comments and critical reading of the manuscript.

References

Bonner, J. T. 1958: The evolution of development. – University Press, Cambridge.

Bourbouloux, A. & Bonnemain, J. C. 1979: The different components of the movement and the areas of retention of labelled molecules after application of (^{3}H) indolyl-acetic acid to the apical bud of *Vicia faba.* — Physiol. Pl. **47:** 260–268.

Braun, A. C. 1956: The activation of two growth substance systems accompanying the conversion of normal to tumor cells in crown gall. – Cancer Res. **16:** 53–56.

– 1978: Plant tumors. – Biochem. Biophys. Acta **516:** 167–191.

Brenner, S., Murray, J. D. & Wolpert, L. (eds.) 1981: Theories of biological pattern formation. – Philos. Trans. Ser. B, **295:** 425–617.

Gersani, M. 1985: Appearance of transport capacity in wounded plants. – J. Exp. Bot. **36:** 1809–1816.

– & Sachs, T. 1984: Polarity reorientation in beans expressed by vascular differentiation and polar auxin transport. – Differentiation **25:** 205–208.

Goodwin P. B., Gollnow B. I. & Letham, D. S. 1978: Phytohormones and growth correlations. – In: Letham, D. S. Goodwin, P. B. & Higgins (eds.), Phytohormones and related compounds, a comprehensive treatise, **2:** 215–250. — Elsevier/North Holland, Amsterdam.

Green P. B. & Poethig. R. S. 1982: Biophysics of extension and initiation of plant organs. – In: Green, P. B. & Subtelny, S. S. (eds.), Developmental order: its origin and regulation (Symp. Soc. Developm, Biol. **40**): 485–509. – New York.

Hooykaas, P. J. J., Ooms, G. & Schilperoort, R. A. 1982: Tumors induced by different strains of *Agrobacterium tumefaciens.* – In: Kahl, G. & Schell, J. S. (eds.), The molecular biology of plant tumors: 373–390. – Academic Press, New York.

Horgan, R. 1986: Cytokinin biosynthesis and metabolism. – In: Bopp, M. (ed.), Plant growth substances 1985: 92–98. – Springer, Berlin.

Jacobs, W. P. 1952: The role of auxin in the differentiation of xylem round a wound. – Amer. J. Bot. **39:** 301–309.
– 1970: Regeneration and differentiation of sieve tube elements. – Int. Rev. Cytol **28:** 239–273.
Jost, L. 1893: Über Beziehungen zwischen der Blattentwicklung und der Gefäßbildung in der Pflanze. – Bot. Zeitung, 2. Abt. **51:** 89–138.
– 1907: Lectures of plant physiology. Translated by R. J. Harvey Gibson. – Clarendon Press, Oxford.
Keeble, F., Nelson, M. G. & Snow, R. 1930: The integration of plant behaviour. II The influence of the shoot on the growth of roots in seedlings. – Proc. Roy. Soc. London, Ser. B., Biol. Sci., **160:** 182–188.
Kende, H. 1965: Kinetin like factors in the root exudate of sunflowers. – Proc. Natl. Acad. USA **53:** 1302–1307.
Kirschner, H. & Sachs, T. 1978: Cytoplasmic reorientation: an early stage of vascular differentiation. – Israel J. Bot. **27:** 131–137.
Larson, P. R. 1975: Development and organization of the primary vascular system in *Populus deltoides* according to phyllotaxy. – Amer. J. Bot. **62:** 1084–1099.
Lyndon, R. F. 1979: The cellular basis of apical differentiation. – In: George, E. C. (ed.), Control of plant development (Monogr. Brit. Pl. Regulator Group **3**): 57–73.
Malacinski, G. M. (ed.) 1984: Pattern formation. A primer for developmental biology. – Macmillan, New York.
Meins, F. jr. 1986: Phenotypic stability and variation in plants. – Curr. Topics Developm. Biol. **20:** 373–382.
Murashige, T. 1974: Plant propagation through tissue culture. – Annual Rev. Pl. Physiol. **254:** 135–166.
Sachs, T. 1968: The role of the root in the induction of xylem differentiation in peas. – Ann. Bot. (London) **32:** 391–399.
– 1969: Polarity and the induction or organized vascular tissues. – Ann. Bot. (London) **33:** 263–275.
– 1972a: The pattern of plasmolysis as a criterion for intercellular relations. – Israel J. Bot. **21:** 90–98.
– 1972b: A possible basis for apical organization in plants. – J. Theor. Biol. **7:** 353–361.
– 1975: Plant tumors resulting from unregulated hormons synthesis. – J. Theor. Biol. **55:** 445–453.
– 1978a: Patterned differentiation in plants. – Differentiation **11:** 63–73.
– 1978b: The development of spacing patterns in the leaf epidermis. – In: Subtelny, S. & Sussex, I. M. (eds.), The clonal basis of development: 161–183. – Academic Press, New York.
– 1981: The control of patterned differentiation of vascular tissues. – Advances Bot. Res. **9:** 151–262.
– 1982: A morphogenetic basis for plant morphology. – Acta Biotheor. **31A:** 118–131.
– 1984: Axiality and polarity in vascular plants. – In: Barlow, P. W. & Carr, D. J. (eds.), Positional controls in plant development: 193–224. – University Press, Cambridge.
– 1986: Cellular interactions in tissue and organ development. – Symp. Soc. Biol. **40:** 181–210.
– & Thimann, K. V. 1967: The role of auxins and cytokinins in the release of buds from dominance. – Amer. J. Bot. **54:** 136–144.
Sattler, R. 1974: A new conception of the shoot in higher plants. – J. Theor. Biol. **24:** 22–34.
Sinnott, E. W. 1960: Plant morphogenesis. – McGraw-Hill, New York.
Skoog F. & Miller, C. D. 1957: Chemical regulation of growth and organ formation in plant tissues cultured in vitro. – Symp. Soc. Exp. Biol. **11:** 118–131.
Snow, R. 1929: The young leaf as the inhibitory organ. – New Phytol. **28:** 345–358.
– 1945: Plagiotropism and correlative inhibition. – New Phytol. **44:** 110–117.

Stewart, R. N. & Derman, H. 1975: Flexibility in ontogeny as shown by the contributionn of the shoot apical layers to leaves of periclinal chimeras. – Amer. J. Bot. **62:** 935–947.

Stubblefield, E. 1986: A theory for developmental control by an program encoded in the genome. – J. Theor. Biol. **118:** 129–143.

Tepper, H. B. & Hollis, C. A. 1967: Mitotic reactivation of the terminal bud and cambium of white ash. – Science **156:** 1635–1636.

Theologis, A. 1986: Rapid gene regulation by auxin. – Annual Rev. Pl. Physiol. **37:** 407–438.

Thimann, K. V. & Skoog, F. 1934: On the inhibition of bud development and other functions of growth substance in *Vicia faba.* – Proc. Roy. Soc. London. Ser. B, Biol. Sci., **114:** 317–339.

Umrath, K. 1948: Dornenbildung, Blattform und Blütenbildung in Abhängigkeit von Wuchsstoff und korrelativer Hemmung. – Planta **36:** 262–297.

Vöchting, H. 1892: Über Transplantation am Pflanzenkörper. – Laupp, Tübingen.

Wang, T. L. & Wareing, P. F. 1979: Cytokinis and apical dominance in *Solanum andigena:* lateral shoot growth and endogenous cytokinin levels in the absence of roots. – New Phytol. **82:** 19–28.

Wareing, P. F. & Al-Chalabi, T. 1985: Determination in plant cells. – Biol. Pl. **27:** 241–248.

Warren, Wilson, P. M., Warren, Wilson, J. & Addicott, F. T. 1986: Induced abscission sites in internodal explants of *Impatiens sultani*: a new system for studying positional control. – Ann. Bot. (London) **57:** 511–530.

Went, F. W. & Thimann, K. V. 1937: Phytohormones. – Macmillan, New York.

Wolpert, L. 1971: Positional information and pattern formation. – Curr. Topics Developm. Biol. **6:** 183–224.

Woolley, D. J. & Wareing, P. F. 1972: The role of roots, cytokinins and apical dominance in the control of lateral shoot form in *Solanum andigena.* – Planta **105:** 33–42.

Address of the author: Professor Tsvi Sachs, Departmant of Botany, The Hebrew University, Jerusalem 91904, Israel.

Berlin and the world of botany

H. W. Lack

Abstract

Lack, H. W. 1988: Berlin and the world of botany. – In: Greuter, W. & Zimmer, B. (eds.): Proceedings of the XIV International Botanical Congress: 261–276. – Koeltz, Königstein/Taunus.

An attempt is made to summarize the contributions to the advancement of knowledge made over three centuries by botanists living in Berlin. The role played by the Royal Academy of Sciences, the Friedrich Wilhelm University and the Kaiser Wilhelm Gesellschaft is lined out. Special emphasis is given to the development of the Botanic Garden and Botanic Museum in Berlin and the *opera magna* of plant taxonomy written or edited by members of staff of this institute – notably Willdenow's Species plantarum, Engler's Pflanzenfamilien, Urban's Symbolae antillanae and Das Pflanzenreich. The historical development of the botanical collections conserved in Berlin and their almost complete destruction during the Second World War is described and the fate of the three major botanical libraries in this city is briefly mentioned. The impact of editing and publishing botanical literature in Berlin is demonstrated on selected examples. From the field of plant physiology the achievements made by Schwendener and Haberlandt are discussed and a short summary on plant genetics und plant breeding in Berlin is also provided.

Those having come from all over the world to Berlin to attend the XIV International Botanical Congress may have got the rash impression of staying in a normal city. If so, I am afraid they were utterly wrong. Berlin is a very special place, a unique place in many respects. It is not only the famous Berlin Wall, that miserable building, that ring of concrete that encircles the western sectors of Berlin; it is not only the fact that in several countries and for good reason the name of this city stands and will stand as a synonym for the most brutal and sophisticated terror ever experienced in the world – here in Berlin the Nazis had their headquarters and brought endless pain, misery and death over most of Europe; it is not only that this city was in turn bombed, set on fire and ruined to a degree almost unbelievable today, and just escaped the more deadly weapons then already at their final test stage. There are other peculiarities.

Many, especially if coming from overseas, may feel in Berlin as being in merry old Europe. Again, they are quite wrong. Berlin is the most American city of Europe, with a very high proportion of immigrants coming from all parts of this continent. Of course Berlin is older than its sister town San Francisco, but not very much so. In 1987, Berlin celebrated its 750th anniversary, thus by European standards it is very young indeed. A fishermen's and merchants' village on the Spree river was the beginning of what eventually became a metropolis of more than 4 million inhabitants. But development was extremely slow – more than two and a half centuries passed until Berlin became the residence of the Electors of Brandenburg, whose territories were called, due to their poor soil and harsh climate, the Holy Roman Empire's box of writing sand. Very much later, as capital of the kingdom of Prussia, this city became one of the main political, economic and cultural centres of Europe. Only 116 years ago did Berlin finally become the capital of the newly founded Deutsches Reich, experiencing from then onwards all stages of rise and fall of this political structure. In a way, it is the story of Europe's Brasilia, but ending in destruction, division and agony.

The time of glory had lasted for a few decades only. For this short period, however, a spectacular concentration of manpower, brainpower and funds took place in the Reichshauptstadt, before being disseminated all over the world.

After the long years of reconstruction and recovery following the Second World War the general situation of this city is still complex and delicate but stabilized by the Four Power Agreement of 1971. Whereas the eastern sector has become the centre of the German Democratic Republic, the western sectors have developed and continue to develop strong links to the Federal Republic of Germany. Still the biggest industrial agglomeration between Paris and Moscow, the Berlin economy, however, has lost momentum, the trendsetters of world trade have moved westwards. Berlin's famous nickname, "Chicago on the Spree" no longer applies. No longer Reichshauptstadt, no longer first address in business, Berlin continues to be a metropolis of outstanding importance for the arts and sciences, and this applies, I may say, for both parts of the city.

The sciences were and are strongly influenced by these more general aspects of the history of Berlin. In this field, too, major developments started late: the Society of Sciences, later called Royal Academy of Sciences and finally Prussian Academy of Sciences, was founded more than a century after the Accademia dei Lincei in Rome. When Alexander von Humboldt returned to his native Berlin from his famous voyage to the Americas and a long stay in France he still had to write "To move from Paris to Berlin, means to move from life to death." And when the Friedrich Wilhelm University finally opened, the universities in nearby Prague, Cracow and Vienna were already more than 450 years old. But then, after this late start, further development was extremely rapid, making Berlin in the 19th century one of the foremost places in the sciences. It is difficult to resist the temptation to list just a few names, but where to start and where to end? The register of the scientific staff of the Friedrich Wilhelm University reads over many years like "Who is who in sciences". At the beginning of this century more than 9000 students were enrolled at this university, then regarded as a colossal number; "to learn German and go to Berlin" was a standard saying among students and young scientists in many parts of Europa.

In several fields the concentration of brainpower was particulary extraordinary – *e.g.* in physics. In the thirties Max Planck, Albert Einstein, Otto Hahn and Lise Meitner worked simultaneously in Berlin and there are photographs of the participants of the seminar for physics, on which you can spot the two or three who later did not receive the Nobel prize. Out of this group of physicists developed the most important experiment ever undertaken in Berlin: the nuclear fission which took place on a wooden laboratory desk standing in the Kaiser Wilhelm Institute for Chemistry in Thielallee 63 and which was first understood in December 1938. Is it only a coincidence or more, that the Four Allies have placed the Berlin Kommandantura, today as in 1945 the Supreme Authority in this city, in a building just opposite this truly historical site? I don't know.

Besides the Royal Academy of Sciences, the Friedrich Wilhelm University and the Kaiser Wilhelm Institutes, a fourth group of scientific institutions grew up in Berlin: the libraries and museums, with famous places like the Royal Library on the boulevard Unter den Linden or the complex of art museums on the Spree island in the very heart of the city. The breath-taking growth of all these institutions was followed by an equally breath-taking collapse and fragmentation of its collections. The colossus is now split leading to the somewhat strange and anomalous situation of most scientific institutions existing in duplicate – starting with the old Friedrich Wilhelm University now split into the Humboldt University in Berlin (East) and the Free University in Berlin (West), the old Preußische Staatsbibliothek split now into the Deutsche Staatsbibliothek in Berlin (East) and the Staatsbibliothek Preußischer Kulturbesitz in Berlin (West), just to name a few examples. This disruption led to curious results: Johann Sebastian Bach's autograph of the Passion According to St. John is split: the music for the voices is in one part of the city, the music for the orchestra in the other. The Natural History collections form a rare exception to this general rule: the very extensive zoological, paleontological, mineralogical and geological material is conserved undivided at the

Museum of Natural Sciences in Berlin (East), whereas the botanical material is kept almost undivided in Berlin (West).

Botany in Berlin was and is based on four pillars: number one – the Electoral, later Royal, finally Prussian State Library; number two – the Royal, later Prussian Academy of Sciences; number three – the Kaiser Wilhelm Institutes, later Max Planck Institutes; number four – the universities. Over the centuries their importance has varied considerably. Today the two academies, the Academy of Sciences of the German Democratic Republic and the new Berlin Academy of Sciences, founded this very year in the western part of the city, have no direct bearing on botany. In the field of biology a single Max Planck Institute is based today in Berlin, specializing in the field of molecular genetics.

Over the centuries the Royal Library has acted as repository for scientific information and houses one of the finest botanical libraries in Europe. An extremely valuable standard bibliography for phytotaxonomic literature is based primarily on its holdings – Pritzel's famous Thesaurus literaturae botanicae, written in Berlin and indispensable even today. The library of the Royal Botanic Garden and Museum and the Horticultural Library housed today by the Technical University form two additional focal points for botanical literature.

The Royal Academy of Sciences acted over the centuries as a communication centre for the sciences, but had little direct involvement with botany. A notable exception is the fact that for about a century the Botanical Garden stood under the administration of the Academy – but these were not golden years: destroyed in the Seven Years' War the institution was in such a bad shape that the proposal was made by the Academy, though not accepted, to sell the land and give up the Botanical Garden altogether. At its bicentenary the Academy launched the "Pflanzenreich" project, aiming at a complete description of the plant kingdom to specific level. No less than 116 volumes were published, but nevertheless the gigantic undertaking remained a torso and was not completed. Otherwise the Academy played a more traditional role – with solemn meetings at its monumental palace Unter den Linden, formal lectures delivered in tails by botanists like Heinrich Friedrich Link, who not only received the famous order Pour le merite but was the only botanist to be elected Rector of the Friedrich Wilhelm University.

In contrast to the traditional Academy the Kaiser Wilhelm Society was a much more modern institution; its foundation was an outstanding achievement in the development of the natural sciences in Berlin. Intended as a "Royal Society for the Advancement of Sciences" by its first president, Adolf von Harnack, professor of theology and director general of the Royal Library, it was soon nicknamed "The thinking factory of the Deutsches Reich"; was there a secret in the thinking factory, which employed scientists like Albert Einstein, Fritz Haber and Otto Warburg?

Yes, I think so; at the turn of this century the professors of the Friedrich Wilhelm University were already complaining about lack of space, the increasing number of students and the heavy burden of administration. Independent from the Academy and the university, the Kaiser Wilhelm Society created pure research institutes, much in the same way as this is done today by its successor, the Max Planck Society in Munich. The aim was to provide ideal working facilities for outstanding scientists without any teaching or administrative commitments. Consequently not projects were financed, but individuals. Since interdisciplinary work was regarded as most promising, the first four institutes of the Kaiser Wilhelm Society were built at a short distance in Dahlem, then a residential suburb far outside the city of Berlin – the institutes for physics, physical chemistry, chemistry and biology, the latter being founded just before the First World War.

The life of Otto Warburg (fig. 1), famous for his research in the fields of respiration, fermentation and photosynthesis and recipient of the Nobel prize in 1931, amply illustrates the possibilities within the Kaiser Wilhelm Society. He had his institute built according to his personal taste – in the form of a Mecklenburg stately home, fine enough to be regarded as suitable for the headquarters of the American Armed Forces in Berlin in 1945. In his Mercedes Warburg drove his dogs every morning to the institute, barely 300 meters away from his villa. Concentrating for his whole life on scientific work only, Warburg, "the Kaiser of Dahlem" as he was called, could tell the Federal President of the Federal Republic of Germany "Just send the decoration by mail", being too busy to travel to Bonn to collect it personally. Such slight eccentricities were regarded as acceptable, in particular since Warburg was suggested a second time for the Nobel prize, but the prize was not conferred – at that time German citizens were not allowed to accept this distinction. Occasionally Warburg was critizied for neglecting the education of his young collaborators. His answer was that three had received the Nobel prize; two of them, however, Otto Meyerhof and Hans Krebs, immortalized in the Krebs cycle, had to emigrate because of the outbreak of Nazi terror, just like Erwin Chagaff, who also had worked in Berlin, famous for the Chagaff rules on the base composition in nucleic acids.

Max Hartmann was another outstanding biologist at the Kaiser Wilhelm Institute for Biology, often regarded as the last general biologist, thus also the

Fig. 1. Otto Warburg at the entrance of the Max Planck Institute for cell physiology in Berlin, 1956. Photograph. – Max Planck Gesellschaft, Archiv, Berlin.

title of his famous textbook “General biology. An introduction to the science of life.“ His research centred on sexuality in plants and animals, thus the title of his second important textbook “Sexuality“, published when the Kaiser Wilhelm Institute for Biology had already been evacuated westwards. The analysis of the tobacco mosaic virus was the research field of Gerhard Schramm, who worked in a laboratory jointly run by the Kaiser Wilhelm Institutes for Chemistry and Biology. Two years before the end of the Second World War he published a paper in “Die Naturwissenschaften“, where also the first note on nuclear fission had appeared, demonstrating that the protein component of the virus did not contain information for replication. In spite of very difficult conditions, excellent research was still done in Berlin, just before this working group was also evacuated westwards. Carl Correns, also at the Kaiser Wilhelm Institute for Biology and one of the three rediscoverers of Mendel's laws, studied cytoplasmatic inheritance, sex determination and sex-inheritance. His successor Fritz von Wettstein summarized the entire mechanism of heredity in the simple formula idiotype = genotype + plastidotype + plasmotype, today we would say nuclear DNA + plastidial DNA + mitochondrial DNA.

After having briefly outlined the role played by the Royal Library, the Royal Academy of Sciences and the Kaiser Wilhelm Institutes and before dealing with the universities, it seems appropriate to look for some common traits.

One peculiarity in the development of the sciences in Berlin may be called the “immigration syndrome“. Extremely few vacancies were filled by people born in this city – the first four holders of the chair for anatomy and physiology of plants at the Friedrich Wilhelm University grew up in Switzerland, the Austro-Hungarian Monarchy, Thuringia and Swabia, and the same holds true for the professors of systematic botany in this period – they came from Hesse, Prussian Silesia (now Poland) and Hamburg. However, for all of them their career ended in Berlin; their positions in the Reichshauptstadt were so attractive that they did not consider changing to another university. A chair in Berlin was then regarded as among the highest academic positions in Central Europe.

The concentration of research activities in a certain number of institutions is another peculiarity. In contrast to London, St. Petersburg or Vienna – just think of the complex relationships between the British Museum and the Royal Botanic Gardens at Kew – there had always been for example only a single place for plant taxonomy in Berlin – the Royal Botanic Garden, later to become the Botanical Garden and Botanical Museum Berlin-Dahlem (this name and the address Königin-Luise-Straße 6–8, are familiar since the

Fig. 2. Friedrich Fedde in Posen (now Poznań, Poland). Undated photograph (Botanischer Garten und Botanisches Museum Berlin-Dahlem, Bildarchiv).

Congress Secretariat was based there). For plant taxonomy, all working facilities, all collections of living and dried specimens were concentrated there, even private scholars like Friedrich Fedde (fig. 2) deposited their collections at this institution. In contrast to London and St. Petersburg the Royal Botanic Garden in Berlin, like the Royal Zoological Museum in Berlin, was led by a succession of scientists, who were simultaneously professors of botany and directors; this concentration of commitments in research, academic teaching and administration was an extremely heavy burden, but the triple construction gave extraordinary results.

A third common trait was the considerable amount of money available for the sciences, coming directly or indirectly from a strong economy. Gustav Krupp and the banker Ludwig Delbrück were the first vice-presidents of the Kaiser Wilhelm Society. The research facilities in Gottlieb Haberlandt's new institute for botany and plant physiology in Königin-Luise-Straße 2–6 were of the latest technology. After the transfer to Dahlem the Royal Botanic Garden covered 42 hectares of ground and thus had become the biggest in Central Europe. Glasshouses of a colossal size were constructed as well as the biggest botanical museum then in existence. The huge red-brick building housed about 4 million herbarium specimens, numerous collections of seeds, fruits, wood specimens, fibers from all over the world and a rich botanical library covering four centuries.

Bearing these three peculiarities in mind – immigration syndrome, concentration of research and opulence of funds – we may now proceed to the fourth pillar of botany in Berlin – the universities. We shall concentrate here on the Friedrich Wilhelm University and mention only briefly work on the more applied aspects of botany as undertaken at the Agricultural Hochschule. Ludwig Wittmack, the founder and director of the Royal Agricultural Museum and specialist on the history of cultivated plants, who was the first to describe the wheat-rye bastard Triticocereale, may be mentioned and Erwin Baur. He will not be forgotten because of his work on snapdragon mutants and his famous "law of homologous series".

At the Friedrich Wilhelm University a basic dichotomy exists in the field of botany: let us deal first with the younger daughter, sometimes called "general botany", more often "anatomy and plant physiology".

Simon Schwendener and Haberlandt, the latter known for the very early use of plant cell cultures and his pioneer studies on phytohormones, may stand here as representatives. Both became very old. Haberlandt, however, writing pessimistically in December 1944 on a card: "It is very uncertain, if it was fortunate for me to see my ninetieth birthday, at a time when such a horrible catastrophe afflicts mankind". Exhausted from his flight from Silesia he died one month later in loneliness in Berlin.

The second daughter, systematic botany or phytotaxonomy, is much older indeed: more than three hundred years ago the Great Elector founded the old botanical garden, which, however, rose to fame only at the beginning of the 19th century. Carl Ludwig Willdenow (fig. 3), the first professor of botany at the Friedrich Wilhelm University, had still taken over "a desert scarcely meriting the name of a garden." And when Alexander von Humboldt returned to Berlin no herbarium of any size existed in Berlin besides Willdenow's small, private collections. The facilities for botanical research were then so much more attractive in Paris that Humboldt decided to have his rich material from the Americas studied there.

However, this situation changed rapidly with the foundation of the university and the creation of the Royal Herbarium. "It seems that every botanist will have to go on a pilgrimage to Berlin – like the Mussulman to the tomb of the prophet – if he wishes to die botanically happy" was the comment of a distinguished visitor from Bavaria a few decades later. A constant flow of living and dried specimens as well as botanical literature from all over the world arrived in Berlin, the scientific output reached several hundred printed pages per year. Among the reasons which made Berlin a Mecca for plant taxonomy in Central Europe two will be discussed here. First, a long succession of travellers and plant collectors gathered material for the Royal Botanic Garden, which was subsequently studied there, and second, a number of *opera magna,* major works, like floras covering a wide area, multivolume conspectuses, were written or edited in Berlin.

It seems appropriate to start with the most important traveller, Alexander von Humboldt, even though he was a private scholar and never worked at the Royal Botanic Garden in Berlin. But his collections were divided between Paris and Berlin, and his chief botanical collaborator, Carl Sigismund Kunth, finally became professor of botany at the Friedrich Wilhelm University. Adelbert von Chamisso, a famous German poet and later curator at the Royal Botanic Garden, is another name to be mentioned here. On the Russian brig "Rurik" he circumnavigated the world thereby discovering on the way from Teneriffe to Brazil metagenesis in tunicates, now a standard topic in first grade courses in biology. His botanical collections were shared between St. Petersburg and Berlin. Friedrich Sellow's material from Brazil, Christian Ehrenberg's and Georg Schweinfurth's collections from Arabia and the coasts of the Red Sea, Rudolf Schlechters's orchids from Malesia, the rich material gathered by various collectors in the German Protectorates, in particular in Tropical Africa – where to end?

Fig. 3. Carl Ludwig Willdenow, director of the Royal Botanic Garden und first professor of botany at the Friedrich Wilhelm University. Copper engraving (Deutsche Staatsbibliothek, Berlin).

Opera magna can be produced only at major institutions with their combination of scientific collections, library holdings and brainpower. In the same way as the Flora Neotropica is produced by the New York Botanic Garden or the Flore du Madagascar by the Muséum National d'Histoire Naturelle in Paris the Royal Botanic Garden in Berlin acted as basis for several *opera magna.*

The last major conspectus of the plant kingdom written in Linnaean tradition, Willdenow's Species Plantarum, was written in Berlin. The last third of the biggest flora ever finished, the Flora brasiliensis in forty volumes, was edited by Ignaz Urban, vicedirector of the Royal Botanic Garden and Museum. "Symbolae Antillanae", also written by Urban, provides in over five thousand pages a miscellany on the flora of the West Indies; Paul Ascherson and Paul Graebner produced an extremely detailed, though incomplete "Synopsis der mitteleuropäischen Flora".

The crown of Berlin plant taxonomy, however, belongs to "Die Pflanzenfamilien", edited by Adolf Engler (fig. 4), for 32 years professor of systematic botany at the Friedrich Wilhelm University and director of the Royal Botanic Garden and Museum. Comprising more than eleven thousand printed pages it is a monument of its own; no fewer than 57 botanists collaborated, the greater part being Engler's students and collaborators in Berlin and Breslau (now Wrocław in Poland), Engler himself wrote about one sixth of the phanerogamic part.

Such an extraordinary work could be produced only by an extraordinary man, and this, no doubt, Engler was. Numerous are the anecdotes about him; he had only two priorities – urgent, meaning tomorrow morning, and very urgent, meaning should have been done yesterday. Engler seems to have had very clear political opinions. Once he was asked why he did not consider Friedrich Fedde, a brilliant taxonomist, as a candidate for a vacancy in his staff. His answer was brief: "I have heard he plays tennis". And this was then regarded as a sign of a doubtful character and leftish, and thus absolutely impossible for a curator at the Royal Botanic Garden. And when saying farewell to members of staff who had been called to arms in the First World War Engler is reported to have said: "I would like to ask the gentlemen to have their pictures taken. There is always such trouble with photographs when writing obituaries." Prussian charm, maybe.

His successor, Ludwig Diels, continued the great tradition of Dahlem plant taxonomy, preparatory work for a big Flora of the Andes was published, the "Pflanzenreich" continued to appear.

Fig. 4. Adolf Engler, director of the Royal Botanic Garden and Museum and professor of botany at the Friedrich Wilhelm University, 1893. Photograph (Botanischer Garten und Botanisches Museum Berlin-Dahlem, Bildarchiv).

All this came to a sudden end, which immediately became known in botanical circles as the Dahlem catastrophe. During the night of March 1st and March 2nd 1943 the peaceful Botanical Museum Berlin-Dahlem was for the most part destroyed in an early air raid on Berlin (fig. 5–6). Four million herbarium specimens of outstanding scientific value annotated by generations of plant taxonomists and one of the finest botanical libraries were reduced to ashes. "Immediately after the first high explosive bomb a second fell on the roof of the herbarium wing, broke it open and set the herbarium on fire. At the same time a number of phosphorus cans fell. Due to the strong wind, which blew that night, the fire very quickly spread. Soon the whole herbarium was in flames . . . After a few hours the tragedy had ended: the fire, which shining over a long distance had produced smoke and heat, died down in a wet and cold night of early spring; the walls of the burnt-out building stood black against the night sky" was an eye-witness's report.

The inferno of the battle of Berlin followed, tanks made their way through the flower beds, trenches were dug. Fortunately the Botanical Garden found

Fig. 5. The herbarium wing of the Botanical Museum, Berlin-Dahlem, 1945. Watercolour, gouache and pencil on paper, signed Daugs (Botanischer Garten und Botanisches Museum Berlin-Dahlem, Bildarchiv).

friends in the Allied troops. "Chozjajstvo podpolkovnikov Ilinskogo i Rodina" was a sign in cyrillic script at the entrance to the botanical garden to inhibit further destruction. And it was the Marshall plan which opened the way to the reconstruction work, which, hard to believe, ended only last Wednesday, when the rebuilt wing was officially opened.

Long was the route to reach this end. At first there was silence. Priorities were totally reversed: potatoes for food, and coal to heat ranked first, not the sciences and botany in particular. Nine years passed until the official journal of the Botanical Garden and Museum reappeared. These were difficult years: the splitting of Berlin, the blockade of the access routes to the western sectors, the workers' revolt in the eastern sector, finally the construction of the Wall – all this brought uncertainty and slowed down recovery. Since the old centres of botanical research are based in the western sectors, new institutions were founded in the eastern sector – thus for example a new, though modest botanical garden forming part of the Humboldt University was created. In zoology, the situation was just reversed: here the new institutions were created

Fig. 6. The Botanical Museum, Berlin-Dahlem, November 1945. Water-colour, gouache, pencil on paper, signed Daugs (Botanischer Garten und Botanisches Museum Berlin-Dahlem, Bildarchiv).

by the Free University in the western sectors, whereas the old remained in the eastern sectors. A very special place, indeed.

A survey of current botanical research undertaken in Berlin is beyond the scope of this paper. To these who want to find out the current state of all these research institutes, I recommend a visit. I am convinced they will be heartily welcome – and, I hope, in the spirit of glasnost, also in the eastern part of this city.

What will the future bring to this very special place? Nobody knows. Let us hope that botany and botanists in Berlin will not experience another catastrophe and that they will be able to continue to work in peace.

Address of the author: Dr. H. Walter Lack, Botanischer Garten und Botanisches Museum Berlin-Dahlem, Königin-Luise-Straße 6–8, D-1000 Berlin 33.

Rarity: a privilege and a threat

V. H. Heywood

Abstract

Heywood, V. H. 1988: Rarity: a privilege and a threat. – In: Greuter, W. & Zimmer, B. (eds.): Proceedings of the XIV International Botanical Congress: 277–290. Koeltz, Königstein/Taunus.

Rarity occurs at many levels. Rare species like rare books command considerable attention from a diverse public – collectors, horticulturalists, taxonomists, conservationists, breeders, educators. This confers on them both a privileged position and in some cases a threat to their continued existence. Rarity may sometimes be an artifact created by the taxonomists who indulge in the unjustified multiplication of species. But more often the phenomenon is real, as documented by a small selection of concrete examples. The IUCN-CMC databank at Kew presently keeps record of over 15,000 vascular plant species that are rare and/or threatened on a world scale; but the estimate is that, if the present trends – particulary forest destruction in the tropics – continue, perhaps 60,000 plant species will have become extinct by the year 2050. Our knowledge and our means are inadequate to prevent this threatened extinction. The "Noah principle" of saving every species is unrealistic. Priorities have to be set – but how? The "usefulness" of species is a hazardous and unsatisfactory criterion. The identification of sites of high diversity and endemism may be one answer. Botanical gardens also have a role to play. Clearly, however, more thought is needed in the very near future.

Introduction

It was Shakespeare who wrote in one of his sonnets, "Sweets grown common lose their dear delight" or in the words of the Spanish proverb "A rose too often smelled loses its fragrance". Thus Man since he learned discrimination soon came to appreciate the unusual, the uncommon, the rare. We tend to view our world from a biased attitude – that we are unique – *Homo sapiens,* the only species that has learned to dominate much of the environment or even destroy it. All our attitudes to rarity, certainly stem from this belief. There would be no such thing as rarity if to collect things was not a widespread human need, shared by rich and poor alike. Rare objects and rare organisms have been avidly sought by collectors and our museums house rare treasures from many cultures and civilizations so that they can be seen by a wide public; our zoos and botanic gardens house rare plants and animals, again for the public to enjoy. Provided it is kept in train there is nothing wrong with the pursuit of rarity in this way. Rarity is one of those concepts that suffuses our culture: it defies precise definition and when used by the scientist it is often given a spurious accuracy to satisfy our need for precision. Yet as biologists we should surely have to agree with rhe de Goncourt brothers that "Nothing is repeated, and everything is unparalleled."

We botanists delight in rarity: rare species, rare cultivars, rare mutations, rare books, rare specimens, and so on. Today, however, rarity tends to have a conservation implication – rarity as has been pointed ourt recently signals species at risk and therefore in need of protection. Alas, the only certainty about rarity in a conservation context is that the list of rare and endangered species grows daily bigger and our ability to stem the tide cannot match it.

My own introduction to rarity in a botanical context was as an undergraduate when my imagination was fired by tales of rare endemics, surviving on a few inaccessible rock pinnacles in mountain holdfasts, and before long I found myself plant collecting in Spain and Turkey where I saw for myself and collected some of these celebrated endemics rarities – the bizarre *Pinguicula vallisneriifolia,* the exquisite *Viola cazorlensis,* and countless others, some of them even described by me and surviving taxonomically to this day! One of the pressures to know about such rare species came from plant geography where endemism was one of the intellectually exciting concepts that taxed our minds – rare epibiotics, hanging on (often literally, on the cliffs) for survival, raising such questions as did species have a fixed life-span? A genetical explanation of epibiotic or relictual endemics was proferred by Stebbins in his classic 1942 paper "The genetic approach to rare and endemic species", which he later modified in the light of recent developments in population genetics. The concept of neo- and palaeo-endemics was introduced and further developed by Favarger & Contandriopoulos (1961), incorporating data on chromosome number and 'ploidy level.

Another pressure came from the horticultural trade and from amateur enthusiasts, such as the bulb collectors, alpine gardeners and others who had the collector's interest in rare plants. Both my first visits to Spain were in fact sponsored by horticultural interests. This is a long standing tradition that in the past has led to many excesses and today is rife in certain rare groups, leading in some cases to a clandestine traffic which conservation legislation has been designed to control, such as CITES, the Convention on International Trade in Endangered Species of Wild Fauna and Flora.

During the tulip mania, in the 17th century, enormous sums were paid for rare tulip bulbs and even two centuries later a prize-winning bulb would change hands at over a hundred pounds (three or four times what a head gardener would earn in a year). Bulbs were even purchased on the instalment system by working men – a weaver of Bethnal Green in London reportedly paying £ 10 for a tulip bulb in weekly instalments. The rich could, of course, afford to buy the rarest plants from the far corners of the earth and often were the patrons of expeditions to find such treasures. Not only did horticulture but botany as a whole benefit from such endeavours.

The meanings of rarity

Quite a sizeable literature has developed around the meaning and definition of rarity in a biological context. The meaning of rarity is the title of a chapter by John Harper in a book "The biological aspects of rare plant conservation" (Synge 1981). Harper notes that rarity as a concept is to be considered a phenomenon both in time and in space. Rabinowitz (1981) describes seven forms of rarity, depending on range, habitat specificity and local abundance, and later develops this concept with reference to the flora of the British Isles (Rabinowitz et al. 1986). Indeed the second volume to be published by Soule with the title "Conservation biology" (1986) has as its subtitle "The Science of scarcity and diversity". He goes so far as to state that just as the genetics of nature conservation is the genetics of scarcity so conservation biology is the biology of scarcity. The conservation biologist is called in like a fireman "when an ecosystem, habitat, species, or population is subject to some artificial limitation – usually a reduction of space and numbers".

Rarity, therefore, requires a particular context before we can apply a definition. It will differ according to intrinsic factors such as geography, ecosystem, climate, or extrinsic such as anthropogenic factors, particularily human modification of the environment, or even modification of taxonomy. In the latter context, rarity can be created at a stroke by taxonomic action.

Taxonomic rarity

Taxonomists have probably made the greatest contribution to the cataloguing of rare and endangered plants as a basis for their conservation. Thus the "List of rare, threatened and endemic plants in Europe" was facilitated by the publication of the five volumes of "Flora europaea". Curiously not many taxonomists have become closely involved in conservation programmes although often they are the people with the most relevant experience in the field of the actual status of populations.

The non-specialist user is at the mercy of the taxonomic profession. While there are, of course, hundreds, possibly thousands of probably valid species kown from only a few localities there are many more which are products of a fertile imagination or simply faulty taxonomy. Naming confers a sense of reality or uniqueness to an organism which may be quite divorced from the biological facts. Differences in taxonomic judgement can all too easily and often quite dramatically chance a situation. Van Balgooy (1971) cites the example of the genus *Canarium (Burseraceae)* which was said to be represented by 45 species in the Philippines, all of them endemic to the region, and today after revision for "Flora malesiana" is considered to have only 9 species of which 4 are endemic. The same species has often in the past been described independently in different parts of the world as separate species due to lack of material for comparison. Such false endemics are quite commonly found, such as *Maytenus senegalensis (Celastraceae)* which formerly contained eleven constitutent "species" from Europe, India and many parts of Africa.

Another revealing example of the dangers of relying on the published taxonomic information is given by Fernández-Pérez (1977): the orchid *Epidendrum paniculatum* is one of the commonest plants in Colombia according to the treatment by Ames et al. (1936) who united 26 different taxa in their study. Subsequently Garay in preparing the orchid flora of Colombia found that *E. paniculatum* is not only distinct from the 25 putative synonyms but does not even occur in Colombia!

The history of taxonomy is full of examples of taxonomists who possessed what has been called "clinical eyesight": some people are better at seeing differences than resemblances between species. I am reminded of the dictum of Van Steenis: the aim should be to see how few species there are in a flora not how many. There are, however, biological reasons underlying some of the splitting found. For example inbreeding populations are the source of many of the so-called species of taxonomists such as Jordan whose name was applied to such evanescent creations. Many such "species" have found their way into lists of endangered species, especially in the Mediterranean area, and for some time they enjoy the privilege of rarity before being consigned to the dustbins of synonymy. This is apt enough since many of these inbreeding populations are shortlived in nature.

The loss of rare species: priorities for conservation

The extinction of rare species is ultimately unavoidable – it is the fate of all species eventually to disappear and to be replaced either in ecological or evolutionary terms by others; some will, however, show remarkable stasigenesis and survive for extremely long periods and outlive their congeners, such as the so-called living fossils, *Ginkgo biloba* and *Metasequoia glyptostroboides.*

Conservation of rare species so as to avoid their imminent extinction is essentially a slowing down process in many cases – either of natural events or of human interference. It has been suggested that we should not be concerned overmuch at the loss of any particular species unless there are good reasons to save it, such as its economic importance to man or its scientific significance. Being rare, it is argued, the contribution of such species to community dynamics is not significant and they can therefore be readily dispensed with. In the words of Main (1982) they are the dross. This is undoubtedly an extreme view and is opposed by another somewhat extreme attitude which is now widespread – the biological diversity approach. The loss of any species, even through natural processes, is to be deplored as a diminution of our natural heritage. Such a "right to exist" approach was prevalent some years ago and was dismissed by most scientists as being based on emotion rather than reason, and was dropped as a cogent argument by conservationists. Its return to favour stems, perhaps from a realization of the scale of the loss of habitat and thereby of species, that we are witnessing today.

Arguments rage as to the rate of, say, the loss or conversion of humid tropical forests to other uses (most of the dry tropical forests having already been destroyed in many parts of the world), with Norman Myers concluding that a present rates of clearance from a third to a half will have gone by the end of this century, with much of the remainder serverely disturbed as to their carrying capacity to support their current rich array of species. This in turn has led to estimates that the number of species at risk in tropical forests alone is of the order of two-thirds to three-quarters of a million. These figures have been challenged on various grounds although other analyses and models, such as those of Newmark (1987) on faunal collapse and Simberloff (1986) on mass extinctions in Latin American rainforests, do tend to confirm Myers' gloomy predictions. We do seem to be faced with mass extinction on a geological scale. But even if this should prove to be a tenfold overestimate, it would still be a terrible indicment of our management of resources.

Biologists themselves (let alone workers in other disciplines, or even politicians) do not seem to have even begun to face up to the possible consequences of losing a major part of the world's natural biological resources – the raw material of botany and zoology. The question has to be asked "How much is society prepared to pay to maintain some of these values?" or put around the other way, "How much would society actually lose if another thousand or ten thousand species became extinct as their habitats are eliminated?" No clear answers can be given to such questions. Yet one has to state unequivocally that while the loss of a single species may be a cause of concern at least to some of us, the loss of up to a million plant and animal species in the next 30–50 years would be a biological disaster without parallel in the whole of evolutionary history to date.

From an ecosystem viewpoint there are those such as Main (1982) who would argue for rare species because they are biologically important, either as a record of the past, or as alternative components of the ecosystems or else as insurance policies. The retention of rare species, he suggests, "gives the ecosystem the possibility of flexibility to respond to environmental change in the future, thereby enabling roles to be filled from the indigenous organisms."

As we shall see, and indeed as I have already implied, the number of rare species is already large and likely to increase very substantially. Some form of selection, or triage as Myers (1983) has termed it, is inevitable since the resources at our disposal for conservation are hopelessly inadequate for the size of the tasks facing us. The preservation or rescue of rare species either *in situ* or *ex situ* involves a considerable amount of prior research into their present status, how they maintain their populations, what factors affect their population size, structure and dynamics, which affect the habitat as a whole. Then there is the question of reserve size, which will often differ from species to species even within the same community, and minimal population size, quite apart from the other problems we mentioned earlier. These and similar questions are highly technical matters that we are not even in a position to answer in many instances. And all this before the actual processes of conservation and management even starts.

The question of selection or priority ranking poses many problems. But before even this can be considered we must attempt to ascertain the size of the problems facing us. In other words, how many "rare" or "endangered" species are there today? For effective conservation we must have the best possible taxonomic, geographical and biological information available. Today this means computerized information systems or databases such as those maintained by the the Nature Conservancy with its state Natural Heritage Programs and the Conservation Monitoring Centre (CMC) of the International Union for Conservation of Nature and Natural Resources (IUCN) which amongst other functions provides a global overview database of rare and endangered species. It is to the IUCN-CMC database that one has to turn for the latest estimates of the numbers of rare and endangered plants species recordet in the world.

Table 1. Statistics of threatened vascular plant species of the world's flora, as recorded in the IUCN-CMC database (July 1987).

Category	Endemics	Non-Endemics	Totals
Extinct (Ex)	361	11	372
Extinct/Endangered (Ex/E)	63		63
Endangered (E)	2867	154	3021
Vulnerable/Rare (V/R)	38	3	41
Rare (R)	5469	590	6059
Endangered/Rare (E/R)	27		27
Rare/not threatened (R/nt)	4270	1301	5571
Totals	13095	2059	15154

Details are, of course, available from national or other databases when these exist.

Table 1 gives the statistics as at July 1987. These figures are based on information obtained or supplied form a wide variety of sources and are from complete but give a factual indication of the numbers of extinct, rare or endangered species for which there are data recorded. It should be noted that there are many examples of species which are rare but not threatened or presently at risk. The categories employed are those devised by IUCN for its red data books and are the most widely employed although they have been criticized on the grounds that the definitions are very subjective. They measure rarity in terms of likelihood of extinction rather than in statistical or numerical terms. The data give good estimates for rarity in Mediterranean ecosystems (except Chile) where there are numerous point endemics, most of which are rare (R), and on oceanic islands where the percentage of endangered (E) and vulnerable (V) species is higher – there are for example nearly as many E species in the Canary Islands as in the whole of continental Europe. The system is less effective in the tropics where rarity in tropical rainforests is often underestimated. A dual coding system has been applied by Leigh et el. (1981) for categorizing rare or threatened Australian plants, comprising a numerical distribution category and an alphabetically coded conservation status, the latter similar to the IUCN system. Holloway (1979) amongst others has suggested a more objective approach so as to lend more precision to conservation activities and programmes. He suggested that it should be possible to calculate a range of theoretical population figures that accord with the conceptual minimal size of the population of a species that is needed to ensure its survival and then use these as a basis for assigning species to various categories of endangerment and vulnerability. Other conservation databases use different systems, laying stress on different factors (see, for example, Ayensu 1981). This makes conversion between one system and another difficult and further efforts towards seeking a more generally acceptable approach are indicated. It has to be said, however, that any system which is dependent on detailed population kowledge is simply not practicable at the present time.

Indeed, for many floras, notably those of the tropics, we simply do not have the knowledge or time to undertake a species-by-species approach. Yet it is precisely in these areas that most plant extinctions are likely to occur. Furthermore, even if rate and threatened species could be ranked according to IUCN criteria, it is clearly not feasible for reserves to be set up for each individual species and not practicable to introduce many thousands of rare and threatened tropical species into cultivation on an adequate scale as a form of *ex situ* conservation. The only workable solution is to identify centres of high diversity and endemism in need of conservation. With this is mind. IUCN is preparing a "Plant Sites Red Data Book" to draw the attention of conservation organizations and aid agencies to about 150 sites or vegetation types considered to be those where most plants can be saved. Plants confined to such areas of high endemism such as the Sinharajah lowland rainforest in Sri Lanka, the Atlantic rainforest of Brazil or the island floras of Madagascar or Hispaniola, may be considered extinction-prone – yet another form of rarity, in this case potential.

Examples of rarity

As I have indicated, it is seldom that one has a precise estimate of population size and it is not always feasible to attempt an assessment in the field especially when the numbers are high or the populations widely dispersed over a large geographical area. Population sizes of rare plants vary widely from species to species, ranging from a single individual to tends of thousands or even higher figures. The "IUCN Plant Red Data Book" (Lucas & Synge 1978) includes a list of 31 rare species with a population size of 20 or fewer individuals. This list is by no means complete and the Threatened Plants Unit of CMC is continually adding to it as data become available.

The following is a selection of some of the more spectacular examples of extremely rare species or those where their rarity poses some problem.

Betula uber – the Virginia round-leaf birch.

This was reduced by 1980 to a single population of 20 individuals in Smyth County, Virginia. Current taxonomic opinion suggests it may not be a good species.

Artemisia granatensis – manzanilla real.

This is a classic example of a species that has been brought to the verge of extinction due to over-collection. It is endemic to the upper heights of Sierra Nevada in S Spain, in an area above 1,800 m in altitude where at least 177 Iberian endemic species occur, 66 of which are endemic to Sierra Nevada. The leaves of the *Artemisia* are used to make an infusion which is highly prized, with the result that the species is now extremely rare.

Stylidium coroniforme.

This is a perennial herb which was discovered in 1963 in the vicinity of the Wongan Hills, Western Australia, where it was fairly common over a small area of about 0.3 ha. In 1980 there was only one flowering plant which in that year failed to set seed, and in 1981 only two plants were found (see Groves 1982).

Persea theobromifolia.

This is an economically important tree that is endemic to the coastal wet forest of western Ecuador. It was formerly the most important timber tree of the region but not described scientifically until 1977 by which time the population was reduced to less than a dozen mature trees in a forest remnant of less than 1 km^2 (Gentry 1986).

Dicliptera dodsonii.

This is another example of an anthropogenic relict endemic species of which there are many in the tropics. It is known only from a single tree in the last remnant of wet coastal Ecuadorion forest at Rio Palenque.

Crescentia portoricensis.

This is an example of a rare tropical island endemic species. It is a taxonomically isolated shrubby tree confined to serpentine outcrops in a small area of Puerto Rico and is known from only four individual specimens and two recently discovered populations not yet at the reproductive stage.

Silene diclinis.

This dioecious annual restricted to the Játiva area of South-east Spain was reported in 1970 as reduced to about 500 individuals showing little sign of regeneration. It was described as falling at the borderline between Endangered, Vulnerable and Rare (Synge 1981). It occurs on the terraced hillsides of a carob plantation *(Ceratonia siliqua).* Recent studies have shown the population size to be very much higher and it appears to be quite a successful an aggressive weed, extending well beyond its 1970 habitat.

Ramosmania heterophylla – café marron.

This monotypic genus is one of the most celebrated members of the endemic flora (40 out of 145 native species) of the island of Rodrigues. It is known from only a single individual in a very poor state of health and despite extensive exploration it does not seem likely that any more plants will be found. The plant was severely damaged by cutting and as a last rescue measure a small cutting was taken in April 1986 and flown to the Royal Botanic Gardens, Kew within 24 hours, through the courtesy of British Airways. The cutting was rooted and is now about 25 cm tall and it is hoped that in turn cuttings will be taken from this propagated plant with a view to raising stock for reintroduction into Rodrigues.

Botanic gardens and rarity

As we have seen from the example of *Ramosmania* just mentioned botanic gardens may often have an important part to play in preserving and propagating rare species. Throughout their history botanic gardens have striven to acquire rare plants in their efforts to build up collections of unusual plants from around the world. Often this has been done with little more motivation than the acquisitive instinct of the garden director or curator, and the occurrence in cultivation in botanic gardens of specimens of rare and endangered or even extinct in the wild species is a consequence of this. Gunatilleke et al. (1987) cite the example of the Royal Botanic Garden, Peradeniya, in Sri Lanka which today contains 72 lowland endemic tree species, 22 of which have not been found in the course of phytosociological inventories of their native area and of which the usually single specimens in the botanic garden are the sole known survivors. Throughout the world there is a considerable number of species that are apparently extinct in the wild and known only from botanic gardens or in cultivation elsewhere.

The part that botanic gardens can play today in the study and rescue of rare and endangered wild species has only recently been recognized. Rate plant conservation in botanic gardens, either in the form of reserve collections or as seeds in gene banks, is a subject that is being very actively considered today with several conferences recently being devoted to this theme. The need to obtain accurate information on which rare and endangered species are in cultivation in botanic gardens led to the establishment by the Threatened Plants Unit of IUCN of a database on the holdings in those gardens that joined what was called the Botanic Gardens Conservation Co-ordinating Body. Altogether 138 botanic gardens cooperated in this venture plus the 116 gardens in the USSR which participate through the Main Botanic Gardens in Moscow. The work of the Body has now been subsumed into the IUCN Botanic Gardens Conservation Secretariat which plans to expand and improve this database of rare plant holdings in as many of the world's 1400 botanic gardens and arboreta as possible. Similar approaches have been adopted by the Center for Plant Conservation, Harvard, for a number of North American botanic gardens.

Recently attention has also been focussed on rare and threatened species or cultivars that are in danger of disappearing from our gardens or from the horticultural trade. The National Concil for the Conservation of Plants and Gardens in Britain is one of the organizations that has been set up in repsonse to this problem.

An important function that gardens and the horticultural trade can perform is to introduce rare plants into cultivation and disseminate them widely. Indeed deliberate propagation of rare plants for introduction through the trade may reduce the pressure on wild populations by collectors. Thus the privilege of rarity is soon lost in the interests of conservation. Usually only one or a few genotypes are brought into cultivation even when there is considerable genetic diversity still to be found in the remaining wild populations. The skills of botanic gardens and the trade in propagating plant material, supplemented today by micropropagation techniques, are being increasingly recognized as a valuable tool in conservation biology.

Some rare individual specimens in cultivation are celebrated such as the Sacred Bo Tree at Anuradhapura in Sri Lanka, derived from a cutting of the original tree under which the Buddah attained enlightenment. This is today one of the most sacred objects of veneration for millions of Buddhists. Its maintenance is the responsibility of the Director of the Royal Botanic Garden, Peradeniya. Another celebrated tree is the "thousand year old" dragon tree *(Dracaena draco)* at Icod in the Canary Island of Tenerife, reputedly seen by von Humboldt and today visited by thousands of tourists every week. Curiously enough, the Canary dragon tree is rare in the wild but is widely cultivated in parks and gardens throughout the islands.

Rare and useful?

If we consider the world totals given above and also the estimate made by Raven (1986) and accepted by IUCN that 60,000 species of higher plant are likely to become extinct by the year 2050 if present trends continue, then we are obliged to apply some system of priority ranking as suggested above. What are the criteria that one should use for this purpose? Two of the obvious criteria are scientific importance and economic value. Various attempts have been made to apply such criteria but they are fraught with difficulties. Addressing this question McMichael (1982) concludes that scientific importance is probably no more important than other factors such as aesthetic appeal and the place that the species occupies in our culture. He writes: "For example, although the Tiger is only one of the larger species of Cate, I do not think the attempts to preserve it are unjustified. It holds such a significant place in human culture that in my view it should be preserved at any cost. The same goes for many of the other great mammals – the rhinos, the elephants, the red kangaroo, chimpanzees, giraffes, oryx, and whales. I believe the same could be said for birds such as the Bald Eagle, the Peregrine Falcon, the Emu, the Ostrich, swans, storks, flamingoes and at the other end of the scale, swallows, nightingales and thrushes, and for some plants, though as far as I am aware, few of the plants which have played any significant role in human history are considered to be endangered."

There is no doubt a lesson to be learned from this observation. Perhaps it is as simple as saying that the plants that have been important in our culture are nurtured by us or are brought into cultivation on a wide scale. Animals are clearly more privileged than plants in this respect. Or perhaps we are taking too narrow a view of our cultural history.

The assignment of an economic value to plant species is complicated by multiple uses. In other cases it may not prove possible to assign an economic value – species included by Ehrenfeld (1976) in the category of "non-resources". He drew attention to the difficulties of applying any ranking system – especially the problems of incomplete knowledge and hence the danger of overlooking value and the need to set one species against another in an unnecessary and unacceptable way. Finally he commented that there was only one account in western culture of a conservation effort on a scale greater than now taking place: the account in the Bible of Noah's Ark in which no species was excluded on the grounds of low priority (and as he observes, no species was apparently lost!). He terms this the Noah Principle. It is, however, idealistic and the fact is that with our limited financial resources for conservation on the one hand and our virtually unlimited capacity on the other to destroy plant (and animals) species and populations, we have to seek some method of concentrating our efforts. It does not seem to me that the botanical community is giving this vital question the attention is deserves. Our children and grandchildren may be unforgiving when they see what they inherit. Rarity may come to have a different meaning for them.

Acknowledgements

I wish to thank my colleagues Steve Davis and Hugh Synge for their help and advice.

References

Ames, O., Hubbard. F. T. & Schweinfurth, C. 1936: The genus *Epidendrum* in the United States and Middle America. – Botanical Museum, Cambridge.

Ayensu, E. S. 1981: Assessment of threatened plant species in the United States. – In: Synge, H. (ed.), The biological aspects of rare plant conservation: 19–58. – Chichester.

Ehrenfeld, D. W. 1976: The conservation of non-resources. – Amer. Sci. **64:** 648–656.

Favarger, C. & Contandriopoulos, J. 1961: Essai sur l'endemisme. – Ber. Schweiz. Bot. Ges. **71:** 384–406.

Fernández-Pérez, A. 1977: The preparation of the endangered species list of Colombia. – In: Prance, G. T. & Elias, T. S. (eds.), Extinction is forever: 117–127. – New York.

Gentry, A. H. 1986: Endemism in tropical versus temperate plant communities. – In: Soulé, M. E. (ed.), Conservation biology: 153–181. – Sunderland.

Gunatilleke, C. V. S., Gunatilleke, I. A. U. N. & Sumithraarachchi, B. 1987: Woody endemic species of the wet lowlands of Sri Lanka and their conservation in botanic gardens. – In: Bramwell, D., Hamann, O., Heywood, V. H. & Synge, H. (eds.), Botanic gardens and the world conservation strategy: 183–194. – London.

Groves, R. H. 1982: Changing directions in research on Australian rare plants. – In: Groves, R. H. & Ride, W. D. L. (eds.), Species at risk: 175–188. – Berlin.

Holloway, C. 1979: IUCN. The red data book and some issues of concern to the identification and conservation of threatened species. – In: Tyler, M. J. (ed.), The status of Australasian wildlife: 1–12. – Canberra.

Leigh, J., Briggs, J. & Hartley, W. 1981: Rare or threatened Australian plants. – Austral. Nat. Park Wildlife Serv. Special Publ. **7.**

Lucas, G. & Synge, H. (eds.) 1978: The IUCN plant red data book. – Morges.

Main, A. R. 1982: Rare species: Precious of dross? – In: Groves, R. H. & Ride, W. D. L. (eds.), Species at risk: 163–274. – Berlin.

McMichael, D. F. 1982: What species, what risk? – In: Groves, R. H. & Ride, W. D. L. (eds.), Species at risk: 3–11. – Berlin.
Myers, N. 1983: A priority-ranking strategy for threatened species. – Environmentalist **3:** 97–120.
Newmark, W. D. 1987: A land-bridge island perspective on mammalian extinctions in western North American parks. – Nature, January: 29.
Rabinowitz, D. 1981: Seven forms of rarity. – In: Synge, H. (ed.), The biological aspects of rare plant conservation: 205–217. – Chichester.
–, Cairns, S. & Dillon, T. 1986: Seven forms of rarity and their frequence in the flora of the British Isles. – In Soulé, M. E. (ed.), Conservation biology: 183–204. – Sunderland.
Raven, P. H. 1986: 60,000 plants under threat. – Threatened Pl. Newsl. **16:** 2–3.
Simberloff, D. 1986: Are we on the verge of a mass extinction in tropical rain forests? – In: Elliott, D. K. (ed.), Dynamics of extinction. – New York.
Synge, H. (ed.) 1981: The biological aspects of rare plant conservation. – Chichester.
Van Balgooy, M. M. J. 1971: Plant geography of the Pacific. – Blumea, Suppl. **6.**

Address of the author: Professor V. H. Heywood, IUCN Plants Office, 53 The Green, Kew, Richmond, Surrey TW9 3AA, UK.

Tropical forests and the botanists' community[1]

N. Myers

Abstract

Myers, N. 1988: Tropical forests and the botanists' community. – In: Greuter, W. & Zimmer, B. (eds).: Proceedings of the XIV International Botanical Congress: 291–300. – Koeltz, Königstein/Taunus.

Tropical forests are by far the richest biome on Earth. They harbour at least 90.000 higher-plant species, or 36% of all such species; and a far greater proportion of animal species. They feature another unique characteristic: they are being depleted faster than any other biome. If present rates of deforestation persist – and they may well accelerate – there will be little left of the forests by early next century except for two large remnant blocs, and these may not persist beyond another few decades. This demise of tropical forests will bring on a mass extinction of species. Already we can suppose that in just three areas – Western Ecuador, Atlantic-coast Brazil and Madagascar – we are losing an aversage of one plant species every three days or so. In other critical sectors of tropical forests, we may witness the elimination of at least 18,000 species in the foreseeable future; and within larger expanses of the forests, some 50,000 species. This could well amount to a greater extinction of plant species than at any time since the first emergence of higher plants. Fortunately we still have time to slow and stem, if not halt, this extinction episode, through a global collaborative effort – an enterprise that will call upon professional contributions from the botantists' community worldwide.

Introduction

This paper presents four main issues.

– Tropical forests are being depleted, both quantitatively and qualitatively, faster than any other biome.

– This is all the more regrettable in that tropical forests are biotically richer, in terms of abundance and diversity of species, than the rest of the biosphere put together.

[1] Adapted from the general Congress lecture that was delivered under the title "The plant world of the Tropics: green hell or paradise lost?"

- At the same time, tropical forests are less explored biologically, and less analysed and understood, than any other biome.
- In light of the mass-extinction episode underway in tropical forests, there is surely a special responsibility for the botanists' community worldwide to engage in more expansive efforts to safeguard this uniquely rich sector of the planetary ecosystem.

Tropical forests: their nature and expanse

For purposes of this paper, let us define tropical forests as evergreen or partly evergreen forests, in areas receiving not less than 100 mm of precipitation in any month for two out of three years, with mean annual temperatures of 24-plus°C, and essentially frost-free. In these forests, some trees may be deciduous. The forests usually occur at altitudes below 1300 m, though often in Amazonia up to 1800 m, and generally in Southeast Asia up to only 750 m. In mature examples of these forests, there are several more or less distinctive strata. Plainly this means that we are dealing with tropical moist forests, including seasonal or monsoonal forests, while tropical dry forests are left out of account. Equally clearly, the definition reflects a highly generalized approach, and does not reflect many significant sub-classifications of tropical forests with their own intrinsic attributes.

Tropical forests are, biologically and ecologically, the most complex and diverse biome on Earth. Justifiably they can be described as the most exuberant expression of nature to have appeared on the face of the planet since the first emergence of life almost four billion years ago. They feature marked diversity of species, especially of trees, woody climbers and epiphytes – a characteristic that appears notably in the wetter forests, with their high atmospheric humidity. The ecological complexity is dynamic, with an exceptional degree of biotic and physiobiotic interactions; and it is stable, able to maintain itself for long periods if not indefinitely. Since environmental conditions are relatively constant (while not necessarily uniform), a high level of dynamic stability can persist within a narrow amplitude of environmental fluctuations.

We should not suppose, however, that stability is inter-related with complexity in a simple and positive fashion. Indeed some theoretical and empirical research suggests that "communities with a rich array of species and a complex web of interactions (the tropical rainforests being the paradigm) are likely to be more fragile than relatively simple and robust temperate ecosystems" (May 1975). Similarly, "The reproductive mechanism of tropical forest plant and animal species are more adapted to biological competition than to largescale environmental disturbance" (UNESCO 1978). Overall we can conclude, so far as we understand the situation, that "The humid tropical environment permits the systems to persist in spite of their fragility, because perturbations are relatively small and restricted to small areas . . . Tropical forests are in a constant turmoil of phasic development . . . and they are stable only within a relatively small domain of parameter space" (UNESCO 1978).

The dynamic relationship between complexity and stability is particularly significant insofar as it appears to depend on one important provision, *viz.* that external forces impinging on the system should not exceed certain threshold values, otherwise distinct and enduring changes may arise. Human intervention – which can be massive both in terms of immediate impact and long-run duration – can readily exceed the capacity of the forests' regulatory processes to maintain the systems.

These considerations are dealt with at some passing length because they carry important implications for both biology in general and botany in particular – also for our strategies and efforts to safeguard the forests with their remarkable arrays of species, both plants and animals.

What is the present expanse of tropical forests? Curiously we do not have a concise figure as yet. By the late 1970s there were remote-sensing data available for arround 65% of the biome, a figure that has risen today to 84 percent (Myers 1988). In all cases of countries where remote-sensing information has come onstream is only the past few years – notably Indonesia, Burma, India, Nigeria, Guatemala and Honduras – we find there is greater deforestation than had been supposed by government forestry agencies. These deficiencies apart, we can note an FAO/UNEP report (1982) that postulated an expanse of around 10,000,000 km^2, while a US National Academy of Sciences survey (Myers 1980, see also Myers 1984) indicated some 9,500,000 km^2, both these estimates being for the late 1970s. Since we have lost a fair amount of tropical forest since the late 1970s, the 1988 figure for largely undisturbed tropical forests, *i.e.* primary or near-primary forests, is likely to be little more than 8,000,000 km^2.

Biotic richness

Tropical forests cover only 6% of Earth's land surface, yet they harbour at least 50% of all Earth's species. Indeed some recent research suggests that just the canopy of tropical forests may contain 30 million (conceivably 50 million) insect species alone (Erwin 1988). Of course these are no more than bald statistics, and do not convey an idea of the extreme biotic diversity in concentrated localities of tropical forests. Let us note then that in 10 one-hectare in Borneo, there have been documented 700 species of trees, or as many as in all of North America (Ashton 1981). A single locality of Costa Rica, the La Selva Forest Reserve, covering a mere 13.7 km^2, features more than 1800 vascular plant species, together with 394 species of breeding birds, 104 mammals, 76 reptiles, 46 amphibians, 42 fish and 143 butterflies – totals to be compared with Great Britain's 233,000 km^2, featuring around 1400 native vascular-plant species, about 240 species of breeding birds, 47 mammals, 6 reptiles, 6 amphibians, 43 fish and 64 butterflies.

In total, systematic taxonomic surveys reveal that tropical forests contain at least 90,000 of Earth's 250,000 identified species of higher plants (another 30,000 could well await discovery in the forests). This total of 90,000 species is to be compared with only 50,000 in the entire northern temperate zone with its vast territories in North America and Eurasia. The Chocó sector of Colombia and Ecuador probably contains at least 8000 vascular plant species, as compared with only 10,000 in all of temperate South America; and Colombia as a whole, where the majority of plant species occur in the Pacific-coast and Amazonian forests, possesses some 25,000 vascular plant species in its 1,100,000 km^2 (or at least 5000 more than the eight-times larger United States). Ecuador is estimated to contain at least 20,000 higher plant species, the bulk of them in its tropical forests – a total almost half es large again as Europe's 13,000 plants in a 31-times greater area.

Each plant species, moreover, represents a unique manifestation of the botanical world's diversity, with its own "genetic fingerprint". So far as we can discern on the basis of limited evidence, and using evaluatory techniques that measure genetic variability through *e.g.* average polymorphism and heterozygosity, there generally appears to be greater genetic diversity in tropical forest species than is the case elsewhere (Ayala 1976, Lewontin 1974).

Tropical forest depletion rates

How fast are the forests being depleted today? There is general agreement (FAO/UNEP 1982, Melillo et al. 1985, Molofsky et al. 1986, Myers 1980, 1985) that between 76,000 and 92,000 km^2 of these forests were being eliminated outright each year in the late 1970s; and at least a further 100,000 km^2 were being grossly distrupted each year. Since the late 1970s, the rates have increased somewhat. This means, roughly speaking, that one percent of the biome is being deforested each year, and rather more than another one percent is being significantly degraded.

By the end of this century of shortly thereafter, there could be little left of the biome in primary state, with full biotic diversity and ecological complexity, outside of two large remnant blocs, one in the Zaire Basin and the other in the western half of Brazilian Amazonia, plus some outlier areas in the Guyana highlands and in New Guinea. While these relict sectors of the biome may well endure for several decades further, they are little likely to last beyond the middle of the next century, if only because of sheer expansion in numbers of smallscale cultivators.

To gain an idea of the extent to which deforestation can proceed at an ultrarapid rate, consider the case of Rondonia in the southern sector of Brazilian Amazonia. Agricultural settlement from other parts of Brazil began to gather pace in the mid-1970s (Fearnside 1985, Malingreau & Tucker 1988). In the total State measuring 243,000 km^2, 1200 km^2 had been cleared by 1975, more than 10,000 by 1982, and almost 17,000 by mid-1986. During the second half of the 1970s, the population was growing at an average rate of 15.8% per year, from 111,000 to almost 500,000; and by mid-1986, the population was well over 1 million.

Three illustrative areas

To illustrate with specific examples, let us look at three geographic areas where deforestation is unusually extensive, where large numbers of extinctions have surely occurred in the recent past – and where we must anticipate there will be a further sizeable fallout of species within the foreseeable future. These areas are western Ecuador, the Atlantic-coast forest of Brazil, and Madagascar. Each of these areas features, or rather featured, exceptional concentrations of species, with high levels of endemism. Western Ecuador is reputed to have once contained some 10,000 plant species, with an endemism rate somewhere between 40 and 60% (Gentry 1986). Since 1960, at least 95% of the forest cover has been destroyed, to make way for banana plantations, oil exploiters and human settlements of various sorts. According to the theory of island biogeography, which is supported by abundant and diversified evidence, we can realistically suppose that when a habitat has lost 90% of its extent, it has lost half its species. If we estimate that 5000 endemic plant species were in question, then 2500 species have already been eliminated or are on the verge of extinction.

Similar baseline figures for species totals and endemism levels, and a similar story of forest depletion (albeit for different reasons, and over a longer period of time), apply to the Atlantic-coast forests of Brazil, where the original 1,000,000 km^2 of forest cover have been reduced to less than 50,000 km^2 (Mori et al. 1981). Parallel data apply also to Madagascar, except that the endemism levels reach 80% or higher (Rauh 1979).

In these three tropical forest areas alone, then, with their documented total of 26,000 vascular plant species, around 12,400 of them being endemic, recent or near-future extinctions could well total some 6200 plant species. If, as is likely, the great majority of these extinctions will have occurred during the last 50 years of this century, there will have been eliminated an average of one plant species every three days. For further details, see table 1.

Table 1: Three hot-spot areas: Madagascar, Atlantic-coast Brazil and western Ecuador.

Area	*Original Forest, km^2*	*Remaining primary forest in 1987, km^2 (% of original)*	*Total of original plant species*	*Total of original plant endemics (and % of original species)*	*Total of plant species eliminated or on verge of extinction*[1]	*Remaining forest area as proportion of Earth's Land Surface*	*Total of original plant species as proportion of all Earth's plant species*
Madagascar	62,000	10,000 (16%)	6,000	4,900 (82%)	2,450	0.00675%	2.4%
Atlantic-Coast Brazil	1,000,000	20,000 (2%)	10,000	5,000 (50%)	2,500	0.0135%	4.0%
West. Ecuador	27,000	2,500 (9%)	10,000	2,500 (25%)	1,250	0.0017%	4.0%
Totals	1,089,000	32,500 (3%)	26,000	12,400 (48%)	6,200	0.02%	10.4%

[1] The number of animal species in a similar situation can be roughly estimated by multiplying the number of plant species by 20, thus supplying a *minimum* estimate. According to the calculations presented here, the total number of animal species in question is 124,000. The actual total could be several times higher. Supposing that most extinctions will have occurred during the last 50 years of this century, the average rate for the period works out at one plant species every three days, and several animal species, perhaps as many as five, every day.

When we consider a further 10 such "hot spot" areas in tropical forests – the Colombian Choćo, uplands of Western Amazonia, Rondonia/Acre in Brazilian Amazonia, the Tanzania/Kenya montane forests, the eastern Himalayas, the Sinharaja forest in Sri Lanka, peninsular Malaysia, north-western Borneo, the Philippines and New Caledonia – we find that rather more than 18,000 endemic higher-plant species (and at least 360,000 endemic animal species) are likely to have been eliminated already or to be on the verge of extinction. These plant species constitute just over 7 % of all plant species on Earth. For further details, see table 2.

Table 2. 13 hot-spot areas in tropical forests: summary.

1. Total extent of original hot-spot areas, 2,554,150 km^2, or 17% of original biome covering 15,000,000 km^2
2. Present extent of primary forest in hot-spot areas, 57,089 km^2, or 2.2% of present biome.
3. Number of endemic plant species in original forests: 36,250.
4. Original endemics as proportion of Earth's total of higher-plant species, 14.5%.
5. Endemic plant species already eliminated or on verge of extinction: 18,125, or 7.25% of Earth's total in 0.4% of Earth's land surface.
6. Endemic animal species already eliminated, or on verge of extinction: 362,500 – minimum estimate, could be several times more.

The mass extinction underway in 13 localities of tropical forests could well amount in itself to a greater extinction of plant species than has occured, so far as we can discern, in the prehistoric past. Generally speaking, terrestrial plant communities have survived with relatively few losses during the mass extinction events at the end of the Cretaceous and in the late Permian (Knoll 1986). How different this time, when there will surely be a "great dying" in tropical forests alone.

Looking at the situation another way, we can reckon, on the basis of what we know about plant numbers and distribution, that almost 20% of all plant species occur in forests of Latin America outside Amazonia; and another 20% in forests of Asia and Africa outside of the Zaire Basin (Raven 1985). That is, some 100,000 plant species altogether. All of the primary forests in which these species occur may well disappear by the end of this century or early in the next. If only half of the species in these forests disappear, this will surely cause the demise of 50,000 plant species.

How about the longer-term future? The immediate extinctions will be far from the entire story. During the next few centuries there will be delayed extinctions, delayed due to the time lag in „equilibriation", or deferred fall-out effects. Consider, for example, the case of Amazonia (Simberloff 1986). If deforestation continues at present rates until the year 2000, then comes to a complete halt (an unlikely prospect), we should anticipate an eventual loss of about 15 percent of the region's plant species; and were Amazonia's forest cover to be ultimately reduced to those areas now set aside as parks and reserves, we should anticipate that 66 percent of plant species will ultimately disappear.

Tropical forests and climatic change

Nor are protected areas likely to provide a sufficient answer, for reasons that reflect climatic factors. In Amazonia, for instance, it is becoming apparent that if as much as half of the forest were to be safeguarded in some way or another (through *e.g.* multiple-use conservation units as well as protected areas), but the other half of the forest were to be "developed out of existence", there could soon be at work a hydrological feedback mechanism that would allow a good part of Amazonia's moisture to be lost to the ecosystem (Salati & Vose 1984). The outcome for the remaining forest would likely be a steady desiccatory process, until the moist forest became more like a dry forest, even a woodland – with all that would mean for the species communities that are adapted to moist forest habitats. Even with a set of forest safeguards of exemplary type and scope, Amazonia's biota would be more threatened than ever.

Responsibility of the botanists' community

This, in brief, is the unfortunate prospect facing tropical forests. Quite properly it is an issue that has merited the attention of the thousands of botanists assembled at the XIVth International Botanical Congress, with its emphasis on forests of the world. While there are severe problems in temperate forests, especially from the threat of acid rain, it is surely in tropical forests that the greatest botanical problems arise. By the time of the next Botanical Congress in 1993, we may well have witnessed the demise of some 1000 plant species in these forests. Worse, we shall have watched the progressive impact of deforestation processes that will have made probable if not inevitable the extinction of many thousand more plant species within the foreseeable future. By the time of the next Botanical Congress after that, we shall be witnessing the full crisis of a greater extinction spasm of plant species than has likely occurred since the first emergence of higher plant species in the distant past.

And unlike occasions in the past, it will all be taking place within a few years: in the twinkling of a geologic eye.

But this supposes that the future continues to be a simple extension of the past. This need not be the case at all. We are only into the opening stages of an unprecedented extinction episode. There is still time – though only just time – to stem and slow, even to halt, this biological debacle. While we face a profound problem, we also face an unmatched opportunity. What an exceptional challenge for the botanists' community around the world, to start to save plant species in vast numbers. Not just to describe and analyse them, but to prevent them from being eliminated from the biosphere.

What is needed, with all due urgency, is a comprehensive plan to safeguard critical plant habitats in tropical forests. To develop such a plan, we shall need to mobilize all the botanical knowledge and understanding we can muster – and to do so in systematized fashion, targetted toward action-oriented goals. It will require the expertise and energies of the entire community of botanists, channelled into a coordinated campaign of global scope. Plainly nothing less will do to measure up to the biotic crisis.

Equally plainly the task does not lie beyond us. For the next Botanical Congress we may suerely look forward to an array of papers that document the accelerating progress of a save-species effort. Within the next few years we have opportunity to engage in a great cooperative endeavour that will elicit the admiration, as well as the gratitude, of generations of botantists into the indefinite future.

References

Ashton, P. S. 1981: Forest conditions in the tropics of Asia and the Far East. – Stud. Third World **13:** 169–179.

Ayala, F. J. (ed.) 1976: Molecular evolution. – Sinauer, Sunderland.

Erwin, T. L. 1988: The tropical forest canopy: the heart of biotic diversity. – In: Wilson, E. O. (ed.), Biodiversity: 105–109. – National Academy Press, Washington.

FAO/UNEP, 1982: Tropical forest resources. – Food and Agriculture Organization, Rome; United Nations Environment Programme, Nairobi.

Fearnside, P. M. 1985: Human-use systems and the causes of deforestation in the Brazilian Amazon. – Paper presented at UNU International Conference on Climatic, Biotic and Human Interactions in the Humid Tropics, San José, dos Campos, Sao Paolo, Brazil, 25th February-1st March, 1985.

Gentry, A. H. 1986: Endemism in tropical versus temperate plant communities. – In: Soule, M. E. (ed.), Conservation biology: the science of scarcity and diversity: 153–181. – Sinauer, Sunderland.

Knoll, A. H. 1986: Patterns of change in plant communities through geological time. – In: Diamond, J. & Case, T. J. (eds.), Community ecology: 126–141. – Harper & Row, New York.

Lewontin, R. C. 1974: The genetic basis of evolutionary change. – Columbia University Press, New York.

Malingreau, M.-P. & Tucker, C. J. 1988: Large-scale deforestation in the southern Amazon Basin of Brazil. – Ambio **17:** 49–55.

May, R. M. 1975: The tropical rainforest. – Nature **257:** 737–738.

Melillo, J. M., Palm, C. A., Houghton, R. A., Woodwell, G. M. & Myers, N. 1985: A comparison of recent estimates of disturbance in tropical forests. – Environmental conservation **12:** 37–40.

Molofsky, J., Hall, C. A. S. & Myers, N. 1986: A comparison of tropical forest surveys. – U. S. Department of Energy, Washington.

Mori, S. A., Bloom, B. M. & Prance, G. T. 1981: Distribution patterns and conservation of eastern Brazilian coastal forest tree species. – Brittonia **33:** 233–245.

Myers, N. 1980: Conversion of tropical moist forests [report to National Academy of Sciences]. – National Research Council, Washington.

– 1984: The primary source: tropical forests and our future. – Norton, New York.

– 1985: Tropical deforestation and species extinction: the latest news. – Futures **17:** 451–463.

– 1988: Tropical deforestation and remote sensing. – Forest Ecol. Managem. **23:** 215–225.

Rauh, W. 1979: Problems of biological conservation in Madagascar. – In: Bramwell, D. (ed.), Plants and Islands: 405–421. – Academic Press, London.

Raven, P. H. 1985: Statement from meeting of IUCN/WWF Plant Advisory Group, Las Palmas, Canary Islands, 24–25th November, 1985. – IUCN, Gland; Missouri Botanical Garden, St. Louis.

Salati, E. & Vose, P. B. 1984: Amazon Basin: a system in equilibrium. – Science **225:** 129–138.

Simberloff, D. 1986: Are we on the verge of a mass extinction in tropical rain forests? – In: Elliott, D. K. (ed), Dynamics of extinction: 165–180. – Wiley, New York.

UNESCO, 1978: Tropical forest ecosystems. – United Nations Educational, Scientific and Cultural Organization, Paris.

Address of the author: Dr. Norman Myers, Upper Meadow, Old Road, Oxford 0X3 8 SZ, UK.

Early land plants – the saga of a great conquest[1]

W. G. Chaloner

Abstract

Chaloner, W. G. 1988: Early land plants – the saga of a great conquest. – In: Greuter, W. & Zimmer, B. (eds.): Proceedings of the XIV International Botanical Congress: 301–316. – Koeltz, Königstein/Taunus.
Fossil evidence for the nature of the earliest vascular land plants and the timing of their appearance is reviewed. Fragmentary specimens of early land plants, large enough to show the association of several adaptive features (vascular tissue, a cuticle with stomata, and spores formed in tetrads) have a record reaching back to the earliest Devonian (c. 410 Ma). Specimens of which the age is more controversial extend this record back to the late Silurian (c. 420 Ma). This evidence is supplemented and extended by small tissue fragments and spores (plant microfossils) obtained by palynological preparations from boreholes and outcrops. Dispersed triradiate spore exines range down to the basal Silurian (c. 435 Ma), with adhering tetrads extending further back to about 440 Ma. Cuticle fragments with stomata do not pre-date the Devonian, but cuticles lacking stomata occur back to c. 440 Ma. Fragmentary tubes resembling tracheids have a record extending back to c. 435 Ma. The fossil record does little at present to bridge the gap in our picture from extant plants, between the simplest archegoniates, and those charophycean algae seen as their most plausible ancestral group. However, recent discoveries of early Devonian gametophytes suggest that they were comparable in size and general structure to the earliest known sporophytes.

Introduction

It has been said that the conquest of the land by higher plants was the single most far-reaching adaptive attainment of the autotrophs following the development of photosynthesis and the carbohydrate-based cell-wall. There are really two sagas involved under this title. One of them is the story of the biological events leading to the adaptation of plant life to survival under the stresses of the terrestrial environment. The other saga is that of the efforts of palaeobotanists to unravel and interpret the evidence which, with its varied imperfections, records what actually took place. The first saga goes back to the Palaeozoic era, perhaps some 450 million years. The second one goes back only to about the middle of the last century. This second saga is still unfinished, although a good account of the story so far is given in Henry Andrews

[1] This Paper is dedicated to Professor Ove Arbo Høeg, in the year of his 90th birthday.

"The fossil hunters" (1980). I shall concentrate here on the earlier, Palaeozoic saga, but elements of the palaeobotanists' saga come into the story.

In this review I shall concentrate principally on the evidence from the fossil record as to the timing and nature of the earliest appearance of vascular plants in the course of the evolution of life. However, recent work on the green algae bears so greatly on the origin of archegoniate plants and the nature of their ancestors that this source of evidence makes a major contribution to our reconstruction of the saga.

There are several core questions that can be directed at the fossil evidence. Some can be answered with conviction, while others may need to have their solutions sought elsewhere:

- What is accepted as incontrovertible evidence of a fossil vascular plant?
- Where and in what form do we first encounter evidence of vascular plants?
- Does the degree of diversity seen in early land plants suggest a single "migration", and a monophyletic origin, rather than a polyphyletic one?
- Does the fossil record give any evidence of the pathway of transition from aquatic green algae to terrestrially adapted forms?
- Was the origin of vascular land plants from green aquatic macrophytes, or from microscopic soil-living unicellular or filamentous algae, already adapted to this form of a terrestrial habitat? In more colloquial terms, this might be rephrased: was the land migration effected by an assault across the marshes or up the beaches – or by a fifth column of microscopic algae, already infiltrated into moist microhabitats on the land surface?
- What was the nature of the earliest alternation of generations? Are the two generations homologous or antithetic in character? And, as a separate question to this, does the earliest evidence of fossil gametophytes favour an antithetic origin or one from algae already showing isomorphic alternation?

The earliest land plants

The earliest record of the occurrence of many groups of organisms in the fossil record is often in varying degree both imperfect and controversial – the earliest acceptable vertebrate, amphibian, mammal and hominid all fit this pattern. Much the same can be said for the recognition of the earliest vascular plants. For this reason it is more satisfactory to start this review with an ancient vascular plant which is generally accepted and uncontroversial, but which is by no means "the oldest", and then to lead back in time onto less secure ground.

The morphologically simple vascular plant *Rhynia* of Kidston & Lang (1917) is still, some seventy years after its discovery, one of the most satisfactory of early vascular plants (see *e.g.* Chaloner & Macdonald 1980). All the plants preserved as fossils in the early Devonian Rhynie Chert were apparently fossilized by being inundated with boiling silica-bearing water associated with extrusive volcanic activity. This retained the microscopic detail of many of the plants growing at this site. As a result we know that the plant *Rhynia gwynne-vaughanii* was small and rather rush-like, with cylindrical upright axes which dichotomized and bore, at some of the apices, terminal sporangia of about the same width as the axis bearing them. The axes have a narrow central protostele with recognizable annular or helically thickened tracheids. There is an intercellular space system in the cortex, which was presumably photosynthetic since it led via stomata in the epidermis to the external atmosphere. Inside the sporangia, spores can be seen formed in tetrads, leaving each spore with a triradiate suture. No-one has had any reason to dispute the interpretation of the original authors of this species in regarding it as a very simple, undifferentiated homosporous vascular plant, reproducing by means of wind-dispersed spores and that this fossil represents the diploid sporophyte generation.

An important revision of our concept of this plant has recently been published by D. S. Edwards (1980), but this does not alter its status in relation to the present review. The same author has also transferred the other species previously included in the genus *Rhynia* to a new genus *Aglaophyton* (D. S. Edwards 1986) on account of its vascular tissue lacking the characteristic cell wall thickenings associated with xylem. Although in the years that have elapsed since the first description of the Rhynie Chert plants we have now come to know of many other late Silurian and early Devonian vascular plant fossils, *Rhynia* remains as one of the best preserved and most fully investigated and understood of early vascular plants. Other relatively undifferentiated leafless plants with terminal sporangia, designated by many authors as *Rhyniophytina,* following Banks's recognition of this group at subdivisional level, are generally known in much less detail. Some fourteen genera assigned to this group have recently been reviewed by Edwards & Edwards (1986). One of the most widely recognized of these other *Rhynia*-like plants from the late Silurian – early Devonian is the genus *Cooksonia,* first described by Lang (1937) from strata close to the Silurian-Devonian boundary in Britain. This genus, now known from a number of widespread localities principally in rocks of latest Silurian age, is preserved only as compression fossils, in which the plant tissue is squashed and coalified so that we do not have the same detail of three-dimensional structure that we have for *Rhynia*. However, we know that *Cooksonia* had a basically similar organization, with dichotomizing axes and terminal sporangia. Lang showed that his material had triradiate spores in the sporangia so that, as with *Rhynia*, we conclude that we are dealing with the products of a meiosis, and that by implication the plant bearing them is the diploid sporophyte. In other axes, of about the same age, not connected to sporangia, Lang also demonstrated the presence of fragments of vascular tissue. Consequently, even with specimens which lack evidence of spores or vascular tissue, but which resemble *Cooksonia* outwardly, palaeobotanists have generally regarded records of that genus as being vascular plants when these come from basal Devonian or Late Silurian rocks. Edwards et al. have recently reported (1986) fragments of *Cooksonia* of earliest Devonian (Gedinnian) age from the Welsh Border, showing stomata in the epidermis and ornamented spores within the sporangia. We therefore have collectively, right at the base of the Devonian, fragmentary plants with spores formed in tetrads in sporangia, with associated axes containing tracheids and others bearing stomata on the outer surface.

As we go back earlier into the Silurian, we have records of other vascular plants of Ludlow age, which are in varying degree less satisfactory than the *Cooksonia* records. Firstly, we have in Australia the *Baragwanathia* flora, containing the lycopsid of that name, which is an undoubted vascular plant, but the age of its earliest occurence remains in some dispute. However, we

Fig. 1. This diagram summarizes the evidence from plant microfossils relating to the time of appearance of certain features of structure relevant to the adaptation of plants to terrestrial existence. It covers only the interval ranging from the mid-Ordovician to the early Devonian. The items shown in "boxes" within each column represent the features in question (spores, stomata and xylem) *in situ* inside plant macrofossil remains or tissue fragments. The spores column shows triradiate spores and tetrads, typical examples being illustrated at each stage. The figures in that column are the numbers of species reported from each time unit, principally from Richardson (1985), Richardson & Ioannides (1973) and Gray & Boucot (1977). This number continues to rise through the Siegenian and Emsian, but reliable totals for those stages are not available. The figures in the "age" column indicate the time in millions of years of the start of each chronostratigraphic unit. The principal literature sources for the records of cuticles and stomata are Edwards et al. (1982), Edwards (1982), Pratt et el. (1978), Gray et. al. (1982), and for xylem and "tubes", Lang (1937), D. S. Edwards (1986), Edwards & Davies (1976), Strother & Traverse (1979) and Niklas & Smocovitis (1983). – The names of Silurian and Devonian time units (in part "series", in part "stages"), given in abbreviated form, are: Llandovery, Wenlock, Ludlow, Pridoli, Gedinnian, Siegenian and Emsian.

▷

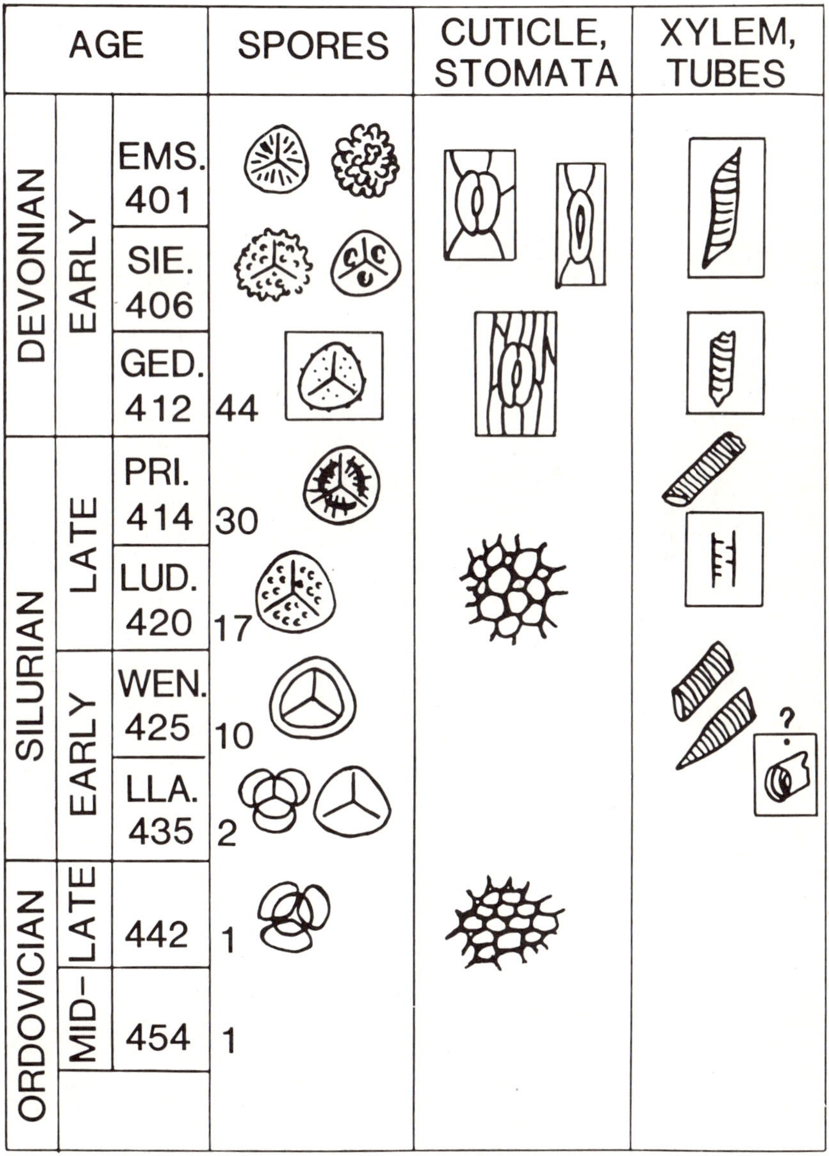
AGE
SPORES
CUTICLE, STOMATA
XYLEM, TUBES
DEVONIAN
EARLY
EMS. 401
SIE. 406
GED. 412
44
SILURIAN
LATE
PRI. 414
30
LUD. 420
17
EARLY
WEN. 425
10
LLA. 435
2
ORDOVICIAN
MID–LATE
442
1
454
1
?

have in the Irish Wenlockian (see fig. 1) *Cooksonia*-like plants which are preserved only as compression fossils, and lack evidence of spores, stomata or vascular tissue (Edwards et al. 1983). These, and similar late Silurian – early Devonian plants, have been designated "rhyniophytoid" (Pratt et al. 1978, Edwards & Edwards 1986). The latter authors offer a redefinition of this term suggesting that "rhyniophytoid be used for plants that look like rhyniophytes, but cannot be assigned unequivocally to that group because of inadequate preservation". Thus the rhyniophytoids, as they use the term, include fossils some of which may be tracheophytes, and others not. It is probably inevitable that in dealing with imperfectly preserved fossils we have to accept designations with a built-in ambiguity of this kind.

In addition to the Australian lower *Baragwanathia* flora, said to be of Ludlovian age, we have from China record of a plant described and very thoroughly documented by Geng (1986), from strata believed to be of Wenlockian age. Geng's genus, *Pinnatiramosus,* shows a distinctive pinnate branching pattern, reminiscent of some higher plant roots, but its anatomy reveals what is apparently conducting tissue, described guardedly by its author as formed of tubes, with scalariform or alternate simple pitting. Some of the fragments illustrated do indeed look extraordinarily like higher plant (?progymnosperm) vascular elements. This is undoubtedly a most significant discovery, but a fuller documentation of the basis for its age would greatly strengthen its status.

In review, the earliest record of plants with an adequate documentation of tracheophyte characters still lies in the basal Devonian (Gedinnian) *Cooksonia.* This may well be pre-dated by the oldest Australian *Baragwanathia,* but the earliest record of that flora is still regarded by some as controversial.

Evidence from microfossils

The impetus given to the study of plant microfossils by their use in hydrocarbon exploration has brought to light a wide range of material which bears on the timing and nature of the land colonization. Microscopic plant fossils in this category include not only spores but fragments of cuticular covering, and individual cells resembling the vascular elements of tracheophytes. Such plant fragments occur, sometimes in considerable numbers, in rock samples obtained in exploratory boreholes, and those most relevant to the present review are from rocks of early Devonian to Ordovician age (fig. 1). Full consideration has been given to the occurrence and significance of such microfossils by Banks (1975), D. Edwards (1982), Gray (1985), Gray & Boucot (1977), Pratt et al. (1978), Richardson (1985) and Strother & Traverse (1979). The most widespread are probably spores, generally regarded as representing exines composed of sporopollenin, and showing evidence of formation in a tetrad either by having a triradiate mark (as in so many living archegoniate plant spores) or by actually adhering together in tetrads, even when in the dispersed state. It may fairly be said that this evidence of tetrad configuration generally separates these spores, (which are accordingly considered, by analogy with living plants, as being the products of a meiosis), from the planktonic cysts of dinoflagellate algae or the more enigmatic acritarchs, which lack evidence of formation in a tetrad. While it has been argued that the majority of Devonian triradiate spores probably represent tracheophytes (Chaloner 1970) the source of the Silurian and late Ordovician spores has been a subject of considerable controversy. Speculation about their likely source is fully explored in the papers cited above. As with any detached organs of plants, there are great difficulties in attributing them to putative parents with any confidence. Where there is close and detailed similarity between dispersed spores and those of about the same age found inside sporangia, a suggested relationship is at least plausible (see Allen 1981, Gensel 1980). When we are dealing with older spores which pre-date any known within sporangia, then the strength of such attribution becomes progressively weaker. Even a very broad-based assignment is not always possible for well-preserved spores with a clear triradiate suture, since such spores are not only produced by many tracheophytes and bryophytes but by other enigmatic fossil plants of unknown affinity. The most often-quoted examples of this are the triradiate spores formed by the late Devonian plant *Foerstia,* also known as *Protosalvinia* (see Gray & Boucot 1979, and Schopf 1978). There is still a wide range of opinions concerning the affinity of this plant, but all authors are agreed that it is not a bryophyte, much less a tracheophyte. This illustrates the hazards in making general attributions of any early Palaeozoic triradiate spores to parental groups on grounds of morphology alone.

The record of dispersed spores (Gray 1985, Richardson 1985, and references there cited) makes it clear (fig. 1) that adhering tetrads of spores first appear in the late Ordovician, and continue in the early Silurian where they are joined by the earliest dispersed triradiate spores. Gray attaches much significance to the biological implications of the fact that such adhering tetrads of spores pre-date those which became separated from the tetrad (presumably prior to dispersal) and show a triradiate slit, representing the site of presumed germination on the proximal surface. She suggests that the close similarity of these (late Ordovician – early Silurian) fossil spores adhering in "obligate" tetrads, to those produced by some extant bryophytes "suggests a non-vascular vegetative grade of organisation for plants of this interval". She further proposes that these tetrads may have involved a sexual segregation, by analogy with some living hepatics, with separate male and female gametophytes developing from the members of each tetrad. If this is true, then unisexual gametophytes would be the primitive state in these precursors of the archegoniates. However, the widespread occurrence of bisexual gametophytes, both in bryophytes otherwise believed to be primitive and certainly in all homosporous pteridophytes, makes this seem improbable. Indeed other authors have argued for very different interpretations of these spores; Smith (1979) for example suggests pre-Downtonian spores might represent aquatic plants. For the present it may be sufficient to acknowledge that spores resembling those of some living archegoniates extend down into the Ordovician, but that very little can safely be said about the habitat, reproductive biology or systematic affinitiy of the plants that produced them.

Fragments of cuticle occur in varying abundance through the same kind of lithologies that have yielded spores from the early Devonian and Silurian. The possession of a water-retaining cuticle has long been acknowledged as one of the more obviously "adaptive" features shown by vascular land plants, and its role seems to be confirmed by the general correlation of plants possessing a thick cuticle with water-stress habitats. Raven (1986) emphasises the combined role of xylem, cuticle, an internal gas distribution system (intercellular spaces) and stomata as a syndrome of adaptive features related to survival on land. Obviously, the cuticle without the pathway for gas exchange represented by stomata would only achieve water retention while enormously restricting the scope of photosynthetic activity. In this connexion it is interesting to note that our record of stomata seen within a cuticle dates only from those macrofossil plants for which we have other evidence of terrestrial adaptation (presence of xylem, structural similarity to known tracheophytes) of early Devonian age (see "boxes" in fig. 1 representing evidence from cuticles *in situ).* We still have no evidence of stomata from Pre-Devonian plants despite the large number of records of cuticle fragments through the Silurian. D. Edwards (1982) has recently reviewed the range of cuticle types seen in this way. In particular she gives consideration to the pieces of cuticle to which Lang had given the name *Nematothallus;* Lang (1937) found an association between pieces of cuticle lacking stomata and some fossil tubes with bands of thickening similar to those of tracheophyte vascular elements. Such "tubes" are considered further below, but in view of their ubiquity in Silurian sediments from many localities, their association with the pieces of cuticle now seems less significant than it appeared to Lang. Indeed, with good justification,

D. Edwards suggests that the evidence for connexion between tubes and cuticle is not sufficient to link them as a single plant, and proposes that the name *Nematothallus* should only be construed as applying to the cuticle alone.

In the early Devonian and Silurian cuticles studied by D. Edwards (1982) there is evidence, of varied quality, that some may have had pores (*i.e.* holes, of about the size of the cells forming the cellular reticulum) in the original cuticle. It is tempting to speculate that these may have had a role analogous to that of stomata, combining a largely water-retaining cuticle with the pores offering a pathway for respiratory and photosynthetic gas exchange. We have in rather later Devonian plants of thalloid organisation *(Spongiophyton* – see Chaloner et al. 1974) evidence of a cuticle with much larger pores, but no stomatal guard cells. This later Devonian plant was apparently terrestrially adapted but was not a vascular plant. It can perhaps only be acknowledged that a range of plants may have produced cuticle as an adaptation to a terrestrial habitat, but that some of the Silurian forms at least did not combine this with the closable gas-exchange structures represented by stomata or any equivalent. The limited value of the microfossil evidence is perhaps only to suggest that the evolution of a cuticle preceded that of stomata by several million years, and apparently predated the development of a vascular system as we see it in tracheophytes.

The occurrence of the earliest record of cells showing the character of xylem elements might be seen as the most acceptable of microfossil evidence for the appearance of tracheophytes, but as in the case of spores and cuticles, various uncertainties and ambiguities blur the picture. Lang prepared some fragments of xylem-like cells by means of a celloidin "pull" preparation from compressions of small plant axes associated with fertile *Cooksonia* axes from the early Devonian of the Welsh Border (Lang 1937). For many years these were regarded as the oldest record of *in situ* vascular tissue. They were undoubtedly not as convincing as the vascular tissue seen *in situ* in *Rhynia gwynne vaughanii* for example, but for many years they were regarded by most palaeobotanists as the earliest acceptable evidence of vascular tissue within a plant fossil. More recently, Edwards & Davies (1976) have illustrated remains of similar fragmentary xylem-like cells within an axis from rather earlier in the Silurian (see item in "box", right-hand column of fig. 1). This probably represents the earliest acceptable evidence of a vascular plant, although the security of the record as unequivocal xylem is perhaps marginally lower than that of Lang's specimen. Again, microfossil evidence extends that from such *in situ* records. We have in Silurian rocks many records of microscopic tubes, of about the dimension of xylem elements, showing either helical or annular bands of thickening on their inner walls. Some authors have regarded their structural similarity to vascular elements as indicating a comparable role (Gray 1985, Taylor 1982), while others (Banks 1975) acknowledge the possibility of their being of entirely different origin, possibly representing remains of marine animals as had been suggested by at least one earlier author. Evidence from the composition of the organic matter contained in the matrix bearing these Silurian tubes suggests the presence of lignin-like residues (Niklas & Pratt 1980). However, as more precisely focussed techniques become available, the composition of the tubes themselves may provide stronger evidence of their ultimate affinity.

Further evidence of the nature of these tubes is offered by an illustration in Niklas & Smocovitis (1983) showing one such tube apparently enclosed within a tissue of non-vascular cells. Their single illustrated fragment is remarkable in also being completely non-flattened, although from a matrix in which other microfossils such as dispersed tubes appear to have collapsed completely. Corroborative evidence of more material of these *in situ* tubes from this significant early Silurian occurrence is needed.

In review, our earliest record of vascular plants combining secure evidence of vascular tissue, spores in sporangia and stomata in the epidermis is still in the earliest Devonian. If, as some maintain, the earliest record of the Australian *Baragwanathia* flora is Ludlovian, then for the combination of acceptable xylem and *in situ* spores, at least (Lang & Cookson 1935), this becomes our oldest record of acceptable tracheophytes (see contrasted views of Hueber 1983 and Garratt et al. 1984). Microfossil records of cuticle lacking stomata extend to the basal Silurian, and spores formed in tetrads to the late Ordovician. Tubes resembling vascular elements also extend back to the earliest Silurian. However, each of these types of microfossil element (spores, cuticles and tubes) stands on its own merit, as limited testimony of one aspect of adaptation to terrestrial life. We have no secure evidence linking them in a single plant before the late Silurian, and this leaves them making only a limited contribution to our record of the saga.

Algal Ancestry

Contemplation of the kind of algal ancestors which may have given rise to the tracheophytes has focussed on the Chlorophyta which share a range of features (photosynthetic pigments, storage products, cellulosic wall) with all archegoniate plants. An admirable review of the current state of thought in this field is given by Stebbins & Hill (1980), while Taylor (1982) reviews the interaction of evidence derived from fossil plants with that from living algae. Some seventy years ago, the Oxford botanist Church (1919) invented a hypothetical ancestral green algal group as a starting point for the "subaerial transmigration", and named them the Thalassiophyta. He pictured their effecting land colonization by what might be called frontal assault, postulating invasion of the terrestrial habitat via the intertidal zone now dominated (in temperate latitudes) by the brown algae. Church suggested that the Thalassiophyta as such became extinct, leaving no direct descendents other than the living archegoniates.

Church's ideas (now largely forgotten or superseded) represent one extreme of a range of possibilities regarding algal ancestry. Thinking on these lines has undergone a radical revision in the last fifteen years with a much fuller exploration of the common ground of biochemical, histological and ultrastructural features shared between the archegoniates and certain green algae. These are reviewed in Mattox & Stewart (1984), Round (1984) and O'Kelly & Floyd (1984). The bulk of the green algae, recognized as the *Chlorophyceae,* differ in a number of seemingly rather fundamental features from those green algae now generally recognized as the subdivision *Charophyceae.* A number of the larger green algae with parenchymatous organization *(e.g. Ulva)* now seem less attractive as possible antecedents to the archegoniates, so reversing the thinking of only a few years ago (*e.g.* Stewart 1983). The *Charophyceae* on the other hand, including *Chara* itself and smaller forms such as *Coleochaete,* sharing a number of features with the archegoniates, have therefore come under renewed scrutiny as possible living representatives of the ancestral group of the tracheophytes.

The *Charales* themselves – and perhaps the whole of the extant *Charophyceae* – are regarded by some as unlikely candidates (Stebbins & Hill 1980, O'Kelly & Floyd 1984, Chapman 1985), being a specialized group distinct from, but sharing important features with, the land plant ancestral stock. However, *Coleochaete* is perhaps a plausible representative of the type of charophycean alga that might have inhabited moist terrestrial habitats pior to the evolution of the archegoniate life cycle. The retention and partial enclosure of the egg (zygote) in that genus and the recent evidence in the form of transfer cells that food is passed from the haploid plant to the zygote are among a number of significant features supporting such a possibility (Graham 1984). Stebbins & Hill (1980) make a vigorous case for the general thesis that microscopic plants must have existed on land in moist and probably impermanent microhabitats from the earliest appearance of such algae (or at least from such time that the incident UV had fallen sufficiently to allow land colonization in that form). Those authors argue convincingly that small soil-living algae comparable in general terms to *Coleochaete* are a more plausible source for archegoniate recruitment than larger aquatic green algae. Further consideration of the consequences of this line of thought takes us into the matter of alternation of generations, and this is dealt with below.

Whether a traditional macrophyte land invasion is postulated, or one accepts a "fifth column takeover" by soil algae, there is still a major gap in our picture of the sequence of land colonization, as it is seen in the fossil record. We have quite an adequate fossil record of supposedly green marine algae, principally calcareous forms, from the Cambrian onwards (for a recent brief review, see Meyen 1987). However, these generally resemble living *Chlorophyceae* rather than *Charophyceae;* the latter group only appears late in the Silurian. We still have no plausible fossil algal links to bridge between the microscopic "terrestrial" *Charophyceae* of the present day and the earliest tracheophytes seen in the fossil record.

Fossil Gametophytes and the origin of alternation

The last twenty years have seen a renewed interest in the question of the gametophyte generation of the earliest vascular plants. Even where the prospects of the preservation of rather delicate plant structures seem good, as for example in the Rhynie Chert, the possible survival of fossil gametophytes had generally been dismissed. Kidston and Lang had attributed the lack of evidence of gametophytes for any of the Rhynie plants, as they construed them, as due to their destruction prior to fossilization. Mercker (1959), Pant (1962) und Lemoigne (1968) drew attention to the possibility, which they developed in slightly different versions, that some of the supposedly sporophytic vascularized axes seen in the Rhynie Chert might in fact represent gametophytes. With increased knowledge of the long-lived, cylindrical, branching subterranean gametophytes seen in the living *Ophioglossales* and the *Psilotales,* this seemed less improbable than it might have appeared in Kidston and Lang's time. Lemoigne figured a plausible sunken archegonium on such a Rhynie axis, but antheridia were less convincing. He went on to suggest that *Rhynia gwynne-vaughanii* might be the gametophyte corresponding to the sporophyte, *Rhynia major.* The subsequent confirmation by D. S. Edwards (1980) that *R. gwynne-vaughanii* did indeed bear sporangia with spore tetrads, as Kidston and Lang had suggested, did not of itself disprove the general case for Rhynie vascularized gametophytes, but merely altered this nomenclatural aspect of the debate. The evidence that some of the Rhynie axes could be gametophytes remains equivocal, but the possibility that these early vascular plants had cylindrical, sporophyte-like gametophytes is still with us.

Renewed interest in fossil gametophytes was generated by Remy & Remy's (1980) recent description of *Lyonophyton,* an upright gametangium-bearing lobed structure from the Rhynie Chert, with suggestive similarities to a marchantialean gametangiophore (see also Remy 1982). This fossil, unlike Lemoigne's gametophytes, bears very convincing antheridia (Remy & Remy 1980: pl. 12), but the evidence of archegonia is rather less persuasive. There is some similitary between *Lyonophyton* and members of the genus *Sciadophyton,* a plant known only as compression fossils from the early Devonian, and recently interpreted as being of similar structure and status as *Lyonophyton* (Remy & Hass 1986, Schweitzer 1983). Schweitzer has illustrated specimens preserved as compressions which suggest that *Sciadophyton* may represent the gametophyte generation of *Zosterophyllum rhenanum,* and possibly of other early vascular plants. While his vigorous reconstructions of stages of the *Zosterophyllum/Sciadophyton* life-cycle are appealing, the state of preservation of the material leaves the relationship between these two genera and the gametophytic status of *Sciadophyton* still open to debate. Remy & Hass (1986) review at length the possible significance of these early Devonian gametophytes, and see in them a measure of support for an origin of tracheophytes from algal ancestors already adapted to a terrestrial existence.

A more cautious view of these gametophyte-like structures might simply be to acknowledge that we have now at least some evidence of Devonian fossils associated with tracheophytes, which bear gametangia. But their relationship with specific sporophyte plants is still speculative, and a real understanding of any diploid/haploid life cycle unproven. However, the gametophyte-like bodies already described from the Rhynie Chert, if taken at their face value, seem to indicate that at this early stage of land plant evolution, gametophytes were present of more or less comparable size and structure to the sporophytes. However this is interpreted, it has important implications for our understanding of the origin of alternation of generations, and emphasizes the need for further search for evidence of Devonian gametophytes.

References

Allen, K. 1981: A comparsion of the structure and sculpture of *in situ* late Silurian and Devonian spores. – Rev. Palaeobot. Palynol. **34:** 1–9.

Andrews, H. N. 1980: The fossil hunters. – Cambridge.

Banks, H. P. 1975: The oldest vascular land plants: a note of caution. – Rev. Palaeobot. Palynol. **20:** 13–25.

Chaloner, W. G. 1970: The rise of the first land plants. – Biol. Rev. Cambridge Philos. Soc. **45:** 353–377.

– & Macdonald, P. 1980: Plants invade the land. – Edinburgh.

–, Mensah, M. K. & Crane, M. D. 1974: Non-vascular land plants from the Devonian of Ghana. – Palaeontology **17:** 925–947.

Chapman, D. J. 1985: Geological factors and biochemical aspects of the origin of land plants. – In: Tiffney, B. H. (ed.), Geological factors and the evolution of plants. – Newhaven.

Church, A. H. 1919: Thalassiophyta and the subaerial transmigration. – Oxford Bot. Mem. **3:** 1–95.

Edwards, D. 1982: Fragmentary non-vascular plant microfossils from the late Silurian of Wales. – Bot. J. Linn. Soc. **84:** 223–256.

– & Davies, E. C. W. 1976: Oldest recorded *in situ* tracheids. – Nature **263:** 494–495.

– & Edwards, D. S. 1986: A reconsideration of the Rhyniophytina Banks. – In: Spicer, R. A. & Thomas, A. B. (eds.), Systematic and taxonomic approaches in palaeobotany. – Syst. Assoc. Special Volume **31:** 199–220.

– & Rayner, R. 1982: The cuticle of early vascular land plants and its evolutionary significance. – In: Cutler, D. F., Alvin, K. L. & Price, C. E. (eds.): The plant cuticle. – Linn. Soc. Symp. Ser. **10:** 341–361.

–, Fanning, U. & Richardson, J. D. 1986: Stomata and sterome in early land plants. – Nature **323:** 438–440.

–, Feehan, J. & Smith, D. G. 1983: A late Wenlock flora from Co. Tipperary, Ireland. – Bot. J. Linn. Soc. **86:** 19–36.

Edwards, D. S. 1980: Evidence for the sporophytic status of the Lower Devonian plant *Rhynia gwynne-vaughanii* Kidston and Lang. – Rev. Palaeobot. Palynol. **29:** 177–188.

– 1986: *Aglaophyton major* (Kidston & Lang) comb. nov., a non-vascular land-plant from the Devonian Rhynie Chert. – Bot. J. Linn. Soc. **93:** 173–204.

Garratt, M. J., Tims, J. D., Rickards, R. B., Chambers, T. C. & Douglas, J. G. 1984: The appearance of *Baragwanathia (Lycophytina)* in the Silurian. – Bot. J. Linn. Soc. **89:** 355–358.

Geng, B.-Y. 1986: Anatomy and morphology of *Pinnatiramosus,* a new plant from the Middle Silurian (Wenlockian) of China. – Acta Bot. Sin. **28:** 664–670.

Gensel, P. G. 1980: Devonian *in situ* spores: a survey and discusion. – Rev. Palaeobot. Palynol. **30:** 101–132.

Graham, L. E. 1984: *Coleochaete* and the origin of land plants. – Amer. J. Bot. **71:** 603–608.

Gray, J. 1985: The microfossil record of early land plants: advances in understanding of early terrestrialization, 1970–1984. – In: Chaloner, W. G. & Lawson, J. O. (eds.), Evolution and environment in the late Silurian and early Devonian. – Philos. Trans., Ser. B, **309:** 167–195.

– & Boucot, A. J. 1977: Early vascular land plants: proof and conjecture. – Lethaia **10:** 145–174.

– & – 1979: The Devonian land plant *Protosalvinia.* – Lethaia **12:** 57–63.

–, Massa, D. & Boucot, A. J. 1982: Caradocian land plant microfossils from Libya. – Geology **10:** 197–201.

Hueber, F. M. 1983: A new species of *Baragwanathia* from the Sextant Formation (Emsian), Northern Ontario, Canada. – Bot. J. Linn. Soc. **86:** 57–79.

Kidston, R. & Lang, W. H. 1917: On Old Red Sandstone plants showing structure from the Rhynie chert bed, Aberdeenshire – Part I, *Rhynia gwynne-vaughani* Kidston and Lang. – Trans. Roy. Soc. Edinburgh **51:** 761–784.

Lang, W. H. 1937: On the plant remains from the Downtonian of England and Wales. – Philos. Trans., Ser. B, **227:** 245–291.

– & Cookson, I. C. 1935: On a flora, including vascular land plants, associated with *Monograptus,* in rocks of Silurian age, from Victoria, Australia. – Philos. Trans., Ser. B, **224:** 421–449.

Lemoigne, Y. 1968: Observation d'archégones portés par des axes de type *Rhynia gwynne-vaughanii* Kidston et Lang. Existence de gamétophytes vascularisés au Dévonien. – Compt. Rend. Hebd. Séances Acad. Sci. **266:** 1655–1657.

Mattox, K. R. & Stewart, K. D. 1984: Classification of green algae: a concept based on comparative cytology. – In: Irvine, D. E. G. & John, D. M. (eds.), Systematics of the green algae. – Syst. Assoc. Special Volume **27:** 29–72.

Merker, H. 1959: Analyse der Rhynien-Basis und Nachweis des Gametophyten. – Bot. Not. **112:** 441–452.

Meyen, S. 1987: Fundamentals of palaeobotany. – London.

Niklas, K. J. & Pratt, L. M. 1980: Evidence for lignin-like constituents in early Silurian (Llandoverian) plant fossils. – Science **209:** 396–397.

– & Smocovitis, V. 1983: Evidence for a conducting strand in early Silurian (Llandoverian) plants: implications for the evolution of the land plants. – Paleobiology **9:** 126–137.

O'Kelly, C. J. & Floyd, G. L. 1984: Flagellar apparatus absolute orientations and the phylogeny of the green algae. – Biosystems **16:** 227–251.

Pant, D. D. 1962: The gametophyte of the Psilophytales. – In: Maheshwari, P., Johri, B. M. & Vasil, I. K. (eds.), Proceedings of the summer school of botany – Darjeeling, 1960: 276–301. – Darjeeling.

Pratt, L. M., Phillips, T. M. & Dennison, J. M. 1978: Evidence of non-vascular land plants from the early Silurian (Llandoverian) of Virginia, USA. – Rev. Palaeobot. Palynol. **25:** 121–149.

Raven, J. A. 1986: Evolution of plant life forms. – In: Givnish, T. J. (ed.), On the economy of plant form and function: 421–476. – Cambridge.

Remy, W. 1982: Lower Devonian gametophytes: relation to the phylogeny of land plants. – Science **215:** 1625–1627.

– & Hass, H. 1986: Das Ur-Landpflanzen-Konzept – unter besonderer Berücksichtigung der Organisation altdevonischer Gametophyten. – Argumenta Palaeobot. **7:** 173–214.

– & Remy, R. 1980: *Lyonophyton rhyniensis* nov. gen. et. nov. spec., ein Gametophyt aus dem Chert von Rhynie (Unterdevon, Schottland). – Argumenta Palaeobot. **6:** 37–72.

Richardson, J. B. 1985: Lower Palaeozoic sporomorphs: their stratigraphic distribution and possible affinities. – In: Chaloner, W. G. & Lawson, J. D. (eds.), Evolution and environment in the late Silurian and early Devonian. – Philos. Trans., Ser. B, **309:** 201–205.

– & Ioannides, N. 1973: Silurian palynomorphs from the Tanezzuft and Acacus formations, Tripolitania, North Africa. – Micropaleontology **19:** 257–307.

Round, F. E. 1984: The systematics of the Chlorophyta: an historical review leading to some modern concepts. – In: Irvine, D. E. G. & John, D. M. (eds.), Systematics of the green algae. – Syst. Assoc. Special Volume **27:** 1–27.

Schopf, J. M. 1978: *Foerstia* and recent interpretations of early, vascular land plants. – Lethaia **11:** 139–143.

Schweitzer, H.-J. 1983: Der Generationswechsel der Psilophyten. – Ber. Deutsch. Bot. Ges. **96:** 483–496.

Smith, D. G. 1979: The distribution of trilete spores in Irish Silurian rocks. – In: Harris, A. L., Holland, C. H. & Leake, B. E. (eds.), The Caledonides of the British Isles – reviewed: 423–431. – London.

Stebbins, G. L. & Hill, G. J. C. 1980: Did multicellular plants invade the land? – Amer. Naturalist **115:** 342–353.

Stewart, W. N. 1983: Paleobotany and the evolution of plants. – Cambridge.

Strother, P. K. & Traverse, A. 1979: Plant microfossils from Llandoverian and Wenlockian rocks of Pennsylvania. – Palynology **3:** 1–21.
Taylor, T. 1982: The origin of land plants: a paleobotanical perspective. – Taxon **31:** 155–177.

Address of the author: Professor William G. Chaloner, Biology Department, Royal Holloway & Bedford New College, Egham Hill, Egham, Surrey, TW20 OEX, UK.

Stability versus change, or how to explain evolution

F. Ehrendorfer

Abstract

Ehrendorfer, F. 1988: Stability versus change, or how to explain evolution. – In: Greuter, W. & Zimmer, B. (eds.): Proceedings of the XIV International Botanical Congress: 317–333. – Koeltz, Königstein/Taunus.

Stability versus change is one of the many aspects of a group's evolutionary success or failure in terms of, *e.g.*, its dominance and biomass, spatial expansion, occupation of different niches and habitats, range of differentation, production of races, species, genera . . ., historical continuity, etc. Generally, evolutionary success or failure depend on a very complex interplay of not clearly separable organismal (endogenous) and environmental (exogenous) parameters. Different individuals, populations, races, species, etc. of a particular group are often affected in a different (positive, neutral, negative) ways by this interplay. Changes which result from this interplay in groups of organisms must necessarily be integrated into their endo- and exogenous parameter syndrome, and are, therefore, limited to a number of available pathways; this leads to characteristic patterns of evolutionary differentation. Examples from the higher plants (*Rubieae, Anthemideae* and *Rutaceae)* illustrate our still very limited understanding of such correlations, integrations and patterns on the individual, population, infra- and supraspecific level.

Introduction

This is one of the key questions concerning evolution and phylogeny: Why are there taxa which have remained stable and uniform over millions of years, and often occupy specialized and regressive habitats, and why are there others which have rapidly changed and differentiated into a multitude of often very polymorphic, ecologically versatile and aggressive descendants?

Obviously, we are confronted here with a very general and at the same time enormously complex problem of biological evolution. This may be the reason why it has attracted relatively little attention, in spite of the current and relevant discussion on gradual versus punctuated evolution. Within the limits of this overview, I can do no more than present a few pertinent examples, mainly from the research work of our team at the University of Vienna, and thus illustrate some of the general aspects of this problem.

First, we have to ask how to approach and possibly quantify stability versus change on the different levels of evolution:

- Individual level: phenotypic and ontogenetic plasticity, etc.
- Population level: gene pool, recombination rate, mutability, etc.
- Infraspecific level: racial differentiation with its cytogenetical, phytochemical, ecogeographical and other aspects.
- Specific and supraspecific levels: rate of origin, transformation and extinction of species and higher taxa.

Clearly, these levels are interrelated and interdependent, but we are still sadly ignorant about such relationships.

Second, there is the question as to what an extent stability versus change in higher plant evolution is influenced by various parameters and syndroms:

- Inside the organisms: heredity, development, ecophysiology, phytochemistry, morphology, reproductive biology, etc.
- Outside the organisms, in the environment: climates, soils, spatial structures and temporal-historical changes of ecosystems, etc.

Here again we are confronted with a complex network of interconnected phenomena which are obviously very difficult to separate and to appreciate as to their individual importance.

Cruciata and *Valantia*

Let me turn first to examples from the *Rubiaceae,* tribe *Rubieae,* and try to illustrate mainly the species and infraspecies level and thus microevolutionary aspects of our problem. The small and closely related genera *Cruciata* and *Valantia* from the Near East, the Mediterranean, and temperate W Eurasia afford good insights into the ways in which different enviroments canalize chromosomal evolution, patterns of racial differentation, changes in reproductive biology, dispersal and distribution, and show how all these parameters are interrelated. *Cruciata* and *Valantia* share the trend from a perennial to an annual life form. Their narrow thyrsus-like inflorescences with few-flowered cymes under or between the leaf whorls and their mericarps borne on recurving peduncles exhibit various specializations for fruit diespersal, particulary in the annuals (Ehrendorfer 1962b, 1965, 1970).

Among the perennial members of *Cruciata* there is a closely related pair of species (Ehrendorfer 1971, 1980): *C. laevipes (=Galium cruciata)* is a rather uniform and consistently diploid member of the temperate deciduous forest flora, ranging widely from the W Himalayas through N Iran, the Caucasus and N Anatolia into C and W Europe. In contrast, *C. taurica* sensu lato *(=C. coronata)* is an enormously polymorphic polyploid complex with a large number of very different but mostly ill-defined 2x, 4x, 6x and 8x races, extending from rock or talus habitats in the Pontic forest zone to alpine and steppe biota in SW Asia, *i.e.* from Turkmenia, Iran und NE Iraq to N Palestine, Anatolia, Georgia, the Crimea, and Euboea in Greece. There is little doubt that with similar evolutionary potentials, *C. laevipes* has changed very little from an assumed mesophilous arcto-tertiary ancestor in a stable habitat, whereas the more or less xerophilous *C. taurica* has been stimulated into a most remarkable cytogenetic differentiation and eco-radiation as a member of various more or less unstable open and dry habitats of the Near East, ultimately strongly modified and expanded under human influence.

For both genera, *Cruciata* and *Valantia,* xeric Near East environments have triggered the origin of annual from ancestral perennial species. In both cases this has been accompanied by remarkable specialisations in their fruit dispersal mechanisms and in their genetic systems. Whereas x = 11 is characteristic for all perennial taxa of *Cruciata* and *Valantia,* descending dysploidy has occurred, down to x = 9 and x = 5 in the former and to x = 10 and x = 9 in the latter genus.

Fruit specialistation in both genera has its start in their common trend for postfloral peduncle and pedicel recurving. In the perennials mericarps drop individually, in the annuals mericarps remain fixed and become incorporated into complex dispersal units, first including only the ± enlarged and reflexed leaves in *Cruciata* or the elaborate penduncles in *Valantia,* but eventually consisting of complete adhesive fruiting individuals. In the outbreeding annual *Cruciata articulata* with 2n = 10, growing in wind-swept Near Eastern semi-deserts, the in frutescence breaks into anemochorous propeller-like whorls, whereas the inbreeding annual *C. pedemontana* with 2n = 18 from submediterranean steppic woodlands has become miniaturized, and whole fruiting plants are adapted to epizoochorous dispersal.

In *Valantia* the only perennial and outbreeding species is *V. aprica* with 2n = 22; it occupies montane to alpine skeleton soils from the Greek mainland to Crete. All other species have become inbreeding annuals: *V. columella,* a tetraploid on the basis x = 10 (2n = 40), has specialized by means of long silky hairs on its peduncles into an anemochorous semi-desert therophyte in N Africa (from Egypt to Tunis). *V. hispida* and the more strongly miniaturized *V. muralis,* both diploids with 2n = 18, have developed bristles and hooks on their peduncles and thus tend to epizoochory. Both have become very successful pioneers and weeds, the former ranging from the

Near East to the Canary Islands, the latter concentrated in the C and W Mediterranean. Further differentiation within this species group has led to the origin of two local neoendemics which exhibit reduction or loss of their epizoochorous adaptations, *i. e., V. deltoidea,* a tetraploid (2n = 36) on Mt. Busambra in Sicily, and *V. calva* (2n = 18) on the little island of Linosa S of Sicily (BRULLO 1979, 1980).

Both in *Cruciata* and in *Valantia,* the annuals, *i.e. C. pedemontana, C. articulata, V. columella* and the *V. hispida-V. muralis* group, have undergone major reorganisations of their karyotypes as compared to their perennial ancestors. This, their deep morphological separation and the total lack of "missing links" strongly suggest that these dysploid annual groups are the result of "catastrophic" selection and very rapid evolution in small populations and under the stress of extreme environmental conditions of the Near East. This pattern of evolution has been demonstrated for several other annual angiosperm groups but is rare in the *Rubieae,* where changes from the basic x = 11 have been recorded only for one other annual species of *Galium (G. spurium,* x = 10) and for three species groups of perennials in *Asperula* (sect. *Cruciana* and sect. *Cynanchicae,* both x = 10) and in Galium (sect. *Aparinoides,* x = 12).

Galium

To what an extent environmental parameters can influence the patterns of evolutionary differentiation, dispersal and distribution, can be further demonstrated by four related perennial and outbreeding groups of *Galium* (cf. Ehrendorfer 1970, 1975, 1976b; Krendl 1967; Teppner & al. 1976). They all exhibit a comparable breadth of morphological diversity, basically the same reproductive biology and the same cytogenetic potential (auto- and allopolypolidy: 2x-4x-6x-8x, etc.), and they all have reached a comparable total area in Europe, the Mediterranean and SW Asia; but of this region as shown in following surrey:

– *G. palustre* group	swamps and marshes,	4 species;
– *G. sylvaticum* group	deciduous forests,	12 species;
– *G. mollugo and G. lucidum* groups	brush and grassland, open soils,	15 species;
– *G. pusillum* group	grassland, rocks and talus,	26 species.

In its rather homogeneous and continuous swamp and marsh habitats, the *G. palustre* group has produced only four recognizable species. In the much more heterogeneous and discontinuous habitats of the *G. pusillum* group, we can register no less than 26 species. The other groups are found in ecosystems that are somewhat intermediate in structure, and they have produced intermediate numbers of species. The more patchy, the more mosaic-like the biota, the stronger the tendency to originate localized, uniform and clearly defined species with small areas; the more interconnected and homogeneous the biota, the more we can expect a small number of widespread and continuously variable, polymorphic species.

Another relevant comparison concerns two alpine groups of *Galium* sect. *Leptogalium* (cf. Ehrendorfer 1958, 1963, 1965) which have evolved as ecological (but not geographical) vicariants in the European mountain systems. The very polymorphic *G. anisophyllon* is closely related to other members of the *G. pulsillum* group at lower elevations, settles in a great variety of open but less extreme habitats, and occupies a large, quite continuous area. It has evolved into an intricate polyploid complex with a large number of ill-defined 2x, 4x, 6x and 8x cytotypes, which all have to be assembled into a single species. The 2x populations are nowadays limited to formerly unglaciated areas, but the 4x, 6x, 8x, and 10x strains that have successively evolved during the Pleistocene have successfully reconquered even the most extensively glaciated territories of the C Alps. On the other hand the *G. baldense* group with its 6 species limited to the alpine zone, is less variable as a whole than *G. anisophyllon* alone. All these species are morphologically quite uniform; each occupies a different and more extreme alpine niche (rock fissures, talus slopes, snowbeds), and a disjunct and geographically reduced area, mostly unglaciated during the Pleistocene.

All these examples clearly demonstrate to what an extent the evolutionary stability and change of individual groups is related to the direction of their initial eco-geographical differentiation and to the structures and historical changes of the biota and ecosystems of which they have become a part.

My last example from the *Rubieae* concerns by far the most successful member of the genus *Galium* with respect to biomass production and worldwide expansion, *i.e.* the agressively weedy *G. aparine.* This belongs to the exclusively annual sect. *Kolgyda,* and we may ask why it has been so successful as compared to the many other, often very inconspicuous and local representatives of this section. Available data suggest a complex answer.

As many other therophytes, most annual species of *Galium* (and of a few satellite genera) have switched to autogamy and to specialisation in fruit dispersal, both of which are advantageous for pioneer plants. Mericarps of *G. aparine* are equipped with hooked hairs, very efficient for epizoochorous dispersal. From a detailed karyosystematic analysis of *G. aparine* and its closest relatives, an allopolyploid origin from two main ancestral sources has become very likely: *G. monachinii,* 2n = 22, a local endemic from limestone gorges in Crete, and *G. spurium,* 2n = 20 and 40, a polymorphic taxon, with the diploid subsp. *spurium* centred in Near Eastern and E Mediterranean therophyte communities, but also including coastal ecotypes along the North and the Baltic Sea and expanding cereal and ruderal ecotypes; the oriental (sub)alpine diploid subsp. *ibicinum;* and the African tetraploid subsp. *africanum* (Puff 1978). Within the allopolyploid *G. aparine* a remarkable array of different polyploid and aneuploid cytotypes has been found (4x with 2n = 42, 44, 48; 6x with 2n = 63, 64, 66, 68; and 8x with 2 n = 86, 88) but so far no clear relationship between chromosome numbers, the considerable morphological and ecophysiological variability and the cosmopolitan distribution is evident. One of the consequences of the extensive genetic component of this variability has been the selection and rapid spread of genotypes unafferted by herbicides used against dicots; this has given *G. aparine* considerable dominance in cereal cultures (cf. Hirdina 1959). Another consequence has been the differentiation into many ecotypes, of which a forest/ruderal and a segetal one were recently studied in detail in C Europe (Groll & Mahn 1966). The forest/ruderal ecotype has a slower, the segetal a faster development capacity. Both ecotypes are strongly modifiable and produce much biomass and seeds on robust plants under good light and reduced competition, less biomass and seeds under shade and increasing competition on small and slender plants. This results in a well balance good seed output under very different (and even unfavourable) conditions, and thus in a remarkable ecological versatility.

Achillea and related genera

With the next examples we turn to the *Asteraceae,* tribe *Anthemidae,* a worldwide tribe of more than 100 genera, and in particular to the generic group of *Achillea* (Meusel & Ehrendorfer 1987). Our attention now moves from the species to the genus level, and from external to internal guiding lines for evolution.

The *Achillea* group consits of five closely related genera with two primary centres of differentation: in NW Africa we find the monotypic and perennial *Heliocauta anthemoides* in the Atlas Mts. and the greatest sprecies diversity of the circum-Mediterranean genus *Anacyclus* (9 species, predominantly annuals). The other centre is located in SE Europe and adjacent SW Asia, with the small genus *Leucocyclus* (2 species) in the mountains of S Anatolia and the greatest sectional and species diversity of *Achillea* (with a total of about 100 species) in the Alps, the Balkan Peninsula, S Russia, Anatolia and the Levant. Finally, this group also contains a semi-shrubby monotypic psammophyte, *Otanthus maritimus,* which has expanded from the shores of the Mediterranean to the Atlantic coasts of England and Ireland. Relationships among this group of genera have been suggested first by phytochemical data, *i.e.* the significant occurrence of amides (= alkamides) (Greger 1978), and were substantiated by fruit anatomy (dorsally compressed cypselas with only two bundles in the pericarp; Humphries 1979). Chromosome numbers are based throughout on $x = 9$, karyotypes and nucleotypes are relatively uniform.

The greatest number of plesiomorphic characters, including less specialized phytochemical profiles, diploidy and medium-sized karyotypes, occur in more or less mesophilous perennials of the mountain zones, *i.e.* in *Heliocauta, Leucoyclus,* and in a few ecologically corresponding members of *Anacyclus* and *Achillea;* today, they appear as relic types. The coastal *Otanthus* exhibits a heterobathmic phenotype with several primitive and some very advanced features (*e.g.* hydrochorous cypselas); it can be regarded as an early specialized member of the group which did not continue to evolve. *Anacyclus* has been relatively successful in producing an array of Mediterranean therophytes progressively more and more specialized in their reproductive biology (out- to inbreeding, increasing anemochory, etc.; Humphries 1979, 1980), phytochemistry, karyotypes, chromosome banding patterns and DNA content (Ehrendorfer & al. 1977; Schweizer & Ehrendorfer 1976, 1983).

In contrast, the much larger genus *Achillea* has conserved and stabilized its perennial life from. Its reproductive biology (outbreeding, miniaturized cypselas) has remained quite constant, and karyotypes (varying to only about 1:2 in total length), DNA and chromosome banding patterns are remarkably little variable for such a high species diversity (Tohidast-Akrad 1981; Schweizer & Ehrendorfer 1983). As a consequence, experimental and natural hybridization – even between morphologically and systematically remote species – is often still possible and may lead to F_1 and further offspring. Differentiation of *Achillea* thus is mainly due to changes in morphology, anatomy, eco-physiology and phytochemistry, mostly of an obvious adaptive nature. In growth form we find a wide spectrum form semi-shrubby to herbaceous pleiocorm perennials with tap roots or ± extensive rhizomes, in leaves from undivided and mesomorphous to deeply segmented and highly xeromorphous, in habitats from open forests to alpine rock and talus, to lowland meadows, steppes and semi-deserts. This multidirectional eco-radiation has mainly been achieved in an allopatric and vicarious fashion, as one rarely encounters more than one species of *Achillea* in the same locality.

An important aspect of this eco-radation in *Achillea* is its remarkable differentiation with respect to phytochemical substances which are obviously of insecticidal and/or fungicidal efficiency (Bohlmann et al. 1973, Greger 1978, 1984 and oral comm., Greger et al. 1987a, b, Hofer & Greger 1984, Hofer et al. 1986, Seaman 1982, Valant 1978, Valant-Vetschera 1981, Wollenweber et al. 1986). The following survey suggests some divergent trends of secondary compound accumulation in different species and species groups of *Achillea:*

A. lingulata:	lignans;
A. nobilis group:	highly unsaturated polyacetylenes;
A. ochroleuca group:	sesquiterpene-coumarine-ethers;
A. falcata group:	alkamides: purely oleifinic, C_{10}, etc.;
A. santolina group:	id.: acetylenic, C_{10}, etc.;
A. millefolium group:	id.: oleifinic and acetylenic, C_{14}, etc.

Let us compare two groups of *Achillea* in more detail in order to demonstrate differences and possible causes of evolutionary stability and success: the stenoecious and regressive *A. clavennae* group (5 species), localized in the E Alps and Balkan mountains, and the euryoecious, expansive *A. millefolium* group (c. 9 species), widespread in the N Hemisphere and even cosmopolitan through some of its weedy members.

Recent biosystematic studies (Franzén 1986, 1987), karyotype analyses (Tohidast-Akrad 1981) and phytochemical surveys of flavonoids (Valant-Vetschera 1981, Franzén 1987) and alkamides (Greger 1984 and oral comm.) have shown that the *A. clavennae* group, well characterized by a silvery-gray indumentum of T-shaped hairs, consists exclusively of diploids with 2n = 18. All are pleiocormic perennials without efficient means of vegetative propagation and with few, relatively large flower heads. They all grow scattered in open (sub)alpine rock, talus and grassland habitats on limestone. *A. ambrosiaca* (Mt. Olympus) and *A. fraasii* (Greece to Montenegro and NW Anatolia) are more isolated. *A. umbellata* (S Greece), *A. pindicola* (with two subspecies in C and NW Greece) and *A. clavennae* (with at least two subspecies, Dinaric Mts. and E Alps) form a closely knit vicarious group which – according to consecutive changes in karyotypes and reduction of flavonoid and alkamide diversity – can be interpreted as a gradual expansion from S to N from the Pliocene through the Pleistocene to the present.

Biometrical analyses and crossing experiments back such an assumption and demonstrate local mosaic-like differentiations and the origin of initial infraspecific reproductive barriers in allpatric Greek mountain populations of *A. umbellata* and *A. pindicola.* Altogether, the *A. clavennae* group appears to be morphologically and ecologically rather conservative, not very competitive and productive, geographically stabilized or even regressive, and only slowly evolving.

A large amount of informations is available on the extremely polymorphic polyploid complex of the *A. millefolium* group, *e.g.* on its modificatory plasticity and the genetics of eco-physiological differentiation (Clausen et al. 1948, Kruckeberg 1951, Hiesey 1953, Hiesey & Nobs 1970, Dabrowska 1977), karyosystematics and the distribution of cytotypes (Ehrendorfer 1953, Tyrl 1969, Gervais 1977, Biste 1978), cytogenetics and microevolution (Ehrendorfer 1959 a, b, c, d, e, 1960, 1961, 1962a, 1963, 1986, Schneider 1958, Tohidast-Akrad 1981), phytochemistry (flavonoids: Valant 1978, Valant-Vetschera 1981, 1984; alkamides: Greger 1984, Greger et al. 1987a, etc.), influences on soil biology (Steubing 1967), etc.

In comparison with the *A. clavennae* group we find a much more advanced, apomorphic set of characters in the (less clearly circumscribed) *A. millefolium* polyploid complex. There is much more variation in all, particularly the vegetative features (stem height, leaf shape and size, indumentum, etc.). Strong development of rhizomes and subterranean runners add to competitive power. The miniaturization of the flower heads has reached a maximum in the genus, but there is over-compensation by numerical increase of heads resulting in conspicuous composite inflorescences (double pseudanthia). The modificational plasticity and potential for regeneration and for the repetitive development of reproductive shoots during one vegetation period are extraordinary, particularly among the polyploids. The variety of eco-physiologically different genotypes and races is enormous and covers a habitat spectrum from the subtropics to the arctic, form xeric to wet and even saline, and from full light to deep shade.

This broad eco-physiological spectrum has been preserved only in part among the rather clearly separated diploid members of the *A. millefolium* group which can be crossed only with some difficulty and may be differentiated into about 5–7 (micro-)species. Examples are the Pannonic *A. aspleniifolia* in wet lowland meadows and fens, *A. setacea* in Pontic steppe communities, and *A. cuspidata* in open W. Himalayan woodlands. All these diploids are relatively uniform, genetically depauperate, more or less relictual and regressive, settle in rather limited areas of Eurasia only, and tend to be competed out by their more widespread and common, more productive and aggressive polyploid descendants. This is exemplified by the tetraploid *A. collina* in continental Europe, for which an allopolyploid origin from *A. aspleniifolia* and *A. setacea* has been experimentally demonstrated. *A. cuspidata* evidently participated in the origin of *A. millefolium* (s.l.) which has spread, mainly with 4x and 6x cytotypes, not only all over Eurasia but most succesfully also into N (and even C) America. In the New World there are no other members of the genus (with the exception of the more hygrophilous *A. ptarmica* group), and the different polyploid races of *A. millefolium* (s. l.) have managed to occupy a multitude of habitats, from the sea shores to the high mountains of the Sierra Nevada and Rocky Mountains.

All polyploids of the *A. millefolium* group suffer little from meiotic disturbances; their predominant bivalent pairing and fertility is near perfect, even in autopolyploids. Hybridization between the different polyploids and even between different ploidy levels is much easier than between the diploids, and evidently has contributed to their enormous genetic diversity and remarkable adaptability.

The combination of genes and characters of the ancestral diploids, in the polyploids, is also evident from phytochemical analyses. In respect to flavonoids one can demonstrate that *A. setacea* and related diploids tend to accumulate C-glycosylflavones, and *A. aspleniifolia* rather flavonol 3-0-glycosides, whereas the polyploids contain both groups of compounds and in addition also flavon 7-0-glycosides. On the leaf surfaces, even more complex and ecologically significant flavonoid aglyca are found. Similar results are obtained from the alkamides and several other groups of secondary compounds, and it is evident that particularly the polyploid members of the *A. millefolium* group are one of the most versatile and diverse Angiosperm groups with respect to their "phytochemical defensive strategy". Their varied phytochemical profile may also have a bearing on their competitive power against other plants and on their association with N_2-binding microorganisms, their positive effects on soil biology, etc.

In conclusion, one can recognize a variety of features in the *A. millefolium* polyploid complex which apparently have been responsible for its becoming morphologically and ecologically so variable, progressive, competitive and productive, geographically extremely expansive, and fast evolving.

Rutaceae

Finally, I would like to illustrate the supraspecific level and the macroevolutionary aspect of our problem of stability versus change by expamples from the *Rutaceae.* This is a large and nearly exclusively woody family with more than 150 genera and 1600 species, centred in the (sub)tropics. A growing body of new information, *i.a.* from morphology and anatomy (*e.g.* Hartl 1957, Gut 1966, Corner 1976, Boesewinkel 1977) and particularly phytochemistry *(e.g.* Fish & Waterman 1973, Waterman 1975, Waterman & Grundon 1983, Waterman & Kholid 1981) clearly demonstrates that the available taxonomic arrangement of the family (Engler 1931) is largely formal and "artificial". Collaboration with Prof. M. Guerra (Recife, Brazil) on aspects of karyology (cf. Guerra 1980, 1984a, 1984b, 1985, 1987, Ehrendorfer 1982) and with Prof. O. R. Gottlieb (Sao Paulo and Rio de Janeiro, Brazil) on chemosystematics (cf. Fernandes da Silva et al. 1988) is now yielding a new and more "natural" alinement of generic groups (the *"Toddalioideae"* are abolished as a polyphletic taxon, etc.). In addition, new palaeobotanical evidence (mainly from fossil seeds: *e.g.* Tiffney 1980, Gregor 1975, 1979, 1984, 1988) clarifies the early history of the family. All this leads to an increasingly better understanding of the remarkable phases in the evolution of the *Rutaceae* (Ehrendorfer 1976a, 1987, 1988).

The family dates back to at least the Upper Cretaceous; some primitive extant groups were already well developed in the Eocene/Oligocene and have little changed since. Examples for this oldest evolutionary phase of *Rutaceae* are the pantropical (rarely warm temparate) *Zanthoxylum* tribe (1–2 genera/220 species), the monotypic tropical African-SE Asiatic *Toddalia* tribe (1/1), the subtropical-temperate E Asiatic *Phellodendron* tribe (1/10), the tropical African *Fagaropsis* tribe (1/2), and the genus *Evodia* (120 species, Australasia to Madagascar). All these groups had their early centre of diversity (and origin?) in the (then humid tropical and geographically still closer) areas of Europe, N Africa and eastern N America *i.e.,* the fragmenting northwestern Gondwana land. During the middle and later Tertiary considerable expansion and diversification occurred. Today these groups are concentrated in tropical refugial areas, particularly in Australasia, and have become relictual. The genera are mostly quite isolated, partly very depauperate, the species often disjunct or with a limited range. With respect to the accumulation of phytochemical substances with a defensive (insecticidal and/or fungicidal) function, these most ancient members of the *Rutaceae* have maintained a variety of benzylisoquinoline alkaloids (comparable to those in the conservative *Magnoliidae)* which have dropped out and become replaced by more advanced substances in all the other more recent evolutionary lines of the family. With regard to karyology, it has been found that *Zanthoxylum* (incl. *Fagara), Toddalia, Phellodendron* and *Evodia* are palaeopolyploid throughout on x = 9, rarely on the 4x, mostly on the 6x und 8x (and sometimes even higher) levels, and have undergone some secondary dysploid changes[1]. There has been differentiation in the structure of interphase nuclei from weakly to deeply stained semi-reticulate and from middle to high DNA 1C mean values per chromosome ($9-17 \times 10^{-2}$ pg).

Quite a different, much more homogeneous pattern of evolutionary differentiation is found in the subfamily *Aurantioideae (= Citroideae;* c. 30/200), a group of 5 rather closeley related tribes with genera and species often difficult to separate. Their fossil record is scanty and suggests a much more recent origin (possibly mid-Tertiary) in SE Asia, where the Aurantioideae still have their centre of diversity, extending with some members into Africa, N Australia and the W Pacific, with little ecological radiation beyond (sub)tropical forests. Accumulation tendencies of phytochemical compounds have clearly switched: There are still some anthranilic acid derived alcaloids (quinolines, acridones), but coumarins, limonoids, flavonoids and essential oils dominate. Evolution has proceeded nearly exclusively on the dipoloid level (x = 9, 2n = 18), without substantial changes in the characteristic areticulate structure of interphase nuclei and in the low DNA 1C mean values per chromosome ($5-7 \times 10^{-2}$ pg). In accordance with this limited karyological differentiation, hybridization is still possible between many species and even genera. The *Aurantioideae* thus represent a group with slow but continuing evolutionary differentiation (mainly based on gene mutations) in saturated and relativley stable ecosystems, with mostly allopatric and often widespread species.

As an example of an aberrant, frost-resistant and herbaceous member of the *Rutaceae* one can introduce the genus *Dictamnus*, isolated in its own tribe, with a fossil documentation going back to the Upper Tertiary and with a disjunct Eurasiatic present distribution ranging from the Iberian Peninsula to Korea. There are several closely related allopatric species (or subspecies), mostly growing in open and relatively dry deciduous forests. The phytochemical profile is characterized by simple and complex anthranilic acid derivatives, some prenylated coumarins, limonoids, etc. Karyological parameters are uniform and aberrant: All taxa are palaeotetraploid (x = 9, 2n = 36), the interphase nuclei exhibit a densely stained and polarized

[1] The karyology *Fagaropsis* has not yet been studied.

reticulum, and the DNA 1C values per chromosomes are very high (c. 19 x 10^{-2} pg). These karyological parameters are in remarkable contrast to those of members of the *Ruta* tribe (in which *Dictamnus* was formerly included), where interphase nuclei are areticulate and DNA 1C values per chromosome very low (c. 0.9–2.6 x 10^{-2} pg). *Dictamnus* thus appears as an advcanced and aberrant northern temperate group, within the *Rutaceae,* which "got stuck" rather early in its evolutionary differentiation.

My last reference is to two tribes within the *Rutaceae* which exhibit a remarkable "explosive" radiation into the dry and nutrient-deficient habitats of sclerophyllous and "mediterranoid" floras of the southern hemisphere, the *Boronia* tribe (= *Boronieae*; c. 12/280), mainly in SE and SW Australia, and the *Diosma* tribe (=*Diosmeae*; c. 21/250) in the Cape region. Together they make up nearly one third of all *Rutaceae* species and one fifth of the genera, and they exhibit an extreme diversity of highly specialized reproductive and more or less xeromorphic vegetative features. It appears that both tribes can be linked independently to tropical-montane members of the relatively primitive palaeopolyploid, but karyologically somewhat variable *Evodia* tribe (c. 12/230; Australasia to E Asia, New Zealand and Madagascar). For the *Diosma* tribe the E to SE African genus *Caldendron* is such a link. Both tribes exhibit an advanced phytochemistry, with only a depauperate set of anthranilic acid derived alcaloids in the *Boronieae* (and none in the *Diosmeae*) and with limonoids in *Calodendron*; but there are coumarins in several genera, widespread essential oils, phenolics etc. While interphase nuclei are semi-reticulate with weakly to medium stained chromatin and with low DNA 1C values per chromosome (c. 6–8.5 x 10^{-2} pg), an enourmous diversity of dysploid and polyploid chromosome numbers has originateted during the evolution of both tribes. In the *Boronieae* the diploid base numers $x = 7, 8, 9, 11$ and 12 are known, together with apparent 4x ($n = 14–20$) and 8x ($n = 28, 32, 36$) derivatives. In the *Diosmeae* no diploids have been found yet, but a comparable range of 4x ($n = 13, 14, 15, 17, 19$) and 6x ($n = 19, 21, 24, 25, 27$) numbers is on record. These two tribes thus give evidence of a relatively recent (probably no older than Upper Tertiary) and rapid evolutionary radiation into unsaturated and relatively unstable ecosystems. This radiation evidently was based *i.a.* on dramatic chromosome structural and genomic mutations and has resulted in large numbers of species, often parapatric or sympatric and mostly with small distributional areas.

References

Biste, C. 1978: Zytotaxonomische Untersuchungen des Formenkreises *Achillea millefolium* (Asteraceae) in der DDR. – Feddes Repert. **88:** 533–613.

Boesewinkel, F. D. 1977: Development of ovule and testa in *Rutaceae: Ruta, Zanthoxylum* and *Skimmia.* – Acta Bot. Neerl. **26:** 193–211.

Bohlmann, F., Burkhard, T. & Zdero, C. (eds.) 1973: Naturally occurring acetylenes. – Academic Press, London & New York.

Brullo, S. 1979: *Valantia calva,* a new species from Linosa, Sicily. – Bot. Not. **132:** 61–64.

– 1980: *Valantia deltoidea* Brullo, sp. nov. from Sicily. – Bot. Not. **133:** 63–66.

Clausen, J., Keck, D. D. & Hiesey, W. M. 1948: Experimental studies on the nature of species. III. Environmental responses of climatic races of *Achillea.* – Publ. Carnegie Inst. Wash. **581.**

Corner, J. H. 1976: The seed of Dicotyledons. – Cambridge University Press, London & New York.

Dabrowska, J. 1977: Effect of soil moisture on some morphological characters of *Achillea collina* Becker, *A. millefolium* L. ssp. *millefolium* and *A. pannonica* Scheele. – Ekol. Polska **25:** 275–288.

Ehrendorfer, F. 1953: Systematische und zytogenetische Untersuchungen an europäischen Rassen des *Achillea millefolium*-Komplexes. (Vorläufige Mitteilung). – Österr. Bot. Z. **100:** 583–592.

– 1958: Die geographische und ökologische Entfaltung des europäisch-alpinen Polyploid-Komplexes *Galium anisophyllum* Vill. seit Beginn des Quartärs. – Uppsala Univ. Årsskr. **1858:** 176–181.

– 1959 a: Spontane Chromosomenaberrationen und andere Meiosestörungen bei diploiden Sippen des *Achillea millefolium*-Komplexes. Zur Phylogenie der Gattung *Achillea,* II. – Chromosoma **10:** 365–406.

– 1959b: Spindeldeffekte, mangelhafte Zellwandbildung und andere Meiosestörungen bei polyploiden Sippen des *Achillea millefolium*-Komplexes. Zur Phylogenie der Gattung *Achillea,* III. – Chromosoma **10:** 461–481.

– 1959c: Zusammenhänge zwischen Sippenstruktur, Lebensraum und Phylogenie bei Formenkreisen der Angiospermen. – Naturwiss. Rundschau **12:** 335–342.

– 1959d: Unterschiedliche Störungssyndrome der Meiose bei diploiden und polyploiden Sippen des *Achillea millefolium*-Komplexes und ihre Bedeutung für die Mikro-Evolution. Zur Phylogenie der Gattung *Achillea,* IV. – Chromosoma **10:** 482–496.

– 1959e: Differentiation-hybridization cycles und polyploidy in *Achillea.* – Cold Spring Harbor Symp. Quant. Biol. **24:** 141–152.

– 1960: Akzessorische Chromosomen bei *Achillea:* Auswirkungen auf das Fortpflanzungssystem, Zahlen-Balance und Bedeutung für die Mikro-Evolution. Zur Phylogenie der Gattung *Achillea,* VI. – Z. Vererbungsl. **91:** 400–422.

– 1961: Akzessorische Chromosomen bei *Achillea:* Struktur, cytologisches Verhalten, zahlenmäßige Instabilität und Entstehung. Zur Phylogenie der Gattung *Achillea,* V. – Chromosoma **11:** 523–552.

– 1962a: Cytotaxonomische Beiträge zur Genese der mitteleuropäischen Flora und Vegetation. – Ber. Deutsch. Bot. Ges. **75:** 137–152.

– 1962b: Notizen zur Systematik und Phylogenie von *Cruciata* Mill. und verwandten Gattungen der *Rubiaceae.* – Ann. Naturhist. Mus. Wien **65:** 11–20.

– 1963: Cytologie, Taxonomie und Evolution bei Samenpflanzen. – In: Turrill, W. B. (ed.), Vistas Bot. **4:** 99–186. – Academic Press, New York.

– 1965: Dispersal mechanisms, genetic systems, and colonizing abilities in some flowering plant families. – In: Baker, H. G. & Stebbins, G. L. (eds.), Genetics of colonizing species: 331–351. – Academic Press, New York.

– 1970: Mediterran-mitteleuropäische Florenbeziehungen im Lichte cytotaxonomischer Befunde. – Feddes Repert. **81:** 3–32.

– 1971: Evolution and eco-geographical differentiation in some South-West Asiatic *Rubiaceae.* In: Davis, P. H., Harper, P. C. & Hedge, C. (eds.), Plant life of South-West Asia: 195–215. – Botanical Society, Edinburgh.

– 1975: Cytosystematik balkanischer Rubiaceae. – Ein Beitrag zur Geschichte und Differenzierung der Flora und Vegetation des Balkans. – In: Jordanov, D. & al. (eds.), Problems of Balkan flora and vegetation: 178–186. – Bulgarian Academy of Sciences, Sofia.

– 1976a: Evolutionary significance of chromosomal differentiation patterns in Gymnosperms and primitive Angiosperms. In: Beck, C. B.. (ed.), Origin and early evolution of Angiosperms: 220–240. – Columbia University Press, New York.

– 1976b: *Rubiaceae.* – In: Tutin, T. G. & al. (eds.), Flora Europaea **4:** 3–38. University Press, Cambridge.

– 1980: Polyploidy and distribution. – In: Lewis, W. (ed.), Polyploidy, biological relevance: 45–60. – Plenum Press, New York.

– 1982: Speciation patterns in woody Angiosperms of tropical origin. – In: Barigozzi, C. (ed.), Mechanisms of speciation: 479–509. – Liss, New York.

– 1986: Chromosomal differentiation and evolution in angiosperm groups. – In: Iwatsuki K., Raven, P. H. & Bock, W. J. (eds.), Modern aspects of species: 59–86. – University of Tokyo Press, Tokyo.

– 1987: Differentiation trends in tropical woody Angiosperms. – In: Urbanska, K. M. (ed.), Differentiation patterns in higher plants: 227–237. – Academic Press, London & New York.

– 1988: Affinities of the African dendroflora: suggestions from karyo- and chemosystematics. – Ann. Missouri Bot. Gard. (in press).

–, Schweizer, D. & Humphries, C. 1977: Chromosome banding and synthetic systematics in *Anacyclus (Asteraceae – Anthemideae).* – Taxon **26:** 357–394.

Engler, A. 1931: *Rutaceae.* – In: Engler, A. & Prantl, K. (eds.), Die natürlichen Pflanzenfamilien, ed. 2, **19a:** 187–359, 458–459. – Engelmann, Leipzig.

Fernandes da Silva, M. F., Gottlieb, O. R. & Ehrendorfer, F. 1988: Major alkaloids and coumarins in *Rutaceae:* suggestions for a natural system and evolutionary interpretation of the family. – Pl. Syst. Evol. (in press).

Fish, F. & Waterman, P. G. 1973: Chemosystematics in the *Rutaceae.* II. The chemosystematics of the *Zanthoxylum/Fagara* complex. – Taxon **22:** 177–203.

Franzén, R. 1986: Taxonomy of the *Achillea clavennae* group and the *A. ageratifolia* group *(Asteraceae, Anthemideae)* on the Balkan Peninsula. – Willdenowia **16:** 13–33.

– 1987: Biosystematics of *Achillea clavennae* and *A. ageratifolia* groups *(Asteraceae).* – Diss., Lund.

Gervais, G. 1977: Cytological investigation of the *Achillea millefolium* complex *(Compositae)* in Quebec. – Canad. J. Bot. **55:** 769–808.

Greger, H. 1978: *Anthemideae* – chemical review. – In: Heywood, V. H., Harborne, J. B. & Turner, B. L. (eds.), The biology and chemistry of the *Compositae.* – Academic Press, London & New York.

– 1984: Alkamides: structural relationships, distribution and biological activity. – Pl. Med. **5:** 366–375.

–, Hofer, O. & Werner, A. 1987a: Biosynthetically simple C_{18}–alkamides from *Achillea* species. – Phytochemistry **26:** 2235–2242.

–, Zdero, C. & Bohlmann, F. 1987b: Pyrrole amides from *Achillea ageratifolia.* – Phytochemistry **26:** 2289–2291.

Gregor, H.-J. 1975: Die Rutaceen aus dem Mittel-Miozän der Oberpfälzer Braunkohle. – Courier Forsch.-Inst. Senckenberg **13:** 119–128.

– 1979: Systematics, biostratigraphy and paleoecology of the genus *Toddalia* Jussieu *(Rutaceae)* in the European Tertiary. – Rev. Paleobot. Palynol. **28:** 311–363.

– 1984: Subtropische Elemente im europäischen Tertiär IV *(Onagraceae, Rutaceae, Vitaceae, Elaeagnaceae).* – Doc. Nat. **16:** 1–37.

– 1988: Aspects of the fossil record and phylogeny of the family *Rutaceae.* – Pl. Syst. Evol. (in press).

Groll, U. & Mahn, E.-G. 1986: Zur Entwicklung ausgewählter Populationen des Kletten-Labkrautes (*Galium aparine* L.). – Flora **178:** 93–110.

Guerra, M. dos. S. 1980: Karyosystematik und Evolution der *Rutaceae.* – Diss., Wien.

– 1984a: Cytogenetics of *Rutaceae.* II. Nuclear DNA content. – Caryologia **37:** 219–226.

– 1984b: New chromosome numbers in *Rutaceae.* – Pl. Syst. Evol. **146:** 13–30.

– 1985: Cytogenetics of *Rutaceae.* III. Heterochromatin patterns. – Caryologia **38:** 335–346.

– 1987: Cytogenetics of *Rutaceae* IV. Structure and systematic significance of interphase nuclei. – Cytologia **52:** 213–222.

Gut, B. J. 1966: Beiträge zur Morphologie des Gynoeceums und der Blütenachse einiger Rutaceen. – Bot. Jahrb. Syst. **85:** 10–247.

Hartl, D. 1957: Struktur und Herkunft des Endokarps der Rutaceen. – Beitr. Biol. Pflanzen **34:** 35–49.

Hiesey, W. M. 1953: Comparative growth between and within climatic races of *Achillea* under controlled conditions. – Evolution **7:** 297–316.

– & Nobs, M. A. 1970: Genetic and transplant studies on contrasting species and ecological races of the *Achillea millefolium* complex. – Bot. Gaz. **131:** 245–259.

Hirdina, F. 1959: Beiträge zur Biologie und Bekämpfung des Kletten-Labkrautes (*Galium aparine* L.). – Z. Acker-Pflanzenbau **109:** 173–197.

Hofer, O. & Greger, H. 1984: Naturally occurring sesquiterpene-coumarin ethers, VI. New sesquiterpene-isofraxidin ethers from *Achillea depressa.* – Monatsh. Chem. **115:** 477–483.

–, Greger, H., Robien, W. & Werner, A. 1986: ^{13}C NMR and ^{1}H lanthanide induced shifts of naturally occurring alkamides with cyclic amide moieties – amides from *Achillea falcata.* – Tetrahedron **42:** 2707–2716.

Humphries, C. J. 1979: A revision of the genus *Anacylus* L. – Bull. Brit. Mus. (Nat. Hist.) Bot. **7:** 83–142.

– 1980: Cytogenetic and cladistic studies in *Anacyclus (Compositae: Anthemideae).* – Nordic J. Bot. **1:** 83–96.

Krendl, F. 1967: Cytotaxonomie der *Galium mollugo*-Gruppe in Mitteleuropa. Zur Phylogenie der Gattung *Galium,* VIII. – Österr. Bot. Z. **114:** 508–549.

Kruckeberg, A. R. 1951: Intraspecific variability in the response of certain native plant species to serpentine soil. – Amer. J. Bot. **6:** 408–419.

Meusel, H. & Ehrendorfer, F. 1987: Pflanzenverbreitung in Raum und Zeit (am Beispiel der Mittelmeerländer). – Nova Acta Leop. ser. 2, **53:** 185–210.

Puff, C. 1978: Chromosome numbers of some southern African *Rubiaceae-Rubieae.* – Linzer Biol. Beitr. **9:** 203-212.

Schneider, I. 1958: Zytogenetische Untersuchungen an Sippen des Polyploid-Komplexes *Achillea millefolium* L. s. lat. Zur Phylogenie der *Achillea,* I. – Österr. Bot. Z. **105:** 111–158.

Schweizer, D. & Ehrendorfer, F. 1976: Giemsa banded karyotypes, systematics and evolution in *Anacyclus (Asteraceae-Anthemideae).* – Pl. Syst. Evol. **126:** 107–148.

– & – 1983: Evolution of C-Band patterns in *Asteraceae-Anthemidiae.* – Biol. Zentralbl. **102:** 637–655.

Seaman, F. C. 1982: Sesquiterpene lactones as taxonomic characters in the *Asteraceae.* – Bot. Rev. **48** (2).

Steubing, L. 1967: Bodenökologische Untersuchungen in der Rhizosphäre. – In: Graf, O. & Satchell, E. J. (eds.), Progress in soil biology: 72–88. – Vieweg, Braunschweig; North Holland Publ. Co., Amsterdam.

Teppner, H., Ehrendorfer, F. & Puff, C. 1976: Karyosystematic notes on the *Galium palustre*-group *(Rubiceae).* – Taxon **25:** 95–97.

Tiffney, B. H. 1980: Fruits and seeds of the Brandon lignite, V. *Rutaceae.* – J. Arnold Arbor. **61:** 1–40.

Tohidast-Akrad, M. 1981: Beiträge zur Karyosystematik und Evolution von *Achillea (Asteraceae-Anthemideae):* Chromosomenzahlen, Feulgen- und Giemsa-C-gebänderte Chromosomen. – Diss., Wien.

Tyrl, J. R., 1969: Cytogeography of *Achillea millefolium* in western Oregon. – Brittonia **21:** 215–223.

Valant, K. 1978: Chatakteristische Flavonoidglykoside und verwandtschaftliche Gliederung der Gattung *Achillea.* – Naturwissenschaften **65:** 437–438.

Valant-Vetschera, K. M. 1981: Vergleichende Flavonoidchemie und Systematik der Gattungen *Achillea* und *Leucocyclus (Asteraceae-Anthemideae).* – Diss., Wien.

– 1984: Laubblattflavonoide der *Achillea millefolium*-Gruppe I: Infraspezifische Variabilität bei *A. setacea* W. & K. und verwandten Arten. – Sci. Pharm. **52:** 307–311.

Waterman, P. G. 1975: Alkaloids of the *Rutaceae:* their distribution and systematic significance. – Biochem. Syst. **3:** 149–180.

– & Grundon, M. F. (eds.) 1983: Chemistry and chemical taxonomy of the *Rutales.* – Academic Press, London & New York.

– & Khalid, S. A., 1981: The biochemical systematics of *Fagaropsis angolensis* and its significance in the *Rutales.* – Biochem. Syst. Ecol. **9:** 45–51.

Wollenweber, E., Valant-Vetschera, K., Ivancheva, S. & Kuzmanov, B. 1986: Flavonoid aglycones from the leaf surfaces of some *Achillea* species. – Phytochemistry **26:** 181–182.

Address of the author: Prof. Dr. F. Ehrendorfer, Botanisches Institut der Universität, Rennweg 14, A-1030 Wien.

Promising new directions in the study of ant-plant mutualisms

D. McKey

Abstract

McKey, D. 1988: Promising new directions in the study of ant-plant mutualisms. – In: Greuter, W. & Zimmer, B. (eds.): Proceedings of the XIV International Botanical Congress: 335–355. – Koeltz, Königstein/Taunus.

Biotic defense by ants is viewed in the context of general hypotheses about plant anti-herbivore defense systems. In facultative interactions, biotic defenses complement the plant's other anti-herbivore defenses. Biotic defense is important for chemically and mechanically vulnerable young tissues, especially those lacking effective phenological defense. It functionally resembles mobile chemical defenses, explaining patterns in the interplay between chemical and biotic defenses in myrmecophytes. Biotic defenses differ from most others in that they are not inherited but must be acquired, a fact with important consequences for the biology of juvenile myrmecophytes. – As the ecology of arboricolous ants is studied in increasing depth, patterns in variation among ant-plant systems, and the functional significance of variation, are becoming clear. Though in its infancy, a comparative biology of protective ant-plant interactions is emerging, similar to that of pollination biology. – Two complementary approaches to studying the evolutionary dynamics of species-specific ant-plant symbioses are briefly presented: study of symbiotic, but facultative associations, probably similar to the evolutionary antecedents of obligate symbioses, and historical ecology, combining comparative field study with phylogenetic analysis of plants and ants.

A nutshell history of the study of ant-plant mutualisms

The study of interactions between ants and plants has had a long and chequered history. Its first flowering was in the last two decades of the 19th century, as naturalists, inspired by Darwin and provided with rich material by the great wave of collections and observations in the tropics that accompanied the colonial era, described a veritable host of phenomena that indicated the diverse and widespread evolutionary interactions between ants and the plant kingdom. Delpino (1886–1889) and others began to document the widespread occurrence of extrafloral nectaries and argued for the importance of protection of plants by ants. Ule (1902) described the ant gardens of the Amazonian region; Treub (1883), Beccari (1884–1886) and others showed the richness of myrmecophytic epiphytes in southeast Asia; Belt (1874) drew attention to the Central American ant-acacias. And in Europe, Sernander

(1906) showed how many plants were dependent on ants for dispersal of their seeds. *Hepatica nobilis,* the symbol of the XIV International Botanical Congress, is one of these.

These phenomena were described, philosophized and argued over, and then discarded, as interest in Darwinism fell into general decline during the early part of this century, where it remained until the first evolutionary synthesis of ecology and genetics that began in the 1930's and 1940's. The phenomena that attracted so much attention from the early naturalists drifted into neglect. Hypotheses about their significance to the plants received ample verbal criticism (*e.g.,* Nieuwenhuis 1907, Zimmermann 1932, Wheeler 1921b, 1942), but as Beattie (1985) points out, with remarkably few exceptions they received no experimental tests. It was not until the 1960's, when Berg (1966) revived interest in seed dispersal by ants and Janzen (1966) conducted his pioneering studies of the ant-acacias, that these interactions again came into the sights of biologists with an evolutionary bent and became the subject of quantitative experimental studies. Following on the heels of Janzen's study of the ant-acacia mutualism, one by one ant/plant phenomena entered the modern age. Bentley (1977) pioneered in studies of the ecological significance of extrafloral nectaries. Rickson (1969, 1971), O'Dowd (1980, 1982) and others began investigating the food bodies produced by plants for ants, turning up some surprises, such as the production of glycogen by an angiosperm (Rickson 1971). Kleinfeldt (1978, 1986), then Davidson (Davidson et al. 1987a), revived interest in ant-gardens. Related phenomena, such as the adaptive significance of acarodomatia, which appear to be important in mite/plant interactions, are only now coming under investigation after decades of neglect.

The focus of studies in the first phase of the modern flowering of ant-plant interactions was the testing of hypotheses that had been posed, in more or less elegant fashion, almost a hundred years ago and lain dormant since. Thus in 1966 Janzen's experiments convincingly demonstrated that acacia-ants were essential to survival of the ant-acacia *Acacia cornigera,* as Belt had surmised in 1874. Bentley (1977) and her successors (*e.g.,* Tilman 1978, Keeler 1980, Schemske 1980, Koptur 1984) demonstrated experimentally that ants attracted to extrafloral nectaries could provide significant benefits through plant protection, as Delpino had argued in 1886. These studies have also shown that many factors, *e.g.,* association with ant-sequestering herbivores (Horvitz & Schemske 1984), influence the effects of ants in an intricate manner. Janzen (1974a), Huxley (1978, 1980), Rickson (1979) and others began to demonstrate the importance of ants to the mineral nutrition of myrmecophytic epiphytes.

In two areas, the early naturalists left us only descriptions of the phenomena, but no hypotheses about their significance to the plant. Thus Janzen (1974a) was the first to propose that the significance of ants to epiphytic ant-plants was the feeding of plants by ants, and subsequent studies (*e.g.,* Rickson 1979, Huxley 1986) support this hypothesis, which applies to ant-garden plants as well (Beattie 1985). Likewise, although Sernander and other workers documented many ant-dispersed plants, only recently have hypotheses been proposed to explain the selective advantages of ant-dispersal. While expanding our vision of the widespread geographic occurrence of ant-dispersed plants (*e.g.,* Berg 1975, Beattie 1983), studies in this area have indicated that ants might provide unique benefits in dispersal of seeds to nutrient-rich microsites (ant nests; Beattie & Culver 1983) and in storage of seeds in sites safe from predators (O'Dowd & Hay 1980, Heithaus 1981), fire (Berg 1975), or competing plant species (Handel 1978).

In all the cases where the early naturalists ventured opinions, they seem to have been mostly right in their speculations about the adaptive significance of ants to plants, and their critics of the early part of this century, such as the American myrmecologist Wheeler – who said of ant-acacias that they have no more need of their ants than a dog has of its fleas (Wheeler 1913) – seem to have been wrong. Though variation in the degree and in the exact nature of the advantages provided by ants becomes apparent as more individual cases are studied (Beattie 1985), the importance of these interactions to plants can be taken as established.

So where do we go from here? Is there anything interesting left to do? I believe that we are on the verge of a second great phase in the modern study of ant/plant interactions, in which for the first time quantitative studies with a *comparative* focus will allow us to move beyond description of individual systems and toward a conceptual framework for studying the evolutionary biology of these interactions. Here, I will review what I perceive as some interesting directions in this current research. Due to limitations of space, I will focus on protective ant/plant interactions. Up-to-date critical syntheses of work on seed dispersal by ants, myrmecotrophic ant-plants, and antgardens may be found in Beattie (1985), Huxley (1986), and Davidson et al. (1987a).

Protective ant-plant interactions: general comments

I will preface this review with two general comments about the study of protective ant/plant interactions. First, through most of the two decades since Janzen's study of ant-acacias, most work with protective interactions, especially quantitative ecological studies, concerned not ant/plant symbioses but facultative, non-symbiotic interactions such as those involving ants opportunistically visiting extrafloral nectaries. It is only fairly recently that ant-plants al have come under renewed intensive study (*e.g.*, McKey 1984, Davidson et al. 1987b, c). There are probably several reasons for this. If I may generalize from personal experience, biologists may have felt after reading Janzen's early work that he had already asked all the interesting questions about myrmecophytes, if not providing all the definitive answers. Second, many biologists have regarded ant-plant symbioses as the natural history of a few weird tropical plants, but of limited importance or interest beyond these relatively few specialized systems. Finally, though I cannot test the supposition, I strongly suspect that many theoretically minded population biologists, especially in the 1970's, when competition and predation dominated community ecology, tacitly regarded mutualisms, especially coevolved ones, as nice pat examples of adaptations, but intellectually rather boring. Thus a theoretical framework for approaching mutualisms was long in coming (Vandermeer 1984) – and in my opinion still lags behind "natural-history", comparative-biology approaches to coevolved mutualisms such as that of Gilbert (1975) in the interest and explanatory power of its results.

None of the reasons for this lack of studies of ant-plant symbioses were valid. Though Janzen's study of ant-acacias is still unsurpassed in its richness and detail, and though a number of other ant/plant symbioses have been described and the benefits to the plants documented, we are just beginning to ask the questions that will provide us with a conceptual framework for understanding their evolutionary biology; and the picture that is emerging is far from simple.

A second observation I would make is that most studies of facultative interactions involving ants attracted to extrafloral nectaries have been devoted to asking the question, for a particular plant, does the presence of ants actually result in protection of the plant, and how does the effect of ants vary in space and time, with ant species, etc.? Surprisingly little work has been aimed at comparative studies that could place protective ant/plant interactions in theoretical perspective. Approaches such as that of Schemske (1983), who has asked under what circumstances symbiotic ant-plant mutualisms have evolved from facultative non-symbiotic interactions, are rare.

What are the new perspectives that can advance the study of ant-plant mutualisms? I will focus on three important aspects:

- Integrating biotic defense into a general theory of the biology of plant anti-herbivore defense systems;

- understanding the diversity of ant/plant interactions in relation to the comparative ecology of different kinds of ants and of plants in different ecological categories; and
- studying the evolutionary dynamics of ant/plant interactions.

Biotic defenses as a component of plant anti-herbivore defense systems

Thanks to the efforts of numerous scientists, a body of theory exists to explain the distribution of chemical anti-herbivore defenses within individual plants (McKey 1979), among plants faced with varying availability of soil nutrients and other resources (Janzen 1974b, Gartlan et al. 1980, Coley et al. 1985), among plants differing in successional stage and ecological strategy (Feeny 1976, Rhoades & Cates 1976), and in traits such as leaf lifespan (McKey 1979, Coley et al. 1985). Biotic defenses, though they are of widespread importance and certainly interact with other kinds of anti-herbivore defenses, have yet to be integrated into this body of theory. In this section I will point to ways in which this might be accomplished.

Biotic and chemical defenses: functional parallels and contrasts

Some of the underlying similarities in the biology of biotic and chemical defenses are readily evident. Maintenance of agents of biotic defense such as ants, for example, imposes a cost to the plant in metabolic terms, and patterns of distribution of ant-related rewards, both among plant species and within individual plants, are subject to the same kind of resource-allocation analysis that has led to concepts useful in interpreting chemical defenses. In fact, certain ideas common to both modes of defense may be more easily tested with biotic defenses. As Janzen (1981) points out, the plant wears its biotic defenses on the outside, and ants are easier to remove experimentally from a plant than are chemical defenses. Likewise, the plant's investment in biotic defense is easier to quantify than is its investment in chemical defenses. However, few studies have taken advantage of this fact, and many fundamental questions remain. What proportion of the plant's resources are allocated to biotic defense (O'Dowd 1979), and how are they distributed among plant parts? Is investment by the plant dependent on presence of ants, as Risch & Rickson (1981) state for the food bodies of *Piper cenocladum*? If extrafloral nectaries, like floral nectaries (Corbet 1978), secrete more nectar when nectar is frequently removed, many plants may be able to so modulate investment in biotic defense.

Just as plants can shunt chemical defenses to those parts being attacked by herbivores (*e.g.*, Carroll & Hoffmann 1980), the recruitment reaction of ants to high-density food resources fortuitously provides the plant with something like an inducible biotic defense (Beattie 1985). Whether increased production of rewards for ants can be induced by herbivore damage (Koptur, pers. comm.) is still an open question.

Some functionally important contrasts are also evident (Beattie 1985): ant-guards may provide a broader-spectrum defense than any single kind of chemical defense, and may be more difficult to overcome evolutionarily; biotic defense involves production of non-toxic ant rewards, rather than toxic metabolites from whose potentially harmful effects the plant itself must be protected (McKey 1979).

These individual points of comparison have not, however, led to testable hypotheses about the role of biotic defenses in relation to the rest of the plant's defensive arsenal. In the following paragraphs I will develop some general hypotheses I believe to be useful approaches to this problem.

Chemical, phenological and biotic defenses of young leaves

According to Feeny's (1976) apparency hypothesis, some kinds of plants, and some kinds of plant parts, are highly "apparent". Because of their abundance in space and time and/or long lifespan, they are "bound to be found" by their specific herbivores and thus require strong chemical and/or mechanical defenses. Other kinds of plants and/or plant parts are unapparent, so patchily distributed in space and time that they are less likely to be found by specialists. They thus require less chemical defenses and must be protected primarily against generalists. Young leaves, for example, are generally more patchily distributed in space and time than are mature leaves. Feeny proposed that young leaves may be primarily defended by their patchy occurrence. The protection that should accrue to plant parts due to intermittent and/or unpredictable occurrence, wich I call phenological defense, is especially important for young leaves because for several reasons these plant parts are difficult to defend chemically or mechanically (McKey 1979). Expanding young leaves cannot be lignified; they have limited storage space for potentially autotoxic allelochemicals, and their high nutrient content means that high concentrations of defenses must be present to discourage herbivores. Dethier (1987) points out, for example, that the deterrent effect of alkaloids on phytophagous insects depends on the concentration of accompanying feeding stimulants such as sugars.

However, young leaves are not uniformly unapparent, and phenological defense will be available to varying degree. The phenology of young-leaf production varies among species in a way that is at least somewhat predictable. Grime (1979) discussed phenology of shoot growth and leaf production as a component of plant strategies, and his remarks are highly relevant to understanding how biotic defenses fit in with other components of the plant's defense system. Grime's strategies include both underground and above-ground components, but only the above-ground component is relevant here. Grime recognizes three distinct strategies; I will focus here on two of these. In Grime's terminology, "competitors" are plants adapted to live in crowded, productive environments. Essential to their success is the ability to rapidly shift the position of their crown in relation to a light environment that is patchy in space and changing over time. The light environment changes rapidly over time because the plant and its neighbors are busy sticking leaves and branches into it at a rapid rate. "Competitors" accomplish rapid shifts in crown position by relatively continuous growth and by rapid and relatively continuous leaf turnover. In such plants, some young leaves are always present. Young leaves of such plants should thus achieve little escape in space and time, because herbivores can easily move from one young leaf to the next. How do such plants protect their young leaves, in the absence of effective chemical and phenological defenses? It is in exactly this category of plants – "competitors" – that *biotic* defense of young leaves by ants attracted to extrafloral nectaries (and to pearl bodies, *e.g.*, O'Dowd 1980) is most important. Bentley (1976) showed that rapidly, continuously growing plants of second-growth more often have extrafloral nectaries than do plants of less productive and crowded environments, such as forest understorey. Another class of plants that can be classed as "competitors" with relatively continuous production of young leaves and growing in crowded environments, vines, are characterized in general by a very high frequency of extrafloral nectaries (Bentley 1981).

In contrast, the success of "stress-tolerators" is due not so much to the capability for rapid shifts in crown position, but to adaptations that enable the plant to survive in conditions of scarcity of resources such as minerals and light (Grime 1979). These plants are characterized by long-lived leaves and other nutrient-conserving adaptations. Many are also characterized by strongly intermittent production of young leaves. This may have originated simply as a way of producing young leaves at a time of year most favourable for growth. This fits well into Grime's scheme, in which stress-tolerators exploit temporal variation in conditions for growth, whereas competitors primarily exploit spatial variation. However, intermittent production of young leaves can have another effect. Just as intermittent production of seeds can result in satiation of seed predators (Janzen 1971), intermittent production of young leaves can result in satiation of young-leaf herbivores. This emergent property of intermittent young-leaf production may have provided selective pressure that exaggerated the pulsed nature of young-leaf production in such plants. The caesalpinioid legume tribes *Detarieae* and *Amherstieae* offer a potential example (McKey 1987). These are tropical trees of mostly nutrient-impoverished sites, with long-lived leaves and highly pulsed young-leaf production, the explosive nature of which has long fascinated naturalists. Such plants, provided with *phenological* defense of young leaves, my require less *biotic* defense. There may be another problem with effective biotic defense of young leaves produced in pulses. The ants must be recruited from somewhere, and since nectaries are often active only on young leaves (*e.g.,* Bentley 1977, Tilman 1978), such plants would offer resources for ants that are discontinuous and unpredictable in time. Attracting enough ants to nectaries on pulsed young leaves may thus be problematical. However, as Beattie (1985) points out, the ability of ants to react quickly to pulses of concentrated food resources should not be underestimated.

I have begun to examine these ideas with a survey of extrafloral nectaries in the family *Leguminosae* (McKey 1987). In the *Papilionoideae,* for example, nectaries are widespread only in the tribe *Phaseoleae,* the group that includes most tropical vines. Biotic defense appears to be very important in the *Mimosoideae,* and less so in the caesalpinioid tribes *Detarieae* and *Amherstieae,* with their pulsed young-leaf production.

In summary, the distribution of traits that encourage biotic defense may be explained by the constraints affecting this type of anti-herbivore defense and alternative defenses open to the plant. Biotic defenses can thus be understood only when they are viewed in this overall functional context. Chemical and mechanical defenses are most effective in mature tissues. Young leaves may in some plants be protected by particularly effective toxin-based defenses, but in most plants phenological or biotic defense may be the main line of protection. Which of these is more important depends on the phenology of young-leaf production, which in turn depends on the plant's competitive strategy. While I have focused here on two extreme phenological patterns (continuous growth *vs.* pulsed flushes synchronous at the level of the individual), in reality the situation will be much more complex. Trees in which individual branches produce pulsed flushes asynchronously over the crown, for example, may be able to maintain both biotic and phenological defenses of young leaves.

What are the effects of enhanced biotic defense on other components of the plant's anti-herbivore defense system?

The evolution of myrmecophytes affords us the opportunity to further explore the relationship between biotic defense and other components of the plant's defense system. What happens to other defenses as biotic defenses become more effective? Janzen and coworkers (Janzen 1966, Rehr et al. 1973) gave us one part of the answer when they showed that neotropical ant-acacias have lost many of the chemical defenses characteristic of non-ant-acacias. This is why they are devastated by herbivores when their ants are removed. As the ant-acacias evolved, their original chemical defenses, costly and now redundant, were by and large evolutionarily replaced by biotic defenses, as the ants took over patrolling of all the plant's leaves. The case of cyanogenic glycosides is best-documented, but other substances must also be involved, since the experiments of Rehr et al. (1973) show that presence of cyanogens cannot account for the avoidance by and poor growth of some insect herbivores on non-ant-acacias. Comparison of contents of fiber and of tannin, substances widespread in the genus, in leaves of ant- and non-ant-acacias would be most instructive.

In other ant-plants, however, we encounter a very different outcome in the relationship between biotic and chemical defenses. For example, in the African legume ant-plant *Leonardoxa africana* we find not that ants have replaced chemical defenses, but instead that there is a division of function between biotic defense and chemical/mechanical defense according to leaf age (McKey 1984). In this plant, concentrations of tannins and lignin in mature leaves are as high as those in mature leaves of the plant's non-myrmecophytic relatives. Mature leaves of *L. africana* are virtually untouched by herbivores, and are not patrolled by ants. The ants patrol exclusively the young leaves, which are tender, nutrient-rich and highly vulnerable to herbivores. Here, biotic and chemical defenses coexist, each restricted to leaves of a different age class.

How do we explain these very different outcomes? Another current in plant defense theory – ideas that relate resource availability to patterns in the distribution of chemical defenses – provides insight here. The decisive difference between ant-acacias and *Leonardoxa,* I will argue, is in the life-span of mature leaves.

The key idea relating resource availability to chemical defenses was elaborated by Janzen (1974b), who argued that when the physical environment is unfavourable to plant growth (*e.g.,* on poor soils or in light-poor understorey), loss of a plant part to herbivores imposes a higher cost to the plant than when conditions for growth are better and tissues lost to herbivores are more easily replaced. Under such conditions, he argued, plants should invest more in anti-herbivore defenses. This is one of a suite of nutrient-conservation adaptations expected in "stress-tolerators". One of the most important of these adaptations is the possession of long-lived leaves.

Life-span of the leaf will in turn impose constraints on the kind of chemical defense employed to protect it. McKey (1979) drew a distinction between reclaimable and non-reclaimable defenses. Non-reclaimable defenses are those such as tannin and lignin that are metabolically inactive. They cannot be reclaimed from senescent leaves. They undergo little turnover, so that their maintenance costs are very low. Since the benefit of this non-reclaimable investment increases with life-span of the plant part, and since few maintenance costs accumulate over time, these defenses are best suited to protection of plants with long-lived leaves, and are in fact characteristic of such plants. In contrast, reclaimable defenses are metabolically active, usually small toxic molecules, that can be removed from senescent leaves before they are shed. In addition, as Coley et al. (1985) point out, the high turnover rate characteristic of these molecules should impose a substantial maintenance cost. Since these defenses are not lost when the leaf is shed, they are suited to the defense of plants with short-lived leaves and rapid leaf turnover. Their high maintenance cost makes them more expensive than non-reclaimable defenses for protection of long-lived leaves, and they are restricted to short-lived leaves, or to temporary phases of leaf development. This same distinction between two kinds of defenses was envisaged by Coley et al. (1985), who re-baptized them "mobile" and "immobile" defenses.

What is the relevance of these considerations for understanding the relationships between biotic and chemical defense in ant-plants? Ants share many features or "reclaimable" or "mobile" defenses. They can be reclaimed from senescent leaves and recycled to other parts of the plant. But because the ants must be fed, biotic defenses impose high maintenance costs. We can expect that biotic defenses in ant-plants will exhibit patterns of distribution similar to those of mobile chemical defenses (McKey 1984).

Ant-acacias are fast-growing plants of second-growth. Leaf turnover is rapid, and mature leaves live no longer than six months. Here, the ants, which are mobile defenses that can be reclaimed from senescent leaves, are cheaper than chemical/mechanical defenses such as tannin and lignin, that would be discarded with the shed leaves. *Leonardoxa,* in contrast, is a shade-tolerant treelet of rainforest understorey. Like other trees of light-poor environments,

Leonardoxa is characterized by long-lived leaves, as I have shown by studies of marked leaves. Their average life expectancy once they mature is 3.5 years. Here, the high maintenance costs of ants would make them more expensive than once-and-for-all investment in tannins and lignin for the defense of mature leaves. Instead, a smaller worker force is maintained that is shunted from one young leaf to the next, and tannins and lignins, with their low maintenance costs, remain the primary defense of mature leaves (McKey 1984).

Though biotic defense has not replaced chemical defense of mature leaves in *L. africana,* there are strong indications that it has replaced *phenological* defense of young leaves. Many members of the tribe *Detarieae,* to which this tree belongs, produce young leaves in large synchronous flushes (McKey 1987). Herbarium specimens of the non-myrmecophytic *Leonardoxa romii* show that it produces flushes of up to internodes 9 plus young leaves at a single terminal, and a non-myrmecophytic close relative of *L. africana* produces flushes 2–3 internodes long (McKey 1988). In contrast, the specialized myrmecophytie *L. africana* studied by McKey (1984) produces single-internode flushes, dribbled out asynchronously over the crown.

I suspect that loss of phenological defenses may have occurred in other ant-plants. In contrast to non-ant-acacias, which in the neotropics are usually deciduous shrubs of open country, the neotropical ant-acacias tend to be evergreen, continuously growing "competitors" in wetter, more crowded, productive habitats (Janzen 1966). The association with vine-clipping *Pseudomyrmex* ants was probably instrumental in enabling ant-acacias to colonize vegetation-choked second-growth in wetter sites that permit continuous growth, and the protection these ants provided against herbivores made up for the loss of phenological defense this strategy entails.

Biotic defenses must be acquired

In the recital of functionally important parallels and contrasts between biotic and chemical defenses preceding this section, one important difference was left out: While chemical defenses are inherited, biotic defenses must be acquired. The plant inherits the information to produce ant attractants, but their protective effect depends on the probabilistic process of establishing an association with mutalistic partners, and is subject to a host of influences on the intensity of the interaction.

This fact is one of the reasons for the natural variation in function and effect that characterizes protective ant-plant interactions (Beattie 1985). Depending on chance and local circumstances, association with ants may not be established, and the effect of the association will depend on the particular assemblage of ant species encountered.

Such variation is minimized in obligate, specific ant-plant mutualisms. However, even in these specialized mutualisms the agents of biotic defense must be acquired. The juvenile phase of the life cycle, when the protective colony has yet to be acquired, and when the young colony is most vulnerable to chance events, predation, and competitive displacement, is likely to be an especially hazardous phase in the life cycle of myrmecophytes.

In *Leonardoxa africana,* for example, extinction of colonies of the mutualist *Petalomyrmex phylax* is a frequent event in young saplings (McKey 1984). Difficulty in establishment of mutualists in juvenile ant-plants provides a foothold for species that reproduce earlier at lower colony size, invest less in workers, have short-lived colonies, and whose effects on the plant are less salubrious than those of the mutualist and usually strongly negative. The *Leonardoxa* x *Petalomyrmex* mutualism is parasitized by *Cataulacus mckeyi* (McKey 1984), the *Acacia* x *Pseudomyrmex* mutualism by *Pseudomyrmex nigropilosa* (Janzen 1975), and the *Tachigali* x *Pseudomyrmex* system by a variety of inquiline ants and by subsocial silvanid beetles (Wheeler 1921a).

Difficulty in acquisition of the mutualist colony by juveniles, frequently exacerbated by the presence of such "cheats", is such a pervasive feature in ant-plant biology that it can be expected to have led to adaptations in both ants and plants to enhance establishment. While this question has received little attention, there are strong indications it is worth pursuing. The lock-and-key system by which the combination of dorsoventrally flattened *Petalomyrmex* alates and slit-like entrance holes exclude *Cataulacus mckeyi* from *Leonardoxa africana* (McKey 1984) would be of most advantage in juvenile plants susceptible to invasion by this parasite. The acceptance of multiple queens in this and a number of other mutualist plant-ants, while of functional importance in many aspects, enables ants to rapidly build up worker numbers in newly colonized juveniles.

The African ant-plant *Barteria fistulosa (Passifloraceae)* offers an intriguing example of a trait that enhances establishment of biotic defense in juveniles. Adult plants have long, hollow horizontal branches in which host-specific *Tetraponera (= Pachysima) aethiops* tend scale insects (Janzen 1972). Adult plants lack extrafloral nectaries. Juvenile plants up to 1.5 m tall lack horizontal branches, and hence domatia, and consist of a single unmodified vertical axis with spirally arranged leaves. In contrast to the swollen horizontal branches, the vertical stem of a juvenile *B. fistulosa* bears 6 to 10 extrafloral nectaries at each node (McKey, unpublished results). This myrmecophyte may thus benefit from facultative association with a variety of ants until it produces its first domatia and can be colonized by its obligate mutualist. Equally important, I suspect, is that the presence of this ready-made food source on juvenile plants reduces the time lag between arrival of the mutualist founding queen and production of a worker force. Encountering abundant food upon her arrival, the founding queen can produce a first crop of workers more quickly, reducing the probability of extinction, than if she first had to establish a colony of scale insects to feed her brood. The case of *B. fistulosa* suggests we should look carefully at juvenile stages of other myrmecophytes whose ants depend primarily on homopteran secretions rather than on food produced directly by the plant. I suspect that the biology of juvenile myrmecophytes holds many surprises for us.

The diversity of protective ant-plant interactions

Though the main avenues by which ants benefit plants are becoming clear, we understand very little about the factors determining variation among different systems in functionally important aspects of the interaction. The comparative biology of protective ant-plant interactions is in its infancy. Though we have hypotheses to explain why and under what circumstances specificity evolves (Schemske 1983), why worker size in specialized plant-ants varies between systems such as *Barteria* × *Tetraponera* (= *Pachysima*) and *Pseudomyrmex* × *Acacia* (Janzen 1972), why some acacia-ants have monogynous and others polygynous colonies (Janzen 1973), and why understorey ant-plants such as *Leonardoxa, Ocotea* (Stout 1979), and *Piper* (Letourneau 1983) differ in many aspects from ant-plants of light-rich sites (McKey 1984), comparison of this small body of hypotheses with the rich literature on comparative biology of pollination interactions (*e.g.*, Real 1983) shows that we have merely scratched the surface.

Why is this? Beattie (1985: 1) identifies the reason precisely. So little is known about the demography of colonies of ant species that interact with plants that he was forced to write his book largely from, as he puts it, "the plant's point of view". Much work in this field has been characterized by this asymmetry, emphasizing impact of ant services on plant fitness, and giving relatively little attention to ant ecology. This lack has prevented us from recognizing much of the diversity that exists in protective ant-plant interactions and understanding its significance.

Fundamental advances will be made as an increasing number of workers bring to this field a sophisticated appreciation of the ecology of ants as well as that of plants. Studies by Davidson and co-workers, for example, are showing us how much more we can learn when we watch ants not only when they are at nectaries or other plant-produced rewards, but broaden our horizons to consider ant-plant interactions in the context of the overall ecology of ants. I will offer two examples of what this work is teaching us.

The behaviour of vine-trimming by ants, first described in acacia-ants (Janzen 1966), provides significant benefits to the plant in removing smothering competitors. In ant-acacias in drier parts of Mexico, the vegetation-trimming activities of acacia-ants can even provide the plant with a dry-season firebreak by removing any plants that could lead to accumulation of litter and fuel (Janzen 1967). Other plant-ants, *e.g.*, *Azteca* in *Cecropia* (Janzen 1969) and *Tetraponera* (= *Pachysima*) in *Barteria* (Janzen 1972), also exhibit this behaviour, which has been presumed to have arisen in coevolution with the plant.

Davidson et al. (1987b), highlighting the importance of interspecific competition for nest-sites, food, and other resources among arboricolous ants, have provided us with a new perspective on this behaviour. Some groups of arboricolous ants, with well-developed chemical defenses, appear to be competitively dominant members of the "tropical ant mosaic". Others, among them the subfamily *Pseudomyrmecinae,* to which acacia-ants belong, lack such defenses and are competitively inferior. The persistence of these ants in a host tree depends on their ability to cut off avenues by which competitively dominant ants can invade their plant. They do this by trimming vines and other vegetation that represent points of contact (Davidson et al. 1987b). Though coevolutionary interactions with ant-plants have doubtless enhanced this behaviour, it seems that its origin may have been in a completely different selective context. This important preadaptation may be one of the reasons why pseudomyrmecine ants are so often the associates of ant-plants in light-rich habitats where competition with vines and other neighbouring plants is severe.

A comprehensive view of ant ecology also leads to recognition of how plant traits other than food rewards and nest-sites are related to ant-plant interactions. Since a variety of ant species may compete for the resources offered by myrmecophytes, and since their effects on plant fitness may vary dramatically, there is selection for plant traits that give the "good" ants a competitive advantage. Davidson et al. (1987c) have shown that trichomes may be one such trait. The density of trichomes on the plant can determine the efficiency with which ant species of different worker size can forage on the plant, and hence their relative competitive success. Further work along these lines is likely to reveal many other traits of both plants and ants whose importance to interactions has gone unnoticed.

Evolutionary dynamics of symbiotic ant-plant mutualisms

Though the *Acacia* × *Pseudomyrmex* symbiosis remains one of the best-documented cases of coevolution, the absence of a chapter on ant-plant interactions in the book on coevolution by Futuyma & Slatkin (1983) says something about conceptual advances in this area since Janzen's pioneering study in the 1960's. One of the main reasons is again our poor appreciation of ant biology. Beattie (1985: 144) notes that in obligate symbiotic ant-plant mutualisms many traits of the plants are more obviously specifically related to the interactions than are traits of the ants, so that "coevolution appears to have been very lopsided": "None of this is to say that more ant traits have not evolved in response to specific mutualistic plant species, but that the detection and analysis of such traits are extremely difficult." When ant demography (and plant growth) receive as much attention as morphology of plants and ants, I believe more evidence for coevolution will be found.

A case study: guilds of ant-plants and plant-ants in South America

Schemske (1983) has pointed to general principles that determine when specific, obligate mutualisms will evolve (and why they evolve so rarely), but the best emerging study of a particular case is that of Davidson et al. (1987b, c). These studies concern a guild of sympatrically occurring, unrelated South American ant-plants that share a guild of arboricolous ants. These symbiotic, but not completely specific interactions probably approach the situation that led to obligate, specific ant-plant symbioses, and the results have many implications for understanding the evolutionary dynamics that have led to obligate symbioses. This work shows the role that multispecies interactions (*e. g.*, interspecific competition of ants for myrmecophytes) may have played in the evolution of pairwise associations. For long we have tacitly viewed obligate ant-plant symbioses as the product of *pairwise* coevolution between the ant and the plant, but the results of Davidson and colleagues indicate that the selective environment is much more complex than this, and that here also, as well as in seed-dispersal and pollination interactions, *diffuse* coevolution has played a very large role.

Historical ecology of ant-plant associations

Field studies of relatively unspecialized ant-plant symbioses offer one avenue for understanding how specific obligate symbioses have evolved. A complementary means to this end is the approach of historical ecology (Mitter & Brooks 1983), which, by combining comparative ecological studies with direct estimates of evolutionary history via phylogenetic analysis, addresses both pattern and process in the evolution of interspecies associations and attempts to understand the interaction between these two aspects of evolution. Ant-plant symbioses, the evolutionary biology of which includes a complex mix of historical constraints, behavioural preadaptations of ants and morphological preadaptations of plants (*e. g.*, for domatia: McKey 1987), as well as strong selective forces, could provide very interesting grist for this mill. Detailed analyses of phylogenetic aspects of coevolution published to date mostly deal with host-endoparasite interactions. One could argue that the parasites, being good parasites, have evolved in such a way as to minimize effects on their hosts. Coevolution in such cases may be lacking a dimension present in systems characterized by strong positive or negative interactions. In mutualisms, for example, selection reinforces the intensity of the interaction. Historical-ecological analysis of ant-plant symbioses and other mutualisms could provide exciting new insights into coevolutionary processes. Two examples of studies in progress will be briefly presented to illustrate the approach and give an inkling of what the future has to offer. The first example concerns an association that seems to have resulted, in Mitter & Brooks (1983) terms, from “colonization“ of an ant-plant mutualism by another ant group, the second concerns a system involving “association by descent“.

The Barteria × Tetraponera (= Pachysima) *association*

Just as the ant-related adaptations of a plant may be a “diffuse“ response to several ant species with which it interacts in contemporary time, they may also be the product of interaction with a sequence of phylogenetically independent ants over evolutionary time: A mutualistic association may be colonized by another mutualist, which may replace the first. One example suggests that an ant-plant has “changed partners“ over evolutionary time – just as slowly-evolving plants interacting with faster-evolving vertebrate seed dispersers have changed partners (Herrera 1985). The example concerns the two species of *Barteria (Passifloraceae)*, ant-plants of tropical Africa. Of these two,

Barteria nigritana is more similar to the closely related genus *Smeathmannia*, and probably resembles more the ancestral *Barteria*. This plant has small domatia that are restricted to the basal internodes of horizontal branches. They are occupied by myrmicine ants of the genus *Crematogaster*, which inhabit a number of other relatively unspecialized ant-plants in the same area (McKey, unplublished results). *Barteria fistulosa*, in contrast, has very large domatia that extend over the entire length of long horizontal branches, and that are occupied by the obligate host-specific ant *Tetraponera aethiops*, from a completely different group, the *Pseudomyrmecinae* (Janzen 1972). It is likely that domatia first evolved in the genus in the context of loose evolutionary interaction with *Crematogaster*, and that at a certain point the ancestor of *Barteria fistulosa* "changed partners" to *Tetraponera*, with which further coevolution produced the highly specialized interaction studied by Janzen (1972) in Nigeria. Incidentally, the evolutionary move to this vine-trimming pseudomyrmecine ant was probably instrumental in enabling *Barteria fistulosa* to become a successful light-gap competitor that colonized the entire Central African rainforest block, a geographic range much larger than in any of its close relatives.

Leonardoxa africana *and its inhabitants; association by descent*

In rainforests on the coast of Cameroon, the caesalpinioid legume tree *Leonardoxa africana* is inhabited by the formicine ant *Petalomyrmex phylax*, described from my collections as a new monotypic genus closely related to *Aphomomyrmex*. *P. phylax* is a highly specialized obligate associate of *L. africana* (McKey 1984). *Aphomomyrmex* appears to be itself monotypic, the sole species *A. afer* being a widespread but apparently rare arboricolous ant in the Central African forest zone (Snelling 1979). Further work has shown that the entity known as *Leonardoxa africana* actually consists of three distinct allopatric taxa occuring in rain-forests from Gabon to eastern Nigeria. One ot these is non-myrmecophytic. Of the two myrmecophytic taxa, one is associated with *A. afer*, the other, more specialized one (*i.e.*, with numerous autapomorphies) with *P. phylax*. Cladistic analysis is clarifying the relationship among the three plant taxa and indicates parallel cladogenesis of plants and ants. A detailed analysis of the evolution of this system will be presented elsewhere (McKey 1988). Incidentally, this system also offers an example of colonization of a mutualism, not by another mutualist, but by an unrelated ant parasitic on the mutualism, the myrmicine *Cataulacus mckeyi* (McKey 1984).

Conclusion

In this brief, selective review I have indicated some little-studied aspects of ant-plant interactions where further work should bring exciting results. It is the mark of a healthy, active field that stock-taking overviews such as this one become quickly out of date. I hope that the readers of this chapter include the graduate students who will advance the study of ant-plant interactions, and that the questions posed here can be answered before the most interesting systems are lost with the world's tropical forests.

Acknowledgements

Financial support for attending the congress was provided by the congress organizers and by the Department of Biology, University of Miami. Discussions with D. Davidson, L. Gilbert, M. Hossaert, and D. Janzen helped clarify the ideas presented here. This study is Contribution No. 274 from the Program in Tropical Biology, Ecology and Behavior, Department of Biology, University of Miami.

References

Beattie, A. J. 1983: Distribution of ant-dispersed plants. – Sonderbeih. Naturwiss. Vereins Hamburg **7:** 249–270.

– 1985: The evolutionary ecology of ant-plant mutualisms. – University Press, Cambridge.

– & Culver, D. C. 1983: The nest chemistry of two seed-dispersing ant species. – Oecologia (Berlin) **56:** 99–103.

Beccari, O. 1884–1886: Malesia. – Instituto Sordo-Muti, Genova.

Belt, T. 1874: The naturalist in Nicaragua. – Dent, London.

Bentley, B. L. 1976: Plants bearing extrafloral nectaries and the associated ant community: interhabitat differences in the reduction of herbivore damage. – Ecology **57:** 815–820.

– 1977: Extrafloral nectaries and protection by pugnacious bodyguards. – Annual Rev. Ecol. Syst. **8:** 407–427.

– 1981: Ants, extrafloral nectaries and the vine life-form: an interaction. – Trop. Ecol. **22:** 127–133.

Berg, R. Y. 1966: Seed dispersal of *Dendromecon*: its ecologic, evolutionary, and taxonomic significance. – Amer. J. Bot. **53:** 61–73.

– 1975: Myrmecochorous plants in Australia and their dispersal by ants. – Austral. J. Bot. **23:** 475–508.

Carroll, C. R. & Hoffman, C. A. 1980: Chemical feeding deterrent mobilized in response to herbivory, and counteradaptation by *Epilachna tredecimnotata.* – Science **209:** 414–416.

Coley, P. D., Bryant, J. P. & Chapin, F. S., III. 1985: Resource availability and plant antiherbivore defense. – Science **230:** 895–899.

Corbet, S. A. 1978: Bee visits and the nectar of *Echium vulgare* L. and *Sinapis alba* L. – Ecol. Entomol. **3:** 25–37.

Davidson, D. W., Epstein, W. W. & Seidel, J. 1987a: Myrmecophilous relations of epiphytes. – In: Lüttge, U. (ed.), Phylogeny and ecophysiology of epiphytes. – Springer, Heidelberg & New York (in press).

–, Longino, J. T. & Snelling, R. R. 1987b: Pruning of host plant neighbours by ants: an experimental approach. – Ecology (in press).

–, Snelling, R. R. & Longino, J. T. 1987c: Competition among ants for myrmecophytes and the significance of plant trichomes. – Biotropica (in press).

Delpino, F. 1886–1889: Funzione mirmecofila nel regno vegetale. – Mem. Reale Accad. Sci. Ist. Bologna, ser. 4, **7:** 215–323; **8:** 601–650; **10:** 115–147.

Dethier, V. G. 1987: Concluding remarks. – In: Labeyrie, V., Fabres, G. & Lachaise, D. G. (eds.), Insects-plants. Proceedings of the 6th international symposium on insect-plant relationships. – Junk, Dordrecht.

Feeny, P. P. 1976: Plant apparency and chemical defense. – In: Wallace, J. W. & Mansell, R. L. (eds.), Biochemical interaction between plants and insects. – Plenum Press, New York & London.

Futuyma, D. J. & Slatkin, M. 1983: Coevolution. – Sinauer, Sunderland, Mass.

Gartlan, J. S., McKey, D. B., Waterman, P. G., Mbi, C. N. & Struhsaker, T. T. 1980: A comparative study of the phytochemistry of two African rainforests. – Biochem. Syst. Ecol. **8:** 401–422.

Gilbert, L. E. 1975: Ecological consequences of a coevolved mutualism between butterflies and plants. – In: Gilbert, L. E. & Raven, P. H. (eds.), Coevolution of animals and plants. – University of Texas Press, Austin.

Grime, J. P. 1979: Plant strategies and vegetation processes. – Wiley, Chichester.

Handel, S. N. 1978: The competitive relationship of three woodland sedges, and its bearing on the evolution of ant dispersal of *Carex pedunculata.* – Evolution **32:** 151–163.

Heithaus, E. R. 1981: Seed predation by rodents on three ant-dispersed plants. – Ecology **62:** 136–145.

Herrera, C. M. 1985: Determinants of plant-animal coevolution: the case of mutualistic dispersal of seeds by vertebrates. – Oikos **44:** 132–141.

Horvitz, C. C. & Schemske, D. W. 1984: Effects of ant-mutualists and an ant-sequestering herbivore on seed production of a tropical herb, *Calathea ovandensis (Marantaceae).* – Ecology **65:** 1369–1378.

Huxley, C. R. 1978: The ant-plants *Myrmecodia* and *Hydnophytum (Rubiaceae)*, and the relationships between their morphology, ant occupants, physiology and ecology. – New Phytol. **80:** 231–268.

– 1980: Symbiosis between ants and epiphytes. – Biol. Rev. Cambridge Philos. Soc. **55:** 321–340.

– 1986: Evolution of benevolent ant-plant relationships. – In: Juniper, B. & Southwood, R. (eds.), Insects and the plant surface. – Arnold, London.

Janzen, D. H. 1966: Coevolution of mutualism between ants and acacias in central America. – Evolution **20:** 249–275.

– 1967: Fire, vegetation structure and the ant-*Acacia* interaction in central America. – Ecology **48:** 26–35.

– 1969: Allelopathy by myrmecophytes: the ant *Azteca* as an allelopathic agent of *Cecropia.* – Ecology **50:** 147–153.

– 1971: Seed predation by animals. – Annual Rev. Ecol. Syst. **2:** 465–482.

– 1972: Protection of *Barteria (Passifloraceae)* by *Pachysima* ants *(Pseudomyrmecinae)* in a Nigerian rain forest. – Ecology **53:** 885–892.

– 1973: Evolution of polygynous obligate acacia-ants in western Mexico. – J. Anim. Ecol. **42:** 727–750.

– 1974a: Epiphytic myrmecophytes in Sarawak: mutualism through the feeding of plants by ants. – Biotropica **6:** 237–259.

– 1974b: Tropical blackwater rivers, animals, and mast fruiting by the *Dipterocarpaceae.* – Biotropica **6:** 69–103.

– 1975: *Pseudomyrmex nigropilosa:* a parasite of a mutualism. – Science **188:** 936–937.

– 1981: The defenses of legumes against herbivores. – In: Polhill, R. M. & Raven, P. H. (eds.), Advances in legume systematics. – Royal Botanic Gardens, Kew.

Keeler, K. H. 1980: The extrafloral nectaries of *Ipomoea leptophylla (Convolvulaceae).* – Amer. J. Bot. **67:** 216–222.

Kleinfeldt, S. E. 1978: Ant-gardens: the interaction of *Codonanthe crassifolia (Gesneriaceae)* and *Crematogaster longispina (Formicidae).* – Ecology **59:** 449–456.

– 1986: Ant-gardens: mutual exploitation. – In: Juniper, B. & Southwood, R. (eds.), Insects and the plant surface. – Arnold, London.

Koptur, S. 1984: Experimental evidence for defense of *Inga (Fabaceae: Mimosoideae)* saplings by ants. – Ecology **65:** 1787–1793.

Letourneau, D. K. 1983: Passive aggression: an alternative hypothesis for the *Piper-Pheidole* association. – Oecologia (Berlin) **60:** 122–126.

McKey, D. 1979: The distribution of secondary compounds within plants. – In: Rosenthal, G. A. & Janzen, D. H. (eds.), Herbivores. Their interactions with secondary plant constituents. – Academic Press, New York & London.

– 1984: Interaction of the ant-plant *Leonardoxa africana (Caesalpiniaceae)* with its obligate inhabitants in a rainforest in Cameroon. – Biotropica **16:** 81–99.

– 1987: Interactions between ants and leguminous plants. – In: Stirton, C. & Zarucchi, J. (eds.), Advances in legume biology. – Monogr. Syst. Bot. – Missouri Botanical Garden, St. Louis (in press).

– 1988: Comparative biology of ant-plant interactions in *Leonardora (Leguminosae: Caesalpinioideae)* I. A revision of the genus, with notes on natural history. – (In preparation).

Mitter, C. & Brooks, D. R. 1983: Phylogenetic aspects of coevolution. – In: Futuyma, D. J. & Slatkin, M. (eds.), Coevolution. – Sinauer, Sunderland, Mass.

Nieuwenhuis von Uexkuell-Guldenbandt, M. 1907: Extraflorale Zuckerausscheidungen und Ameisenschutz. – Ann. Jard. Bot. Buitenzorg **21:** 195–327.

O'Dowd, D. J. 1979: Foliar nectar production and ant activity on a neotropical tree, *Ochroma pyramidale.* – Oecologia (Berlin) **43:** 233–248.

– 1980: Pearl bodies of a neotropical tree, *Ochroma pyramidale:* ecological implications. – Amer. J. Bot. **67:** 543–549.

– 1982: Pearl bodies as ant food: an ecological role for some leaf emergences of tropical plants. – Biotropica **14:** 40–49.

– & Hay, M. E. 1980: Mutualism between harvester ants and a desert ephemeral: seed escape from rodents. – Ecology **61:** 531–540.

Real, L. A. (ed.) 1983: Pollination biology. – Academic Press, New York & London.

Rehr, S. S., Feeny, P. P. & Janzen., D. H. 1973: Chemical defence in Central American non-ant-acacias. – J. Anim. Ecol. **42:** 405–416.

Rhoades, D. F. & Cates, R. G. 1976: A general theory of plant antiherbivore chemistry. – In: Wallace, J. W. & Mansell, R. L. (eds.), Biochemical interaction between plants and insects. – Plenum Press, New York.

Rickson, F. R. 1969: Developmental aspects of the shoot apex, leaf, and beltian bodies of *Acacia cornigera.* – Amer. J. Bot. **62:** 913–922.

– 1971: Glycogen plastids in Mullerian body cells of *Cecropia peltata* – a higher green plant. – Science **173:** 344–347.
– 1979: Absorption of animal tissue breakdown products into a plant stem – the feeding of a plant by ants. – Amer. J. Bot. **66:** 87–90.
Risch, S. J. & Rickson, F. R. 1981: Mutualism in which ants must be present before plants produce food bodies. – Nature **291:** 149–150.
Schemske, D. W. 1980: The evolutionary significance of extrafloral nectar production by *Costus woodsonii (Zingiberaceae)*: an experimental analysis of ant protection. – J. Ecol. **68:** 959–967.
– 1983: Limits to specialization and coevolution in plant-animal mutualisms. – In: Nitecki, M. H. (ed.), Coevolution. – University of Chicago Press. Chicago.
Sernander, R. 1906: Entwurf einer Monographie der europäischen Myrmekochoren. – Kongl. Svenska Vetenskapsakad. Handl. **41:** 1–410.
Snelling, R. R. 1979: *Aphomomyrmex* and a related new genus of arboreal African ants (Hymenoptera: *Formicidae).* Contr. Sci. Nat. Hist. Mus. Los Angeles County **316:** 1–8.
Stout, J. 1979: An association between an ant, a mealy bug and an understory tree from a Costa Rican rainforest. – Biotropica **11:** 309–311.
Tilman, D. 1978: Cherries, ants and tent caterpillars: timing of nectar production in relation to susceptibility of caterpillars to ant predation. – Ecology **59:** 686–692.
Treub, M. 1883: Sur le *Myrmecodia echinata* Gaudich. – Ann. Jard. Bot. Buitenzorg **3:** 129–159.
Ule, E. 1902: Ameisengärten im Amazonasgebiet. – Bot. Jahrb. Syst. **30:** 45–52.
Vandermeer, J. 1984: The evolution of mutualism. – In: Shorrocks, B. (ed.), Evolutionary ecology. 23rd Symposium of the British Ecological Society. – Blackwell, Oxford.
Wheeler, W. M. 1913: Observations on the Central American *Acacia* ants. – In: Transactions of the 2nd International Entomological Congress, Oxford (1912), **2:** 109–139.
– 1921a: A study of some social beetles in British Guiana and of their relations to the ant-plant *Tachigalia.* – Zoologica **3:** 35–183.
– 1921b: A new case of parabiosis and the "ant gardens" of British Guiana. – Ecology **2:** 89–103.
– 1942: Studies of neotropical ant-plants and their ants. – Bull. Mus. Comp. Zool. **90:** 1–262.
Zimmermann, J. G. 1932: Über die extrafloralen Nektarien der Angiospermen. – Beih. Bot. Centralbl. **49:** 99–196.

Address of the author: Dr. Doyle McKey, Department of Biology, University of Miami, P. O. Box 249118, Coral Gables, FL 33124, USA.

Ecophysiology of photosynthesis: performance of poikilohydric and homoiohydric plants

O. L. Lange

Abstract

Lange, O. L. 1988: Ecophysiology of photosynthesis: performance of poikilohydric and homoiohydric plants. – In: Greuter, W. & Zimmer, B. (eds.): Proceedings of the XIV International Botanical Congress: 357–383. – Koeltz, Königstein/Taunus.

The objecitve of plant ecophysiology is to explain processes in plant ecology, such as plant performance, survival, and distribution in physiological, biophysical, and biochemical terms. Analysis of photosynthesis has to combine field investigations of plant responses to natural and experimental conditions with laboratory experiences. Great progress has been made over the last decade concerning equipment for gas exchange measurement with poikilohydric as well as homoiohydric plants in the field. Two examples demonstrate actual work. – Primary production of lichens is governed through their thallus water content and time periods of metabolic activity are restricted. Species with green algae are even able to make use of high air humidity without hydration by liquid water. With higher plants, CO_2-assimilation is limited by stomatal conductance as well as by mesophyll carboxylation activity. Interplay of these characteristics determine seasonal photosynthetic performance and adaptions to stress of leaves and canopies of Mediterranean sclerophylls.

Introduction

Plant ecophysiology attempts to explain in physiological, biophysical, and biochemical terms ecological processes of plant performance, survival, and distribution. Thus, ecophysiological research on photosynthesis investigates the photosynthetic primary production of plants in the field by determining the fixed range of internal responses, of specific plant adaptions, and actual plant performance in response to transient external conditions.

Ecophysiological data on photosynthetic primary production must therefore be gathered in several stages at different levels of plant research (fig. 1). First, photosynthesis has to be measured in the plant's natural environment under natural conditions so that events of primary production can be correlated with external climatic factors. From this complex response to environment, the specific influences of single factors must then be identified by altering measurement conditions experimentally in the field. Next, the performance of the intact plant must be analyzed under better controlled conditions *e.g.* in the growth chamber. The biochemical and biophysical reactions of the photosynthetic apparatus may then be determined through fractions of plant material in vitro.

In this manner, each study provides additional information towards explaining plant performance at succesively more complex levels. For example, single leaf responses may be integrated so as to simulate whole plant responses and thereby describe and explain canopy performance. The integration of all available information into different models may finally result in interpretation and prediction of plant photosynthetic behaviour under natural conditions.

Experimental ecology is still far from having achieved an ideal linkage between various research levels, but the progress made in the last few decades especially through the development of sophisticated instrumentation for field measurements has been considerable.

Instrumentation for field measurement of plant gas exchange

Equipment for photosynthetic measurements under natural conditions in the field has improved enormously over the last few years. Equipment now

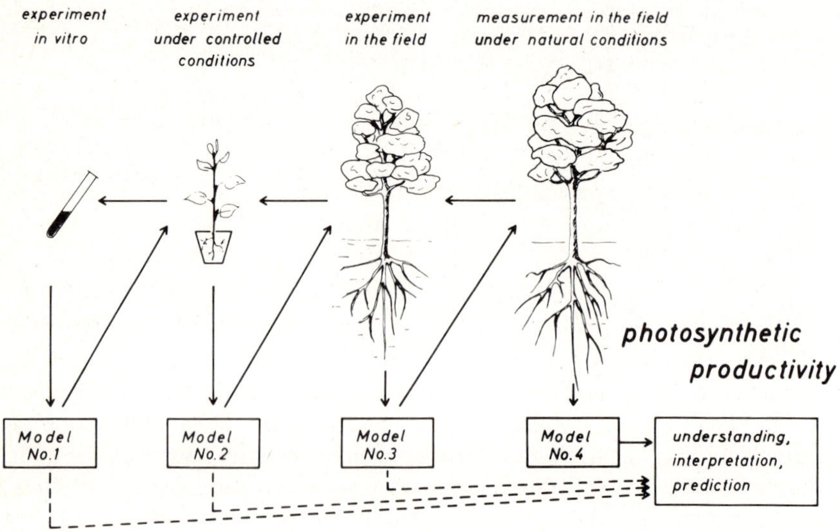

Fig. 1. Diagram of investigations required to explain photosynthetic productivity of plants in the field.

available usually allows for measuring CO_2 exchange as well as transpirational water loss and the relevant weather parameters.

To continually monitor gas exchange, fully conditioned plant cuvettes can be used (*e.g.* Lange et al. 1969, Beyschlag et al. 1986). A leaf or short twig with leaves is enclosed in a transparent gas exchange cuvette, the interior of which is climatized. The system is controlled to track outside air temperature and humidity. In addition, carbon dioxide and oxygen concentrations can also be controlled in the cuvettes. With this gas exchange method, long-term monitoring can be conducted so that large data sets can be generated for statistical evaluation. However, enclosing leaves in a fully conditioned system for periods of up to a week may influence response characteristics. In addition, measurements are restricted to single individuals of leaves or twigs and a considerable amount of technical support instrumentation is required to continue such measurements. Thus, field work with fully conditioned cuvettes is restricted to selected measuring sites.

In contrast, porometer-type gas exchange instruments increase flexibility and allow quick spot measurement of plant CO_2 assimilation and transpiration (Schulze et al. 1982, Lange & Tenhunen 1985). They allow measurements on a

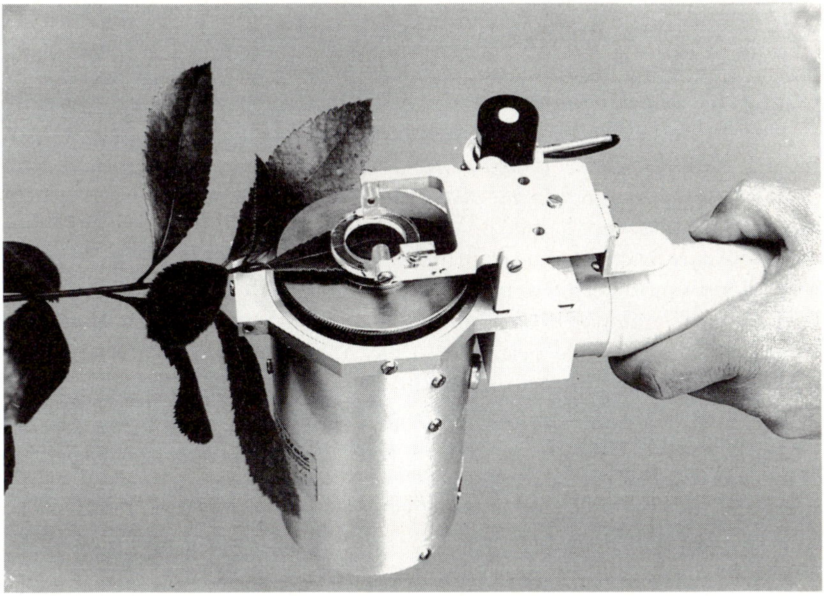

Fig. 2. H_2O/CO_2-porometer with clamp adaptor. The lower leaf face *(Arbutus unedo)* is exposed to the interior of the porometer cuvette.

variety of leaves of the same plant individual or within a plant stand. Fig. 2 shows one such porometer cuvette. The opening is clamped at the lower side of a hypostomatous leaf so that the other side is fully exposed to the natural environment. Within one to two minutes the porometer indicates photosynthetic CO_2 exchange and stomatal diffusive conductance under natural conditions. The total instrumentation is portable and can be used at remote sites.

For an optimal combination of methods, fully conditioned plant chambers in combination with CO_2 porometers can be used to obtain reliable ecophysiological information on natural gas exchange performance.

Once data on natural responses have been obtained, field measurements must be followed by field experiments conducted under controlled external environmental conditions. A portable minicuvette system consists of a small temperature and humidity conditioned cuvette and a light source mounted to provide controlled illumination (Lange & Tenhunen 1984). With this portable instrumentation, temperature, humidity, light, and CO_2 response curves are easily generated for plants in the field (Lange et al. 1985).

Types of diurnal time courses of CO_2 exchange

One of the first lessons learned when diurnal courses of gas exchange are followed under natural conditions in the field is the fact that photosynthetic patterns of different plant types may differ strikingly and characteristically even at the same site. This is shown in fig. 3 in which four different species from a desert habitat (Negev, Israel) are schematically depicted. The poikilohydric lichen *Ramalina maciformis* is photosynthetically active only after hydration of the thallus. As the lichen is dry during most of the day, its metabolism is reactivated by water vapour and dew uptake during the night. This allows a short period of high photosynthesis in the early morning hours. On the other hand, plants with homoiohydric structure are metabolically active over more extended periods of time. Varied response patterns of net photosynthesis are reflected in the different types of homoiohydric structures. The C_4 plant *Hammada scoparia* shows a plateau-shaped curve for net photosynthesis during the entire day. The mesophytic *Prunus armeniaca* cultivated in a run-off farm system closes its stomata under stressful conditions and so exhibits a pronounced midday depression of CO_2 uptake. The CAM plant *Caralluma negevensis* shows dark CO_2 fixation with its stomata closed during most of the daylight hours.

These varied response patterns of net photosynthesis, both poikilohydric and homoiohydric, demonstrate different strategies of carbon gain and water use under arid habitat conditions. The typical daily courses of CO_2 exchange may vary as the plant responds to transient climatic conditions and undergoes developmental changes to meet long-term environmental influences.

Several categories of factors which determine daily carbon gain under field conditions are illustrated by these examples. The first consists of fundamental constituent differences between plants with poikilohydric and homoiohydric

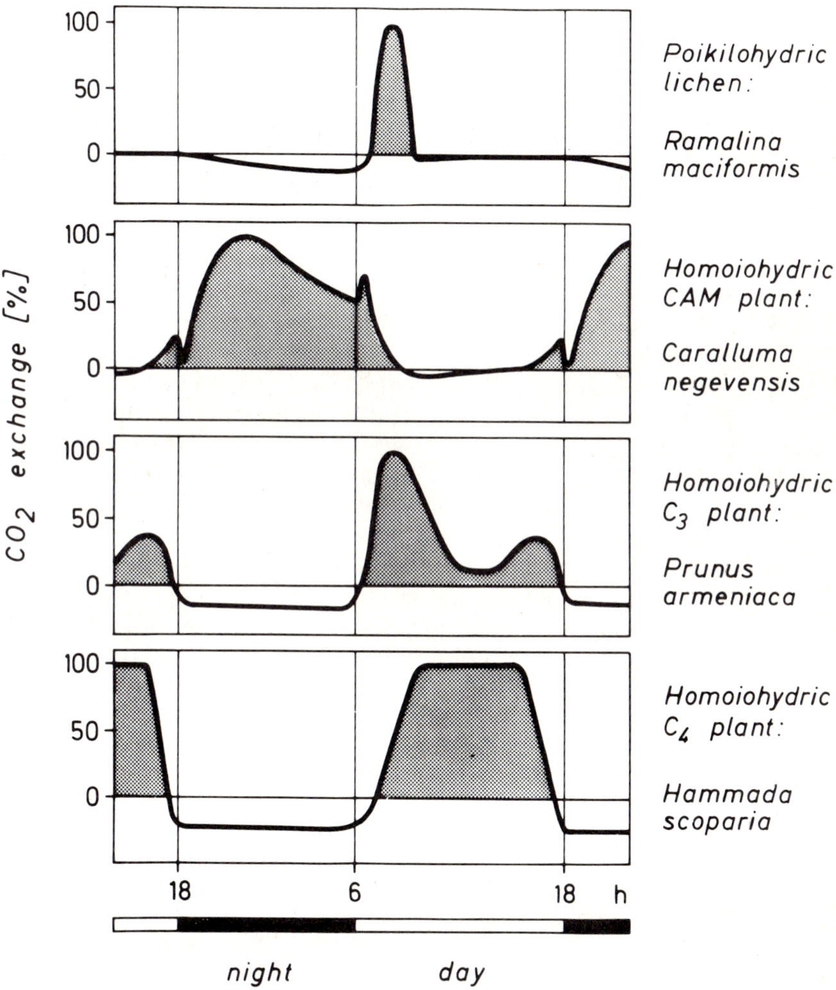

Fig 3. Diurnal courses of CO_2 exchange (CO_2 uptake positive) of different plant types growing in the same arid habitat in the Negev Desert, Israel. (After Lange et al. 1987.)

structure, various leaf types, and carbon fixation pathways. The second category of factors concerns the variations in photosynthetic capacity and performance with age and the seasonal stage of plant development. At any given time, net photosynthetic rate is determined by actual environmental conditions such as light, water availability, and nutrition. These external factors may eventually modify the functional properties of the plant and thus its range of possible responses. This is the case when acclimation to certain environmental conditions occurs.

As a result of so many interactions between various internal and external factors, control of photosynthesis at the level of the whole plant is enormously complicated. However, the differences in leaf photosynthetic performance of different plant types as well as variations during plant development and in response to external conditions can be characterized in terms of either of two photosynthetic mechanistic processes.

The first process is the actual net photosynthesis of a leaf as determined by the state and capacity of the biochemical and biophysical potential of its photosynthetic apparatus. Secondly, net photosynthesis is controlled and limited by diffusion resistances in the plant organs when carbon dioxide has to be supplied at the sites of carboxylation. The magnitude of and changes within these resistances depend both on the water status and the water requirements of the plant. As each CO_2 uptake is inevitably connected with concomitant water loss, the evolution of diffusion barriers to prevent excessive water loss in terrestrial plants has the danger of dehydration by unavoidably restricting the CO_2 supply at the same time. This is further complicated by the fact that hydrated tissue itself hampers CO_2 diffusion. Strategies for meeting the need to secure sufficient CO_2 assimilation under dry atmospheric conditions are illustrated by the two contrasting plant types, poikilohydric and homoiohydric.

Photosynthesis of poikilohydric lichens

Lichen photosynthetic metabolism is governed through the transient water conditions of the thallus. Fig. 4 illustrates how the water content of the thallus of *Ramalina maciformis* determines net photosynthesis and shows the plant's four characteristic phases of response to different degrees of hydration. At very low water contents (A), the dry lichen has no metabolic activity. In the range between 20 and 50% of dry weight related water content (B), there is a steep increase of net photosynthesis. Subsequently (C), an optimal water content for photosynthesis is reached. Finally (D), thalli are fully saturated with liquid water, and CO_2 assimilation is depressed.

Thus, in phases B and D, water conditions restrict primary production of the activated plant. In the case of B, biochemical activity of the photosynthetic apparatus is limited because of insufficient hydration. In the case of D, the depression of primary production results from the increased diffusion resistances for CO_2 when the capillary system in the thallus is filled with water (Lange & Tenhunen 1981). This is illustrated in the series of experiments depicted in fig. 5. At time zero, the lichen thalli were fully saturated with water, as is the case after a rain storm in the field. Subsequently, they were allowed to slowly dry out, and their net photosynthesis was monitored. At natural ambient CO_2 concentration around 350 ppm (second panel from above), initial CO_2 assimilation was severely depressed for as long as 10 hours. Then, with decreasing water content, CO_2 uptake increased to an optimum value, and later the lichen became fully dry. That the initial depression is actually caused by hampered gas diffusion processes, rather than by metabolic features, becomes clear from the lower panels of the diagram. When the ambient CO_2 level was increased to 800 and 1000 ppm, the initial depression gradually decreased until it was completely absent when photosynthesis was saturated at 1600 ppm external CO_2.

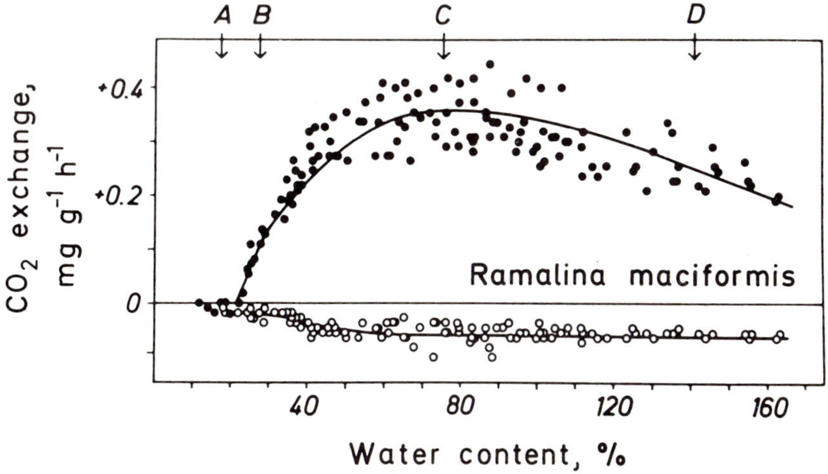

Fig. 4. Net photosynthesis of the lichen *Ramalina maciformis* in the light (solid circles) and its dark respiration (open circles) as a function of thallus water content (related to dry weight). Ordinate: CO_2 uptake (positive) and CO_2 release (negative). (After Lange 1980.)

The depression of potential CO_2 assimilation because of difficulties in CO_2 transport seems to be a general problem for plants with poikilohydric structure. It is critical for plants of this evolutionary stage of development to keep CO_2 diffusion pathways open under conditions of high hydration. One can therefore appreciate what evolution has achieved in generating the leaf of a higher homiohydric plant with its stomata and its intercellular air space system. These ensure free diffusion of CO_2 in the gas phase directly from the atmosphere near to the sites of carboxylation.

Diffusion resistances for CO_2 are small in the lichen thallus at low degrees of hydration (see phase B in fig. 4). However, the water potential at which lichens still perform CO_2 assimilation is almost unbelievably low. This is demonstrated in fig. 6 for the extreme case of *Dendrographa minor,* a species from California. The first significant CO_2 uptake is detected at a thallus water potential of −380 bar, and −180 bar already enables 50% of potential photosynthesis. This ability to absorb CO_2 at low degrees of hydration is a special ability of lichens (and some algae), unique in terrestrial plants.

Lichens are also unique in that they do not even need liquid water to reactivate their photosynthetic apparatus. For the experiment depicted in fig. 6, dry lichen material had been subjected without any water condensation to constant levels of air humidity. The thalli took up water vapour and after several hours had reached water potential equilibrium with the water vapour partial pressure of the surrounding air. Both the relative humidity at 15°C, and the air/thallus water potential in equilibrium with the ambient vapour partial pressure are indicated on the abscissa of fig. 6. As photosynthesis had begun in equilibrium with a relative humidity between 74% and 75%, the lichen reached high rates of photosynthesis without any liquid water through water vapour uptake only. This ability to utilize extremely low degrees of hydration for photosynthetic primary production determines the ecological potential of a species like *Dendrographa minor.*

However, lichens differ greatly in their ability to reactivate their photosynthesis by humidity uptake. A fundamental difference exists between lichen species with blue-green algae (Cyanobacteria) and those with green algae as their phycobionts (Lange & Kilian 1985, Lange et al. 1986). In the experiment

Fig. 5. Time courses of net photosynthetic CO_2 uptake (NP, in percent of maximal rates) of drying thalli of *Ramalina maciformis* (17°C and 750 $\mu Em^{-2}s^{-1}$ photosynthetic active radiation). The lichens were initially moistened to maximal water holding capacity (beginning of each individual curve). Subsequently they lost water. The response curves have been synchronized according to the time of maximal rates of CO_2 uptake (arrows). Experiments were conducted at different ambient CO_2 partial pressures, which are indicated in ppm. (After Lange & Tenhunen 1981.) ▷

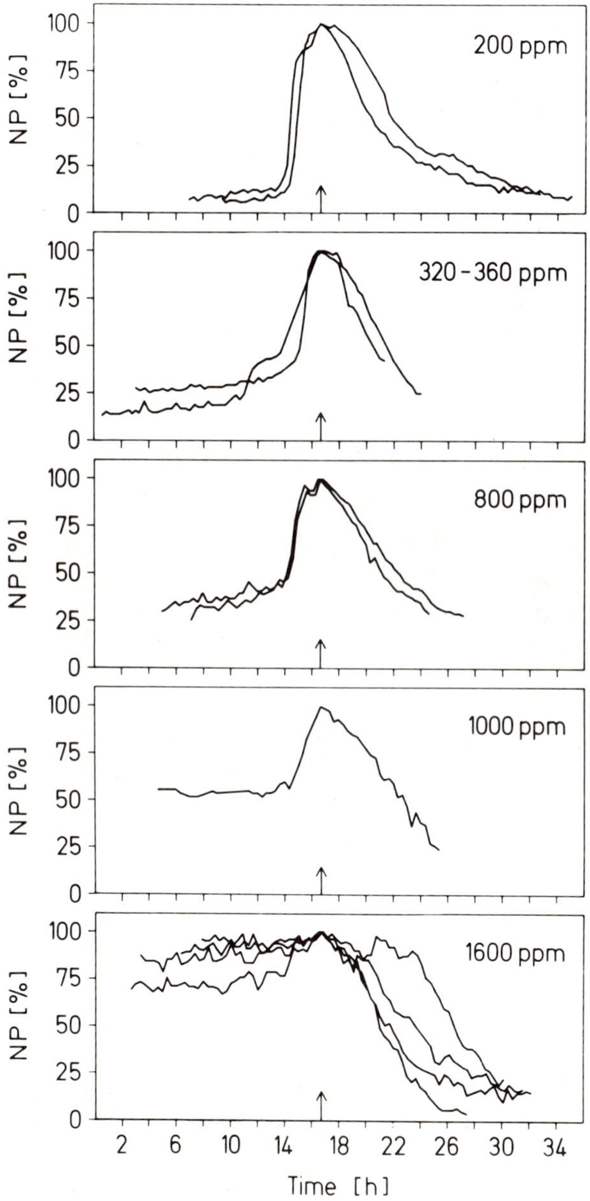
NP [%]
100
75
50
25
0
200 ppm
320 - 360 ppm
800 ppm
1000 ppm
1600 ppm
2
6
10
14
18
22
26
30
34
Time [h]

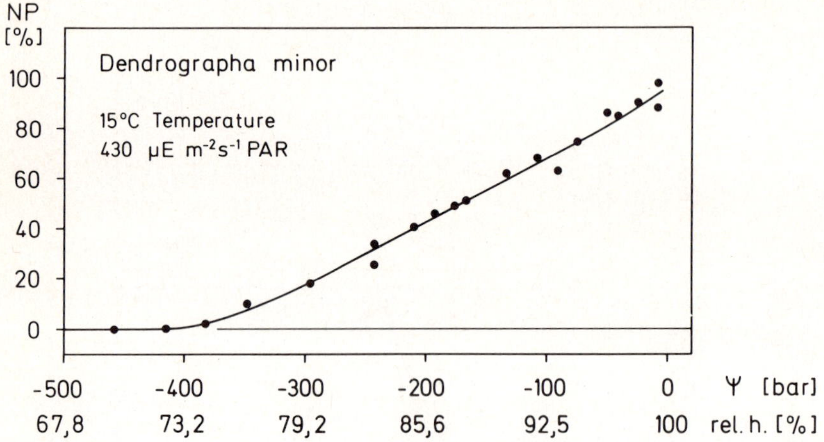

Fig. 6. Net photosynthesis (NP, in percent of maximal rates) of *Dendrographa minor* as a function of thallus water potential (ψ) and ambient relative air humidity (rel. h.), respectively, under equilibrium conditions (temperature and photosynthetic active radiation indicated).

shown in fig. 7, dry lichen thalli of three different species were subjected to a high relative air humidity near saturation beginning with time zero. Their water content (upper panel) increased and reached a constant level at equilibrium conditions after c. 40 hours of exposure. The green algae of *Peltigera leucophlebia* became photosynthetically activated and attained high rates of net CO_2 assimilation. In contrast, the blue-green algae of the two other species, the gelatinous *Collema auriculatum* and the foliaceous, heteromerous *Nephroma resupinatum,* did not become activated. Requiring liquid water for net photosynthesis, they immediately took up CO_2 after being sprayed (arrows in fig. 7). This fundamental difference in performance also exists when closely related lichens with different phycobionts are compared with each other, such as species within one genus (*e.g. Lobaria* spp., *Peltigera* spp., *Nephroma* spp., *Sticta* spp.) or even photosymbiodemes with lobes of green or blue-green algae within the same individual thallus (*e.g. Pseudocyphellaria* spp.: Lange et al. 1988.

Fig. 7. Time courses of water content (in percent of dry weight) and CO_2 exchange in the light (CO_2 uptake positive) of lichens during water vapour uptake. At starting time 0, the dry lichen thalli were treated with air of a relative humidity of 97% (15°C) and were later sprayed with liquid water (arrows): *Peltigera leucophlebia* (green phycobiont), *Nephroma resupinatum* (blue-green phycobiont), and *Collema auriculatum* (blue-green phycobiont). (From Lange et al. 1986.) ▷

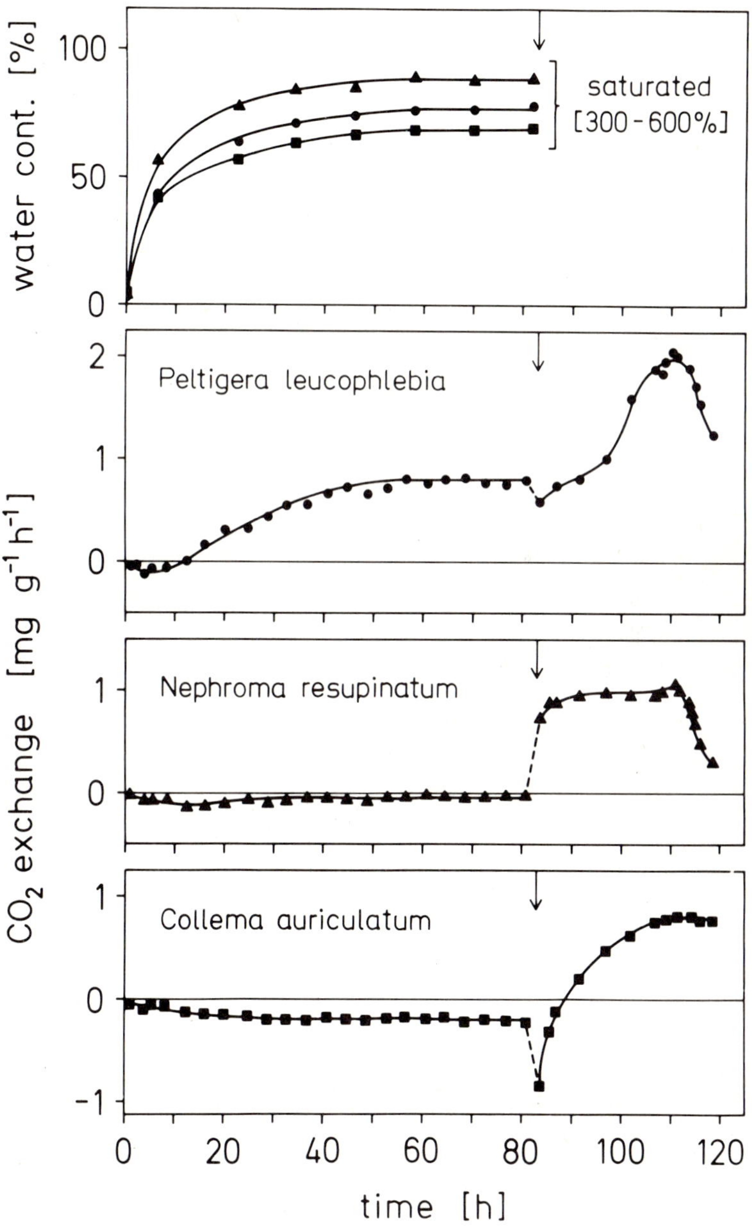
water cont. [%]
100
50
0
saturated
[300 - 600%]
Peltigera leucophlebia
Nephroma resupinatum
Collema auriculatum
CO_2 exchange [mg g^{-1} h^{-1}]
2
1
0
-1
0
20
40
60
80
100
120
time [h]

This contrast in lichen performance probably reflects fundamental differences in the structure of the photosynthetic apparatus of green and blue-green algae. This may yet be shown in current studies investigating low-temperature as well as room-temperature fluorescence of lichens at different degrees of hydration.

Ecologically, these functional differences may be very important determinants of lichen habitat as illustrated in the stable carbon isotope ratios of lichen thalli containing green or blue-green algae (Lange & Ziegler 1986, unpublished data of H. Ziegler). When $\delta^{13}C$ discrimination ratios for lichen species of genera with both green and blue-green algae were examined (table 1), all species with green algae had low $\delta^{13}C$ values, indicating a high degree of discrimination. In contrast, the species with blue-green algae had much higher $\delta^{13}C$ values and lower rates of discrimination. These differences must be caused through differences in CO_2 diffusion resistances in the thalli during assimilation, resulting in differences in carbon dioxide partial pressure at the sites of carboxylation (H. Ziegler, unpublished; see Osmond et al. 1982). The lichens with blue-green algae exhibit high diffusion resistances during CO_2 fixation because they can only assimilate CO_2 when they are moistened by liquid water. In contrast, the lichens with green algae also photosynthesize after water vapour uptake and thus with small diffusion resistances.

Carbon isotope composition is an integration of total carbon-gain life history of the different lichen thalli. Thus, the laboratory response differences between the two groups of lichens also reflect the water content related photosynthetic performance of the lichens under natural conditions. It can therefore be understood why lichens with blue-green algae need habitats where at least periodically liquid water is available for photosynthetic activity. The ability to activate photosynthesis through humidity uptake, on the other hand, enables lichens with green algae to colonize special habitats with a minimum or even lack of liquid precipitation. Such lichens are found under extremely dry conditions, for instance in the fog oases in Chilean deserts (Redon & Lange 1983), at the surface of overhanging rocks in talus slopes or on cliffs where liquid water is not available but the relative humidity is high.

Table 1. $\delta^{13}C$ discrimination values ($^0/_{00}$ PDP as standard, see Osmond et al. 1982) in the thallus material of different folious heteromerous lichen species with different phycobionts within single genera. (After Lange & Ziegler 1986 and unpublished data from H. Ziegler, Munich; lichen samples provided by U. Buschbom, T. G. A. Green, J. Poelt, R. Türk, V. Wirth.) ▷

Species (origin) phycobiont:	green	blue-green
Lobaria adscripta (New Zealand)		−23.84
– *scrobiculata* (Austria)		−24.56
– *laetevirens* (Portugal)	−30.45	
– *pulmonaria* (Austria, 1)	−31.85	
– *pulmonaria* (Austria, 2)	−32.15	
– – *var. meridionalis* (Tenerife)	−32.93	
Nephroma resupinatum (Austria, 1)		−22.36
– *bellum* (Austria, 1)		−23.95
– *bellum* (Austria, 2)		−24.07
– *resupinatum* (Austria, 2)		−24.30
– *laevigatum* (France)		−24.70
– *parile* (Austria)		−24.72
– *arcticum* (Sweden)	−29.38	
Peltigera canina (Austria)		−20.72
– *rufescens* (Germany, Franconia)		−21.32
– *praetextata* (Austria)		−22.55
– *neckeri* (Austria)		−23.40
– *polydactyla* (Germany, Franconia)		−24.24
– *nana* (New Zealand)		−26.75
– *leucophlebia* (Alaska)	−29.43	
– *aphtosa* (Austria)	−32.19	
– *leucophlebia* (Austria)	−32.95	
Sticta fuliginosa (Austria)		−19.63
– spec. (Brasilia, 1)		−22.40
– *latifrons* (New Zealand, 1)	−22.73	
– *dufourei* (Tenerife)		−22.80
– *latifrons* (New Zealand, 2)	−22.89	
– spec. (South Africa)	−23.46	
– *fuliginosa* (New Zealand)		−23.52
– spec. (Brasilia, 2)		−23.63
– *subcaperata* (New Zealand)	−25.75	
– *limbata* (New Zealand)		−26.53
– *filix* (New Zealand)	−30.66	
– spec. (Brasilia, 3)	−31.69	
– spec. (Brasilia, 4)	−32.20	
– *latifrons* (New Zealand, 3)	−32.89	

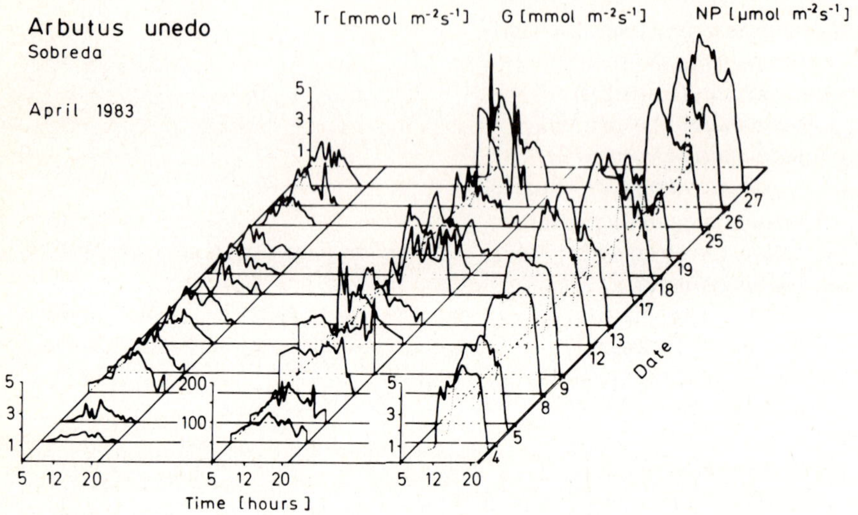

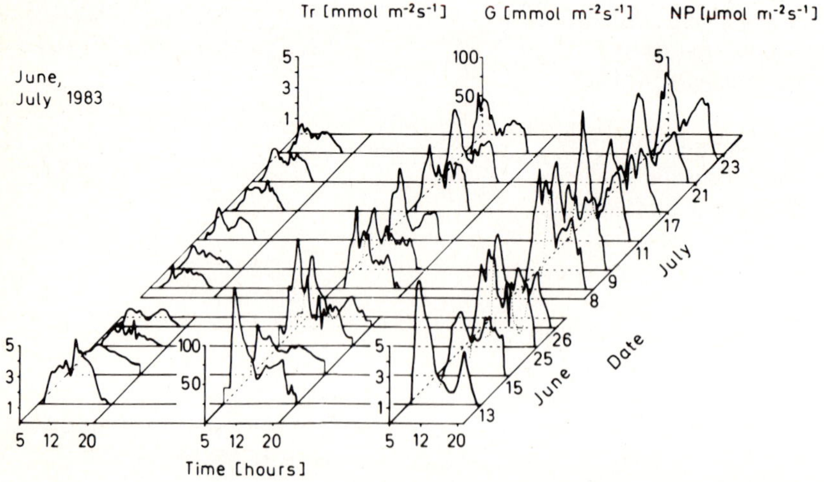

Fig. 8. Daily time courses of net photosynthesis (NP), transpiration (Tr), and leaf conductance (G) in April (above) and in June/July (below) for leaves of *Arbutus unedo* under natural conditions in an evergreen macchia, Quinta São Pedro, Sobreda, Portugal. (After Beyschlag 1984.)

Photosynthesis of homoiohydric higher plants

In contrast to poikilohydric lichen performance, the homoiohydric structure of higher plants requires a strictly controlled gas exchange that ensures sufficient CO_2 uptake without risk of dehydration of the mesophyll through excessive water loss. This regulation is performed by stomata which are controlled and finely tuned by various internal and external influences. Such interrelationships between plant water relations and carbon gain in a daily and seasonally fluctuating environment are illustrated by Mediterranean-type sclerophyllous shrubs. As the Mediterranean environment is characterized by a change from a mild and moist winter to a hot and dry summer, the same leaves of the evergreen vegetation are exposed to extreme seasonal phases. Therefore, environmental influences on stomatal controlling mechanisms can be well illustrated by studying these shrubs.

Arbutus unedo, a shrub or small tree of the *Ericaceae,* is characteristic of the more mesic sites of the Mediterranean-type macchia. Throughout the year, its leaves experience great seasonal changes in water status. At a study site in Portugal (Beyschlag 1984), soil water supply is favourable in the winter. However, leaf predawn water potential decreases in the summer to around −50 bar under extreme water stress. In combination with atmospheric stress, these seasonal fluctuations in soil and plant water relations determine primary production, as clearly reflected through seasonal changes in the daily patterns of CO_2 assimilation.

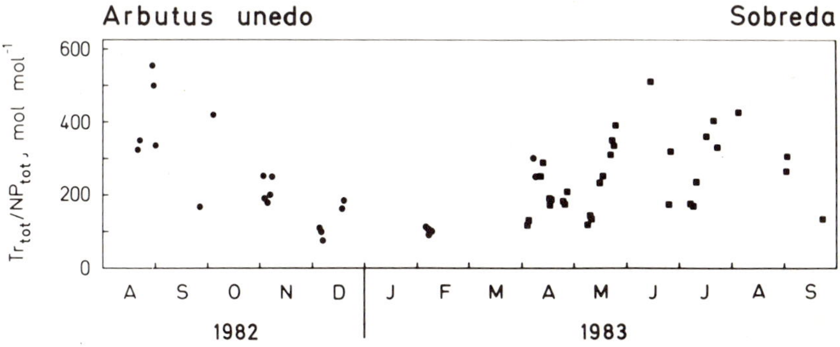

Fig. 9. Annual course of the transpiration ratio (Tr/Np: sum of water loss per sum of CO_2 uptake for the daylight period between the two light compensation points of net photosynthesis in the morning and in the evening) for leaves of *Arbutus unedo.* Months are indicated at the abscissa. (See fig. 8; after Beyschlag 1984.)

In fig. 8, diurnal courses for net photosynthesis, transpiration, and leaf conductance are plotted. During the first part of April, stomata are open most of the day and the diel courses of photosynthesis in general have a bell-shaped appearance with the highest rates around noon. This high rate of photosynthesis is associated with low rates of water loss, because the water vapour deficit of the air is low. At the end of the month, the stomata tend to decrease conductance after the highest values in the early morning hours so that two-peaked curves of gas exchange appear. This tendency becomes more pronounced in June and July, when stomatal conductance and net photosynthesis are characterized by a steep midday depression which appears every day and which illustrates a sensitive control of water use. The effectiveness of the stomatal regulation of gas exchange becomes obvious in fig. 9 where the transpiration coefficient, the ratio between total daily transpiration and the daily total of CO_2 fixation, is plotted. Water use efficiency is highest in winter but does not show much difference from April through September, even if climatic differences are extreme between spring and late summer. Such a successful regulation of water use is a result of the midday depression of leaf conductance by which CO_2 uptake is restricted to the morning and the afternoon hours and stomata close during the hottest and driest part of the day. From the calculations of Cowan (1982), we can infer that this is a strategy which plants of arid habitats have adopted to optimize water use.

It is possible to reproduce this typical plant response under simulated conditions in a growth chamber. Potted plants of *Arbutus unedo* were subjected to artificial light, temperature, and humidity conditions mimicking standard Portuguese climatic summer conditions (Tenhunen et al. 1981). Two different experiments are plotted in fig. 10. In both, the plants exhibited typical midday depression of net photosynthesis and transpiration with a concomitant increase in stomatal resistance. This response of *Arbutus unedo* under simulated conditions is identical to that under natural conditions in the field.

With this experimental design, the factors which control midday depression can be further evaluated. Because the experiment depicted in fig. 10 was conducted with plants well hydrated with soil water at field capacity, the triggering factors for midday stomatal closure were atmospheric in nature and not

Fig. 10. Daily time courses of leaf temperature (T_L), air-to-leaf water vapour concentration difference (WD), transpiration rate (Tr), leaf diffusion resistence (R), and net photosynthesis rate (NP), measured for leaves of well-watered potted *Arbutus unedo* plants under simulated natural conditions in an environmental chamber. Measurements with two different leaf samples. (From Tenhunen et al. 1981.) ▷

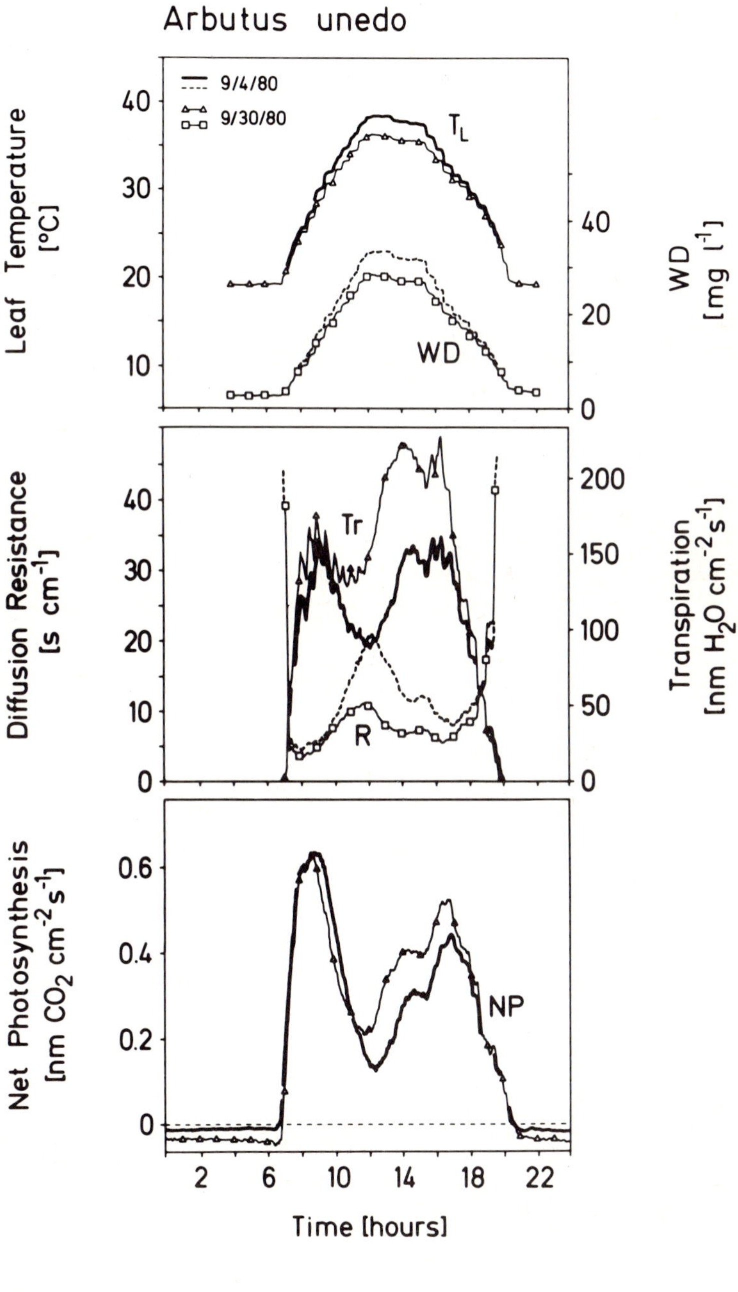
Arbutus unedo
9/4/80
9/30/80
T_L
WD
Leaf Temperature [°C]
WD [mg l⁻¹]
Tr
R
Diffusion Resistance [s cm⁻¹]
Transpiration [nm H₂O cm⁻² s⁻¹]
NP
Net Photosynthesis [nm CO₂ cm⁻² s⁻¹]
Time [hours]

primarily related to soil drought. This becomes still clearer in the series of experiments shown in fig. 11. On four consecutive days, a small tree of *Arbutus unedo* was subjected to increasing atmospheric stress in the growth chamber. Standard day conditions were applied at constant water vapour dewpoint while the level of air temperature was raised from day to day. Under cool and humid conditions, stomata opened in the morning and closed gradual during the remainder of the day; net photosynthesis and transpiration exhibited one-peaked courses. This changed with increasing atmospheric stress when maximum leaf conductance dropped slightly and stomata closed at noon. Thus, water loss as well as CO_2 assimilation was exhibiting an increasingly two-peaked performance curve.

This experimental was also conducted under unlimited soil water supply, and it clearly domonstrates the control of diurnal response patterns by atmospheric conditions alone, probably in connection with photoinhibitory processes (see Demmig-Adams et al. 1988). In addition, this reaction is mediated by the internal water stress of the plant as evidenced by fig. 12. An evergreen oak, *Quercus suber,* was experimentally subjected to increasing soil water stress under the same standard day conditions. Predawn water potential dropped from −23 to less than −37 bar on consecutive days. This resulted in

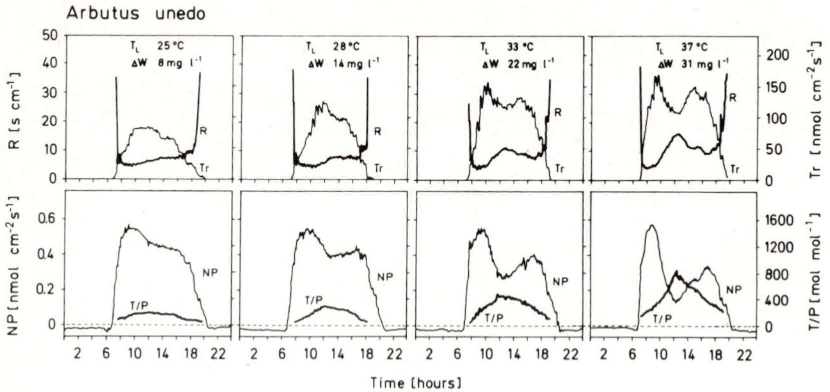

Fig. 11. Diurnal time courses of leaf resistence (R), rates of net photosynthesis (NP) and transpiration (Tr), and of the ratio Tr/NP (T/P), with potted, well watered plants of *Arbutus unedo* on four consecutive days. Plants were subjected to simulated natural time courses of temperature, air humidity, and light. Air temperatures (T_L) and leaf-to air water vapour concentration differences (ΔW) were increased each day to the maximal value shown in the body of the figure. (From Tenhunen et al. 1987.)

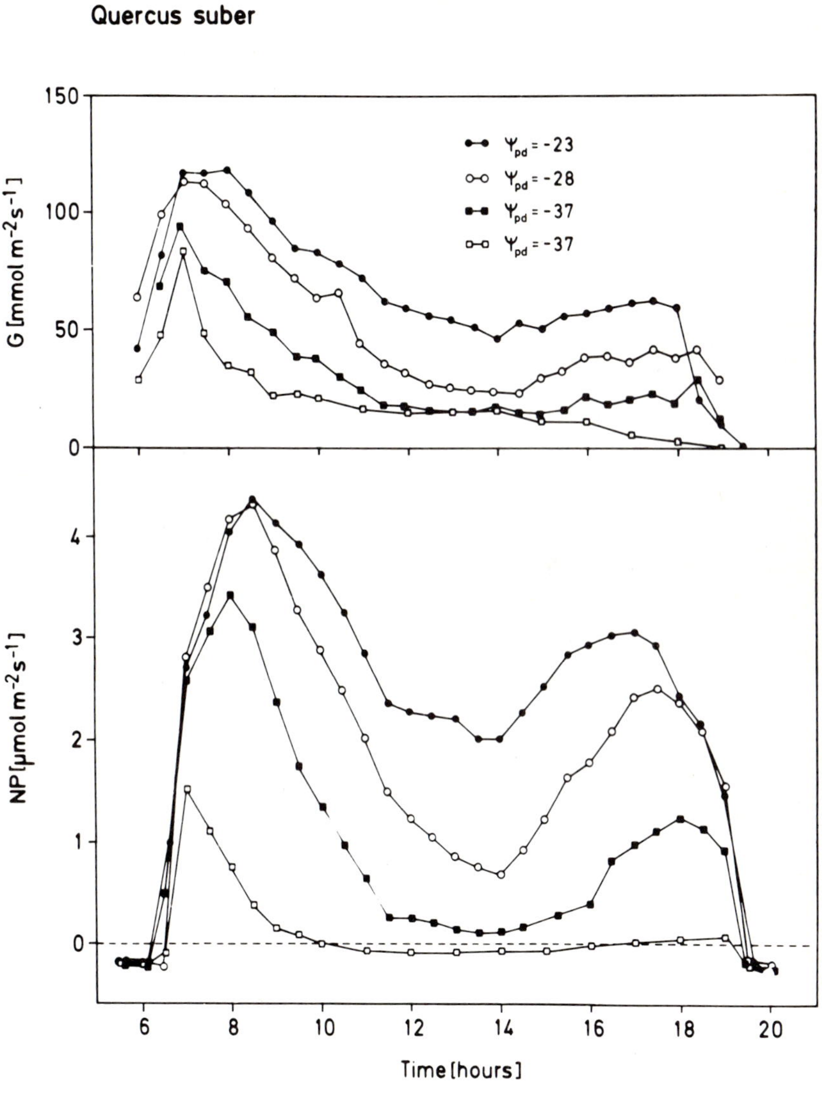

Fig. 12. Diurnal time courses of leaf conductance (G) and rate of net photosynthesis (NP) in leaves of potted *Quercus suber* plants on the third (solid circles), fourth (open circles), fifth (solid squares), and sixth (open squares) day of a soil drying cycle under standard summer day atmospheric conditions (see text) simulated in a growth chamber. Pre-dawn leaf water potential (ψ_{PD}) indicated in the top panel. (After J. Gebel, Diplomarbeit, University of Würzburg.)

successively more intense midday depressions, until at last only the morning peak of net photosynthesis subsisted under high soil-water stress. Midday depression of gas exchange is thus apparently induced by atmospheric stress, but the extent of the depression greatly depends on actual soil water stress. Furthermore, a plant's stress history, *e.g.* during the course of the dry season, is also an important determinant of its stomatal and photosynthetic performance (see Schulze & Hall 1982).

An important feature of plant performance, in fig. 12, is that the morning depression of leaf conductance occurs earlier with increasing soil moisture stress. This indicates changes in the temperature and/or humidity sensitivity of the stomata. The same phenomenon is seen under natural conditions. In winter and spring, maximal daily photosynthesis takes place around noon, with one-peaked diurnal response patterns. In the summer, however, predominantly two-peaked diurnal courses of gas exchange occur so that the maximum net photosynthesis is shifted to the early morning hours, *i.e.* towards the cooler part of the day with high relative air humidity. Fig. 13 illustrates this for measurements made on *Arbutus unedo* through the course of the seasons (Beyschlag 1984). While in winter the maximal daily air-to-leaf vapour pressure gradients are low for *Arbutus unedo* in the macchia when the days are humid, the summer gradients are steep when the weather turns hot and dry. Nevertheless, the air-to-leaf vapour pressure gradient, at the time of day when maximum photosynthesis occurs, surprisingly stays at the same level of about 15 mbar and at similar leaf temperatures all the year round. This indicates an enormous homoeostasis is seasonal behaviour of photosynthesis. In spite of extreme differences in temperature and humidity conditions throughout the year, the period of maximal primary production is always shifted to the time of day that is best suited for maximal carbon gain with minimal water loss, namely noon in the winter and the morning hours in the

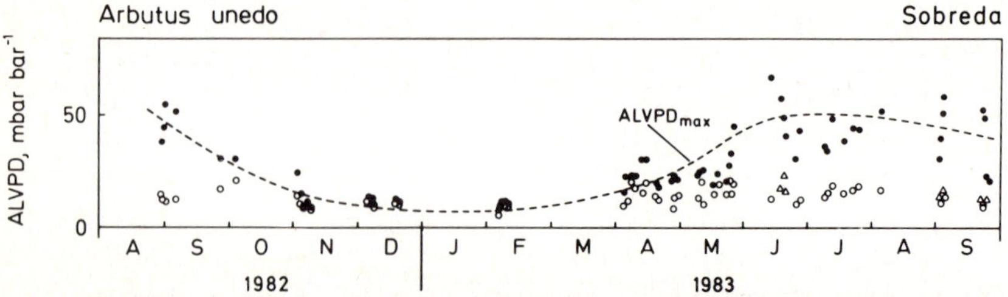

Fig. 13. Annual courses of the daily maximum of the air-to-leaf water vapour partial pressure difference (ALVPD $_{max}$, closed symbols) and ALVPD measured at the time of the daily maximum of net photosynthesis (open symbols) for leaves of *Arbutus unedo.* Months are indicated at the abscissa. (See fig. 8, after Beyschlag 1984.)

summer. This demonstrates the success of stomatal action in optimizing water use efficiency.

Midday depression of stomatal conductance is a central feature in achieving such homoeostatic control. However, further analysis has shown that

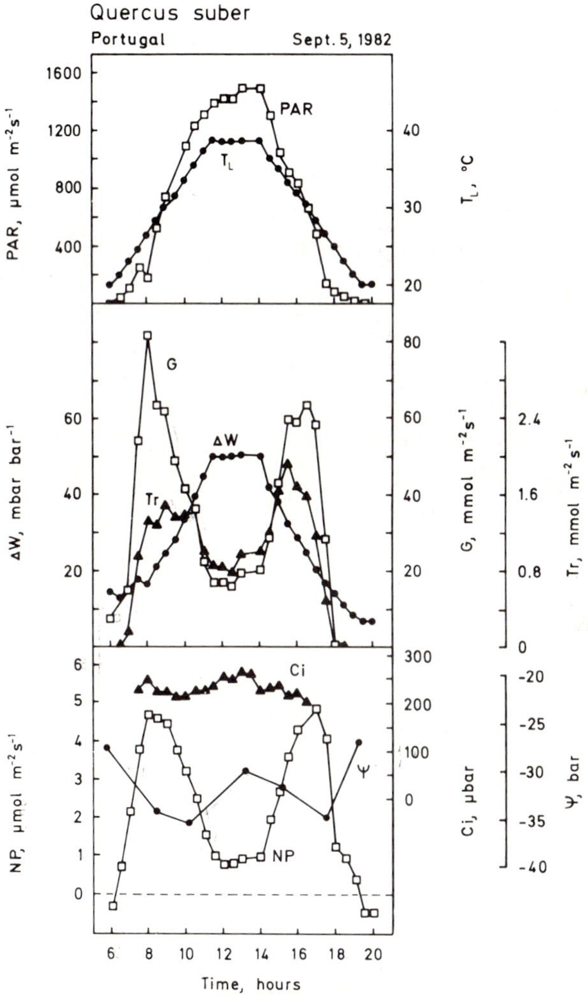

Fig. 14. Daily time courses of photosynthetic active radiation (PAR), leaf temperature (T_L), air-to-leaf water vapour partial pressure difference (ΔW), leaf conductance (G), rate of transpiration (Tr), rate of net photosynthesis (NP), leaf xylem water potential (ψ), and leaf internal CO_2 partial pressure (C_i) of *Quercus suber* leaves in a Portuguese evergreen macchia. (From Tenhunen et al. 1984.)

stomatal responses to atmospheric and/or soil water-stress are insufficient explanations for the total phenomenon. This becomes clear from fig. 14 which shows a typical diurnal course of the gas exchange performance of the cork oak, *Quercus suber*, during summer in Portugal. Leaf conductance exhibited a deep midday depression which resulted in greatly reduced rates of transpiration at noon. Concurrently, net photosynthesis also strongly decreased. However, this drop in CO_2 assimilation could not have been caused by a restricted supply of CO_2 to the mesophyll through increased stomatal diffusion resistance at midday. In examining the diurnal course of internal CO_2 partial pressure in the mesophyll (Ci), as calculated from stomatal conductance and net photosynthesis, it was found that Ci stayed remarkably constant at about 230 mbar during the day and even increased somewhat at noon. Thus, a drop of leaf internal CO_2 could not explain decreased net photosynthesis. The only explanation is a drop in biochemical and biophysical capacity of the photosynthetic apparatus at midday. Only this would account for a decrease in net photosynthesis under conditions of constant internal levels of CO_2.

In order to characterize the metabolic capacity of the mesophyll, photosynthesis has to be related to leaf internal CO_2 partial pressure. The basis for such

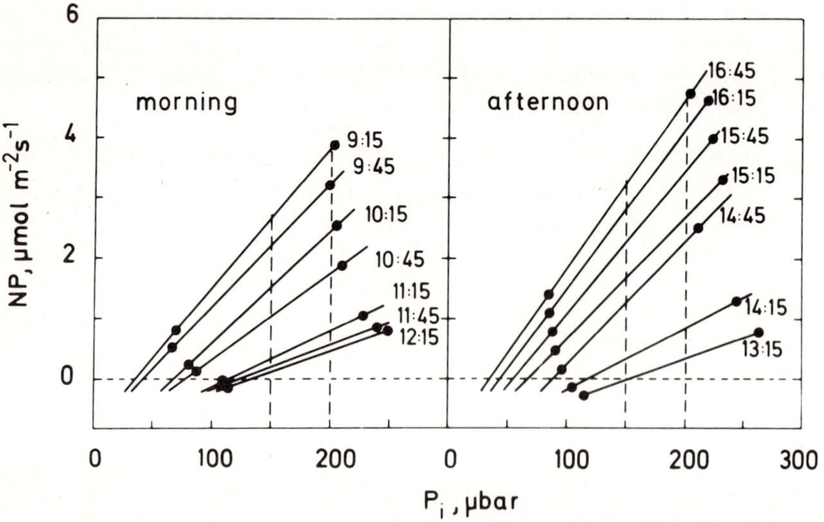

Fig. 15. Initial slope of the CO_2 response curves (carboxylation efficiency) obtained at different times of the day with leaves of *Quercus suber*; data corresponding to those of fig. 14. Ordinate: net photosynthesis (NP); abscissa: leaf internal CO_2 partial pressure (P_i). (From Tenhunen et al. 1984.)

analyses are CO_2 response curves of intact leaves taken in the field. These exhibit saturation-type characteristics. Maximal photosynthetic capacity of CO_2 assimilation at saturating CO_2 concentration is thought to be determined by maximal possible regeneration of the CO_2 acceptor. The initial slope of the CO_2 response curve, the carboxylation efficiency, is believed to be determined mainly through the kinetics of the carboxylating enzyme system (see Farquhar & Caemmerer 1982). These two parameters are important for characterizing the status of the photosynthetic apparatus of a leaf.

As an example, fig. 15 shows the diurnal patterns of the initial slope of the CO_2 response curves of *Quercus suber* leaves. With changes in temperature, light and air humidity, carboxylation efficiency changes drastically during the course of a summer day. Carboxylation efficiency is strongly depressed during the midday hours, and at the same time maximal photosynthetic capacity is also decreased. This explains the natural midday depression of photosynthesis as a result of diurnal changes in the activity of the photosynthetic apparatus, indepedent of stomatal control. It should be mentioned that photosynthetic characteristics not only differ with the time of day but also change over time with the course of the seasons (Tenhunen et al. 1985). Actual changing weather conditions can not only affect photosynthesis transiently but may induce absolute changes in the characteristics of the photosynthetic apparatus as a result of phenological changes and long-term effects of water stress and seasonal acclimation. Together with stomatal responses, the combination of these influences explains the seasonal changes in diurnal patterns of photosynthetic responses.

As shown in fig. 1, such studies will not be complete unless research is directed towards a deeper physiological and biochemical understanding of plant responses. One direction is to determine the involvement of phytohormonal control (Burschka et al. 1985, Raschke & Hedrich 1985) for answering the question of how the reactions of the stomata and the photosynthetic mesophyll capacity are coordinated within the leaf to produce constant internal CO_2 partial pressure. Another is to explain why leaves exposed to high solar radiation apparently tend to avoid low internal CO_2 partial pressure, perhaps in terms of an increased danger of midday photoinhibition (see Demmig-Adams et al. 1988).

In addition to further physiologically analyzing single factors and dependencies, we also need to integrate results in order to understand plant responses to the total complex of enviromental factors. This step involves mathematical models which describe and simulate the multivariate relationships of leaf CO_2 assimilation (see Caemmerer & Farquhar 1981, Harley et al. 1986,

Tenhunen et al. 1987). Such models are efficient tools for integrating the available information in order to better quantify the physiological and environmental mechanism which determine leaf primary production.

Canopy photosynthesis

Photosynthesis models are a basic requirement for addressing questions of photosynthetic behaviour of whole plants and whole plant stands. In plant canopies, not only do the physiological characteristics vary at different locations, for instance between those leaves in the sun and those in the shade, but also between portions of foliage elements as the microenvironment of one is altered by the presence of others.

Assessing these canopy relationships of structure and function is not at all trivial as the relevant parameters cannot be measured simultaneously. Thus, detailed simulations must be combined with selected measurements to elucidate the various interactions in canopies by producing very complicated model systems. Fig. 16 illustrates the general structure of such a canopy model for the sclerophyllous shrub *Quercus coccifera* under Mediterranean condi-

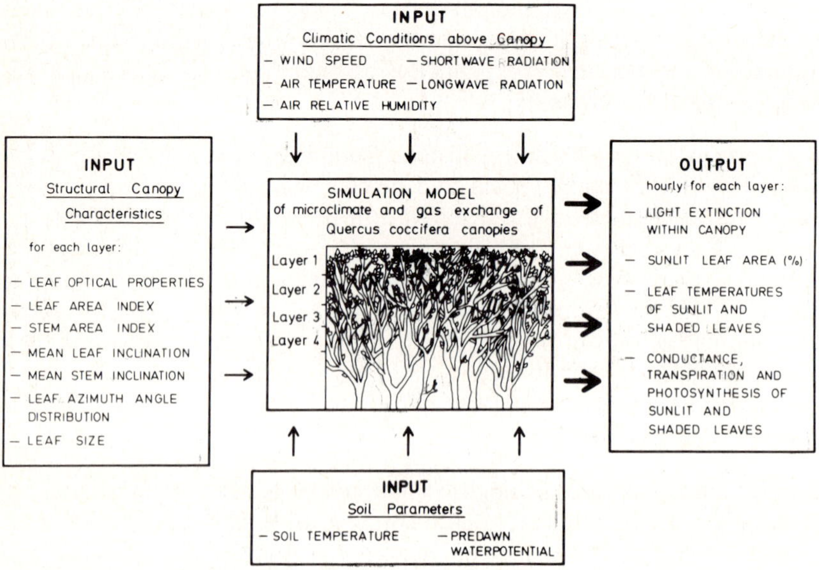

Fig. 16. Input and output parameters of an canopy microclimate and gas exchange model for *Quercus coccifera*. (From Meister et al. 1987.)

tions (Caldwell et al. 1986, Meister et al. 1987). Input data include soil temperature, water potential and the climatic conditions above the canopy. The canopy itself is characterized by such structural parameters as optical leaf properties, leaf area index, and leaf angle distribution. Together, these data are fed into a microclimatic model which simulates light and temperature conditions for the individual leaf classes at the different canopy levels. Its output is used in a stomatal and a gas exchange model which simulates leaf conductance, transpiration, and leaf net photosynthesis on an hourly basis and subsequently integrates the resulting data for days and seasons.

Such models can assess important ecological problems such as the extent to which different canopy layers contribute to the total carbon gain. This assessment might be essential to interpret different plant growth forms and their competetive potential. Specific physiological functions of single leaves, such as the midday depression of gas exchange, can be considered for the whole plant or for whole plant stands. Quantitative estimates of the significance of sun versus shade leaf adaptations for total canopy photosynthesis and water use efficiency are also possible (see Meister et al. 1987).

While canopy models are highly complex and difficult to construct, they are potential tools to learn more about the control of CO_2 assimilation at the level of the organism and ultimately to physiologically understand the whole plant and canopy functioning of photosynthetic production. This goal is important not only for theoretical ecology but also for applications in agriculture and forestry.

Acknowledgements

The research was supported by the “Deutsche Forschungsgemeinschaft“ and by the “Fonds der Chemischen Industrie im Verband der Chemischen Industrie e.V.“ The author is grateful to Dr. H. Ziegler (Munich), Dr. T. Nash III (Tempe, Arizona), and Dr. T. G. A. Green (Hamilton, New Zealand) for their agreement to use unpublished data. Ms Rebecca Richards (Logan, Utah) carefully edited the English rendition of the manuscript.

References

Beyschlag, W. 1984: Photosynthese und Wasserhaushalt von *Arbutus unedo* L. im Jahreslauf am Freilandstandort in Portugal. Gaswechselmessungen unter natürlichen Bedingungen und experimentelle Faktorenanalyse. – Diss., Würzburg.

–, Lange, O. L. & Tenhunen, J. D. 1986: Photosynthese und Wasserhaushalt der immergrünen mediterranen Hartlaubpflanze *Arbutus unedo* L. im Jahreslauf am Freilandstandort in Portugal. I. Tagesläufe von CO_2-Gaswechsel und Transpiration unter natürlichen Bedingungen. – Flora **178:** 409–444.

Burschka, C., Lange, O. L. & Hartung, W. 1985: Effects of abscisic acid on stomatal conductance and photosynthesis in leaves of intact *Arbutus unedo* plants under natural conditions. – Oecologia (Berlin) **67:** 593–595.

Caemmerer, S. von & Farquhar, G. D. 1981: Some relationships between the biochemistry of photosynthesis and the gas exchange of leaves. – Planta **153:** 376–387.

Caldwell, M. M., Meister, H.-P., Tenhunen, J. D. & Lange, O. L. 1986: Canopy structure, light microclimate and leaf gas exchange of *Quercus coccifera* L. in a Portuguese macchia: measurements in different canopy layers and simulation with a canopy model. – Trees **1:** 25–41.

Cowan, I. R. 1982: Regulation of water use in relation to carbon gain in higher plants. – In: Lange, O. L., Nobel, P. S., Osmond, C. B. & Ziegler, H. (eds.), Physiological plant ecology II (Encyclopedia of plant physiology, new series, **12B**): 589–613. – Springer, Berlin etc.

Demmig-Adams, B., Adams III, W. W., Winter, K., Meyer, A., Schreiber, U., Pereira, J. S., Krüber, A., Czygan, F.-C. & Lange, O. L. 1988: Photochemical efficiency of PSII, photon yield of O_2 evolution, photosynthetic capacity, and carotenoid composition during the „midday' depression" of net CO_2 uptake in *Arbutus unedo* growing in Portugal. – Planta (submitted).

Farquhar, G. D. & Caemmerer, S. von 1982: Modelling of photosynthetic response to environmental conditions. – In: Lange, O. L., Nobel, P. S., Osmond, C. B. & Ziegler, H. (eds.), Physiological plant ecology II (Encyclopedia of plant physiology, new series, **12B**): 549–587. – Springer, Berlin, etc.

Harley, P. C., Tenhunen, J. D. & Lange, O. L. 1986: Use of an analytical model to study limitations on net photosynthesis in *Arbutus unedo* under field conditions. – Oecologia (Berlin) **70:** 393–401.

Lange, O. L. 1980: Moisture content and CO_2-exchange of lichens. I. Influence of temperature on moisture-dependent net photosynthesis and dark respiration in *Ramalina maciformis.* – Oecologia (Berlin) **45:** 82–87.

–, Beyschlag, W. & Tenhunen, J. D. 1987: Control of leaf carbon assimilation – input of chemical energy into ecosystems. – In: Schulze, E.-D. & Zwölfer, H. (eds.), Potentials and limitations of ecosystem analysis (Ecol. Stud., **61**): 149–163. – Springer, Berlin, etc.

–, Gebel, J., Schulze, E.-D., Walz, H. 1985: Eine Methode zur raschen Charakterisierung der photosynthetischen Leistungsfähigkeit von Bäumen unter Freilandbedingungen – Anwendung zur Analyse "neuartiger Waldschäden" bei der Fichte. – Forstwiss. Centralbl. **104:** 186–198.

–, Green, T. G. A. & Ziegler, H. 1988: Water status related photosynthesis and carbon isotope discrimination in species of the lichen genus *Pseudocyphellaria* with green or blue-greeen photobionts and in photosymbiodemes. – Oecologia (Berlin) **75:** 494-501.

– & Kilian, E. 1985: Reaktivierung der Photosynthese trockener Flechten durch Wasserdampfaufnahme aus dem Luftraum: artspezifisch unterschiedliches Verhalten. – Flora **176:** 7–23.

–, – & Ziegler, H. 1986: Water vapour uptake and photosynthesis of lichens: performance differences in species with green and blue-green algae as phycobionts. – Oecologia (Berlin) **71:** 104–110.

–, Koch, W. & Schulze, E.-D. 1969: CO_2-Gaswechsel und Wasserhaushalt von Pflanzen in der Negev-Wüste am Ende der Trockenzeit. – Ber. Deutsch. Bot. Ges. **82:** 39–61.

– & Tenhunen, J. D. 1981: Moisture content and CO_2 exchange of lichens. II. Depression of net photosynthesis in *Ramalina maciformis* at high water content is caused by increased thallus carbon dioxide diffusion resistance. – Oecologia (Berlin) **51:** 426–429.

– & – 1984: A minicuvette system for measurement of CO_2-exchange und transpiration of plants under controlled conditions in field and laboratory. – Walz, Effeltrich.

– & – 1985: CO_2/H_2O porometer for the measurement of CO_2 gas exchange and transpiration of plants under natural conditions. – Walz, Effeltrich.

– & Ziegler, H. 1986: Different limiting processes of photosynthesis in lichens. – In: Marcelle, R., Clijsters, H. & van Poucke, M. (eds.), Biological control of photosynthesis: 147–161. – Nijhoff, Dordrecht.

Meister, H.-P., Caldwell, M. M., Tenhunen, J. D. & Lange, O. L. 1987: Ecological implications of sun/shade-leaf differentiation in sclerophyllous canopies: assessment by canopy modeling. – In: Tenhunen, J. D., Catarino, F. M., Lange, O. L. & Oechel, W. (eds.), Plant response to stress – functional analysis in Mediterranean ecosystems: 401-411. – Springer, Berlin etc.

Osmond, C. B., Winter, K. & Ziegler, H. 1982: Functional significance of different pathways of CO_2-fixation in photosynthesis. – In: Lange, O. L., Nobel, P. S., Osmond, C. B. & Ziegler, H. (eds.), Physiological plant ecology II (Encyclopedia of plant physiology, new series, **12B):** 479–547. – Springer, Berlin etc.

Raschke, K. & Hedrich, R. 1985: Simultaneous and independent effects of abscisic acid on photosynthesis and stomatal resistance. – Planta **163:** 105–118.

Redon, J. & Lange, O. L. 1983: Epiphytische Flechten im Bereich einer chilenischen "Nebeloase" (Fray Jorge). I. Vegetationskundliche Gliederung und Standortbedingungen. – Flora **174:** 213–243.

Schulze, E.-D. & Hall, A. E. 1982: Stomatal responses, water loss and CO_2 assimilation rates of plants in contrasting environments. – In: Lange, O. L., Nobel, P. S., Osmond, C. B. & Ziegler, H. (eds.), Physiological plant ecology II (Encyclopedia of plant phsyiology, new series **12B):** 181–230. – Springer, Berlin etc.

–, – & Lange, O. L., Walz, H. 1982: A portable steady-state porometer for measuring the carbon dioxide and water vapour exchange of leaves under natural conditions. – Oecologia (Berlin) **53:** 141–145.

Tenhunen, J. D., Lange, O. L. & Braun, M. 1981: Midday stomatal closure in mediterranean type sclerophylls under simulated habitat conditions in an environmental chamber. II. Effect of the complex of leaf temperature and air humidity on gas exchange of *Arbutus unedo* and *Quercus ilex*. – Oecologia (Berlin) **50:** 5–11.

–, –, Gebel, J., Beyschlag, W. & Weber, J. A. 1984: Changes in photosynthetic capacity, carboxylation efficiency, and CO_2 compensation point associated with midday stomatal closure and midday depression of net CO_2 exchange of leaves of *Quercus suber*. – Planta **162:** 193–203.

–, Harley, P. C., Beyschlag, W. & Lange, O. L. 1987: A model of net photosynthesis for leaves of the sclerophyll *Quercus coccifera*. – In: Tenhunen, J. D., Catarino, F. M., Lange, O. L. & Oechel, W. C. (eds.), Plant response to stress – functional analysis in Mediterranean ecosystems: 339-354. – Springer, Berlin etc.

–, Meister, H. P., Caldwell, M. M. & Lange, O. L. 1985: Environmental constraints on productivity of the Mediterranean sclerophyll shrub *Quercus coccifera*. – Options Medit. **84 (1):** 33–53.

Address of the author: Prof. Dr. Otto L. Lange, Lehrstuhl für Botanik II der Universität Würzburg, Mittlerer Dallenbergweg 64, D-8700 Würzburg, FRG.

Plant root systems and competition

M. M. Caldwell

Abstract

Caldwell, M. M. 1988: Plant root systems and competition. – In: Greuter, W. & Zimmer, B. (eds.): Proceedings of the XIV International Botanical Congress: 385–404. – Koeltz, Königstein/Taunus.

The function of roots under field conditions as this relates to resource competition between neighbouring plants is considered from several perspectives: the importance of root competition vis-à-vis competition for light and above-ground space; theoretical considerations on the "zones of influence" exerted by individual roots; the implications of different dispersion patterns of individual roots, including microscale distributions; the relative importance of root geometry and physiological absorption capacity in competitive effectiveness; and the efficiency of resource-rich patch exploitation by roots and their associated mycorrhizae. Experiments are portrayed that address competition for water, a mobile resource the uptake of which is largely a physical process, and for phosphate, an immobile resource the uptake of which is primarily a physiological process, often facilitated by mycorrhizae. Some of these experiments indicate an immediate competition for essentially the same resource molecules. Evidence is presented for a phenomenon termed hydraulic lift, which involves water transfer between soil layers by roots. Implications of this process including water parasitism are discussed.

Introduction

Soil resources are often the object of intense competition between neighbouring plants. An enormous literature exists on the phenomena of competition both for light and for belowground resources. Yet, comparatively little is known about the process of competition – how root systems of neighbouring plants confront one another and interfere with the acquisition of soil resources and why some species seem to have a decided advantage under certain conditions.

A physical displacement of one plant's roots by those of its neighbours is unlikely. Even at extremely high rooting densities of 20 cm root cm^{-3} soil volume, roots only occupy about 2% of the soil pore volume (assuming roots of 0.1 mm diameter and a pore volume of 30%). Thus, roots interfere with one another primarily by the zones of influence they exert in the soil. The zones may be areas of water or nutrient depletion or areas where root exudates have permeated. These zones may be considered on the scale of entire root systems or the scale of individual roots. Although these zones will usually disadvantage other roots in their acquisition of soil resources, occasionally they can have a positive influence.

What is known about these zones of influence comes largely from the study of roots under laboratory or greenhouse conditions and a body of theory which predicts how materials diffuse toward or away from roots under various soil conditions. Little direct evidence exists about how roots interact in nature. Indirect assessments of root activity by isotope uptake (*e.g.,* Currie & Hammer 1979) or water extraction from soils (Rambal 1984) have been used in some ecological and agricultural studies. For the most part, however, ecologists working in the field have been able to do little more than excavate, weigh and map root systems.

This essay will address a few facets of root system function under field conditions, primarily from the perspective of water and nutrient competition. Other aspects of this subject have been addressed in recent papers (Caldwell & Richards 1986, Caldwell 1987a, 1987b).

Arrangement of competing roots

The fine roots of neighbouring plants may explore a common soil volume and be in very close proximity to one another. There is, however, very little empirical evidence to support or refute this contention. Even the degree to which large, principal roots of neighbouring plants intermix is quite dependent on shoot density (Atkinson et al. 1976), site (Cannon 1911, Richards 1986) and other factors (Caldwell 1987a). Root profile trench observations and detailed hydraulic excavations suggest substantial mixing of roots in grassland prairie (Clements et al. 1929), chaparral woodland (Kummerow et al. 1977) and in experimental field plots (Caldwell & Richards 1986). Following the trajectory of the very fine roots and distinguishing the fine roots of different species is difficult under the best of circumstances. Nevertheless, since competition and other interactions between neighbouring plant roots most likely involve the fine roots, which are the most active in absorption and exudation, determination of their spatial relationships is of obvious value.

Since the diffusion of water and nutrients in soils toward roots can be rate limiting in the uptake of soil resources by a plant, the pathlength for diffusion through the soil and, therefore, the distance between roots is clearly important (Nye & Tinker 1977, Passioura 1985). If the root systems of plants were suspended in a well mixed nutrient solution, competition for nutrients would be determined by the amount of actively absorbing root surface of each plant and the absorption kinetics of those roots. Such a situation can be approached in some aquatic systems and has led to the thesis that competitive plants are those that can deplete resources to levels unavailable to competitors (Tilman 1982). In soil, the comparatively slow movement of water and nutrients toward roots, particularly when soil moisture and/or nutrient concentrations are low, means that greater root absorption capacity (per surface area of root) will not necessarily lead to greater uptake of soil resources. Indeed, species adapted to infertile soils have low nutrient absorption capacities (Chapin 1980).

Numerous models have been developed to quantitatively describe the movement of moisture or nutrients toward and into individual roots (Gardner 1960, Cowan 1965, Nye & Tinker 1977, Cushman 1979). Most are solutions to a common cylindrical continuity equation describing radial movement from a cylinder of soil toward a central axis representing the root (fig. 1). Considerable attention has been directed to the average distance between roots as a particularly sensitive parameter in such models. These distances are calculated from the average root density. The models simplify this further to

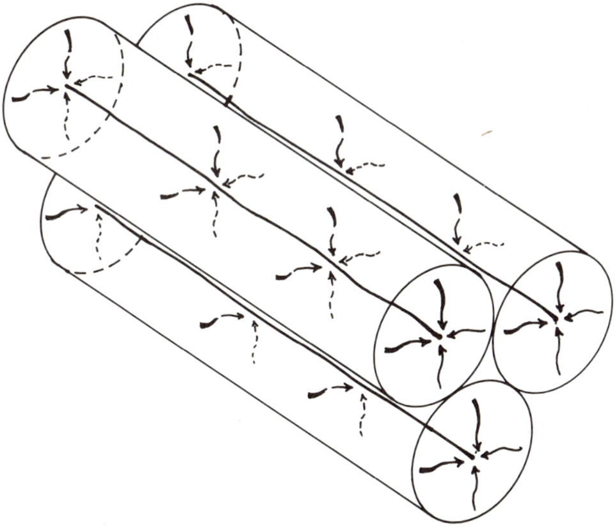

Fig. 1. Patterns of water and nutrient flow toward individual roots following assumptions of single-root uptake models. The roots are parallel, absorbing uniformly along their length, and each root has exclusive access to resources in a soil cylinder surrounding the root. The radii of the soil cylinder represent half the distance between roots.

a system of parallel roots that are more or less evenly spaced. Each root absorbs uniformly along its length and has exclusive access to the resources in the surrounding cylinder of soil. Despite the assumptions and simplifications of such models, they can be useful tools in examining the potential importance of different root and soil characteristics for water and nutrient acquisition.

An example is provided by the results of a sensitivity analysis for potassium uptake. A few parameters of the model have been increased or decreased about an average value while other parameters are held constant (fig. 2). The rate of water uptake is insensitive in the model since the amount of potassium carried by water flow toward the root is trivial compared to the diffusion

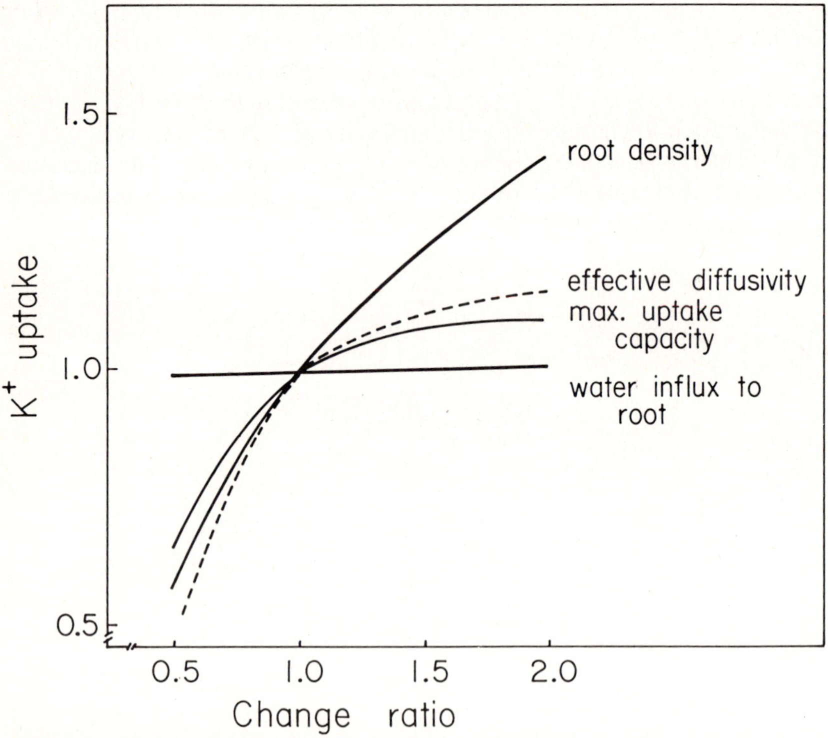

Fig. 2. Plant potassium uptake as affected by changes in a few individual parameters of an ion uptake model while other parameters are held constant. Root density is varied while both total soil volume and root volume (and mass) are held constant. (Adapted from Silberbush & Barber 1983). The value 1.0 represents the average value for these parameters determined experimentally for potted soybeans growing in an agricultural silt loam soil.

of potassium. It is perhaps not surprising that any of the other characteristics can limit uptake when they are individually decreased in the model. Of greater interest is the degree to which ion uptake can be enhanced by increases of each parameter. Increased maximum potassium absoprtion capacity of the root or a greater diffusivity of potassium in the soil is much less important in increasing potassium uptake than an increase of root density. In this example, root density has been changed without changing the total soil volume available to

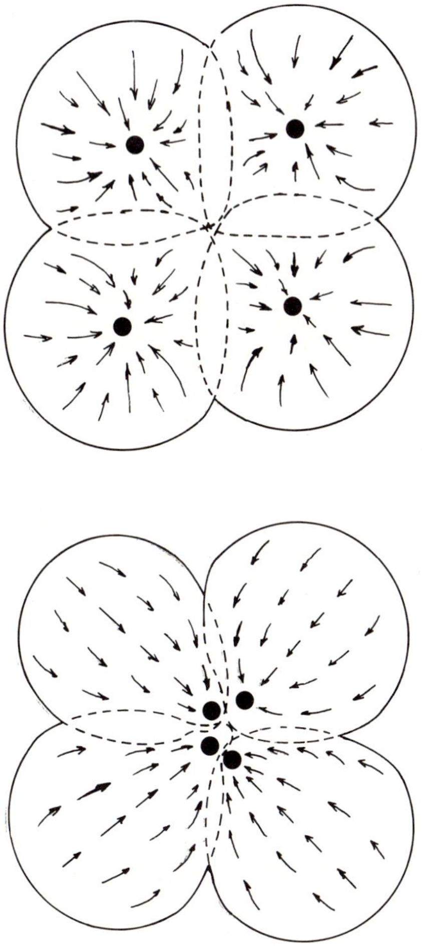

Fig. 3. Patterns of water and nutrient movement toward evenly spaced (top) or clustered (bottom) roots which have the same available soil volume.

the plant nor the total root mass. In other words, root density has been changed by simply altering root thickness. The importance of the average distance between individual roots is emphasized by this example.

The interroot distance determines the average pathlength that ions must diffuse through the soil to reach the roots and is calculated in the models with the assumption that roots are more or less evenly distributed in the soil. However, when actual root distributions in the soil are determined, very uneven patterns often emerge. Rather than being randomly or evenly distributed in the soil, roots are often clustered in their distribution as groups of roots follow cracks and crevices, or pores in the soil left by decaying roots of by earthworms. This variability has been shown on scales ranging from decimetres to millimetres (Fusseder 1985, Passioura 1985, Tardieu & Manichon 1986, Wang et al. 1986, Caldwell et al. 1987). The clustered distribution of roots results in a much longer average pathlength for ion diffusion than would be calculated with the assumption of an even distribution (fig. 3). This diminishes the effective rooting density and magnifies the limitation to ion uptake imposed by diffusion in the soil (Baldwin et al. 1972, Passioura 1985, Tardieu & Manichon 1986). Although data concerning average root density are abundant, information on root dispersion patterns in soils is quite limited. Similarly, root branching patterns and their significance for an effective exploitation of the soil have received little attention (Fitter 1987). More concerning root distribution patterns will appear later in this paper.

Depletion zones

Apart from developing perspective on root characteristics that should contribute to greater soil resource acquisition, root models also suggest other features of specific relevance to resource competition. This is discussed in greater detail elswhere (Barley 1970, Baldwin 1976, Nye & Tinker 1977, Caldwell & Richards 1986) but a cursory overview follows.

The degree to which the absorption process ist limited by movement of water and nutrients in the soil as opposed to limitation by root uptake capacity depends in part on the diffusion rates of specific substances in soil. These diffusion rates of water and various ions vary greatly (fig. 4). If diffusion in the soil is limiting water or ion movement toward the root, a localized depletion can develop near the root surface. For materials that diffuse very slowly, there is a tendency for the depletion zones to be very close to the root surface and very pronounced. Actively absorbing root hairs can effectively extend the outer radius of the root and thus the radius of the depletion shell. In the absence of other changes in the soil environment or root activity, the depletion shells will slowly broaden through time. The depletion zones of evenly spaced neighbouring roots will overlap when the distance between the roots is less than $(Dt)^{1/2}$, where D is diffusivity and t is time.

For relatively mobile nutrients such as NO_3^-, depletion shells will be broad and shallow such that the effective concentration at the root surface will not differ appreciably from that in the bulk soil (fig. 5). However, an ion such as phosphate diffuses so slowly that depletion shells are expected to be very narrow (fig. 5) and the shells of adjacent roots should only overlap if they are very close. The implications for immediate nutrient competition between individual roots, whether of the same plant or neighbouring plants, should be clear. For example, in order for phosphate competition to ensue, roots of neighbouring plants would need to be within a few millimetres of one another (Nye & Tinker 1977).

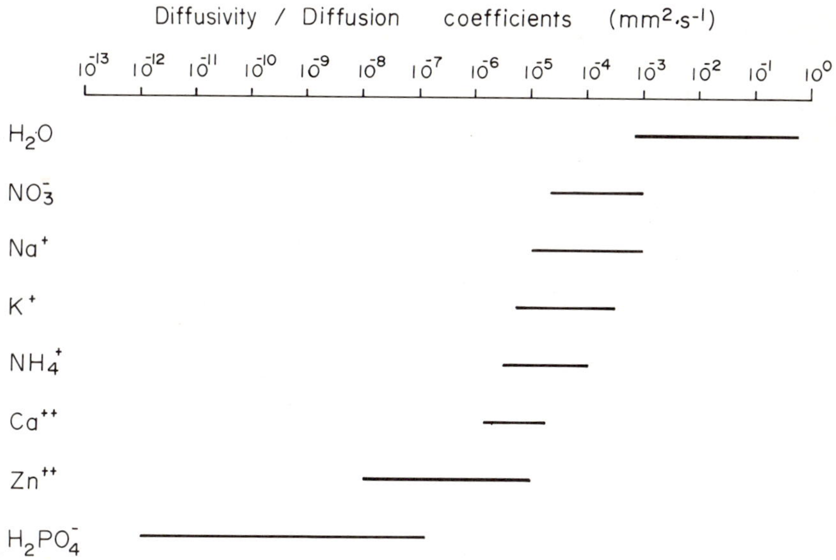

Fig. 4. Diffusivity and diffusion coefficients for water and various ions in soil. Diffusivity for water includes both liquid and vapor phase movement over a range of soil water contents from saturation to approximately 0.02 (cm^3 water/cm^3 soil). Compiled from Fried & Broeshart (1967), Rose (1968), Barber (1974) and Fitter & Hay (1981). Diffusion coefficients for ions are shown for moist soils. As soils dry, these coefficients decrease rapidly.

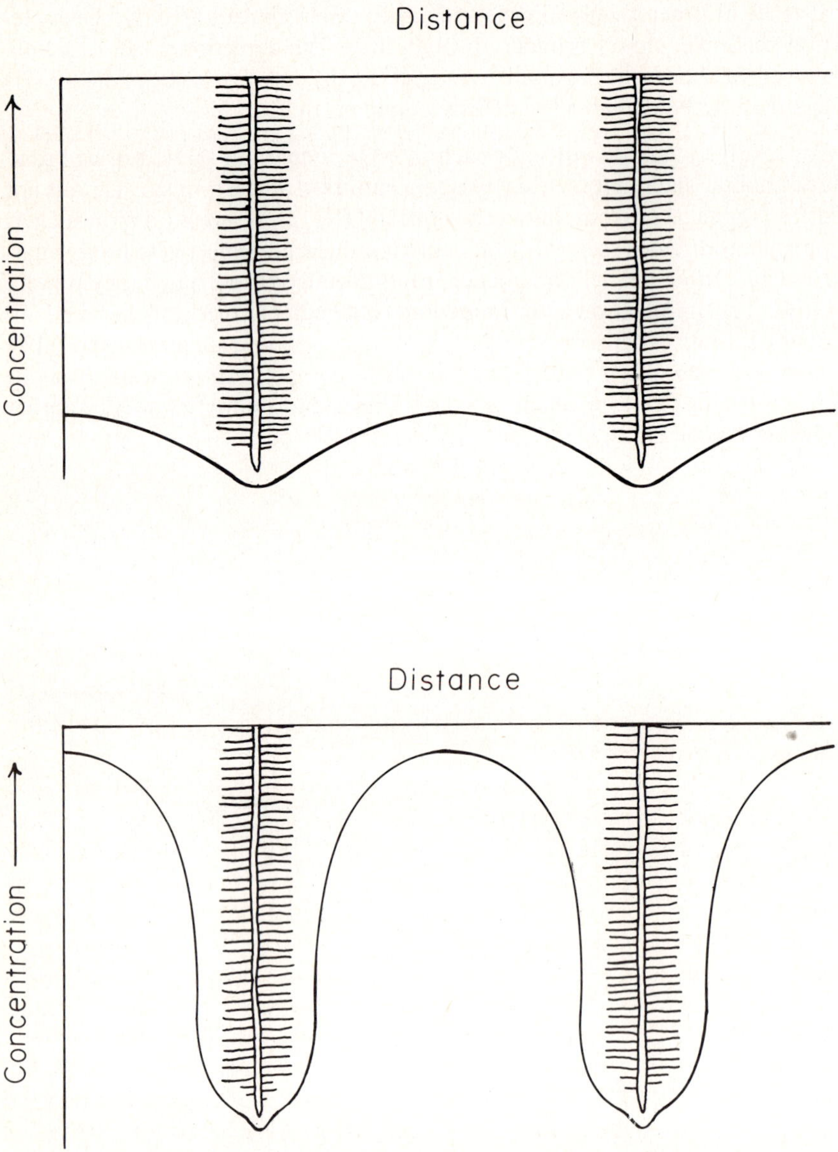

Fig. 5. Depiction of ion concentration gradients near neighbouring roots with root hairs for relatively mobile (top) and immobile (bottom) ions. Diagrams of the roots are superimposed on the concentration profiles. (Near the root tips concentrations would be greater than indicated here.) Adapted from Nye & Tinker (1977).

Mycorrhizae, especially with hyphae external to the root, can complicate this picture. Hyphae, for example of vesicular-arbuscular mycorrhiza (VAM) associations, may range more than 5 cm from a root. Furthermore, a network of hyphae may infect roots of neighbouring plants thus forming conduits between plants (*e.g.* Read et al. 1985). Transport of ^{32}P between neighbouring plants has been clearly demonstrated (*e.g.*, Chiariello et al. 1982). The suggestion has even been made that a competitively superior plant merely taps into a mycorrhizal network that interconnects several root systems and exerts a greater demand on resources than do neighbouring plants (Read et al. 1985). Although mycorrhizae can clearly facilitate phosphate uptake under certain circumstances (Nye & Tinker 1977), the quantitative significance of transport between plants by mycorrhizal bridges remains to be demonstrated.

An active root system continues growing and some of this growth reexplores the same soil volume. Over a period of weeks, depletion zones of even the very immobile ions will become gradually blurred. Thus, the combination of mycorrhizal activity and continued root growth results in more competition for immobile ions among widely spaced roots than is predicted by the root uptake models as portrayed in fig. 5.

Exploiting rich patches

As discussed earlier, roots should not be expected to be evenly distributed in the soil. Apart from root aggregation due to physical discontinuities in the soil, root dispersion patterns may also reflect patchiness in the distribution of soil nutrient ions. Patches enriched in nutrients may result from the activity or the decay of soil organisms or the uneven incorporation of surface litter into the soil. If a plant's roots proliferate to achieve a much higher density in nutrient-rich patches, a competitive advantage might obtain. Such localized root proliferation has been demonstrated with root systems divided into nutrient solution compartments of differing concentration (Drew & Saker 1975). This has also been reported in agricultural soils when fertilizer has been placed in strips at a particular depth (Passioura & Wetselaar 1972). Root growth response to very localized injection of liquid fertilizer in a field experiment is shown in fig. 6. The degree to which roots of differing species respond to enriched soil patches and the implications of localized root proliferation for nutrient competition among plants has not been well explored. A competitive plant may be one which can quickly detect and exploit small fertile patches. Such resource preemption may be more important than the ability to generally deplete nutrients in soil to very low concentrations.

Other zones of influence

Apart from depletion shells created by roots, other zones of influence may develop. Roots deposit a variety of materials into the soil by sloughing and exudation. Materials potentially detrimental to the survival or function of other roots, commonly termed allelochemics, are included among these substances. The inhibitory nature of such compounds has been clearly demonstrated (*e.g.*, Newman & Miller 1977) and reports of circumstantial evidence for allelopathic effects in the field are abundant (Rice 1984). However, the effects of allelochemics are difficult to separate experimentally from direct resource competition under field conditions (Newman 1983). Many compounds of allelopathic potential are degraded rather rapidly in the soil (*e.g.*, Oleszek & Jurzysta 1987). Tracing the fate and evaluating the quantitative effectiveness of these compounds in the soil remain a challenge for future work in this area. Allelopathy is a subject which attracts both devoted adherents and strong critics. Hopefully, future research will help to separate potential and realized consequences of allelochemics under field conditions.

Other rhizosphere exudates may have positive effects for neighbouring plant roots. Proton or phosphatase efflux may facilitate uptake of nutrients such as phosphate or iron otherwise not readily available to the plant (Marschner et al. 1986). Recently, Horst & Waschkies (1987) demonstrated enhanced phosphate uptake by roots of wheat when they were allowed to intermix with roots of lupine, a plant known to release compounds that facilitate phosphate uptake. There was no effect if roots of the two species were growing in the same soil but separated by fine stainless steel mesh. Thus, very close association of the two species' roots was necessary for the phenomenon to be evident.

There are many other avenues by which roots of one plant may influence those of another. These may, for example, be mediated by soil microorganisms (Christie et al. 1978). Although the potential for root exudates to influence roots of neighbouring plants, both positively and negatively, is considerable, the importance of these effects under field conditions vis-à-vis direct resource competition is not clear. As suggested by the title of this paper, the emphasis is on competition rather than other forms of

Fig. 6. Root length appearing next to glass tubes (m of root per m^2 glass surface area) installed in the soil near tussocks of *Agropyron desertorum* in the field. The maps are plotted on axes of depth in the soil and azimuth angle around the tubes without (top) and with (bottom) localized fertilizer injections. The glass tubes were 119 mm in circumference. The bold arrows indicate the location of liquid fertilizer injections 28 days before the observations. From Eissenstat & Caldwell (1988).

▷

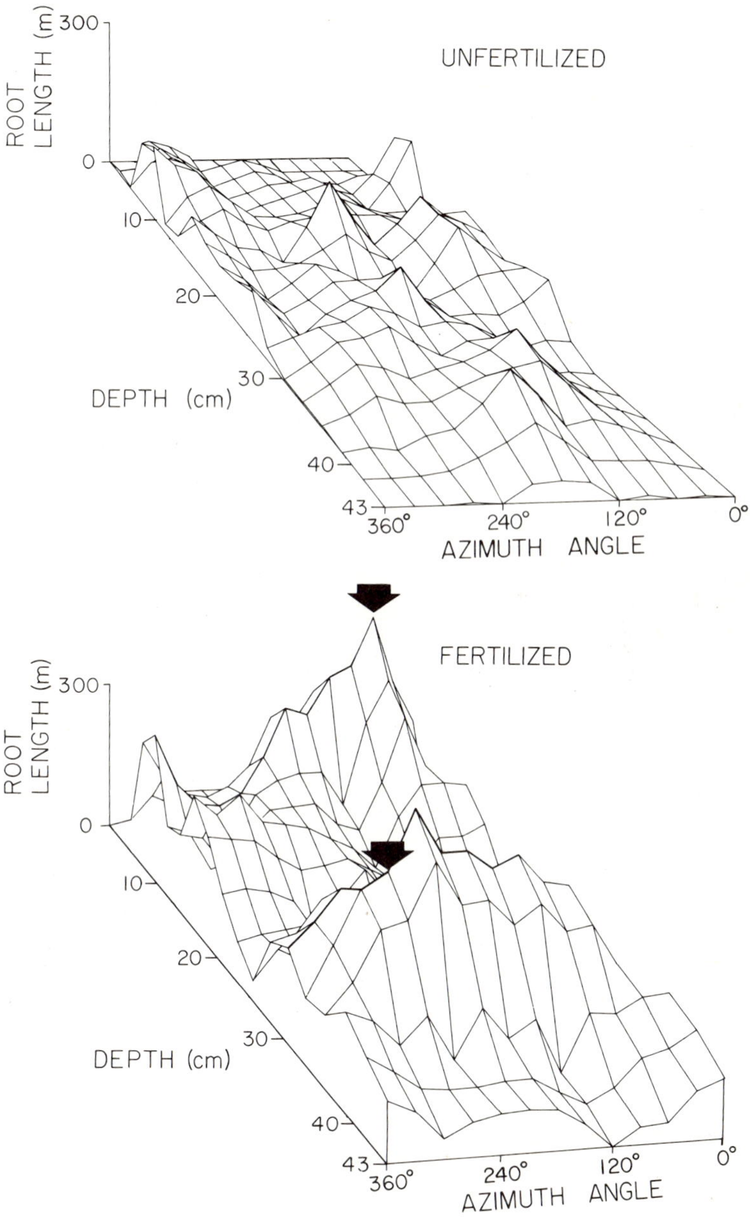
ROOT LENGTH (m)
300
0
UNFERTILIZED
10
20
30
DEPTH (cm)
40
43
360°
240°
120°
0°
AZIMUTH ANGLE
FERTILIZED
ROOT LENGTH (m)
300
0
10
20
30
DEPTH (cm)
40
43
360°
240°
120°
0°
AZIMUTH ANGLE

interference or influence between neighbouring plant roots. Nevertheless, mention should be made of the potential importance of these phenomena.

The immediacy of interference

Although there is little doubt that interference between root systems of neighbouring plants takes place in the field, the immediacy of the interaction is not often apparent. One approach to the study of plant interactions in the field has been to remove plants and to observe the response of remaining neighbours. When this has been done with plants in water-limited environments, the remaining individuals have exhibited improved water status, growth and development of reproductive structures (Fonteyn & Mahall 1981, Robberecht et al. 1983, Ehleringer 1984). Such experiments are sometimes interpreted to indicate the nature of competition among the plants before the removals were conducted. However, the responses were reported only after several months following the neighbour removals. Therefore, it is not clear how much of the response reflects the nature of interference at the time of removal or how much of the response was due to growth of roots into the soil space previously occupied by the missing plants. Such experiments may be limited by the ability to detect immediate change in the remaining individuals reflecting their release from interference by neighbours.

In a somewhat different approach to this question, changes in phosphate acquisition have been measured following clipping (rather than removal) of a neighbouring plant.

An indicator plant, *Artemisia tridentata,* surrounded by tussock grasses *(Agropyron desertorum,* in one experiment, *A. spicatum* in another), was given access to phosphorus radioisotopes by placing labelled orthophosphoric acid in the soil on opposite sides of the *Artemisia.* The labelled P was placed halfway between the indicator plant and a neighbouring tussock grass using ^{32}P on one side of the indicator plant and ^{33}P on the other. One of the tussock grasses sharing a labelled soil interspace was clipped at the time of labelling to reduce the competitive effectiveness of the grass. The ratio of P isotopes appearing in the indicator plant was measured through time. Within two weeks there was a sizeable shift in phosphate acquisition by the indicator plant such that as much as 6 times more P was obtained in competition with the clipped plant as with the unclipped neighbour (fig. 7). This rapid shift suggests an immediate competition for the same phosphate resource at the same time and location in the soil rather than major changes in root distribution due to root exploration of released soil space. Micro-scale mapping of the roots of the neighbouring plants in this experiment showed that individual roots of neighbouring plants were close enough to be in immediate competition for soil P according to theoretical expectations discussed earlier. Still, this experiment does not necessarily preclude alternative explanations of the mechanisms involved. The competition may have been directly in the soil, or part of the shift in phosphate acquisition may have involved a change in phosphate demand exerted by the clipped plants on a mycorrhizal network interconnecting all of the plants. Conceivably even allelopathy could have been involved. No matter what the exact route, the immediacy of the interaction among the plants is clear.

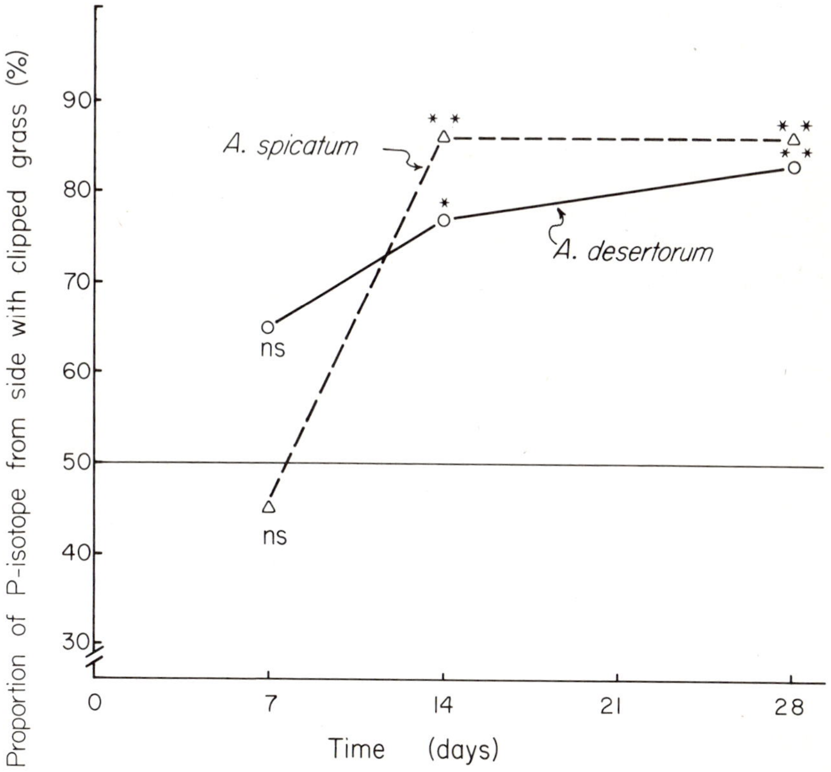

Fig. 7. The amount of phosphate isotope acquired from the soil interspace between an indicator plant and a clipped neighbour plant as as proportion of the total phosphate isotope acquired from labelled soil interspace on opposite sides of the indicator plant. From Caldwell et al. (1987).

Hydraulic lift

If the roots of various species in a community extend to different depths and appear to be seeking their own region of the soil profile in which to expand, this is often taken as evidence that species are minimizing interspecific competition (*e.g.*, Cody 1986, Davis & Mooney 1986). While this may be correct in many respects, it may be misleading in others. Some species may be profiting at the expense of others.

Evidence has been recently published that a considerable efflux of water from plant roots can occur (Richards & Caldwell 1987). This flow of water is thought to follow a water potential gradient from moister, and usually deeper, soil layers through the roots to dryer, and usually shallower, parts of the soil profile where it is released. Such flow had been postulated to occur under field conditions (Schippers et al. 1967, Hansen & Dickson 1979, Mooney et al. 1980, van Bavel et al. 1984), and laboratory experiments (van Bavel & Baker 1985, Baker & van Bavel 1986) have clearly demonstrated significant water

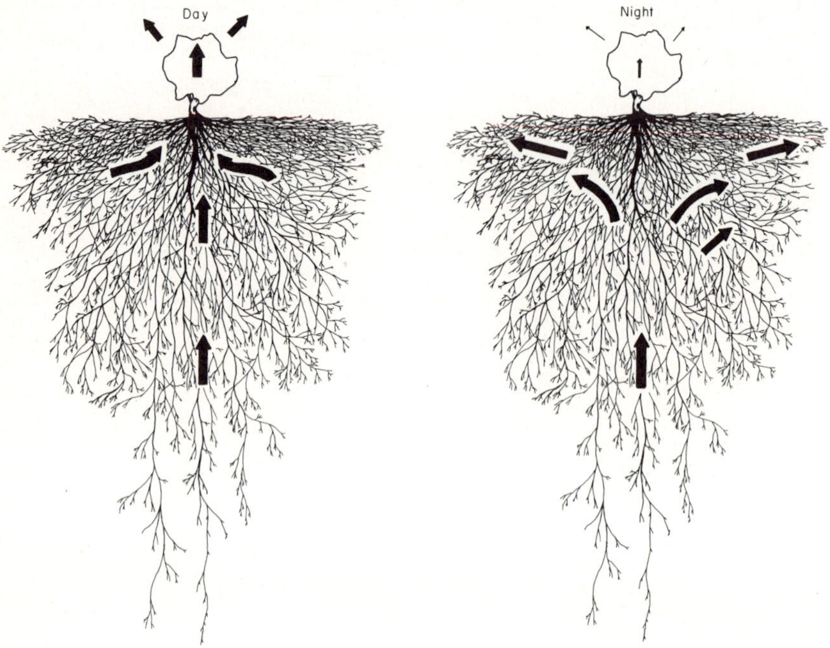

Fig. 8. Pattern of water movement according to the hydraulic lift hypothesis during day and night periods.

flow from moist to dry soils held in separate compartments but interconnected by a common root system.

A cycle is envisaged whereby water is absorbed by deeper roots, flows to roots in shallower layers and into the soil at night. During the day when the plant is transpiring, the water potential gradient is toward the atmosphere and water is absorbed both by deeper roots and those in shallower layers (fig. 8). The water absorbed by the shallower roots may in large part be the water that flowed out of those roots during the previous night. The phenomenon is termed hydraulic lift (Richards & Caldwell 1987). The field evidence for this phenomenon comes primarily from pronounced diel fluctuations in soil water potential in the upper part of the soil profile during long drying cycles (fig. 9). Water potential declines during the day and rises at night, or continues to rise for several days if the plants are covered to stop transpiration (fig. 10). Other lines of evidence are being developed to more directly support this concept.

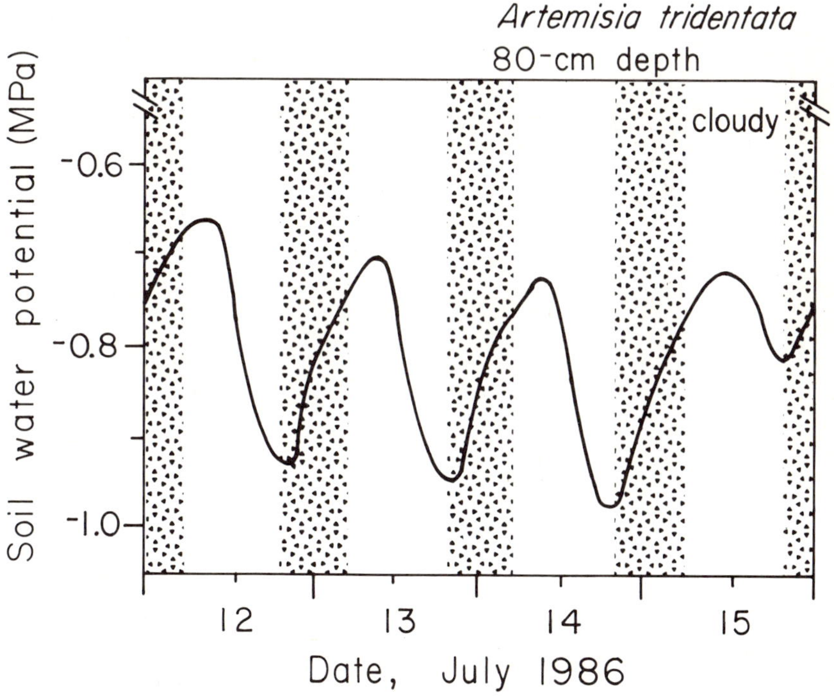

Fig. 9. Diel fluctuations of soil water potential at 80 cm depth in an experimental plot planted with *Artemisia tridentata* shrubs. The stippled zones indicate nighttime. Adapted from Richards & Caldwell (1987).

During long drying cycles when the soil moisture reserve in the upper part of the profile is largely exhausted, hydraulic lift can have numerous implications. Even though roots of many plants may reach several metres depth, roots in deeper soil tend to be very sparse compared to roots of the upper profile in most ecosystems (Richards 1986). Hydraulic lift may facilitate water uptake by deep roots since these roots can absorb and transport water both day and night while the plant uses the shallower soil as a temporary moisture reservoir (fig. 8). The daily irrigation of roots and their rhizosphere in the upper soil profile provided by hydraulic lift should promote soil nutrient mineralization and absorption by roots and their associated mycorrhizae in otherwise dry soil (Nambiar 1977, Richards & Caldwell 1987). Another implication of hydraulic lift is that moisture depletion zones that might develop around roots of the upper profile during the day could be obliterated overnight. Although hydraulic lift may be most easily envisaged in environments where long drying cycles prevail, it may also operate at a less dramatic level even in

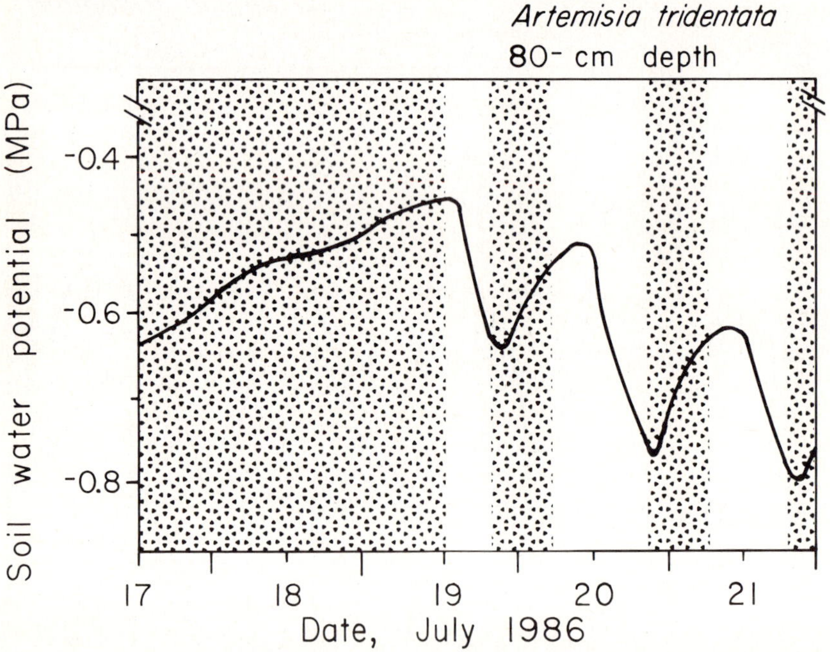

Fig. 10. Fluctuations in soil water potential as in fig. 9 showing a two-day period in which the shrubs were covered with opaque plastic to suppress transpiration (indicated by broad stippled zone). Overheating of the shrubs during the transpiration suppression was prevented by applying moistened layers of cloth over the plastic coverings. Adapted from Richards & Caldwell (1987).

environments where only the upper soil layers experience temporary dryness. In this case, maintenance of an active fine root system by hydraulic lift during short-term dryness may allevirate the amount of fine root death that could otherwise take place.

Although these and several other advantages can accrue for plants with hydraulic lift, there may be disadvantages too. Neighbouring plants with only shallow roots may parasitize plants conducting hydraulic lift by using some of the water released into upper soil layers. These neighbours may also benefit from the increased soil nutrient mineralization resulting from hydraulic lift. Even though the deeper roots of several species in a community may extend to different depths and appear to avoid interspecific competition, they can have overlapping root systems in shallower soil layers. If hydraulic lift proves to be a general and significant phenomenon, a more complicated picture of species interactions in plant communities may emerge.

Concluding remarks

The soil is a complicated and patchy milieu in which roots of neighbouring plants intermesh and interact. Field experiments addressing mechanisms of root competition will seldom lead to unequivocal interpretation, at least with present technology. Therefore, it is necessary to employ several experimental approaches, both in the field and laboratory, with available theory to approach questions of root system interaction. Despite the deficiencies of each individual experiment or technique, by using a sufficiently diverse set of experiments a more refined understanding of how root systems interact in the field will emerge. The questions are compelling and ecologists should not be content with merely excavating and mapping roots.

Acknowledgements

I thank my colleague J. H. Richards for thoughtful comments on this manuscript. Some of the information and concepts contained in this paper resulted from research supported by the National Science Foundation (BSR 8207171 and BSR 8705492) and the Utah Agricultural Experiment Station. Reviewing by S. D. Flint, typing by R. Hart and drafting of figures by C. W. Warner are gratefully acknowledged.

References

Atkinson, D., Naylor, D. & Coldrick, G. A. 1976: The effect of tree spacing on the apple roots system. – Hort. Res. (Edinburgh) **16:** 89–105.

Baker, J. M. & Baval, C. H. M. van, 1986: Resistance of plant roots to water loss. – Agron. J. **78:** 641–644.

Baldwin, J. P. 1976: Competition for plant nutrients in soil; a theoretical approach. – J. Agric. Sci. **87:** 341–356.

–, Tinker, P. B. & Nye, P. H. 1972: Uptake of solutes by multiple root systems from soil. II. The theoretical effects of rooting density and pattern on uptake of nutrients from the soil. – Pl. & Soil **63:** 693–708.

Barber, S. A. 1974: Influence of the plant root on ion movement in soil. – In: Carson, E. W. (ed.), The plant root and its environment: 525–564. – Charlottesville.

Barley, K. P. 1970: The configuration of the root system in relation to nutrient uptake. – Advances Agron. **22:** 159–201.

Bavel, C. H. M. van, & Baker, J. M. 1985: Water transfer by plant roots from wet to dry soil. – Naturwissenschaften **72:** 606–607.

–, Lascano, R. J. & Stroosnijder, L. 1984: Test and analysis of a model of water use by sorghum. – Soil Sci. **137:** 443–456.

Caldwell, M. M. 1987a: Competition between root systems in natural communities. – In: Gregory, P. J., Lake, J. V. & Rose, D. A. (eds.), Root development and function: 167–185. – Cambridge.

– 1987 b: Plant architecture and resource competition. – In: Schulze, E. D. & Zwölfer, H. (eds.), Potentials and limitations of ecosystem analysis: 164–179. – Berlin.

– & Richards, J. H. 1986: Competing root systems: morphology and models of absorption. – In: Givnish, T. J. (ed.), On the economy of plant form and function: 251–273. – Cambridge.

–, –, Manwaring, J. H. & Eissenstat, D. M. 1987: Rapid shifts in phosphate acquisition show direct competition between neighbouring plants. – Nature **327:** 615–616.

Cannon, W. A. 1911: The root habits of desert plants. – Publ. Carnegie Inst. Washington **131.**

Chapin, F. S., III. 1980: The mineral nutrition of wild plants. – Annual Rev. Ecol. Syst. **11:** 233–260.

Chiariello, N., Hickman, J. C. & Mooney, H. A. 1982: Endomycorrhizal role for interspecific transfer of phosphorus in a community of annual plants. – Science **217:** 941–943.

Christie, P., Newman, E. I. & Campbell, R. 1978: The influence of neighbouring grassland plants on each others' endomycorrhizas and root-surface microorganisms. – Soil Biol. Biochem. **10:** 521–527.

Clements, F. E., Weaver, J. E. & Hanson, H. C. 1929: Plant competition: an analysis of community function. – Publ. Carnegie Inst. Washington **398.**

Cody, M. L. 1986: Structural niches in plant communities. – In: Diamond, J. & Case, T. J. (eds.), Community ecology: 381–405. – New York.

Cowan, I. R. 1965: Transport of water in the soil-plant-atmosphere system. – J. Appl. Ecol. **2:** 221–239.

Currie, P. O. & Hammer, F. L. 1979: Detecting depth and lateral spread of roots of native range plants using radioactive phosphorus. – J. Range Managem. **32:** 101–103.

Cushman, J. H. 1979: An analytical solution to solute transport near root surfaces for low initial concentration: I. Equations development. – Soil Sci. Amer. J. **43:** 1087–1090.

Davis, S. D. & Mooney H. A. 1986: Water use patterns of four co-occurring chaparral shrubs. – Oecologia (Berlin) **70:** 172–177.

Drew, M. C. & Saker, L. R. 1975: Nutrient supply and the growth of seminal root system in barley. II. Localized, compensatory increases in lateral root growth and rates of nitrate uptake when nitrate suppy is restricted to only part of the root system. – J. Exp. Bot. **26:** 79–90.

Ehleringer, J. R. 1984: Intraspecific competitive effects on water relations, growth and reproduction in *Encelia farinosa.* – Oecologia (Berlin) **63:** 153–158.

Eissenstat, D. M. & Caldwell, M. M. 1988: Seasonal timing of root growth in favorable microsites. – Ecology **69:** 870–873.

Fitter, A. H. 1987: An architectural approach to the comparative ecology of plant root systems. – New Phytol. **106:** 61–77.

– & Hay, R. K. M. 1981: Environmental physiology of plants. – London.

Fonteyn, P. J. & Mahall, B. E. 1981: An experimental analysis of structure in a desert plant community. – J. Ecol. **69:** 883–896.

Fried, M. & Broeshart, H. 1967: The soil-plant system in relation to inorganic nutrition. – New York.

Fusseder, A. 1985: Verteilung des Wurzelsystems von Mais im Hinblick auf die Konkurrenz um Makronährstoffe. – Z. Pflanzenernähr. Bodenk. **148:** 321–334.

Gardner, W. R. 1960: Dynamic aspects of water availability to plants. – Soil Sci. **89:** 63–73.

Hansen, E. A. & Dickson, R. E. 1979: Water and mineral nutrient transfer between root systems of juvenile *Populus.* – Forest Sci. **25:** 247–252.

Horst, W. J. & Waschkies, C. 1987: Phosphatversorgung von Sommerweizen (*Triticum aestivum* L.) in Mischkultur mit Weißer Lupine (*Lupinus albus* L.). – Z. Pflanzenernähr. Bodenk. **150:** 1–8.

Kummerow, J., Krause, D. & Jow, W. 1977: Root systems of chaparral shrubs. – Oecologia (Berlin) **29:** 163–177.

Marschner, H., Römheld, V., Horst, W. J. & Martin, P. 1986: Root-induced changes in the rhizosphere: Importance for the mineral nutrition of plants. – Z. Pflanzenernähr. Bodenk. **149:** 441–456.

Mooney, H. A., Gulmon, S. L., Rundel, P. W. & Ehleringer, J. 1980: Further observations on the water relations of *Prosopis tamarugo* of the northern Atacama desert. – Oecologia (Berlin) **44:** 177–180.

Nambiar, E. K. S. 1977: The effects of water content of the topsoil on micronutrient availability and uptake in a siliceous sandy soil. – Pl. & Soil **46:** 175–183.

Newman, E. I. 1983: Interactions between plants. – In: Lange, O. L., Nobel, P. S., Osmond, C. B. & Ziegler, H. (eds.), Encyclopedia of plant physiology, **12C:** Physiological plant ecology III. Responses to the chemical and biological environment: 679–710. – Berlin.

– & Miller, M. H. 1977: Allelopathy among some British grassland species. II. Influence of root exudates on phosphorus uptake. – J. Ecol. **65:** 399–411.

Nye, P. H. & Tinker, P. B. 1977: Solute movement in the soil-root system. – Oxford.

Oleszek, W. & Jurzysta, M. 1987: The allelopathic potential of alfalfa root medicagenic acid glycosides and their fate in soil environments. – Pl. & Soil **98:** 67–80.

Passioura, J. B. 1985: Roots and water economy of wheat. – In: Day, W. & Atkin, R. K. (eds.), Wheat growth and modelling: 185–198. – New York.

– & Wetselaar, R. 1972: Consequences of banding nitrogen fertilizers in soil. II. Effects on the growth of wheat roots. – Pl. & Soil **36:** 461–473.

Rambal, S. 1984: Water balance and pattern of root water uptake by a *Quercus coccifera* L. evergreen scrub. – Oecologia (Berlin) **62:** 18–25.

Read, D. J., Francis, R. & Finlay, R. D. 1985: Mycorrhizal mycelia and nutrient cycling in plant communities. – In: Fitter, A. H., Atkinson, D., Read, D. J. & Busher, M. (eds.), Ecological interactions in soil: 193–217. – Oxford.

Rice, E. L. 1984: Allelopathy. – Orlando.

Richards, J. H. 1986: Root form and depth distribution in several biomes. – In: Carlisle, D., Berry, W. L., Kaplan, I. R. & Watterson, J. R. (eds.), Mineral exploration: biological systems and organic matter: 82–97. – Englewood Cliffs.

– & Caldwell, M. M. 1987: Hydraulic lift: Substantial nocturnal water transport between soil layers by *Artemisia tridentata* roots. – Oecologia (Berlin) **73:** 486–489.

Robberecht, R., Mahall, B. E. & Nobel, P. S. 1983: Experimental removal of intraspecific competitors-effects on water relations and productivity of a desert bunchgrass, *Hilaria rigida.* – Oecologia (Berlin) **60:** 21–24.

Rose, D. A. 1968: Water movement in porous materials. III. Evaporation of water from soil. – Brit. J. Appl. Phys. **1:** 1779–1791.

Schippers, B., Schroth, M. N. & Hildebrand, D. C. 1967: Emanation of water from underground plant parts. – Pl. & Soil **49:** 533–550.

Silberbush, M. & Barber, S. A. 1983: Sensitivity analysis of parameters used in simulating K uptake with a mechanistic mathematical model. – Agron. J. **75:** 851–854.

Tardieu, F. & Manichon, H. 1986: Caractérisation en tant que capteur d'eau de l'enracinement du mais en parcelle cultivée. II. – Une méthode d'étude de la répartition verticale et horizontale des racines. – Agronomie **6:** 415–425.

Tilman, D. 1982: Resource competition and community structure. – Princeton.

Wang, J., Hesketh, J. D. & Woolley, J. T. 1986: Preexisting channels and soybean rooting patterns. – Soil Sci. **141:** 432–437.

Address of the author: Professor Martyn M. Caldwell, Range Science Department and the Ecology Center, Utah State University, Logan, Utah 84322-5230, USA.

Pollution and forest ecosystems

P. D. Manion

Abstract
Manion, P. D. 1988: Pollution and forest ecosystems. – In: Greuter, W. & Zimmer, B. (eds.): Proceedings of the XIV International Botanical Congress: 405–421. – Koeltz, Königstein/Taunus.

Region wide air pollution impacts on forest ecosystems are a major ecological and political issue of today. The complexities of evaluating effects of man-made pollutants in relation to naturally occurring "pollutants" complicates the topic. "Natural" changes produced by the interaction of biological and physicl factors on forest ecosystems are very difficult to separate from unnatural changes. Value judgements have, to some degree, ignored these points in suggesting ecosystem level impacts for four forest systems in North America. It is appropriate to recognize the limitations of our concepts and data and to generate a better foundation for addressing this critical issue.

Introduction

The local impact of pollutants on forest ecosystems has been observed for centuries. The gradient of dying and dead trees around an ore smelting facility or a highly industrialized city was and still is quite obvious. Some of these point source pollutant problems have been reduced by changes in technology but others have been "solved" locally by dilution of the pollutant output with taller exhaust stacks.

The collective regional impact of pollutants from diluted point sources combined with the pollutants from large numbers of small, sometimes mobile, point sources (small factories, residential heating, automobiles, trucks and aircraft) presents a potential but less obvious threat to forest ecosystems.

This topic of regional impacts of air pollutants on forest ecosystems is one of the critical ecological issues of today. The extensive literature on the topic, produced in the last two decades, provides a basis for developing opinions on the magnitude and solution of the problem. For an objective scientist it is appropriate to continually evaluate the foundations of concepts and conclusions being used.

Westman (1980), in his opening remarks before a pollution and forest ecosystem symposium, noted that the social context of the issues may affect the emphasis of the presentation. He further suggested that as scientists we need to clearly differentiate between empirical observations and value judgements. The social context has clearly affected many of the presentations in the air pollution/forest decline topic. To emphasize this point, value judgements will be noted throughout my remarks.

This presentation will critically evaluate past and present understanding of regional air pollutant impacts on forest ecosystems. I will emphasize the four regional forest problems in North America that have received the most attention. These are the red spruce, eastern white pine and sugar maple problems of the east and the ponderosa pine problems of California. The emphasis is on these as individual species, but it is important to recognize that each species occurs in an array of compositional mixtures to form numerous natural ecosystems.

I will start with the key points of a recent article by Woodman & Cowling (1987). They suggest that "regional losses in productivity of whole forests (as a result of pollution) has not been proven." They further believe that "by 1990 there will still be no rigorous proof that one or several air pollutants cause decreases in the productivity of commercial forests in any region of North America." These statements are made even though injury in the field due to air pollutants has been rigorously ducumented in the case of eastern white pine, and ponderosa pine. The statements are also in direct contradiction to the cental value judgement of a number of review articles on ecosystem response to pollution (Bormann 1985, Kozlowski & Constantinidou 1986, Smith 1984). Why is it not presently possible to document a regional ecosystem level response to air pollution? To answer this question we will examine the foundations of the concepts of pollutants, the theories of ecosystem responses to stress, and the presently available data on the four systems of concern.

Concepts of pollutants

Pollutants are characterized in a number of different ways: for example, primary and secondary pollutants, oxidants and reductants, trace metals, growth regulators, etc. The term pollute implies unclear or contaminant. It is appropriate to recognize that some unusual man-made products may be unnatural contaminants but other pollutants are natural products (Finlayson-Pitts & Pitts 1986, Smith 1981) that plants have had to contend with throughout their evolutionary development. Nitrogen oxides, sulfur oxides, and ozone are pollutants of regional concern, but they are also natural products of the environment. Their synthesis, breakdown and biological importance are interwoven into the physical and biological forces that provide stability to the biosphere of this planet.

Nitrogen oxide, for example, is produced by bacterial action under anaerobic conditions. Nitrogen oxide is also produced during lightning storms and when fuels are combusted under high temperature and pressure. The naturally produced nitrogen oxide of this planet was estimated to be 10 times the amount produced by combustion of fuels (Robinson & Robbins 1970). Others suggest that about half the nitrogen oxide is of natural origin (Becker et al. 1985). The man-made nitrogen oxide is concentrated in the densely populated areas. In North America, combustion of fuel increases the nitrogen oxide 3 to 13 times the natural levels (Finlayson-Pitts & Pitts 1986). These various figures suggest some level of disagreement on the relative amounts of anthropogenic and natural nitrogen oxide. Naturally produced nitrogen oxide varies in relation to the activity of the anaerobic bacteria which, in turn, are affected by temperature, moisture, and other environmental variables. The oxidation of nitrogen oxide in the atmosphere to nitrogen dioxide and nitrate are influenced by the concentration of various oxidants and reductants, ultraviolet light, and moisture.

These are just few of the many relationship affecting nitrogen oxide which in turn affect the levels of ozone and the pH of the rain. This brief introduction stresses that there are both natural and man-made sources of the pollutant nitrogen oxide, there are variations in the output of the pollutant, and there are a series of factors that affect rates of chemical reactions. The key to understanding the impacts of nitrogen oxide on forest ecosystems will require a recognition of the range of variation at each of these stages, an understanding of the controlling factors and feedback mechanisms between stages, and an understanding of the acceptable limits of variation to the total ecosystem.

I do not wish to belabour the point but it is important to recognize that other pollutants have natural sources. It has been estimated that about half the sulfur in the atmosphere is from human activity (Postel 1984). Others suggest different proportions. Disagreement over the magnitude of the naturally produced sulfur oxides is similar to the nitrogen oxide situation.

Annual average ozone levels of 0.01 to 0.07 ppm are considered natural background resulting from injection of stratospheric ozone into the troposphere and photochemical reaction of naturally occurring nitrogen oxide (Corn et al. 1975, Becker et al. 1985). Ozone peaks exceeding 0.20 ppm may occur from stratospheric origin on mountain tops and in clean areas. These peaks are not obvious in urban areas because of the destruction of ozone by nitrogen oxide in the polluted air. Polluted air can contribute precursors for production of secondary pollutants but may also function as a scavanger for destruction of pollutants.

The natural levels of sulfur, nitrogen, carbon, and other compounds in the atmosphere also contribute to a base line level of pH which has been estimated to be between 4.5 and 5.6 (Galloway et al. 1984).

Ecosystem responses to stress

Is it correct to assume that normal plant succession processes evolved in an unpolluted atmosphere (Kozlowski & Constantinidou 1986, Smith 1980) or is it appropriate to recognize that there are background levels of the pollutants of primary concern, that these chemicals can stimulate as well as suppress growth, and that there are many environmental and biological factors that affect plant response to pollutants (Winner 1986)? Are pollutants toxicants that follow the rules of toxicology and exhibit differential dose dependent responses (Smith 1987)? Pollutant concepts directly affect the interpretation ecosystem responses to stress.

Woodwell (1970) set the foundations of present day concepts of ecosystem responses to pollution. He suggested that changes in natural ecosystems caused by many different types of disturbance are similar and predictable. Disturbances strip away layers of the ecosystem starting with the trees and progressing down through taller shrubs, to shorter shrubs, to herbs, and finally to lichens and mosses.

The Woodwell concept of a disturbance factor included radiation, air pollutants, fire, clearcutting, pathogens, and any other stress inducing factor. His primary example of ecosystem responses was derived from data on vegetation sampling along a stress gradient from a point source of radiation. He also relied on the gradient analysis study of Gorham & Gordon (1960) around the Sudbury, Canada, nickel smelter. These two systems both responded to stress with a reduction in community diversity and a shift to earlier successional generalists.

The spatial gradient response system of Woodwell is based on a dose difference with distance from a point source. Extension of the concept suggests that a dose response gradient can be developed in an ecosystem over time with low levels of accumulative pollutant stress.

Smith (1974) suggested a classification system for pollution impacts on plant systems. At low pollutant levels, Class I, the vegetation acts as a sink but does not directly respond. Class II involves higher levels of pollutants that may induce subtle chronic effects such as reduced productivity, shifts in species, or increasing secondery effects such as insect and disease epidemics. Class III involves high enough levels of pollutants to produce acute leaf symptoms or plant mortality.

Most of our understanding on ecosystem effects of pollutants is at the Class III level. This is the pollutant level on which the Woodwell community diversity responses are founded. The question today is with Class I and Class II in which subtle effects may occur in the absence of direct symptom expression. The simple ecosystem poisoning model of Woodwell does not necessarily apply especially when the pollutants of concern are normal components. In this case the response model needs to cover the range of concentrations and timing that appropriately predict sensitivity, both promotive and adverse effects, and secondary effects of individual and community responses.

I would like to suggest a few general principles as a foundation for the ecosystem response model for low level chronic stress:

- The ecosystem has structure and function (Odum 1962) but value judgements should not be used to prioritize levels of structure or function;
- the forest ecosystem succession model has no beginning or end but rather relies on disturbance to generate long term cycles (Runkel 1985);
- no species encounters optimum conditions for all of its functions (Dansereau 1956);
- most ecosystems cannot maintain an ideal balance of resources for primary producers, decomposers, and consumers and therefore internal adjustments take place over time (Waring 1985);
- there is redundancy of function within an ecosystem;
- optimum development for a community does not necessarily occur in the absence of stress (Dyne 1981);
- fast growth or size are not necessarily of selective advantage for a species (Lugo & McCormick 1981);
- and the effects of long-term chronic stress may not be obvious but they may contribute to predisposition and subsequently to significant losses associated with unique catastrophic events (Sharpe & Scheld 1986).

Ecosystem modelling involves the measurement of three kinds of variables:

- driving variables such as precipitation, wind, temperature, and other meteorological phenomena;
- system-state variables such as dynamics of soil water, vegetation, animal populations and other variables over time;

– rate-process variables such as decomposition, photosynthesis and other rate-of-change variables (Dyne 1981).

As originally presented these variables may seem like three different groups of factors but in fact they should be considered as different states of the same factors. For example, nitrogen oxide could be considered as a driving variable affecting other variables, it has a dynamic system state over time, and it has a rate of change in relation to source outputs and reaction rates. The populations of plants, animals, pests, and pathogens within the ecosystem also have different states from driving variables, to system state variables, to rate process variables.

Each pollutant such as ozone or sulfur oxide needs to be considered as a driving variable, a system-state variable, and as a rate process variable. A simple model of forest response to a pollutant like the FORET model (West et al. 1980) is based on the assumption that the pollutant functions as a driving variable that induces growth reductions. In this model individual species change in importance in relation to differential responses to pollutant and competition. A more realistic model would need to consider the other system state and rate process components of the pollutant.

Ecosystem level pollutant systems in North America

Using the concepts of pollution and ecosystem responses to stress as a foundation let us critically look at four North American forest situations that value judgements have characterized as examples of regional pollution problems. Value judgements also have suggested that the pollution based forest problems of North America are less serious than in Germany (McLaughlin 1985). We will examine the red spruce, eastern with pine, sugar maple and ponderosa pine problems.

Red spruce

The European concern on the effects of acid rain developed a terrestrial focus in North America initially with the red spruce. Value judgements suggested that, since simple explanations were not apparent for the decline and death of red spruce, acid rain was a responsable suspect. Although the early speculation centred on acid rain (Vogelmann 1982), the current concern has expanded to include ozone, lead, changes in freezing tolerance associated with nitrate fertilization, and combinations of these and other factors.

The symptoms of the red spruce problem are varied and diverse. Increased mortality was one of the first observations. Crown thinning resulting from loss of recent needles, overall chlorosis, chlorosis of older needles, death of twigs and branches, tufting of foliage, reduced annual increment, reductions in stand basal area, changes in species composition, and lack of reproduction are some of the other symptoms of the spruce problem today.

There is no clear progression of symptoms over time. There is no agreement on the primary symptoms nor on the spatial distribution of the symtoms within or among trees. There are no recognized symptoms for direct effects of any of the pollutants of interest.

The mixture of symptoms and lack of progression of symptoms suggest the possibility that different factors may be involved in different places (Manion 1985). This symptom mixture also introduces non-random variation into statistical analysis thereby making it very difficult to identify important components of the problem. The lack of direct effects symptoms limits the design and possible interpretation of the array of pollution exposure experiments presently in progress.

The USDA Forest Service coordinated aerial photographic and field surveys of red spruce and balsam fir decline and mortality in the northeastern states of New York, Vermont and New Hampshire (Weiss et al. 1985). They found that decline and mortality occurred widely in both the upper elevation slopes where spruce and fir predominate and at lower elevations where spruce and fir are mixed with hardwood species. From aerial photographs of the Adirondack region of New York they classified 28% of the slope type and 11% of the mixed wood type as having heavy mortality. Field surveys of sample areas determined that heavy mortality classification on the photographs represented sites with 16 to 17% mortality of red spruce. They estimated that at least 10% standing dead trees was a normal condition for the forest but recognize that this figure may be considerably higher at upper elevations because the rate of deterioration of dead trees is slower. Field surveys also noted that spruce regeneration increased with increasing mortality.

The tabulation and conclusions for other areas of the survey may differ slightly, but the judgement that a substantial portion of the mortality is "normal" begins to set the stage for a better understanding of the problem. Runkel (1985) suggests that average annual mortality rates for canopy level trees of temperate forests range from 0.05 to 2.0%. The relationship of mortality to regeneration likewise provides some perspective and suggests that natural disturbance may be contributing to successional recycling.

The documentation of growth reduction in red spruce beginning in the early 1960's suggested that something was affecting the trees (Hornbeck & Smith 1985). By 1980 the increment was 13 to 40% less than the increment of 1960. Current analysis suggests that growth reductions are in part due to forest ageing that can be predicted based on yield data published in 1929 (Federer & Hornbeck 1987). The spruce forest appears to be developing like an even-aged stand that originated early in this century due to extensive cutting and insect problems.

Detailed stem analysia based on destructive sampling of trees from Whiteface Mountain identified two periods of growth reduction of red spruce, the mid 1960's and the 1975 to 1983 period (LeBlanc et al. 1986). The 1960's growth reduction coincide with a very severe regional drought. The more recent growth reduction may be associated with a collection of unusual winters. There have been three unusually warm winters and three unusually cold winters during the period. Johnson et al. (1986) likewise suggest unfavourable temperature involvement based on recent and historic spruce mortality records.

Value judgements have suggested that air pollutants may contribute to the decline and death of red spruce. The air pollution data base for testing these judgements has been recently installed. A coordinated network of six monitoring stations in the red spruce forests are now in their second season of sampling (Bradow & Mohnen 1987). The current sampling network plus information from some previous sampling, provide preliminary information on the upper elevation pollution system. For example, concentrations of ozone at the summit were generally 0.01 to 0.02 ppm higher than in nearby valley sites. The amount of cloud cover may affect the ozone concentration. The maximum concentration occurred during the night and early morning hours at high elevations and during the afternoon hours in the lower elevations. Annual average hourly ozone concentrations were lesse than 0.05 ppm (Mueller 1987, Burgess 1984). (Consider this figure in relation to the figures for natural stratospheric inputs presented above.)

At high elevations, hydrogen peroxide and other oxidants rapidly oxidize sulfur and nitrogen oxides to acids. Acidic cloud and fog moisture collecting on vegetation substantially increase the acid inputs.

The meteorological data base suggests that the higher elevations may be good sites for assessing the impacts of air pollutants. Unfortunately the physical conditions of ice and wind are very damaging and complicate any evaluation of pollution injury. The stratospheric inputs of ozone and other oxidants also complicate the system. The relationships of concentrations of pollutants to physiological receptivity of the trees need to be properly considered. The possibility that genetic selection for tolerance to the conditions, including pollution, has occured at high altitudes needs to be considered in designing experiments or in interpreting experimental data. There is an opportunity to derive reliable empirical observations to substantiate or refute the value judgements that started the issue of spruce decline.

In summary: The red spruce ecosystem is experiencing a level of mortality that can not be clearly defined as excessive; a collection of poorly characterized symptoms that cannot be accounted for by a single factor; growth reductions that can be accounted for, to some degree, by ageing and climatic factors; and reasonable regeneration that suggests natural disturbances. The proper analysis procedures for establishing causality for any of the suspected pollutants have not been accoumplished. These procedures first require that injury or dysfunction symptoms of trees in the forest must be constantly associated with the presence of the suspected causal factor (Woodman & Cowling 1987). The data base for assessment of association is just beginning to emerge.

These statements should not be interpreted as indicating that pollutants are not affecting the red spruce ecosystem. What they should suggest is that claims of pollution impacts should be critically examined to determine the level of empirical observations and value judgements involved. A level of thinking and analysis that incorporates a more thorough understanding of the system will be required to properly characterize the impact, if any, of pollutants on the red spruce ecosystem. An array of current research, coordinated under the USDA Forest Service Spruce-Fir Research Cooperative, is attempting to address these and other questions.

Eastern white pine

The origins of concern with pollution problems on eastern white pine date to early in this century but the foundations of the ozone involvement can be traced to a collection of research activities in the 1960's.

White pine ecosystem level impacts have been suggested in various review articles (Kozlowski & Constantinidou 1986, Smith 1987). These are value judgements based on a correlation of growth reduction with foliar ozone sensitivity, growth reductions with controlled fumigations and growth reductions in ambient air of open-top-chambers as compared to charcoal filtered open-top-chambers.

Benoit et al. (1982) demonstrated radial increment reductions in natural populations of trees from ten plots scattered along a 446 km section of the Blue Ridge Mountains. At each plot three trees were selected to represent high, medium, and low severity of pollution symptoms. The 10 trees classified as sensitive to pollution, based on foliar symptoms, had significantly less mean annual increment than 10 correspondingly symptomless trees. These data, plus information from previous surveys that found that 4 to 11% of the white pine in the area were sensitive (Skelly et al. 1979), suggested the possibility of ecosystem level impacts. In addition, Benoit et al (1982) provided the value judgement that it was unlikely that ozone sensitive dominant and codominant white pines could have developed to their present state under conditions of high ozone and therfore ozone stress has most likely increased in recent years.

These data and value judgements provide a basis for inclusion of white pine in an ecosystem level pollution discussion, but further examination of other references provides appropriate perspective. The growth reductions with symptomatic trees in the above study are in contrast to another study in the same general area which identified no significant growth differences between symptomatic and asymptomatic trees in response to a fluctuating air pollution level (Phillips et al. 1977), but this study contrasts with an earlier study which found growth reductions correlated with pollution levels (Stone & Skelly 1974). A composite of all these studies does not logically lead to the conclusion that ecosystem level impacts are occurring.

A consideration not discussed in the above studies is the possible involvement of other factors. The root pathogen *Verticicladiella procera* was associated with roots of dying white pines of the Blue Ridge Mountains (Skelly 1980). The interaction of pathogens and pollutants is often incorrectly simplified to suggest that pollutant injured trees are less resistant to pathogens.

Another consideration in assessing growth reductions is the effects of drought. Puckett (1982) did an analysis of growth of white pine and a number of other tree species in eastern New York state and concluded that the relationship of climate to tree growth had been altered in recent years. Acid rain or other pollutants were suggested as a possible cause. More recent analysis using Kalman filter techniques that remove the influences of weather found no apparent residual growth effect that could be attributed to pollutants (Visser 1987).

In summary: The relationship of air pollutants to specific symptoms in white pine have been well established; growth reductions associated with controlled fumigations are well documented; air-pollutant-like symptoms occur in the field; growth reductions associated with symptoms in the field are not well documented; and other biological and physical factors may play a significant role in growth and development. Therefore only premature value judgements would suggest that ozone pollution is damaging the white pine ecosystem. At the present time there is no coordinated effort of various agencies to document the magnitude or role of pollutants in the eastern white pine ecosystem.

Sugar maple

The interest in the sugar maple ecosystem in relation to air pollutants has been gaining some momentum in recent years. The USDA Forest Service has recently funded a number of studies under an Eastern Hardwoods Cooperative Research Program to investigate the problem even though there is a long history of maple decline (McIlveen et al. 1986) and a number of cases that do not involve air pollutants (Gregory et al. 1986, Sinclair 1966). The value judgements of two provincial research groups in Canada provide most of the foundation for the current interest in acid rain and maple decline (McLaughlin et al. 1987, Carrier 1986). In contrast to the provincial judgements, personnel of the federal Canadian Forest Insect and Disease Service suggest that their inventories find no unusual damage has occurred in maple and any dieback can be attributed to normally occuring factors like insects, diseases, climate, soil, or management practices (Kondo & Taylor 1986, Lachance et al. 1984). The issue is extremely political since the maple decline is the only Canadian tree problem that has been considered as similar to the forest decline situations of Europe and the United States (Addison & Linzon 1986). The problem will require highly sophisticated field research and data analysis to separate the natural factors from anthropogenic pollutants in this very complex forest ecosystem.

Ponderosa pine in California

Conspicuous needle damage was noticed on ponderosa pine of the San Bernardino Mountains east of Los Angeles, California, in 1953. By 1961 it was reasonably well documented that oxidant air pollution was a primary cause of the symptoms observed in the field. The research has continued to the present time and provides the data base for one of the most thoroughly studied air pollution problems of forests anywhere in the world. The literature on this problem is extensive and often cited when concepts of air pollution impacts on forest ecosystems are developed. Critical evaluation is needed to separate the empirical observations from the value judgements, and again such an evaluation demonstrates some questionable foundations to some of the more readily accepted conclusions. If this is the model system for assessing ecosystem impacts, it is appropriate to recognize the weaknesses of the foundation and then, understanding these weaknesses, design a better model. In this connection, I will address the involvement of disease and insects and the assessment of changes in growth and mortality in relation to pollution.

The impression that disease agents and insects increase in importance in pollution stressed tree populations is often suggested even though review articles (Heagle 1973, Treshow 1980) clearly indicate both positive and negative effects. James et al. (1980) suggested that oxidant air pollution increased the susceptibility of pines to colonization by *Heterobasidion annosum.* This value judgement is based on the relationship of pollution symptoms (needle retention) to natural infection in the field, the colonization of roots following inoculation in the field, and exposure chamber studies in which inoculated seedlings exposed to ozone were evaluated for infection. The results indicate that only one of four field study plots had a significant correlation of needle retention to natural infection, there was no overall difference in amount of colonization in relation to needle retention following inoculation but there was a significant difference in proximal colonization, and that fumigation experiments with seedlings showed no increased infection percentage for ponderosa pine but did show a significant difference with Jeffrey pine. The data neither support nor refute the often cited conclusion that oxidant air pollution increases the susceptibility to colonization by the root rot pathogen. Value judgements must be used to develop a model from these results to conclude that *Heterobasidion annosum* root disease would increase by seven fold under pollution injury (Taylor 1980).

It is widely assumed that the impact of western pine bettle is increased by pollution injury, but the data of Dahlsten & Rowney (1980) indicate that insect attack density was significantly lower on oxidant damaged trees and the number of brood output did not differ between oxidant injured and non injured trees. The data can clearly be interpreted in a number of different ways.

The assessment of impact of more than thirty years of pollution on the forest can likewise be interpreted and presented in a number of different ways. Modelling the effects of pollution on a southern California forest generates a number of outputs but a worst case scenario projects destruction of the conifer component of the forest and conversion to hardwood brush (Kickert & Gemmill 1980). In this projection, the more fire tolerant pines are more severely affected by pollutants, while the more pollutant tolerant white fir and incense cedar are severely affected by the fires which should increase in response to increased fuel provided by the dying pines.

This model is based on the assumption that pollution increases mortality and reduces growth rate in the pines. Experience has demonstrated mortality rates of 2–3% per year (Miller et al. 1982) and that the combined effect of pathogens and insects produced most of the mortality (Taylor 1980). Outside the areas of highest pollution there was a low correlation of mortality with pollution injury (Taylor 1980).

The relationship of pollution to growth rate involves exposure studies on seedlings and field observations in relation to pollution levels and symptom expression. Exposure to 0.45 to 0.80 ppm ozone reduces photosynthesis by 10% but a 24 hour exposure to 0.05 to 0.06 ppm produces injury symptoms on ponderosa pine (Miller et al. 1982). Ozone levels in the field range from typical mountain background levels of 0.03 to 0.04 ppm to maximums of 0.10 to 0.12. The hightest levels do not occur at night, as with most mountains, but occur during the mid afternoon (Miller et al. 1982). Although the symptoms of ozone injury are generally well recognized and used to identify the problem in the field, Vogler & Pronons (1980) identified severe pollution injury in the southern Sierra Nevada mountains of California even though ozone levels were low. The combined effect of temperature and moisture status on symptom expression complicates the assessment of symptoms in relation to ozone levels (Miller 1986). Increment analysis of asymptomatic and symptomatic Jeffrey and ponderosa pines identified no apparent growth reduction in ponderosa pine. Symptomatic Jeffrey pine trees grew faster than asymptomatic trees in the 1928 to 1964 period but since 1965 have been growing slower (Peterson & Arbaugh 1987).

The above examples are just a few of many conflicting considerations that need to be moved beyond the level of value judgements if the ponderosa pine system is to be used as a model system for considering pollution impacts. A better integration of the historical factors that set the stage for the forests of today (Miller & McBride 1975, Miller et al. 1987) would contribute to a better understanding of the changes in ecosystem being observed. There is still confusion regarding pollutant interactions in mechanistic processes. Basic physiological data are needed to integrate beyond simple cause-effect models (Sharpe & Scheld 1986).

Summary

Major impacts of pollutants on forest ecosystems have been identified for pollutants originating from point sources. Regional impact of pollutants represents a potential but less obvious threat. Simple toxicant type models are not necessarily appropriate to characterize regional pollutant impacts. A more realistic understanding of the relationship of natural and anthropogenic levels of pollutants to plant systems is needed. An ecosystem framework that builds on an understanding of structure and function, but avoids value judgements for characterizing position or direction of succession, recognizes that disturbance is a normal part of the process and that compromise and redundancy affect short term as well as long term ecosystem relationships.

This presentation has pointed out a number of problems with our understanding of four region-wide pollution problems in North America. The approach deviates from the usual presentation that builds hypotheses of concern and cites examples where pollutants are destroying our environment. I share the concern but suggest that empirical observations rather than value judgements should be the driving force of science. Value judgements have established concepts that need to be seriously reexamined before meaningful forest ecosystem level impacts of air pollutants can be properly identified and assessed.

Acknowledgements

The author thanks D. J. Raynal, R. A. Zabel, R. A. Bruck, and H. Fuernkranz for their technical review and comments on this manuscript.

References

Addison, P. A. & Linzon, S. N. (eds.) 1986: Assessment of the state of knowledge on the long-range transport of air pollutants and acid deposition – part 4 – terrestrial effects. – Federal/Provincial Research and Monitoring Coordinating Committee, Canada.

Becker, K. H. et al. 1985: Formation, transport and control of photochemical oxidants. – In: Guderian, R. (ed.), Air pollution by photochemical oxidants: 3–111. – Springer, Berlin.

Benoit, L. F. et al. 1982: Radial growth reductions in *Pinus strobus* L. correlated with foliar ozone sensitivity as an indicator of ozone-induced losses in eastern forests. – Canad. J. Forest Res. **12:** 673–678.

Bormann, F. H. 1985: Air pollution stress on forests: an ecosystem perspective. – In: Air pollution effects on forest ecosystems: 39–57. – Acid Rain Foundation, St. Paul, Minnesota.

Bradow, R. & Mohnen, V. 1987: Mountain cloud chemistry project (MCCP). In: NAPAP terrestrial effects task group (V) peer review summaries: B127–B165.

Burgess, R. L. (ed.) 1984: Effects of acidic deposition on forest ecosystems in the northeastern United States: an evalution of current evidence. – State University of New York College of Environmental Science and Forestry, Syracuse, N. Y.

Carrier, L. 1986: Decline in Québec's forest, assessment of the situation. – Ministère de l'Energie et des Ressources, Québec.

Corn, M. et al. 1975: Photochemical oxidants: sources, sinks and strategies. – JAPCA **25:** 16–18.

Dahlsten, D. L. & Rowney, D. L. 1980: Influence of air pollution on population dynamics of forest insects and on tree mortality. (In: Effects of air pollutants on mediterranean and temperate forest ecosystems.) – USDA Forest Serv. Gen. Techn. Rep. PSW **43:** 125–130.

Dansereau, P. 1956: Le coincement, un processus écologique. – Acta Biotheor. **11:** 157–178.

Dyne, G. M. van 1981: Responses of shortgrass prairie to man-induced stresses as determined from modelling experiments. – In: Barret, G. W. & Rosenberg, R. (eds.), Stress effects on natural ecosystems: 57–70. – Wiley, New York.

Federer, C. A. & Hornbeck, J. W. 1987: Expected decreases in diameter growth of even-aged red spruce. – Canad. J. Forest Res. **17:** 266–269.

Finlayson-Pitts, B. J. & Pitts, J. N. 1986: Atmospheric chemistry: foundations and experimental techniques. – Wiley, New York.

Galloway, J. N. et al. 1984: Acid precipitation: natural versus anthropogenic components. – Science **226:** 829–831.

Gorham, E. & Gordon. A. G. 1960: Some effects of smelter pollution northeast of Falconbridge, Ontario. – Canad. J. Bot. **38:** 307–312.

Gregory, R. A. et al. 1986: Proposed scenario for dieback and decline of *Acer saccharum* in northeastern U.S.A. and southern Canada. – IAWA Bull., ser. 2, **7:** 357–369.

Haegle, A. S. 1973: Interactions between air pollutants and plant parasites. – Annual Rev. Phytopathol. **11:** 365–388.

Hornbeck, J. W. & Smith, R. B. 1985: Documentation of red spruce growth decline. – Canad. J. Forest Res. **15:** 1199–1201.

James, R. et al. 1980: Effects of oxidant air pollution on the susceptibility of pine roots to *Fomes annosus*. – Phytopathology **70:** 560–563.

Johnson, A. H. et al. 1986: Recent and historic red spruce mortality: evidence of climatic influence. – Water, Air & Soil Pollut. **30:** 319–330.

Kickert, R. N. & Gemmill, B. 1980: Data-based ecological modeling of ozone air pollution effects in a southern California mixed conifer forest ecosystems. (In: Effects of air pollutants on mediterranean and temperate forest ecosystems.) – USDA Forest Serv. Gen. Techn. Rep. PSW **43:** 181–188.

Kondo, E. W. & Taylor, R. G. 1986: Forest insect and disease conditions in Canada 1985. – Forest Insect and Disease Survey, Canadian Forestry Service, Ottawa.

Kozlowski, T. T. & Constantinidou, H. A. 1986: Responses of woody plants to environmental pollution. – Forest Abstr. **47:** 5–132.

Lachance, D. P. et al. 1984: Insectes et maladies des arbres, Québec – 1983. – Forêt Conservation **50,** suppl.: 15–17.

LeBlanc, D. C. et al. 1986: Characterization of historical growth patterns in decline red spruce trees. – In: LRTAP workshop No. 6, forest decline workshop: 190–201. – Federal LRTAP Liaison Office, Environment Canada.

Lugo, A. E. & McCormick, J. F. 1981: Influence of environmental stressors upon energy flow in a natural terrestrial ecosystem. – In: Barret, G. W. & Rosenberg, R. (eds.), Stress effects on natural ecosystems: 79–102. – Wiley, New York.

Manion, P. D. 1985: Factors contributing to the decline of forests, a conceptual overview. – In: Air pollution effects on forest ecosystems. 63–73. – Acid Rain Foundation, St. Paul, Minnesota.

McIlveen, W. D. et al. 1986: A historical perspective of sugar maple decline within Ontario and outside Ontario. – Ontario Ministry of the Environment, Ottawa.

McLaughlin, D. S. et al. 1987: Sugar maple decline in Ontario. – In: Proceedings of the NATO advanced research workshop on the effects of acidic deposition and air pollutants in forests, wetland and agricultural ecosystems: 101–116. – Springer, New York & Berlin.

McLaughlin., S. B. 1985: Effects of air pollution on forests: a critical review. – JAPCA **35:** 512–534.

Miller, P. R. 1986: Effect of tree development and environment on response to airborne chemicals. (In: Proceedings of a workshop on controlled exposure techniques and evaluation of tree responses to airborne chemicals.) – NCASI Techn. Bull. **500:** 55–63.

– & McBride, J. R. 1975: Effects of air pollutants on forests. – In: Mudd, J. B. & Kozlowski, T. T. (eds.), Responses of plants to air pollution: 195–235. – Academic Press, New York.

– et al. 1982: Oxidant air pollution effects on a western coniferous forest ecosystems. – EPA-600/D-82-276, USEPA, Corvallis, Oregon.

– et al. 1987: Investigating the effects of acid deposition and gaseous air pollutants on forest tree physiology. – In: Bicknell, S. H. (ed.), California forest response program planning conference proceedings: 77–89. – Natural Resources Institute, Humboldt State University, Arcata, California.

Mueller, S. F. 1987: Pollutant deposition budgets for red spruce on Whitetop Mountain, Virginia. – In: NAPAP terrestrial effects task group (V) per review summaries: B166–B174.

Odum, E. P. 1962: Relationships between structure and function in the ecosystem. – Jap. J. Ecol. **12:** 108–118.

Peterson, D. L. & Arbaugh, M. J. 1987: Effects of ozone injury on pine growth in the southern Sierra Nevada. – In: NAPAP Terrestrial effects task group (V) peer review summaries: A194–A201.

Phillipps, S. O. et al. 1977: Eastern white pine exhibits growth retardation by fluctuating air pollution levels: interaction of rainfall, age and symptom expression. – Phytopathology **67:** 721–725.

Postel, S. 1984: Air pollution, acid rain and the future of forests. – Worldwatch Paper No **58.**

Puckett, L. J. 1982: Acid rain, air pollution and tree growth in southeastern New York. – J. Environmental Qual. **11:** 376–380.

Robinson, E. & Robbins, R. C. 1970: Gaseous nitrogen compounds from urban and natural sources. – JAPCA **20:** 303–306.

Runkel, J. R. 1985: Disturbance regimes in temperate forests. – In: Pickett, S. T. A. & White, P. S. (eds.), The ecology of natural disturbance and patch dynamics: 17–33. – Academic Press, New York.

Sharpe, P. J. H. & Scheld, H. W. 1986: Role of mechanistic modeling in estimating long-term pollutant effects on natural and man-influenced forest ecosystems. – (In: Proceedings of a workshop on controlled exposure techniques and evaluation of tree responses to airborne chemicals.) – NCASI Techn. Bull. **500:** 76–82.

Sinclair, W. A. 1966: Decline of hardwoods: possible causes. – Proc. Int. Shade Tree Conf. **42:** 17–32.

Skelly, J. M. et al. 1979: Impact of photochemical oxidant to white pine in the Shenandoah Blue Ride Parkway and Great Smoky Mountain National Park. – In: Abstracts, Second Conference of Scientific Research in the National Parks: 131. – San Francisco, California.

– 1980: Photochemical oxidant impacts on mediterranean and temperate forest ecosystems: real and potential effects. – (In: Effects of air pollutants on mediterranean and temperate forest ecosystems.) – USDA Forest Service Gen. Techn. Rep. PSW **43:** 38–50.

Smith, W. H. 1974: Air pollution – effects on the structure and function of the temperate forest ecosystem. – Environmental Pollut. **6:** 111–129.

– 1980: Air pollution – a 20th century allogenic influence on forest ecosystems. (In: Effects of air pollutants on mediterranean and temperate forest ecosystems.) – USDA Forest Serv. Gen. Techn. Rep. PSW **43:** 79–87.

– 1981: Air pollution and forests interaction between air contaminants and forest ecosystems. Springer, New York & Berlin.

– 1984: Ecosystem pathology: a new perspective for phytopathology. – Forest Ecol. Managem. **9:** 193–219.

– 1987: Assessment of the influence atmospheric deposition on forest ecosystems: the challenge of differential effects of local, regional and global scale pollutants. – In: Bicknell, s. H. (ed.), California forest response program planning conference Proceedings: 64–76. – Natural Resources Institute, Humboldt, State University, Arcata, California.

Stone, L. L. & Skelly, J. M. 1974: The growth of two forest tree species adjacent to a periodic source of air pollutants. – Phytopathology **64:** 773–778.

Taylor, O. C. (ed.) 1980: Photochemical oxidant air pollution effects on a mixed conifer forest ecosystem. – EPA 600/3–80–002, USEPA, Corvallis, Oregon.

Treshow, M. 1980: Interactions of air pollutants and plant diseases. (In: Effects of air pollutants on mediterranean and temperate forest ecosystems.) – USDA Forest Serv. Gen. Techn. Rep. PSW **43:** 103–109.

Visser, H. 1987: Analysis of tree ring data using the Kalman filter techniques. – IAWA Bull. ser. 2, **7:** 289–297.

Vogleman, H. W. 1982: Catastrophe on camel's hump. – Nat. Hist. **91:** 8–14.

Vogler, D. R. & Pronos, J. 1980: Ozone injury to pine in the southern Sierra Nevada of California. (In: Effects of air pollutants on mediterranean and temperate forest ecosystems.) – USDA Forest Serv. Gen. Techn. Rep. PSW **43:** 253.

Waring, R. H. 1985: Imbalanced ecosystems – assessment and consequences. – Forest Ecol. Managem. **12:** 93–112.

Weiss, J. J. et al. 1985: Red spruce and balsam fir decline and mortality in New York, Vermont and New Hampshire 1984. – USDA Forest Service, NA-TP-11.

West, D. C. et al. 1980: Simulated forest response to chronic air pollution stress. – J. Environmental Qual. **9:** 43–49.

Westman, W. E. 1980: Opening remarks and summary of panel audience discussion. (In: Effects of air pollatants on mediterranean and temperate forest ecosystems.) – USDA Forest Serv. Gen. Techn. Rep. PSW **43:** 215–220.

Winner, W. E. 1986: Measures of plant responses to airborne chemicals. (In: Proceedings of a workshop on controlled exposure techniques and evaluation of tree responses to airborne chemicals.) – NCASI Techn. Bull. **500:** 64–71.

Woodman, J. N. & Cowling, E. B. 1987: Airborne chemicals and forest health. – Environmental Sci. Technol. **21:** 120–126.

Woodwell, G. M. 1970: Effects of pollution on the structure and physiology of ecosystems. – Science **168:** 429–433.

Address of the author: Professor Paul D. Manion, State University of New York, College of Environmental Science and Forestry, Syracuse, New York 13210, USA.

Deterioration of forests in Central Europe

H. Ziegler

Abstract

Ziegler, H. 1988: Deterioration of forests in Central Europe. – In: Greuter, W. & Zimmer, B. (eds.): Proceedings of the XIV International Botanic Congress: 423–444. – Koeltz, Königstein/Taunus.

During the last 10 years many reports have been made of forest decline ("Waldsterben") in Central Europe and other regions in the northern hemisphere. In a number of cases the responsible factors are well defined, *e.g.* identified intoxication by air pollutants, infection diseases, deficiency in mineral nutrition, climatic conditions, game feeding. The remaining cases are called "novel forest decline" ("Neuartige Waldschäden") and many explanations are offered. In this group, too, it is necessary to distinguish different types of damages and to look for different causal agents or complexes of agents. There is no single cause for the "Waldsterben", it is doubtful whether a greater number of the damage types belonging to the "novel forest decline" is really novel. In some cases, however, an imbalance in mineral nutrition through airborne nitrogen input (increasing amounts of NO_X or – in some regions – NH_3 in the atmosphere), notably in areas where the soils are naturally poor in potassium and or magnesium, may be held responsible for a new type of forest damage.

According to Napoleon Bonaparte there is only one convincing rhetoric phrase: the repetition. The repeated statement gets so much fixed in the brains that it is finally considered to be the proved truth.

Introduction

There is a great number of different opinions on the causal agents for the intensively discussed forest decline and about the probable fate of the forests in Central Europe. I can therefore be confident that only a few of the experts in this field will agree with my statements. Since, however, the XIV International Botanical Congress has been placed under the Motto "Forests of the World", it is not surprising that the organizers considered it necessary to have general reviews of this problem.

The limited time available forces me to concentrate on the situation in the Federal Republic of Germany and on our most important forest tree, Norway spruce *(Picea abies* [L.] Karsten). There are similar phenomena and intensive

Table 1. Criteria for characterization of damage classes used in the inventory of Forest Decline in the FRG. They are based on needle/leaf – loss + degree of yellowing (Bundesminister für Ernährung, Landwirtschaft und Forsten 1986).

Damage class	*Loss of needles/leaves*
0: without damage	up to 10%
1: slightly damaged	11–25%
2: medium damaged	26–60%
3: strongly damaged	60%
4: dead	dead

Damage class based on loss of needles/leaves	Degree of yellowing (% of yellow needles/leaves)		
	1 (11–25 %)	2 (26–60 %)	3 (61–100 %)
	definitive damage class		
0	0	1	2
1	1	2	2
2	2	3	3
3	3	3	3

scientific efforts also in Switzerland and Austria (which are well-known to us), and especially strong damages in the forests of the GDR, Poland and ČSSR. Other tree species also show damage symptoms, some of them *(e. g. Abies alba* Miller) even more severe ones than spruce. But Norway spruce is much more important economically. It is perhaps the plant species with the highest total biomass on earth.

Since I have to assume that not all readers are familiar with details of our present topic, I will briefly explain some terms and definitions wich are "daily bread" to the colleagues involved in this field of research.

Fig. 1. Forest decline census 1986 (FRG). All tree species, damage classes 2 to 4. Damaged area in percent of forest area in the different regions (Bundesminister für Ernährung, Landwirtschaft und Forsten 1986). ▷

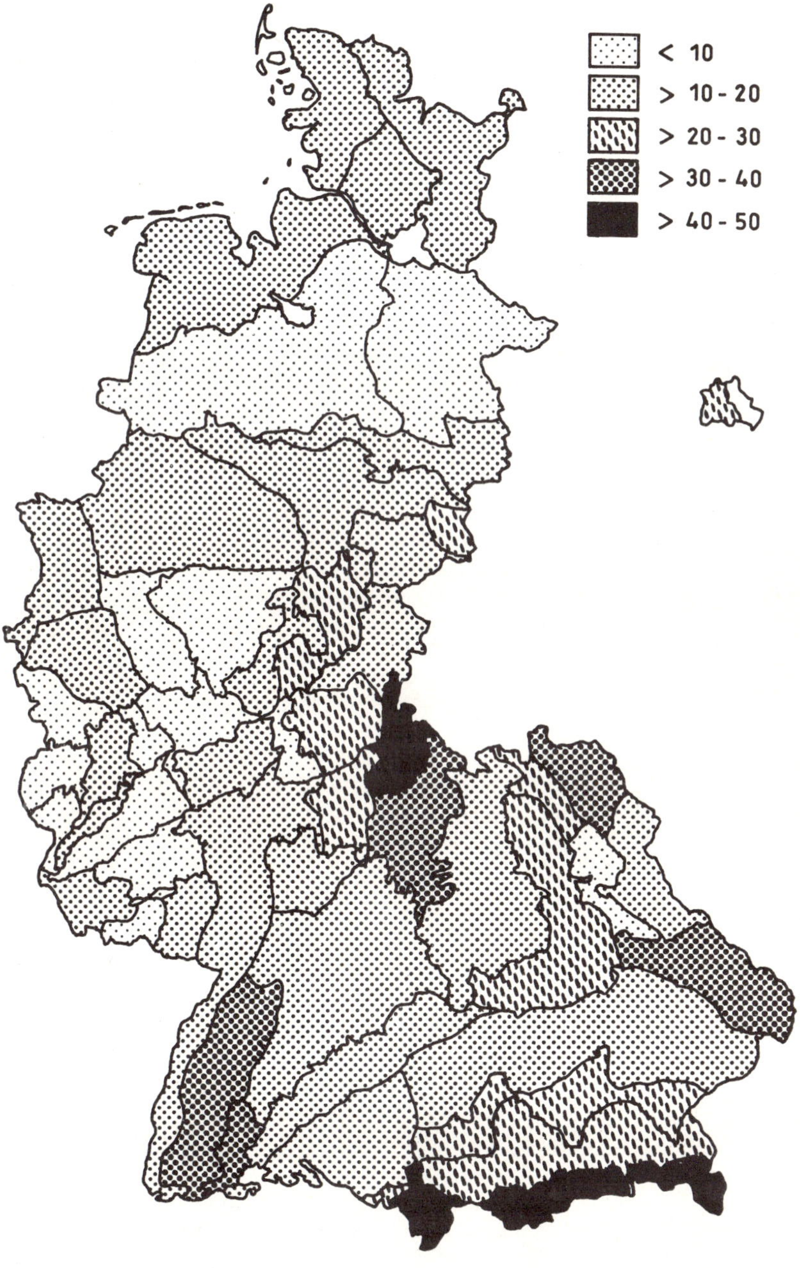
< 10
> 10 - 20
> 20 - 30
> 30 - 40
> 40 - 50

Forest damages are, of course, not new in principle, but have been discussed ever since forestry developed into a scientific discipline. Absolutely new in the present discussion was to my opinion the sensitivity of the public to this problem, when it arose around 1980. And the interest of Man makes a problem really a problem. This has been recognized more than 2000 years ago: "Man is the measure of all things. If they exist to him, they are; if not, they are not" (Protagoras of Abdera, born 480 B.C.).

The inventory of forest deterioration in the FRG, published annually since 1983 by the "Bundesminister für Ernährung, Landwirtschaft und Forsten" (last issue 1986) is mainly based on an estimate of foliage loss of the important forest trees as compared to regional "standard" conditions. Understorey vegetation is not considered in this context, and only limited information is available on this topic.

Forests are classified as healthy (category 0) as long as they carry at least 90% of the estimated normal amount of foliage. Weakly damaged stands (category 1) have foliage losses from 11 to 25%, moderately damaged ones (category 2) from 26 to 60%. Heavily damaged stands (category 3) carry less than 40% of the "normal" amount of leaves. The damage class is raised when needles are yellowing to an important degree (table 1).

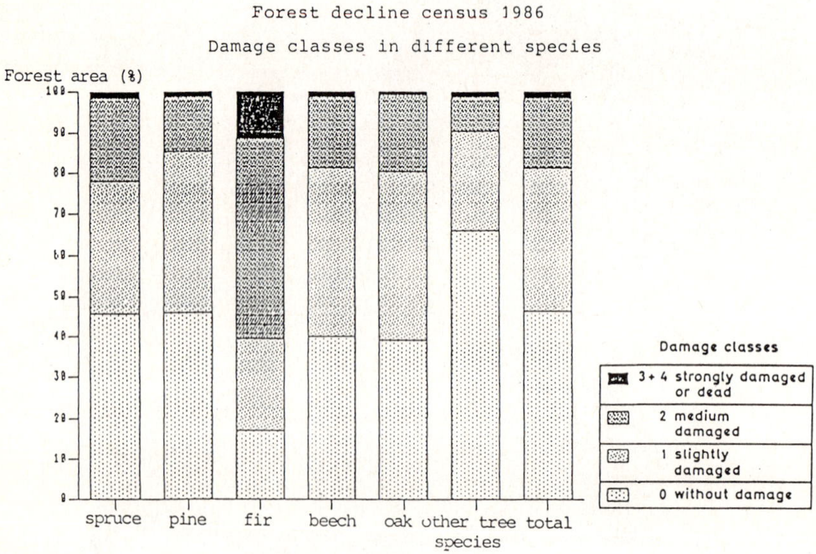

Fig. 2. Forest decline census 1986 (FRG). Damage classes in different species (Bundesminister für Ernährung, Landwirtschaft und Forsten 1986).

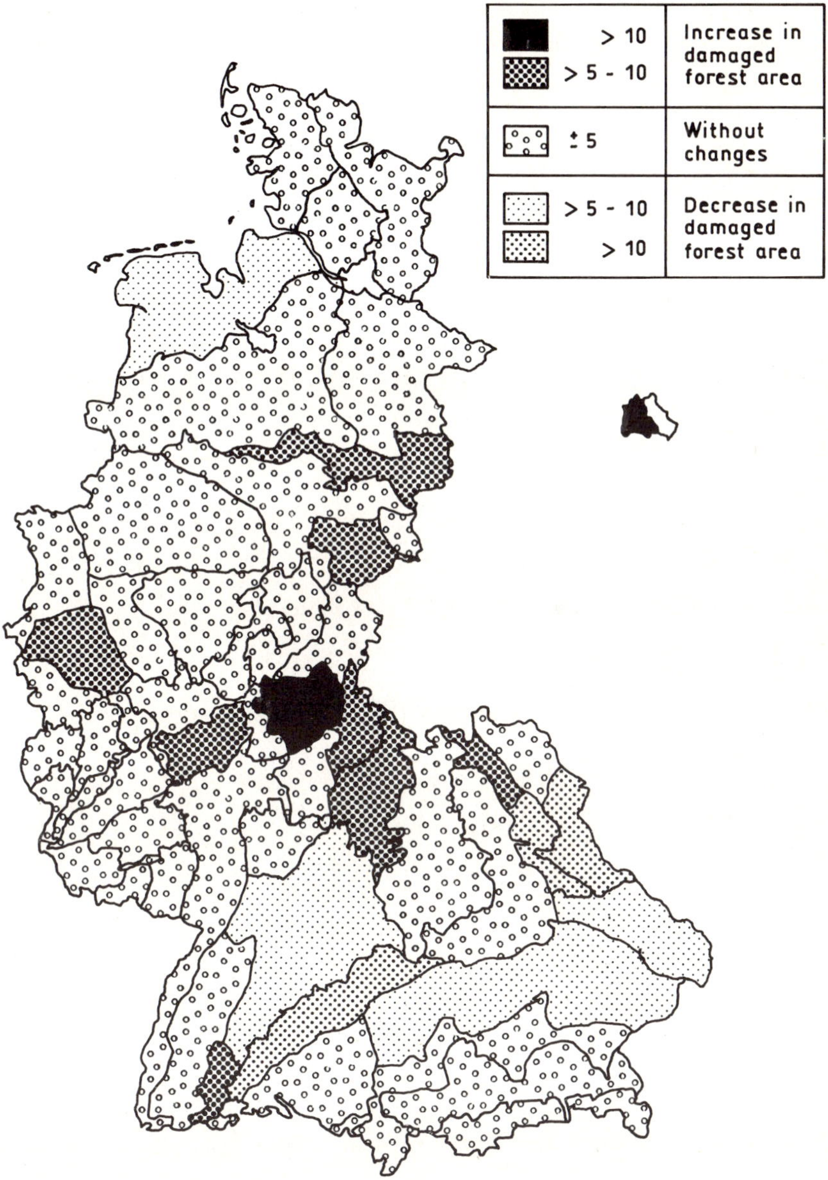

Fig. 3. Forest decline census 1986 (FRG). All tree species. Changes compared with 1985 in % of forest area in the different regions (Bundesminister für Ernährung, Landwirtschaft und Forsten 1986).

However, the standard situation, which is the basis for comparison, has until now been but inaccurately defined. Damage class 1, as a consequence, is quite questionable. It is therefore advisable to consider damage classes 2–4 only.

In the last census (1986) the classes 2–4 for all tree species together, diminished by 0.3% as compared to 1985, to cover about 20% of the total forest area. There are much higher damages in South- than in North-Germany (fig. 1), although the concentration of most pollutants in the atmosphere is generally lower in the south. Damage classes in different species are shown in fig. 2.

The inventory comes to the conclusion that the overall damage, for all tree species together, increased from 1982 to 1985 and first came to a standstill in 1986, admittedly at a high level (fig. 3, 4). Conifers, in particular, recovered considerably (pine and fir) or moderately (spruce) in 1986 (fig. 5). This development, of course, is not indicative of a continuous and inevitable dying of their stands.

The recovery in 1986 is mostly interpreted as a consequence of the cool and wet weather in that year, interestingly enough also by people who categorically deny a promoting influence of hot and dry sommers on the degree of damage.

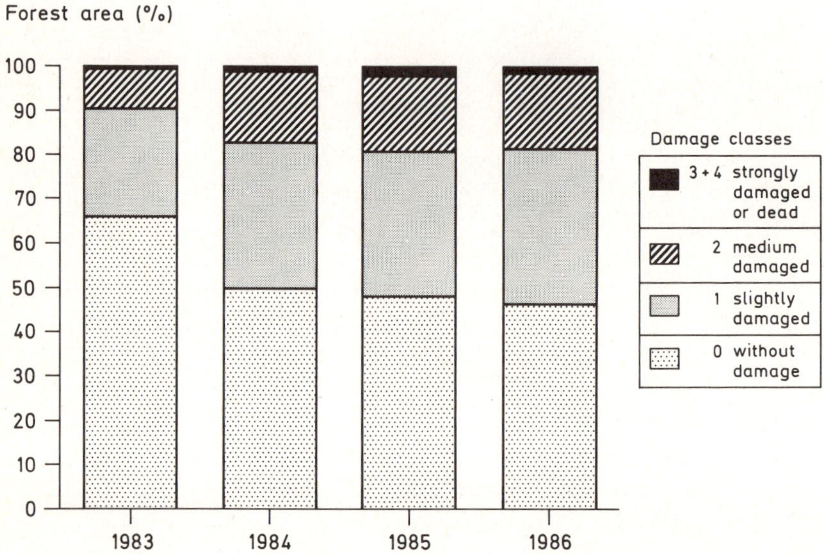

Fig. 4. Forest decline census 1986 (FRG). Change in the percentage of different damage classes (all tree species) in the years 1983–1986. (The census 1983 is comparable to the later ones only to a limited degree.) (Bundesminister für Ernährung, Landwirtschaft und Forsten 1986).

If we use radial growth as an indicator of vitality (Kenk 1987) we get a quite different picture: With the exception of trees with strong needle loss we find in the last years a considerable increase in fir (less in spruce) as well as in beech (fig. 6).

A careful comparison of the various disease – or crown damage – phenomena in Norway spruce, silver fir, Scots pine, European beech and in oaks in the FRG and in the neighbouring countries reveals an extreme diversity of symptoms and growth responses. Therefore, it is plausible to assume that we have to deal with quite different damages. They are partly injuries, partly diseases and partly declines according to the definitions of Manion (1984), partly really new and partly long known.

In the forest inventory for the FRG for 1985 (Forschungsbeirat Waldschäden, 1986), therefore, 5 different damage types were distinguished for Norway

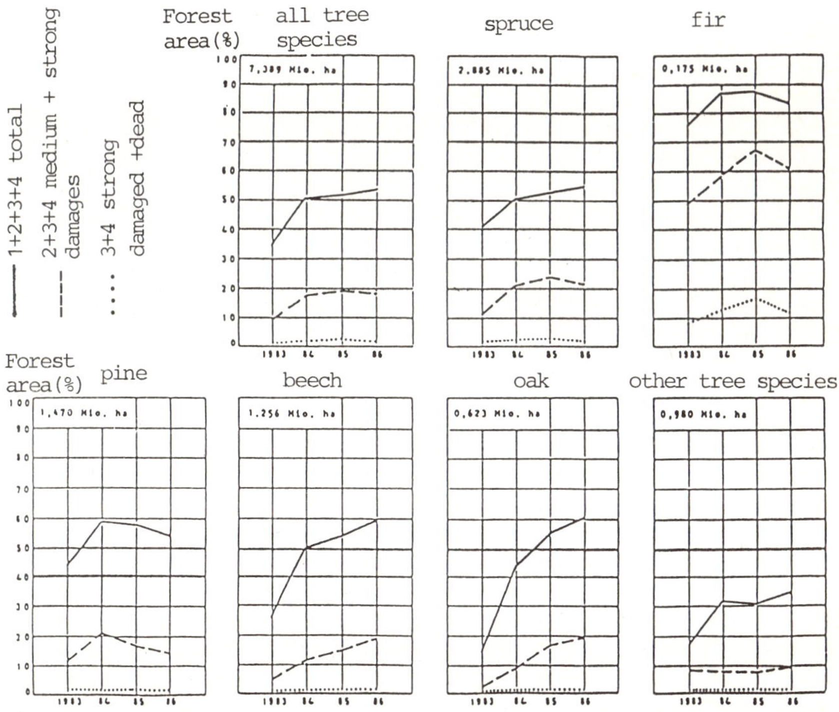

Fig. 5. Forest decline census 1986 (FRG). Changes in the damage classes between 1983 and 1986 for different tree species (Bundesminister für Ernährung, Landwirtschaft und Forsten 1986).

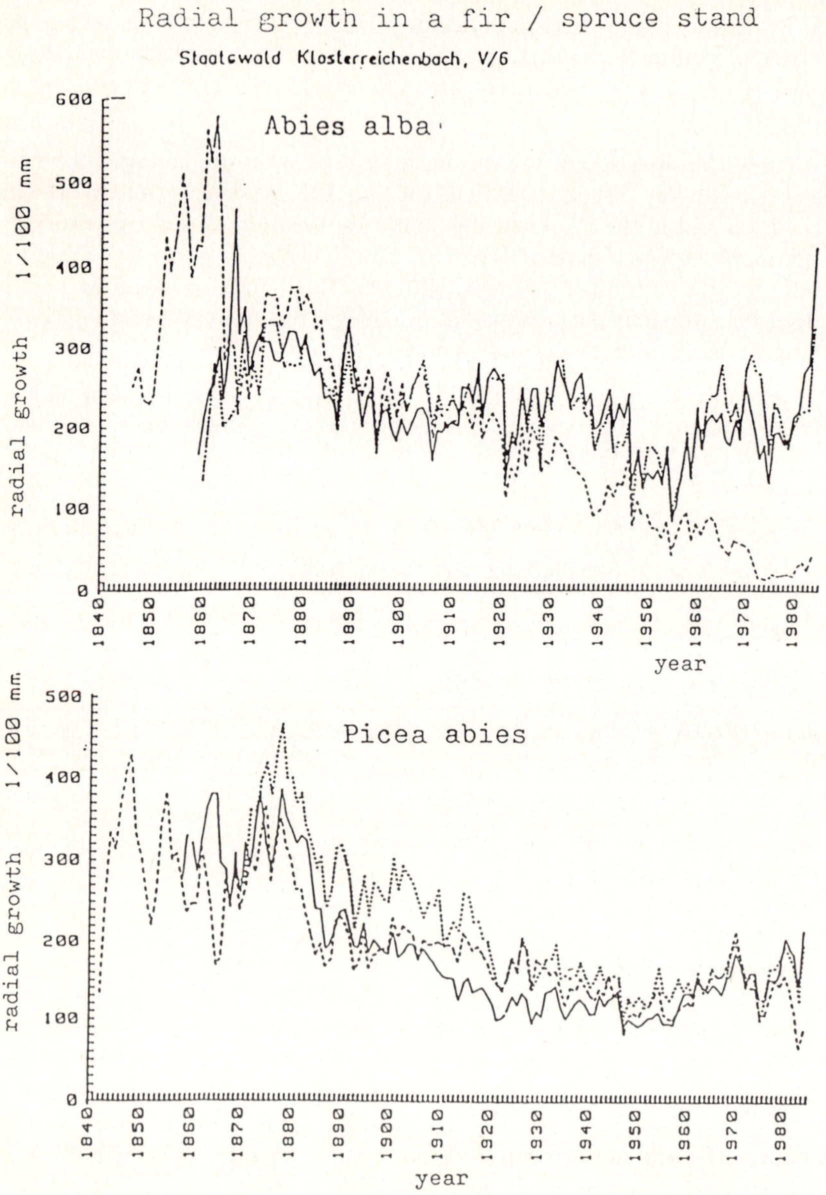

Fig. 6. Radial growth of trunks in a fir/spruce stand in the Black Forest (Staatswald Klosterreichenbach). – *Abies alba*: —— : Trees with 11–25% needle loss; ⋯⋯: 26–60%; ------: > 60%. – *Picea abies*: —— : 0–10% needle loss; ⋯⋯: 11–25%; ------: > 60%. (Kenk 1987.)

spruce. Interestingly, only the supposed “novel“ damages were considered in detail, while the well defined ones were mentioned only shortly or not at all. I therefore added some of them myself (table 2).

Table 2. Different damage types in the forests of the FRG. Typ I to V according to the Forschungsbeirat Waldschäden/Luftverunreinigungen (1986). Type VI to VIII added by the author.

I.	Needle yellowing at higher altitudes of the German Mittelgebirge.
II.	Thinning of tree canopies at medium altitudes of the Mittelgebirge.
III.	Needle reddening of older stands in southern Germany at lower altitudes.
IV.	Needle yellowing at higher altitudes of the calcareous Alps.
V.	Thinning of tree canopies in the vicinity of the sea cost.
VI.	Defined infection diseases.
VII.	Defined effects of air pollutants.
VIII.	Damages by game.

Let me discuss these different types of damage backwardly; the reason will be understandable at the end of my considerations.

Damages caused by game

These damages (injuries) are at present by far the most serious danger for the most important forests in our country, namely the mountain forests in the Alps with their protective function. Since many politicians are hunters – hunting is fashionable – it is very hard to reduce the number of deers and other tree-injuring animals. In some regions of the Alps and of the prealpine regions foresters need three fence generations to raise the palatable species like spruce, mountain maple *(Acer pseudoplatanus)*, fir *(Abies alba)*, mountain ash *(Sorbus aucuparia)* and beech *(Fagus sylvatica)* to a size that allows them to survive the attacks.

Defined effects of air pollutants

These damages belong to the diseases. They are well known since ancient times: The greek geographer Strabo (born c. 63 B.C.) reports from Spain that smelting furnaces for silver sulfides were to be established on higher elevations to distribute the produced noxious gas (SO_2) to a wider area at lower concentrations. Damages by air pollutants, especially SO_2, were studied scientifically since more than 100 years, especially in Germany (compare the reprint "Waldsterben im 19. Jahrhundert" by VDI-Verlag; see under Wislicenus, 1906–1908 in references).

Well known are the catastrophic effects of SO_2 (a real "Waldsterben") in the Ore Mountains and Iser Mountains along the border between Czechoslowakia and the GDR. There the average annual concentration auf SO_2 in the air amounts to 80–90µg m^{-3}, whereas the threshold for damage of Norway spruce is assumed to be c. 50µg m^{-3} (Wentzel 1978). About 30 000 ha of spruce forests have been really killed, mainly during the past 10 years, and have been cleared (Rehfuess 1987). Even the youngest spruce plantations are severely affected, while beech and mountain ash look quite normal. Proton deposition in the Ore Mountains is estimated to amount to 8–10 kmol ha^{-1} a^{-1}; its influence on soil fertility is not considered to be a decisive factor (Rehfuess 1987).

It seems that in the regions of high SO_2 immission quite frequently the combined effect of SO_2 (preconditioning) and frost causes the damage, which would thus belong to the decline phenomena. This is true for the Ore Mountains as well as for regions in northeastern Bavaria (Fichtelgebirge), where we have average annual SO_2 concentrations between 30 and 50 µg m^{-3}. The damaged spruce needles (the youngest needles only) had sulphur contents between 1400 and 1800 µg g^{-1} dry weight; 1250 µg is the upper limit for the natural "normal" S-content, and a content above 1500 µg is considered to be potentially damaging per se (above 2000 µg damage is inevitable).

In these regions with predominant SO_2 damage lichens are absent on the trunks and branches of the trees. Only *Lecanora conizaeoides* occurs, which is known to resist high concentrations of SO_2.

Defined infection diseases

There are many well defined diseases of forest trees caused by pathogens. It is reported that of the c. 450 infection diseases of North American forest trees 97% are caused by fungi, 2% by bacteria and 1% by viruses.

As an example of a fungal disease the spectacular yellowing of spruces in parts of the Alps (mainly South Tyrolia) in the year 1983 (and again 1987) should be mentioned. It was caused by the rust fungus *Chrysomyxa rhododendri.* An obligatory change of host is typical for many rust fungi. From spring to autumn *C. rhododendri* grows as the haploid generation in the youngest needles of spruce, producing numerous aecidia and inducing yellowing and finally shedding of the infected needles (fig. 7). In winter the fungus lives as a diplont in the leaves of *Rhododendron* species.

Therefore, it can only occur in regions where spruce and *Rhododendron* grow together. The fungus develops best in years with a wet spring and a hot and dry summer, which happened in 1983. In 1984 the yellow needles were not present any more (fig. 8), the new needles did not show any symptoms (Ziegler 1986).

It is certain that the course of such infection diseases is influenced not only by climatic factors, but also by anthropogenic impact. It has, *e.g.*, been

Fig. 7 a. *Picea abies* in the Rain valley (South Tyrolia). Youngest needles infected by the rust fungus *Chrysomyxa rhododendri.* In the background the second host of the fungus, *Rhododendron ferrugineum.* August 1983.

demonstrated experimentally that relatively low doses of SO_2 or $SO_2 + NO_X$ inhibit the development of certain pathogenic fungi on the leaf surface, *e.g.* of mildew on wheat (Roberts et al. 1983). On the other hand, the development of aphids is promoted by polluted air (Flückinger & Oertli 1978, Dohmen et al. 1984). This is of special relevance in connection with the transmitting function of these insects for systemic pathogens (*e.g.* mycoplasms, viruses, viroids).

We will now discuss such forest damages, which are included in the extensive report of the Research Council for Forest Decline (Forschungsbeirat Waldschäden 1986) and have therefore been suspected to be real "new damages".

Thinning of tree crowns in the vicinity of the sea coast

This is certainly not a new event. As far as spruce is concerned, the stands are not autochthonous in these regions and they do not tolerate the special soil conditions and the impact of salt load from the air.

Spruce needle yellowing at higher altitudes in the calcareous Alps

This type of damage is mainly manifest in spruce stands on shallow soils on steep southerly slopes between 800 and 1500 m (Rehfuess 1987). The symptoms are needle yellowing and needle loss. Quite often the spruce trees are

Fig. 7b. Aecidia of *Chrysomyxa rhododendri.* Lavazei-Joch (South Tyrolia). September 1987. Bar = 100 μm. Scanning electron micrograph by Jutta Künne.

deficient in potassium and/or manganese. Also nitrogen, phosphorus and iron may be wanting. Normally, there is ample supply of calcium and magnesium. There ist no indication at all that SO_2, acid rain (1–1.5 kmol H^+ ha^{-1} a^{-1} on average; Rehfuss 1987) or NO_x immission plays a role in these stands. Ozone (70–90 µg m^{-3}; Reiter et al. 1983) is not high enough to be responsible alone for the damage, but too high to be neglected.

This air pollution situation is mirrored by the lichen flora in these stands. Köstner & Lange (1986) have identified no less than 76 taxa of lichens, including some species known to be very SO_2 sensitive *(e.g. Lobaria pulmonaria).*

Since these sites show high precipitation (1500–2200 mm a^{-1}), leaching of the limiting elements is also discussed.

Fig. 8. Branch from *Picea abies* from the same site as in fig. 7a, in September 1984. The infected needles from the year 1983 are shed. The new needle generation (1984) shows no damage symptoms.

Intensive efforts are presently made to gain a better insight in this typical "decline" phenomenon. It is not clear whether there is something really new in this damage type. The sites are very poor indeed and would not seem to allow a normal spruce growth at any time. New and very serious, in these regions, is the extent of injuries caused by game (see above).

Needle reddening of older spruce stands in South Germany at lower altitudes

This disease type seems to be the most widespread in older Norway spruce stands at lower altitudes all over southern Germany (Rehfuess & Rodenkirchen 1984, 1985). The stands look healthy during summer. In autumn, several age classes of older needles turn red. This happens with trees on fertile soils and on poor sites alike. The red needles are regularly infected by fungi, mostly *Lophodermium piceae* and *Rhizosphaera kalkhoffii* (fig. 9). The infected needles are shed in early winter.

In the spruce type whose secondary twigs hang down from the main branch, the so-called "silver-tinsel symptom" ("Lametta-Symptom") appears. The vertical position of the twigs itself is genetically fixed ("Kammfichten"), not a damage symptom (Magel & Ziegler 1987) as has been assumed by many authors.

Fig. 9. Pycnidium of the fungus *Rhizosphaera kalkhoffii,* emerging from a spruce needle stoma and carrying the wax plug of the stomatal chamber. Scanning EM, bar = 10 μm. Site: Glashütte near Rottach-Egern, 1983.

As the needle reddening and shedding process repeats itself, the crowns become more and more transparent. Since only stands older than 60 years are affected by this disease, damaged trees are found side by side with healthy (younger) plantations. No long-lasting growth reduction has been detected in such spruce stands.

Needle reddening and fungal infection are not dependent on the nutritional status of the tree. For most stands one can exclude SO_2, NO_x, photo-oxidants and acid deposition as responsible factors.

Rehfuess (1987) describes the needle reddening as a "conventional epidemic" needle cast disease such as was already described by Hartig (1889) about 100 years ago. Rehfuess considers that a series of frost shocks in the late seventies and early eighties was most probably involved in the development of this disease.

Fig. 10. Upper side of a branch of *Picea abies* with "needle yellowing" damage. Only the older needles are yellow, the youngest are normal green, as are all needles on the lower side. Bavarian Forest (photograph by W. Oßwald).

A pronounced recovery process has been observed within these spruce stands in the very last years, especially in the region between the Alps and the Danube: On the upper side of the main branches new shoots are formed which obviously compensate for the severe foliage losses (Rehfuess 1987).

Thinning of spruce crowns at medium altitudes in the Mittelgebirge

This damage type is characterized mainly by a loss of older needles, while needle yellowing is negligible. The symptoms occur mostly on poor sites at an altitude of 400–600 m. The SO_2 load and the (wet) proton deposition are relatively high, the nitrogen and ozone concentrations in the air are medium.

This type is not well defined. It is very doubtful whether it is a real damage, and the symptoms are quite probably not new.

Needle yellowing at higher altitudes of the German Mittelgebirge

This last damage type is especially interesting, since it appears to be the only one that is really new and had not been described before. This is the reason why I treat this type at the end.

The symptoms are observed since about 1980 (Prinz et al. 1982) in spruce stands on acidic soils. It occurs in the Bavarian Forest as well as in the Black Forest, generally at elevations of more than 900 m above sea level.

The decline starts with an pronounced yellowing of the older needles (fig. 10), mostly followed by necrosis and shedding, and leads to a marked thinning of the crowns. It is characteristic that only the older foliage is involved, the youngest shoots ar normally green. The discoloration starts at the tips of the older needles and is particulary evident when the new needles are developed. The golden-to-yellow colour is generally much more intensive on the upper side of the needles than on the lower. This indicates that light is involved in the disappearance of chlorophyll. The decline affects all age classes of spruce trees. A pronounced tree-to-tree and stand-to-stand variation of damage intensity is reported.

It was conclusively demonstrated by several authors (Zech & Popp 1983, Zöttl & Mies 1983, Krause et al. 1983, review in Rehfuess 1987) that deficiency in magnesium and/or potassium was involved in this decline. It is reasonable to assume a restricted uptake of these essential elements on soils poor in these minerals. The amounts in the needles may be additionally diminished due to accelerated leaching from the soil by increased acid deposition. The proton

Table 3. Ratios chlorophyll a/chlorophyll b and chlorophyll/carotene in the youngest needle generation of branches of 40–100 year old spruce trees under different treatments with ozone and different radiation intensities (Senser et al. 1987).

	O_3 $\mu g m^{-3}$	*Chlorophyll a*[1] / *Chlorophyll b*			*Chlorophylls* / *Carotenes*		
		26.3.	14.4.	16.6.	26.3.	14.4.	16.6.
Picea abies I	0	2.67	3.34	2.79	14.16	14.55	16.81
Bot. Garden	300	3.35	3.29	2.53	15.20	15.37	15.42
Munich	900	3.58	3.20	2.65	17.80	14.95	18.74
(100–80 klx)	2000	2.56	3.60	–	28.21	30.85	–
(<80 klx)	2000	3.02	2.76	2.16	19.56	15.05	19.99
Picea abies II	0	3.21	3.25		20.19	14.75	
Bavarian	300	2.84	3.40		21.80	14.70	
Forest	900	2.87	3.07		15.18	18.25	
(100–80 klx)	2000	2.52	2.94		32.52	35.27	
(<80 klx)	2000	3.02	2.96		15.99	16.85	

[1] Chlorophyll a and pheophytin a.

input in throughfall at these sites may be as high as 2 kmol $ha^{-1}a^{-1}$ (Rehfuess 1987). High contents of exchangeable aluminium may additionally hinder the root (mycorrhiza) uptake of both elements.

Prinz et al. (1982) suggest that ozone in combination with acid rain and especially acid mist may cause damages to the cell membranes and promote leaching of Mg^{2+} and/or K^+ out of the needles. The assumption that ozone in realistic concentrations also influences the permeability of the cuticle was not confirmed by experiments of Lendzian and Kerstiens (unpublished) in our department.

The decisive role of ozone in this scenario is doubtful for other reasons too: As was shown by Senser et al. (1987), ozone fumigation causes a destruction of the pigments in spruce needles in the order carotene > chlorophyll a > chlorophyll b > xanthophyll (table 3). This means that the chlorophyll/carotene ratio increases under ozone influence. This is not the case of the yellowing needles from sites in the Bavarian Forest (Kandler et al. 1987; table 4). They behave more or less like extreme sun-needles.

Table 4. Pigment content and ratios of 12–15 years old spruce with different stages of "needle yellowing". Site: Spiegelhütte (Bavarian Forest). Date of sampling: 15. 5. 86 (before flushing, 1986). All date in μ/g needle fresh weight (Kandler et al. 1987).

Condition of the tree	*needle year*	*Chlorophyll a+b*[1]	*Carotenes*	*Chlorophylls / Carotenes*	*Chla / Chlb*
	1985 green	698	57.8	12.1	3.5
healthy	1984 green	782	77.0	10.1	2.4
	1983 green	802	70.1	11.0	2.1
since 1984	1985 green	734	60.0	12.2	3.26
progressive	1984 yellow	423	32.5	13.1	3.45
needle yellowing	1983 yellow	264	23.3	11.3	3.40
yellowing 1984,	1985 green	762	56.0	13.6	2.7
since 1985	1984 green	731	51.4	14.2	2.8
at a standstill	1983 yellow	508	32.3	15.8	2.8
yellowing 1984,	1985 greeen	749	45.5	16.5	3.2
1986 regreening	1984 green	811	53.5	15.0	3.1
	1983 green	861	55.4	15.5	2.0

[1] Including pheophytine a + b.

An especially interesting hypothesis has resulted from intensive discussions in the very recent past (*e. g.* Schulze et al. 1987; Mohr 1986): The interaction of a surplus of nitrogen with the deficiency of magnesium and/or potassium (fig. 11).

A huge and increasing input of nitrogen (NH_4^+, NO_3^- presently occurs via dry and wet deposition. In clean-air regions of North America or Scandinavia the annual N deposition is less than 1 kg ha^{-1}. In the Netherlands, however, more than 60 kg ha^{-1} a^{-1} and in the Black Forest up to 40 kg ha^{-1} a^{-1} are deposited (Houten 1983). The annual requirement on a spruce forest is only 5 to 8 kg (Schulze et al. 1987). As long as the other essential elements are available in sufficient amounts, the higher N inputs increase the growth of the trees dramatically.

As soon as an essential element (*e.g.* magnesium or potassium) is lacking, the needles turn yellow (Schulze et al. 1987). In the Fichtelgebirge the present deposition of N should be diminished to about 25% for the trees to cope with the low actual (natural) magnesium supply (or the magnesium amount should be increased proportionately).

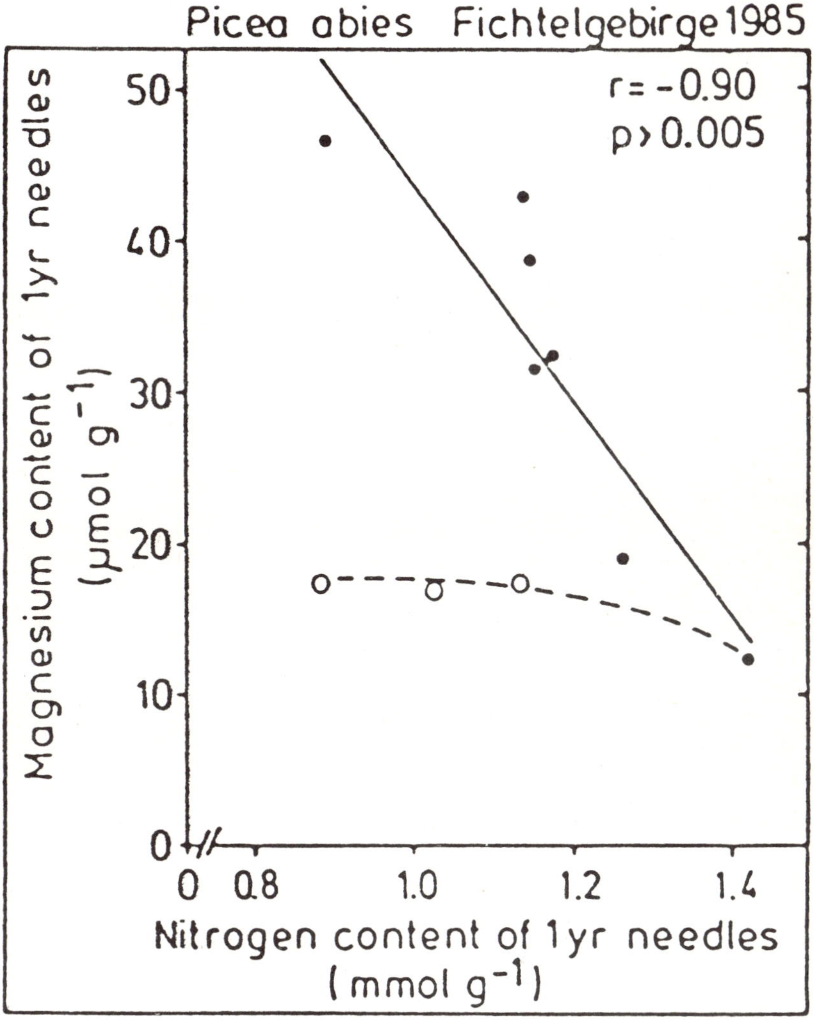

Fig. 11. Relations between magnesium and nitrogen contents of *Picea abies* needles in the Fichtelgebirge (NE Bavaria). The highest nitrogen content correlates with the lowest magnesium content, the needles being yellow. The dashed line shows a special situation in which lower magnesium contents occur at low nitrogen contents; in this case the needles remain green (Schulze et al. 1987).

The effect may be reinforced by the fact that NO_x permeates not only through stomata, but also quite easily through the cuticle (Lendzian 1986). In addition, NO_2 nitrifies some compounds in the cuticle (Kisser-Priesack et al. 1987) and considerably increases water permeability and also leaching. There are also indications that this high N impact may influence the function of mycorrhizas (Meyer 1987).

Conclusions

The main conclusion to be drawn from the intensive and extensive studies of forest deterioration in Central Europe in the last few years seems to be that there is an urgent need of carefully distinguishing the different symptoms of tree damages on the different sites. Detailed causal analysis should not be started before this first step has been made. This plea sounds trivial, but was in fact ignored by most research workers – else they would not have looked for a single cause for *the* "Waldsterben".

Acknowledgements

I am indepted to many colleagues for intensive discussions and partly also for providing illustrations (especially colour slides during the lecture). Particular thanks are due to: O. Kandler, O. L. Lange, W. Oßwald, K. G. Rehfuess, G. D. Schulze and M. Senser.

Experimental work of the author and his coworkers, in this field, was promoted by grants from the "Bundesminister für Forschung und Technologie", the "Deutsche Forschungsgemeinschaft", the Commission of the European Communities, the "Projektgruppe Bayern zur Erforschung der Wirkung von Umweltstoffen" (PBWU) and the "Bayerische Forschungsgruppe Forsttoxikologie".

References

Bundesminister für Ernährung, Landwirtschaft und Forsten, 1986: Waldschäden in der Bundesrepublik Deutschland. Ergebnisse der Waldschadenserhebung 1986. – Schriftenreihe Bundesminist. Ernähr., Reihe A: Angew. Wiss., **334.**

Dohmen, G. P., McNeill, S. & Bell, J. N. B. 1984: Air pollution increases *Aphis fabae* pest potential. – Nature **307:** 503.

Flückinger, W. & Oertli, J. J. 1978: Observations on an aphid infestation on hawthorn in the vicinity of a motorway. – Naturwissenschaften **65:** 654–655.

Forschungsbeirat Waldschäden / Luftverunreinigungen der Bundesregierung und der Länder; 1986: 2. Bericht. – Kernforschungszentrum, Karlsruhe.

Hartig, R. 1889: Lehrbuch der Baumkrankheiten, ed. 2. – Springer, Berlin.

Houten, J. G. ten 1983: Biological indicators of air pollution. – Environmental Monit. Assessment **3:** 257–261.

Kandler, O., Miller, W. & Ostner, R. 1987: Dynamik der "akuten Vergilbung" der Fichte. Epidemiologische und physiologische Befunde. – Allg. Forstz. **27/28/29:** 715–723.

Kenk, G. 1987: Referenzdaten zum Waldwachstum. Der Vergleich von Wachstumsdaten aufeinanderfolgender Waldbestandsgenerationen. – PEF-Ber. Freiburg 1987.

Kisser-Priesack, G. M., Scheunert, I., Gnatz, G. & Ziegler, H. 1987: Uptake of $^{15}NO_2$ and ^{15}NO by plant cuticles. – Naturwissenschaften **74:** 550–551.

Köstner, B. & Lange, O. L. 1986: Epiphytische Flechten in bayerischen Waldschadensgebieten des nördlichen Alpenraumes: Floristisch-soziologische Untersuchungen und Vitalitätstests durch Photosynthesemessungen. – Ber. Akad. Naturschutz Laufen **10:** 185–210.

Krause, G. M. H., Jung, K. D. & Prinz, B. 1983: Untersuchungen zur Aufklärung immissionsbedingter Waldschäden. – Verein Deutsch. Ing. Ber. **500:** 257–266.

Lendzian, K. J. 1986: Der Einfluß von "saurem Regen" auf die Barriere-Eigenschaften pflanzlicher Kutikeln. [Statusseminar Wirkungen von Luftverunreinigungen auf Waldbäumen und Waldböden.] – Kernforschungsanlage Jülich. Spezielle Ber. **369:** 59.

Magel, E. & Ziegler, H. 1987: Die "Lametta-Tracht" – ein Schadsymptom? – Allg. Forstz. **27/28/29:** 731–733.

Manion, P. D. 1981: Tree disease concepts. – Prentice-Hall, Inglewood Cliffs.

Meyer, F. H. 1987: Das Wurzelsystem geschädigter Waldbestände. – Allg. Forstz. **27/28/29:** 754–755.

Mohr, H. 1986: Die Erforschung der neuartigen Waldschäden. – Biol. Unserer Zeit **16:** 83–89.

Prinz, B., Krause, G. H. M. & Stratmann, H. 1982: Waldschäden in der Bundesrepublik Deutschland. – Ber. Landesanst. Immissionsschutz Nordrhein-Westfalen **28.**

Rehfuess, K. E. 1987: Perceptions on forest diseases in Central Europe. – Forestry **60:** 1–11.

– & Rodenkirchen, H. 1984: Über die Nadelröte-Erkrankung der Fichte *(Picea abies* Karst.) in Süddeutschland. – Forstwiss. Centralbl. **103:** 248–262.

– & – 1985: Über die Nadelröte-Erkrankung der Fichte *(Picea abies* Karst.) in Süddeutschland. – Forstwiss. Centralbl. **104:** 381–390.

Reiter, R., Munzert, K., Sladkovic, R., Poetzl, K. & Kanter, H.-J. 1983: Basiserarbeitung zum Problem "Waldschäden im Alpenvorland". – Bayerisches Staatsministerium für Landesentwicklung und Umweltfragen, München.

Roberts, T. M., Darrall, N. M. & Lane, P. 1983: Effects of gaseous air pollutants on agriculture and forestry in the U.K. – Advances Appl. Biol. **9:** 1–142.

Schulze, E.-D., Oren, R. & Zimmermann, R. 1987: Die Wirkung von Immissionen auf 30jährige Fichten in mittleren Höhenlagen des Fichtelgebirges auf Phyllit. – Allg. Forstz. **27/28/29:** 725–730.

Senser, M., Höpker, K.-A., Peuker, A. & Glashagen, B. 1987: Wirkungen extremer Ozonkonzentrationen auf Koniferen. – Allg. Forstz. **27/28/29:** 709–714.

Wentzel, K. F. 1978: Immissionsgrenzwerte für den Wald. – Schweiz. Z. Forstwesen **5:** 368–380.

Wislicenus, H. (ed.) 1908–1916: Sammlung von Abhandlungen über Abgase und Rauchschäden unter Mitwirkung von Fachleuten. – Parey, Berlin [Reprint 1985: Waldsterben im 19. Jahrhundert. – Verband Deutscher Ingenieure, Düsseldorf.]

Zech, W. & Popp, E. 1983: Magnesiummangel, einer der Gründe für das Fichten- und Tannensterben in NO-Bayern. – Forstwiss. Centralbl. **103:** 50–55.

Ziegler, H. 1986: Pflanzenphysiologische Aspekte der Waldschäden. – Vortr. Rhein.-Westfäl. Akad. Wiss. **347.**

Zöttl, H. W. & Mies, E. 1983: Die Fichtenerkrankung in den Hochlagen des Südschwarzwaldes. – Allg. Forst- Jagdzeitung **154:** 110–114.

Address of the author: Prof. Dr. Hubert Ziegler, Institut für Botanik und Mikrobiologie, Technische Universität München, Arcisstr. 21, D-8000 München 2, FRG.

Canopy dieback and ecosystem processes in the Pacific area

D. Mueller-Dombois

Abstract

Mueller-Dombois, D. 1988: Canopy dieback and ecosystem processes in the Pacific area. In: Greuter, W. & Zimmer, B. (eds.): Proceedings of the XIV International Botanical Congress: 445–465. – Koeltz, Königstein/Taunus.

Forest dieback in the form of canopy collapse or stand-level mortality is common to a number of Pacific island forests that are not affected by industrial air pollution or epidemic diseases. The symptomology of canopy dieback was defined as it became manifested initially in the indigenous Hawaiian *Metrosideros polymorpha (Myrtaceae)* dominated rain forest in the mid 1960s. The subsequent search for the causes and the accumulated facts are summarized. Comparisons are made with other Pacific forest diebacks in New Zealand, Papua New Guinea, Japan and the Galapagos Islands. A theory for natural dieback is proposed involving as predisposing factors (1) the cohort structure of the dieback stands, (2) habitat relationships of the affected species and (3) frequently recurring environmental perturbations. Chronic stress brings the canopy stand eventually into a state of physiological senescence, at which time the third factor becomes the dieback trigger. Biotic agents can assume a subsidiary role. They may become involved only as decomposers. Some conclusions are drawn with regard to the forest dieback in industrially polluted countries of the Atlantic area.

Introduction

During the mid 1960s, various segments of the native Hawaiian rain forest were observed to have shed much of their crown foliage for no obvious reason, a feature considered to be unusual in an evergreen tropical rain forest. The phenomenon was called forest decline by some (Papp et al. 1979) and canopy dieback by others (Mueller-Dombois et al. 1980).

To this day, the terms are used interchangeably. The symptomology includes trees that lose their foliage over the whole crown suddenly and others that lose their foliage slowly from the top down. Trees with slow or gradual loss of foliage are often recognized as stag-headed trees with adventitious branches coming from their trunks. Both types of dieback trees grow intermixed in the same stands that are affected by canopy dieback or decline. The term stand-level dieback is used for the same phenomenon when forest segments or whole stands rather than isolated trees are affected.

Another symptom is that dead and dying trees remain in upright position for a very long time. In fact in nearly all dieback stands sampled, trees were found that died a longer time ago, others that died recently and still others that were in the process of dying.

Initially, the native rain forest decline or canopy dieback was thought to be a uniquely Hawaiian problem. Then it was found that similar canopy diebacks occur in indigenous forests of other Pacific islands. This led to a symposium at the XVth Pacific Science Congress in New Zealand in 1983, where canopy dieback and associated dynamic processes were discussed from a Pacific-regional perspective (Mueller-Dombois 1983). In spite of local differences, it was possible to find underlying commonalities. These related to the structure of the dieback stands, to the research approaches taken and the causative relationships elucidated.

In this paper, the Hawaiian dieback research will first be summarized as it developed from a pathological into an ecological problem. Then other Pacific examples will be briefly referred to and an attempt will be made towards a unifying theory. Finally, the Pacific forest diebacks will be set in perspective to the Atlantic forest diebacks.

Dieback research in Hawaii

The disease hypothesis

The Hawaiian rain forest is dominated over large areas by a single endemic canopy species, *Metrosideros polymorpha (Myrtaceae),* locally known as the 'ōhi'a lehua tree. There are many more native tree species in the undergrowth (about 20 to 25 per hectare), but these do not usually join the upper canopy. A further general characteristic is the presence of tree ferns (*Cibotium* spp.) which often form a distinct layer in the undergrowth. The largest, continuous area of this forest occurs on the island of Hawaii, the youngest and biggest in the archipelago (fig. 1). It was here that canopy or stand-level dieback

Fig. 1. *Metrosideros* rain-forest occurs in all high islands (Kauai, Oahu, Molokai, Maui and Hawaii) on their north-east (trade-wind exposed) sides and mountainous interiors. Dieback stands are currently present and have been recorded in the past on all islands. The currently largest dieback area is a patchy mosaic extending from Kilauea Volcano north along the east slopes of Mauna Loa and Mauna Kea and west upslope (10 km) from Hilo on Hawaii. Island ages in millions of years, as determined by the K/Ar method, are shown on the upper left side of each larger island. The climate diagram of Hilo is representative of the Hawaiian rain-forest climate. (Reproduced with permission from Geogr. Rundschau.) ▷

increased rapidly in the early 1970s. An analysis of aerial photographs covering an 80,000 ha territory across the adjoining eastern slopes of Mauna Loa and Mauna Kea revealed that heavy dieback was restricted to an area totalling 120 ha in 1954; in 1965 the dieback area had increased to 16,000 ha and in 1972

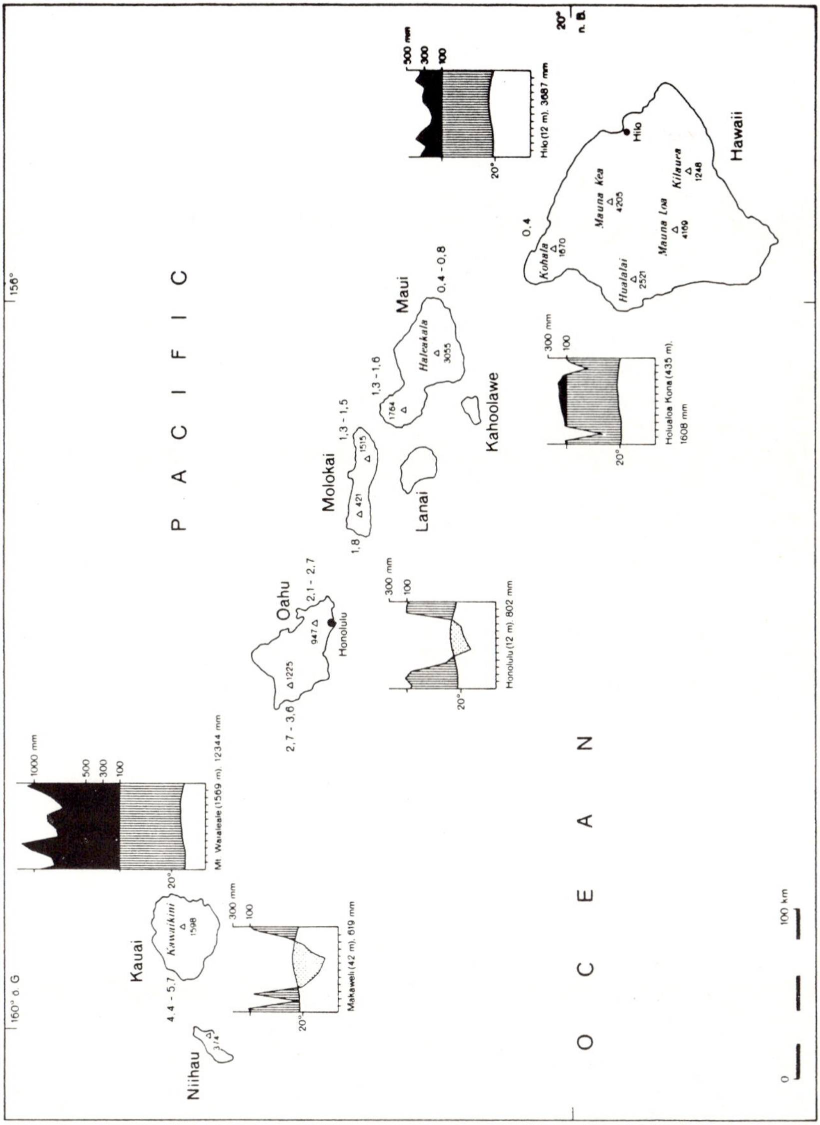

to a cumulative area of 34,500 ha (Petteys et al. 1975). It was then predicted that the Hawaiian rain forest in this 80,000 ha territory was struck by an alien disease and that the whole forest would die in the next 15 to 25 years.

Intensive research on the disease hypothesis during the 1970s revealed that several disease causing agents (fungi, nematodes, bark beetles) were in the rain forest area. Two of these, the root fungus *Phytophthora cinnamomi* and the *Metrosideros* wood borer *Plagithmysus bilineatus,* appeared to have some causative involvement in the dieback. However, thorough sampling in dieback stands failed to reveal consistent relationships. For example, *Phytophthora cinnamomi* was present in some dieback stands but not in others (Hwang & Ko 1978), and the *Metrosideros* borer likewise was not consistently abundant in all dieback stands. Experimental inoculation of healthy *Metrosideros polymorpha* trees with *Plagithmysus bilineatus* failed to cause dieback (Papp et al. 1979). Experimental application of fungicide in a partial dieback stand revealed no improvement. Instead fertilization with NPK resulted in the refoliation of some trees and a partial recovery (Kliejunas & Ko 1974). Finally it was concluded that the canopy dieback was not caused directly by a biotic disease, but rather by an abiotic stress or a combination of factors (DLNR 1981, Hodges et al. 1986).

The abiotic stress hypothesis

The positive response of dieback trees to fertilization was an indication that nutrient limitation, particularly nitrogen deficiency, was part of an environmental stress-complex. Earlier observations revealed large areas of dieback stands to occur on poorly drained substrates. This led to the idea that stands had died from lack of oxygen, since their root systems were partially or wholly submerged in these habitats.

Dieback stands on poorly drained pāhoehoe lava were often sharply separated from healthy stands on 'a'ā lava with better drainage. Rainfall analyses (Doty 1982, Evenson 1983) done on the assumption that a sequence of heavier than normal rainfall years may have caused dieback on poorly drained sites, failed to reveal any significant relationship. Doty (1982) discovered a slight long-term drying trend over the dieback area, but this trend was so small that it was dismissed as biologically insignificant. A study of water table fluctuations (Doty 1980) in adjacent poorly and better drained sites showed frequent and substantial fluctuations in the root zone in relation to current rainfall events. No rise in the surface water table could be detected that would suggest a prolonged submersion of the root systems of dieback stands.

Extension of the survey revealed dieback stands also on well drained sites, on nutritionally rich deep soils from volcanic ash, as well as on excessively drained young lava flows. In fact, dieback stands were found over an entire spectrum of habitats from relatively young lava flows to old deeply weathered, well-drained soils and to bogs, and over a range of soil reactions from weakly acid (pH 6.0) to strongly acid (pH 3.5) (Balakrishnan & Mueller-Dombois 1983).

These site differences suggested dieback relationships to soil water and nutrient regimes. Soil nutrients, when in low supply or imbalanced form, can be expected to exert a stress which may increase with stand development. In contrast, soil water can be expected to exert a fluctuating stress, unless there is evidence for a gradual and permanent change.

An analysis of month to month rainfall variations (Mueller-Dombois 1986) revealed that extreme rainfall events in terms of floods and droughts occur at irregular, but quite frequent intervals. There were nine severe floods and six severe droughts recorded at Hilo over an 85 year timespan (fig. 2). At times, these extreme rainfall events are clustered in successive years. A severe drought was followed by a severe flood at the start of the main dieback period in the mid 1950s. The severe drought caused by the giant “El Niño“ in the early 1980s did not result in any significant expansion of the dieback area, presumably because most of the vulnerable stands already had entered the dieback stage.

There is little doubt that such extreme climatic disturbances cause physiological upsets in the forest vegetation. Yet, since such extreme events occur repeatedly throughout the normal lifespan of a *Metrosideros* tree (estimated to be between 200 to 400 years), and since they are a natural component of the dynamic environment of this native forest, they can not be considered as the only cause of stand-level mortality.

Environmental stresses other than extreme rainfall events were considered early in the research, such as air pollution from volcanic fuming, animal damage from feral pigs, cold air drainage, a fire in the past, lightning damage, but none of these possibilities would correlate well with the spatial pattern of the dieback stands. Another factor, substrate vibration from seismic activity, still needs further study. Substrate vibrations are a frequent phenomenon in this volcanic environment. Seismic events attaining between 6 to 7 on the Richter scale occur from twice a year to 11 year intervals on the island of Hawaii. An event reading 7.2 on the Richter scale occurred in November 1975 in the area south of Hilo at Kalapana (A. S. Furamoto, pers. comm.). However, this was after most of the dieback had occurred, and *Metrosideros* rain forests in this volcanically most active area along the south-east rift zone of the Kilauea volcano have not yet gone into dieback.

The succession hypothesis

A literature review revealed that dieback in the Hawaiian rain forest ist not a new phenomenon. Nineteen century authors (Clarke 1875, Miller ined.) refer to the "decadence" of Hawaiian forests, and a major stand-level dieback was discovered and researched early in this century in the lower montane rain forest on the island of Maui (Lyon 1909, 1918, 1919, Lewton-Brain 1909). This dieback was initially also thought to be caused by a biotic disease agent but later diagnosed as caused by soil ageing, which in Hawaiian volcanic soils is associated with soil toxicity, particularly the development of abundant reducing iron under poor drainage conditions (Clements et al. 1974). The Maui dieback can be characterized as a "stand-reduction dieback" occurring in the retrogressive phase of primary succession. It did not result in the disappearance of *Metrosideros* from the area, but in poor recovery of the same species (Holt 1983). The vegetation stature changed from closed forest to patchy scrub. A similar "stand-reduction dieback" was noted in the 1950s by Fosberg (1961) on poorly drained flats in the rain forest on Kauai. Selling (1948), who encountered dieback of *Metrosideros* stands in the late 1930s, suggested to Lyon that the trees may have died as a result of climatic change. Selling's bog-pollen cores showed drastic fluctuations in *Metrosideros* pollen. Recent carbon datings (Juvic, pers. comm.) indicate an age of 10,000 years for the 325 cm deep bog profile on Molokai, which brings the fluctuations to a time scale that could reflect generation turnovers of *Metrosideros* stands (fig. 3).

Thus, stand-level dieback appears to be a recurring phenomenon in the Hawaiian rain forest. Here, primary succession begins on new lava flows, where *Metrosideros polymorpha* is the first tree species to invade usually after four years (Mueller-Dombois & Smathers 1975). Depending on seed source availability, lava flows are often stocked by open grown cohorts of *M.*

Fig. 2. Extreme rainfall events at Hilo from 1900 to 1984. Months with < 50 mm rain are shown left of drought index and those with > 750 mm right of flood index. Asterisks on the left indicate severe droughts, less than 10 mm/month or two or more sequential months with < 50 mm rain. Asterisks on the right indicate severe floods, with > 1250 mm rain/month or two more sequential months with > 750 mm. The shaded central zig-zag line represents the mean monthly rainfall for each year. Compare to diagram of average climate for Hilo on fig. 1. (Reproduced with permission by Annual Reviews Inc. from Annual Rev. Ecol. Syst., vol. **17**: 1986.) ▷

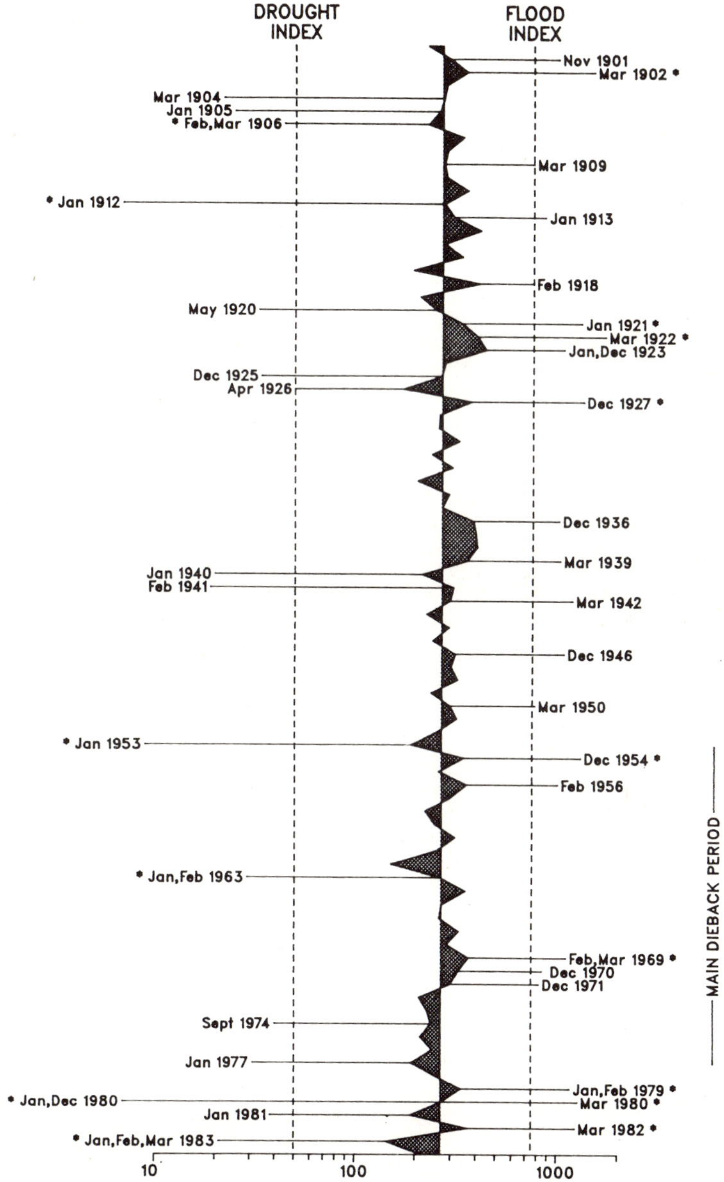
DROUGHT INDEX
FLOOD INDEX
Nov 1901
Mar 1902 *
Mar 1904
Jan 1905
* Feb,Mar 1906
Mar 1909
* Jan 1912
Jan 1913
Feb 1918
May 1920
Jan 1921 *
Mar 1922 *
Jan,Dec 1923
Dec 1925
Apr 1926
Dec 1927 *
Dec 1936
Mar 1939
Jan 1940
Feb 1941
Mar 1942
Dec 1946
Mar 1950
* Jan 1953
Dec 1954 *
Feb 1956
* Jan,Feb 1963
Feb,Mar 1969 *
Dec 1970
Dec 1971
Sept 1974
Jan 1977
Jan,Feb 1979 *
* Jan,Dec 1980
Mar 1980 *
Jan 1981
Mar 1982 *
* Jan,Feb,Mar 1983
MAIN DIEBACK PERIOD
10
100
1000

polymorpha, which after 20 years can be from 1 to 2 m tall. Succession is associated with floristic changes, which include after the initial barren state, fields of the lichen *Stereocaulon volcani* with pockets of the sword fern *Nephrolepis exaltata.* Later, after 50–100 years, the lichens become displaced by the creeping and mat-forming fern *Dicranopteris linearis,* when the *Metrosideros* cohort stands may reach 5 to 8 m in height. Eventually, the trees develop into closed forests, which can be 15 to 25 m tall when mature, depending on site. With development of canopy closure, *Dicranopteris* retracts and tree ferns of the genus *Cibotium* begin to form a separate layer in the undergrowth. The process from barren lava flow to mature *Metrosideros-Cibotium* rain-forest has been estimated by Atkinson (1970) to take about 400 years. On ash substrates, the process appears to proceed much faster, taking only about 200 years.

At this time, saplings of several other native tree species can be found under the *Metrosideros* canopy, but rarely are there saplings of the canopy species.

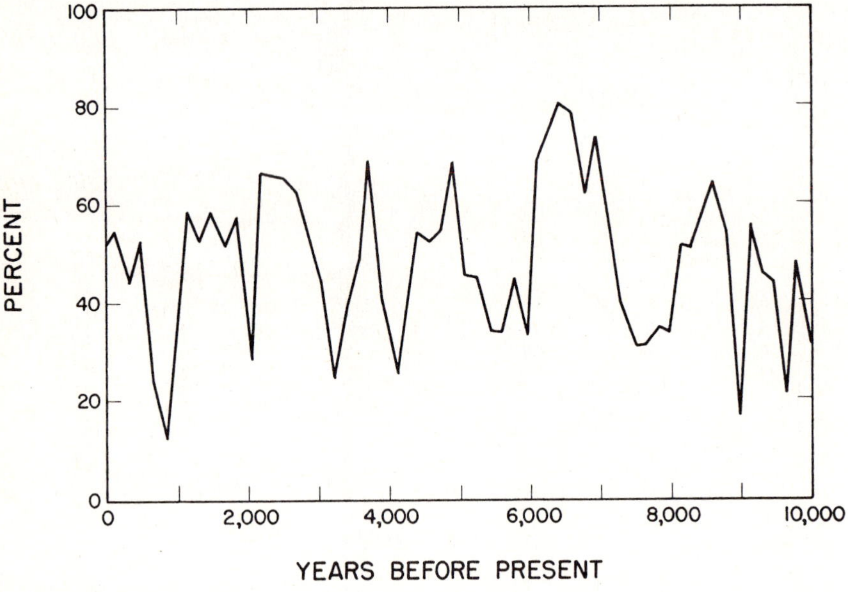

Fig. 3. *Metrosideros* pollen fluctuations in a 325 cm deep bog-soil core from a rain-forest bog on Molokai (after Selling 1948, modified). The bottom of the pollen core was recently carbon dated as 10,000 years old. Percent abundance is based on tree fern spores and tree pollen. *Metrosideros* displays a continuous but oscillating presence. (Reproduced with permission by Annual Reviews Inc. from Annual Rev. Ecol. Syst., vol. **17**: 1986).

Therefore, one can speak of a "sapling gap". Absence of *Metrosideros* saplings is a characteristic feature in many mature non-dieback stands. In some stands, *Metrosideros* seedlings are found on logs and tree ferns. With further maturation of the first generation cohort, a breakdown of the canopy stand can be expected to occur. When this happens, the existing seedlings start to grow into saplings supported also by new seedlings that appear after canopy opening (Burton & Mueller-Dombois 1984).

Most stand-level dieback was found to be restricted to mature stands, while immature stands and seedlings under dieback stands appeared less or not at all affected. Upgrowth of new saplings was recorded following dieback (Jacobi et al. 1983) in two of the main dieback types recognized, the so-called "wetland dieback" (on poorly drained lava flows) and the „dryland dieback" (on well to excessively drained flows). These two dieback types were structurally recognized as "replacement diebacks". In contrast, on nutritionally rich (eutrophic), relatively young (1000 to 2000 years old) soils from volcanic

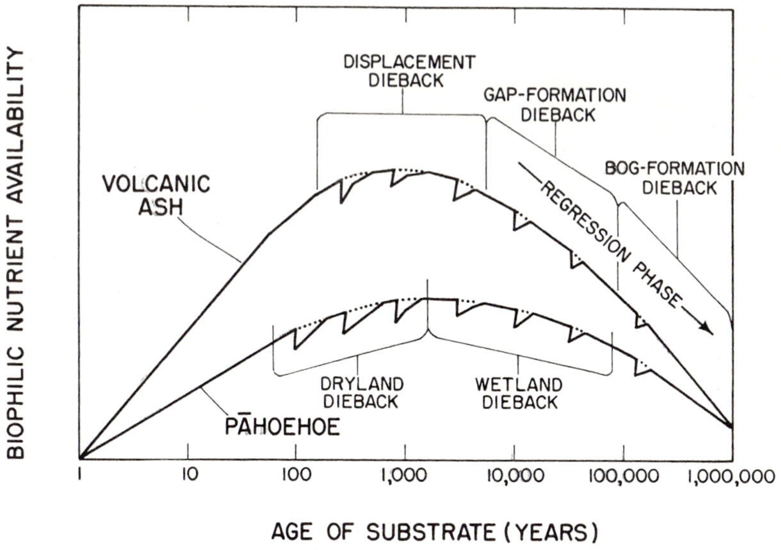

Fig. 4. An idealized relationship of five site-related dieback types over a soil-age sequence on the two major volcanic substrates. The curves indicate an increase of plant biomass associated with biophilic nutrient content reaching a maximum at approximately 1000–2000 years. Primary succession enters a regressive phase after about 3000 years, when soils become increasingly acid and nutrients imbalanced. (Reproduced with permission by Annual Reviews Inc. from Annual Rev. Ecol. Syst., vol. **17**: 1986.)

ash we recognized what we call "displacement dieback" on account of the strong quantitative exclusion after dieback of *Metrosideros* saplings by the dense tree fern layer (Mueller-Dombois et al. 1980). A fourth major type, called „bog-formation dieback", structurally recognized as "stand-reduction dieback", which occurs on permanently water-logged and boggy sites, was already mentioned. A fifth type called "gap-formation dieback" has been recognized as a smaller area (usually <1 ha) stand-level dieback occurring on better drained knolls and ridges in physiographically older terrain. In some gap-formation stands *Metrosideros* reproduction is displaced by undergrowth competition, others show *Metrosideros* replacement depending on the availability of "safe" sites, such as logs and tree fern trunks that do not become completely overgrown by other species. Fig. 4 shows an idealized arrangement of the five dieback types along a primary successional gradient relating to soil age sequences of the two main substrate types, volcanic ash and lava rock.

Metrosideros polymorpha, as mentioned before, is the pioneer tree on new lava flows. It is replaced by later-successional native species in seasonal environments. In the rain-forest environment, however, it retains its canopy dominance over the soil age sequence. It is still the dominant canopy species on the island of Kauai with over three million year-old soil substrates. Periodic canopy dieback may have contributed to its successful maintenance.

Conspecific succession

Another factor promoting persistence appears to be the evolution of successional races in *Metrosideros polymorpha.* It has been observed only recently that there is a pattern of dominantly pubescent-leaved *Metrosideros* to appear as the early lava flow variety, while on older substrates glabrous-leaved varieties prevail (Stemmermann 1983, 1986). Pubescent-leaved varieties which occur on the older islands also may have a similar relationship by responding positively to larger area disturbances and to invading harsher sites, whereas the glabrous-leaved forms are better adapted to mesic rainforest sites. While the latter relationship still needs further study, it is clear now that *Metrosideros polymorpha* has become adapted to a wide range of habitats by racial segregation. It is very well adapted to new volcanic habitats, while it may be differentially adapted to others. The tree can hold on to periodically poorly drained sites, but in bogs it becomes dwarfed and totally displaced in extreme cases. It is, moreover, probable that *Metrosideros* is triggered more easily into dieback by climatic disturbances on poorly drained sites than on well drained deep soils or on lava flows.

The successional hypothesis has led to the recognition of two other causes in the Hawaiian dieback etiology, namely to the involvement of stand dynamics in form of cohort development, and the involvement of site as a medium of differential adaptation for *Metrosideros polymorpha.* In other words, the *Metrosideros* rain forest can be recognized as an irregular mosaic of cohort stands of spatially quite variable dimensions. This cohort mosaic extends over a spectrum of habitats from new lava flows to bogs, and in spite of subspecific segregation, the canopy species does not appear to be equally well adapted to all habitats over its range.

Dieback in other Pacific forests

Stand-level dieback occurs in a number of Pacific forests. Examples are found in Australia and several Pacific islands including New Zealand, Papua New Guinea, Japan and the Galápagos.

New Zealand

Large area dieback has been reported in New Zealand to occur in the *Metrosideros umbellata – Weinmannia racemosa* forest type, which is widely distributed on the west side of the South Island of New Zealand. Here stand-level dieback has been attributed to the Australian brush-tailed opossum *(Trichosurus vulpecula),* which has successfully multiplied and is feeding on the foliage of *Metrosideros umbellata* (Batcheler 1983). However, the animal depredation thesis as the sole cause of *Metrosideros* dieback in New Zealand was challenged by Veblen & Stewart (1982) and Stewart & Veblen (1982), who pointed out convincingly that stand dynamics is another important cause. They noted that opossum feeding occurs primarily in older stands and much less so in immature stands. They also discovered that stands occur in even-aged segments or cohort stands. They consider overmaturity or senescence of such stands as an important predisposing factor to stand-level dieback. Similarly, various *Nothofagus* forests in New Zealand exhibit stand-level dieback. In the monospecific southern mountain beech *(Nothofagus solandri* var. *cliffordioides)* forest, dieback has also been related to several factors, involving stand dynamics with the formation of cohort stands and subsequent high density of juvenile stands or development to overmaturity as a predisposing factor, while snow storms and snow-pack breakage have been considered as trigger factors (Wardle & Allen 1983). Reproduction of the canopy species is not impeded by this type of dieback. To the contrary,

Nothofagus regeneration is stimulated, and again in form of cohorts, similarly as in some of the Hawaiian *Metrosideros* dieback types.

Recently, Hosking (1986) and Hosking & Hutchinson (1986) have recognized three forms of stand-level dieback in New Zealand's *Nothofagus* forests:

- stands in which there is extensive decline of the old canopy but adequate regeneration;
- stands with re-establishment problems following breakdown of the old canopy; and
- stands in which both old and young trees show clear symptoms of decline.

Papua New Guinea

Nothofagus stand dieback has also been reported as common in mountain beech forests of this tropical island (Paijmans 1976, Ash 1981). Arentz (1983) found *Phytophthora cinnamomi* in dieback stands of *Nothofagus pullei,* but also in non-dieback stands. As in Hawaii, the root fungus was initially believed to be a new introduction, but it was later recognized as a normal and longstanding member of *Nothofagus* forest communities on account of its wide distribution. Moreover, dieback stands have been recognized as cohort stands consisting largely of overmature individuals. *Nothofagus* reproduction in the same stands is common and not impeded by *Phytophthora cinnamomi.*

Japan

Wave regeneration dieback in fir *(Abies veitchii* and *A. mariesii)* dominated subalpine forests of central Japan has been recognized likewise as a totally natural stand-level dieback (Kohyama & Fujita 1981, Kohyama 1984). Here it is called the "Shimagare" phenomenon according to its development on Mt. Shimagare, which literally means the mountain with the white stripes of dead trees. As in the Adirondac mountains of upstate New York, where wave regeneration dieback in *Abies balsamea* stands has been reported and researched by Sprugel (1976), emphasis was placed on the combination of canopy dieback with reproduction of the same species. In the dieback etiology of both of these fir dominated forest ecosystems, endogenous or demographic factors relating to stand dynamics are recognized as causatively important together with abiotic environmental stresses, such as frosty and desiccating winds. The latter are recognized as dieback triggers that operate perpendicular to the exposed stand edges. These stand edges collapse in direction of the stand interior in "domino" fashion leaving behind striped dieback stands in which *Abies* regeneration comes up in waves, *i.e.* in form of cohorts.

Galápagos

One of the major canopy species in the Galápagos islands is *Scalesia pedunculata,* a tree that grows to 20 m height and apparently evolved from a forb belonging to the family *Compositae. Scalesia pedunculata* is the dominant canopy tree of the moist zone, which is well represented at mid to upper elevation (from 500–700 m) on the windward side of Santa Cruz island. This *Scalesia* forest also goes into periodic canopy dieback (Kastdalen 1982, Itow 1983), which is followed usually by abundant regeneration in form of cohorts. The mature stands likewise are recognized as cohort stands. Their dieback is clearly a natural phenomenon. *Scalesia* dieback appears to be related in part to the stand demography resulting in ageing cohort stands and in part to environmental triggers, such as heavy rains associated with "El Niño" or perhaps other physiological upsets, such as a drought. Certainly environmental disturbances associated with *Scalesia* dieback cannot be considered as the only causes of mortality, since they merely trigger the dieback of the mature or overmature stand with the result of subsequent stand rejuvenation.

In contrast to *Metrosideros polymorpha* in Hawaii, which is a slow-growing pioneer species initially adapted to exploit new volcanic habitats, *Scalesia pedunculata* is a fast-growing species that may complete its life cycle in 20 years. It compares well to fast-growing pioneer trees typical of secondary tropical rain forests.

Towards a unifying theory for stand-level dieback

The decline disease theory

Canopy and stand-level dieback can be fitted into the three-stage etiology developed by Houston (1973, 1981, 1984) for the decline disease theory (see also Manion 1981). This can be considered a unifying theory for all stand-level diebacks or decline that have no obvious single cause. The theory consists of three generic cause factors, which act in chain reaction. They are:

- predisposing factors;
- precipitating (inciting) or trigger factors; and
- accelerating or hastening factors.

A number of specific causes can be considered as *predisposing*, such as air pollution, climatic change, soil toxification, nutrient imbalances, genotype of host and old age. *Precipitating* causes are considered to be periodically fluctuating or recurring environmental stresses such as frost damage, storms, salt spray, logging damage, drought, floods, cold air drainage, and periodically fluctuating doses of air pollution. Epidemic diseases and insect pests could also act as dieback precipitating or trigger factors rather than as sole causes of tree death. The latter would apply only to the classic biotic diseases, which are attributable mostly to pest organism alien to an indigenous ecosystem. However, more commonly, disease-causing organisms assume the third causal role in the typical decline disease theory, that of dieback *accelerating* agents.

It is often assumed that stand-level dieback is the result of anthropogenic causes. The decline disease theory exemplifies anthropogenic causes particularly among the predisposing and precipitating ones. If they relate to anthropogenic causes, such as industrial air pollution and logging damage, it follows that the dieback accelerating biotic disease agents also find suitable conditions through certain management practices. However, causes of decline or dieback can also be completely natural, *i.e.* not influenced by human agency or management. This applies to several of the Pacific forest diebacks studied in recent years, including the native Hawaiian rain forest dieback.

In attributing dieback causes to new human-induced stresses it is important to also understand the extent and mechanisms of natural dieback causes. In this respect, the etiology as currently understood for the native Hawaiian rain forest may be used as a paradigm.

An etiology for natural dieback

Dieback structural types. – Vegetation and stand structure can always be used as a first indicator of function. Whenever the dieback causes are elusive, structure should receive special attention. As mentioned before, Hosking (1986), recognized three dieback structural types in New Zealand's *Nothofagus* forests which are similar to those recognized in the Hawaiian *Metrosideros* rain forests. The latter were characterized as *replacement, displacement* and *stand-reduction* diebacks.

The first two are typical canopy diebacks relating to the older stand component, one associated with adequate regeneration, the other with reestablishment problems of the canopy species. In Hawaii, such reestablishment problems refer to the successional displacement of the canopy species by competing species. The third structural dieback type, the stand-reduction dieback, relates to a rather limited form of stand recovery which is associated with site degradation for *Metrosideros* and other indigenous tree species. It occurs on permanently water-logged old sites with soil toxification problems, where the canopy stand of *Metrosideros polymorpha* and associated tree species have died. Here, recovery is only in the form of low-stature trees and shrubs. Moreover, regrowth of *Metrosideros* appears to be mostly from vegetative resprouts, and formerly more continuous stands are replaced only by disjunct patches of stunted regrowth. This also is a natural process in Hawaii, in this case related to physiographic landscape ageing.

Functional attributes. – Factors that are considered to play an important role in the Hawaiian dieback can be stated as the following five:

- The cohort stand structure (s) of the forest, which appears to result in an irregular spatial mosaic of relatively even-structured (and probably even-aged) stands. (The term cohort is here applied in the original sense to individuals belonging to the same life stage or age state. It is not restricted to "birth cohorts", *i.e.* individuals of the same calender age monitored through time).
- The edaphic site spectrum (e) from recent lava flows to ancient bogs which, in terms of forest development, is dominated largely by a single canopy species *(Metrosideros polymorpha).* Across this wide habitat spectrum, there are sites which are less favourable for the development of *M. polymorpha* stands. It follows that the species is not equally well adapted to all sites in its distribution range.
- Superimposed on this soil-site spectrum and life-stage mosaic of cohort stands are occasional perturbations (p) in form of climatic extremes, about 2 to 3 per decade. They include flash floods associated with gusty storms and periods of low solar radiation and/or extreme droughts that may last for 1 to 2 months or more. They may also include recurring substrate vibrations from seismic activity. Cohort stands in a vigourous life stage suffer physiological set-backs from which they can recover afterwards, but others on less buffered sites or those in a more advanced life stage may be stressed sufficiently into a state of more or less permanently reduced vigour. Such senescing cohort stands are eventually triggered into stand-level dieback.

- Following such drastic loss of tree vigour, biotic agents (b) which are weakly parasitic may then become aggressive in certain localities, where they can hasten the dieback process. Moreover, the decomposer biota can accelerate stand breakdown in moist and warm environments, while their activity may be very slow in cool, montane environments, a factor that can significantly prolong the dieback stage.
- Time (t) is an overriding dimension. It is involved in all other factors, and plays a role in ageing stands which become more vulnerable to external perturbations on account of increasing chronic stresses, *i.e.* natural loss of energy with advancing age.

The above role factors can be seen as functionally related in fig. 5.

The predisposing factor complex is of special interest, since it addresses the loss of stand vigour prior to dieback. The cohort structure and its postmature life stage (senescence) has been considered the most important predisposing factor in the Hawaiian dieback etiology (Mueller-Dombois 1986). There is little doubt that the range extension of *Metrosideros polymorpha* into harsh habitats, *i.e.* those to which it is less well adapted, is of similar importance. Cohort stands growing on nutritionally imbalanced or poorly drained soils may become forced into a low-vigour state sooner than under optimal conditions. This has been referred to as "premature senescence" (Mueller-Dombois 1986). The third factor, p (= environmental perturbations), which operates as a periodically growth-disrupting or pulsating environmental stress, will no doubt contribute further to the loss of vigour in stand development when the stand's ability to recover decreases with advancing age. This factor may thus also contribute to premature or physiological senescence. It seems reasonable to use this term as long as the contributing factors are identified. Future research should aim at analysing each factor individually as much as possible, whereby ontogenetic senescence in polycarpic trees and responses to climatic and seismic perturbations should receive special attention. Other less generally operating factors that may contribute to loss of stand vigour, such as autotoxicity or insect depredation without subsequent dieback, should, however, not be included in the concept of physiological senescence.

The Hawaiian and other Pacific island examples show that canopy or stand-level dieback extending over large areas (from a few hectares to several square kilometres) can result purely from natural factors that operate in contemporary ecosystems over relatively short time spans, such as decades. Changing climates, which historically had great effects on vegetation changes, perhaps also involving large-area stand-level diebacks, do not need to be involved.

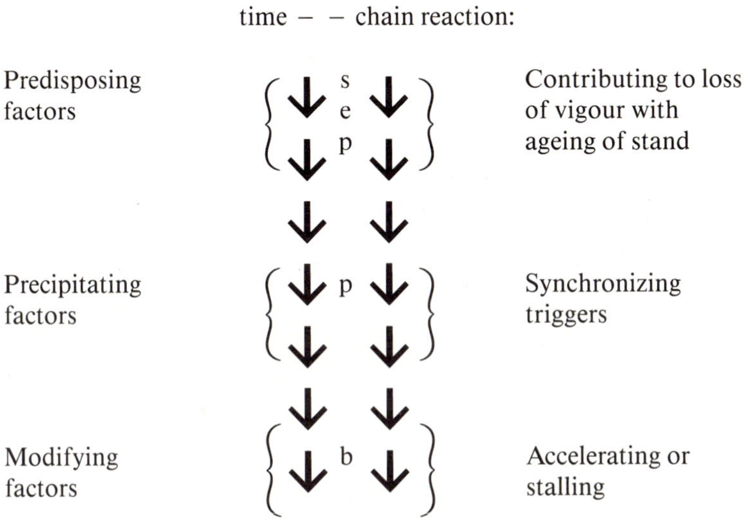

s = stand structure (senescence of canopy cohorts as related to forest growth cycle and past disturbances)

e = edaphic factors (stressful for the species in certain areas of its site spectrum)

p = environmental perturbations (droughts, floods, and vibrations of substrate from seismic activity)

b = biotic agents (appearance of parasites as accelerators vs. absence of parasites, and/or slow rate of decomposition as stalling factors)

time = overriding dimension for all factors; s, e, and p define spatial dimension.

Fig. 5. Factors involved in natural canopy dieback.

Such changes, however, should not be excluded from an etiology of natural dieback that pretends to include all possibilities. It is entirely possible to witness extensive forest dieback in areas where such events have never occurred before. This adds to the difficult task of separating natural from anthropogenic causes of dieback.

Conclusions: Pacific versus Atlantic diebacks

Forests in industrially polluted areas on both sides of the Atlantic have come under a new anthropogenic stress. Much of the European dieback in particular has been attributed to industrial air pollution (Schütt & Cowling 1985). The same new stress factor has fortunately not yet spread across the Pacific area, and the Pacific island forests of Hawaii, New Zealand, Papua New Guinea and the Galápagos may perhaps serve functionally as comparative dynamic models.

There is little doubt that large-area stand-level dieback occurs as a natural phenomenon in these island forests, including the subalpine fir forests of Japan. Thus forest dieback is not always a disease or abnormality related to new anthropogenic stresses. In all cases discussed in this paper there is a demographic basis that plays an important role.

Stand-level dieback occurs naturally in forests with few canopy species, as is typical for islands and mountains. These often display a uniform size or age structure over large areas. Such forests, as a rule, lack saplings of the canopy species and, in some cases, seedlings under closed stands. However, when their canopy breaks down, sapling stands are often found to grow up after dieback. In some cases, these forests lack successional tree species. Therefore, saplings of the same species grow up following dieback. They behave according to Hosking's (1986) type 1 dieback pattern in New Zealand's *Nothofagus* forests which corresponds to the Hawaiian replacement dieback. In several cases the species are typical pioneers. They are adapted to invade open areas. When pioneer site conditions change due to subsequent invasion and competition by other species, the maintenance strategies of the pioneer species may become ineffective, resulting in their successional displacement by other species. Disturbances which produce open sites, however, can permit these pioneer species to remain in the area.

In all cases of natural dieback referred to here, the etiology cannot be reduced to a single cause. This may also apply to the Atlantic forest dieback in spite of its strong anthropogenic cause. Air pollution and its associated complexities, such as acid rain, dry deposition of sulfur dioxide, nitrous oxide, heavy metal deposition, and synergistic effects, may have been added to the other predisposing factors. These other factors include the establishment of artificial cohort stands, sometimes in large-area plantations of uniform age structure. Also included may be the range extension of single canopy species over habitats to which the planted species are not well adapted. In addition, fluctuating and growth-upsetting climatic disturbances may be components also of the natural dynamics of Atlantic forest ecosystems, whereas seismic disturbances may be more common in the Pacific area.

It appears therefore that the origin of stand-level breakdown now under investigation in the Atlantic dieback forests is likely to include a combination of anthropogenic and natural causes. Moreover, among the new anthropogenic stresses, of which air pollution is the major concern, are probably some that simulate conditions which are natural in forests of isolated islands, namely few canopy species extending over a wide spectrum of habitats. Such a spectrum usually includes areas of habitat to which the species in question may respond favourably only during a restricted time span of its life cycle.

References

Arentz, F. 1983: *Nothofagus* dieback on Mt. Giluwe, Papua New Guinea. – Pacific Sci. **37**: 453–458.

Ash, J. E. 1981: The *Nothofagus* Blume *(Fagaceae)* of New Guinea. – In: Gressitt, J. L. (ed.), Biogeography and Ecology of New Guinea. – Monogr. Biol. **42**: 355–363.

Atkinson, I. A. E. 1970: Successional trends in the coastal and lowland forest of Mauna Loa and Kilauea Volcanoes, Hawaii. – Pacific Sci. **24**: 387–400.

Balakrishnan, N. & Mueller-Dombois, D. 1983: Nutrient studies in relation to habitat types and canopy dieback in the montane rain forest ecosystem, Island of Hawaii. – Pacific Sci. **37**: 339–350.

Batcheler, C. L. 1983: The possum and rata-kamahi dieback in New Zealand: a review. – Pacific Sci. **37**: 415–426.

Burton, P. J. & Mueller-Dombois, D. 1984: Response of *Metrosideros polymorpha* seedlings to experimental canopy opening. – Ecology **65**: 779–791.

Clarke, F. L. 1875: Decadence of Hawaiian forests. – All About Hawaii (Thrum's Hawaiian Annual) **1**: 19–20.

Clements, H. F., Putman, E. W., Suehisa, R. S. & Yee, Y. L. 1974: Soil toxicities as causes of sugar cane leaf freckle, macadamia leaf chlorosis (Keaau) and Maui sugar cane growth failure. – Hawaii Agric. Exp. Sta. Univ. Hawaii, Techn. Bull. **88.**

DLNR, 1981: A biological evaluation of the ohia decline on the island of Hawaii. – USDA Forest Service, Pacific SW Region Forest Pest Management, & State of Hawaii Department of Land and Natural Resources (DLNR), Division of Forestry and Wildlife.

Doty, R. D. 1980: Groundwater conditions in the 'ōhi'a rain forest near Hilo. – In: Smith, C. W. (ed.), Proceedings of the third conference in natural sciences, Hawaii Volcanoes National Park: 101–111. – Cooperative Park Resource Studies Unit, University of Hawaii, Manoa.

– 1982: Annual precipitation on the Island of Hawaii between 1890 and 1977. – Pacific Sci. **36**: 421–425.
Evenson, W. E. 1983: Climate analysis in 'ōhi'a dieback area on the island of Hawaii. – Pacific Sci. **37**: 375–384.
Fosberg, F. R. 1961: Guide to Excursion III. Tenth Pacific Science Congress. – Tenth Pacific Science Congress & University of Hawaii. (Ed. 2, 1972, University of Hawaii.)
Hodges, C. S., Adee, K. T., Stein, J. D., Wood, H. B. & Doty, R. D. 1986: Decline of ohia *(Metrosideros polymorpha)* in Hawaii: a review. – USDA Forest Serv., Pacific SW Forest & Range Exp. Sta., Gen. Techn. Rep., **PSW–86.**
Holt, R. A. 1983: The Maui forest trouble: a literature review and proposal for research. – Hawaii Bot. Sci. Pap., **42.**
Hosking, G. P. 1986. Beech death. What's new in forest research. – Government Printer, Wellington. (Art. **140**).
– & Hutchinson, J. A. 1986. Hardbeech *(Nothofagus truncata)* decline on the Mamaku Plateau, North Island, New Zealand. – New Zealand J. Bot. **24**: 263–270.
Houston, D. R. 1973: Diebacks and declines: diseases initiated by stress, including defoliation. – Int. Shade Tree Conf. Proc. **49**: 73–76.
– 1981: Stress triggered tree diseases: the diebacks and declines. USDA Forest Service (NE-INF–41–81), Washington, DC.
– 1984: Stress related diseases. – Arborical J. **8**: 137–149.
Hwang, S. C. & Ko, W. H. 1978: Quantitative studies of *Phytophthora cinnamomi* in decline and healthy ohia forests. – Trans. Brit. Mycol. Soc. **70**: 312–315.
Itow, S. 1983: The Galápagos Islands: cradle of evolutionary theory. – Chuoukouron-sha, Tokyo (in Japanese).
Jacobi, J. D., Gerrish, G. & Mueller-Dombois, D. 1983: *Metrosideros* dieback in Hawaii: vegetation changes in permanent plots. – Pacific Sci. **37**: 327–337.
Kastdalen, A. 1982: Changes in the biology of Santa Cruz, 1935–1965. – Not. Galápagos **35**: 7–12.
Kliejunas, J. T. & Ko, W. H. 1974: Deficiency of inorganic nutrients as a contributing factor to ohia decline. – Phytopathology **64**: 891–896.
Kohyama, T. 1984: Regeneration and coexistence of two *Abies* species dominating subalpine forests in central Japan. – Oecologia (Berlin) **62**: 156–161.
– & Fujita, N. 1981: Studies on the *Abies* population of Mt. Shimagare. – Bot. Mag. (Tokyo) **94**: 55–68.
Lewton-Brain, L. 1909: The Maui forest trouble. – Hawaiian Pl. Rec. **1**: 92–95.
Lyon, H. L. 1909: The forest disease on Maui. – Hawaiian Pl. Rec. **1**: 151–159.
– 1918: The forests of Hawaii. – Hawaiian Pl. Rec. **20**: 276–281.
– 1919: Some observations on the forest problems of Hawaii. – Hawaiian Pl. Rec. **21**: 289–300.
Manion, P. D. 1981: Tree disease concepts. – Prentice-Hall, Englewood Cliffs, New Jersey.
Miller, L. H. (ined.): Collecting trips, 1900. – (Verbatim copy of journal, the original of which was donated to the Bancroft Library at U.C. Berkeley. Xeroxed copy in Sinclair Library, University of Hawaii, Honolulu.)
Mueller-Dombois, D. 1983: Canopy dieback and dynamic processes in Pacific forests: introductory statement/concluding synthesis. – Pacific Sci. **37**: 313–316, 483–488.
– 1986: Perspectives for an etiology of stand-level dieback. – Annunal Rev. Ecol. Syst. **17**: 221–243.
–, Jacobi, J. D., Cooray, R. G. & Balakrishnan, N. 1980: 'Ohi'a rain forests study: ecological investigations of the 'ōhi'a dieback problem in Hawaii. – Hawaii Agric. Exp. Sta. Misc. Publ. **183**.

– & Smathers, G. 1975: Sukzession nach einem Vulkanausbruch auf der Insel Hawaii. – In: Schmidt, W. (ed.), Sukzessionsforschung: 159–188. – Cramer, Vaduz.

Paijmans, K. 1976: New Guinea vegetation. – CSIRO & Australian National University Press, Canberra.

Papp, R. P., Kliejunas, J. T., Smith, R. S. jr. & Scharpf, R. F. 1979: Association of *Plagithmysus bilineatus (Coleoptera: Cerambycidae)* and *Phytophthora cinnamomi* with the decline of 'ōhi'a-lehua forests on the Island of Hawaii. – Forest Sci. **25**: 187–196.

Petteys, E. Q. P., Burgan, R. E. & Nelson, R. E. 1975: Ohia forest decline: its spread and severity in Hawaii. – Pacific SW Forest & Range Exp. Sta., US Forest Serv. Res. Pap., **PSW–105**.

Schütt, P. & Cowling, E. B. 1985: Waldsterben, a general decline of forests in Central Europe: symptoms, development and possible causes. – Pl. Dis. **69**: 548–558.

Selling, O. H. 1948: Studies in Hawaiian pollen statistics. Part III. On the late Quarternary history of the Hawaiian vegetation. – Bernice P. Bishop Mus. Spec. Publ. **39**.

Sprugel, D. G. 1976: Dynamic structure of wave-regenerated *Abies balsamea* forests in the northeastern United States. – J. Ecol. **64**: 889–911.

Stemmermann, L. 1983: Ecological studies of Hawaiian *Metrosideros* in a successional context. – Pacific Sci. **37**: 361–373.

– 1986: Ecological studies of 'ōhi'a varieties *(Metrosideros polymorpha, Myrtaceae),* the dominants in successional communities of Hawaiian rain forests. – Ph D Diss., University of Hawaii, Honolulu.

Stewart, G. H. & Veblen, T. T. 1982: Regeneration patterns in southern rata *(Metrosideros umbellata)*-kamahi *(Weinmannia racemosa)* forest in central Westland, New Zealand. – New Zealand J. Bot. **20**: 55–72.

Veblen, T. T. & Stewart, G. H. 1982: The effects of introduced wild animals on New Zealand forests. – Ann. Assoc. Amer. Geogr. **72**: 372–397.

Wardle, J. A. Allen, R. B. 1983: Dieback in New Zealand *Nothofagus* Forests. – Pacific. Sci. **37**: 397–404.

Address of the author: Professor D. Mueller-Dombois, Botany Department, University of Hawaii at Manoa, 3190 Maile Way, Honolulu, Hawaii 96822, USA.

Plants in the city – door-mats or pampered kids?

Reinhard Bornkamm

Abstract

Bornkamm, R. 1988: Plants in the city – door-mats or pampered kids? In: Greuter, W. & Zimmer, B. (eds.): Proceedings of the XIV International Botanical Congress: 467–476. – Koeltz, Königstein/Taunus.

Metropolitan life has two conflicting aspects: On the one hand, social problems and environmental risks are particularly pronounced in big cities, on the other hand, the urban setting has many facets that are most attractive to Man. Plants in a city are subject to a similar polarity. Many species that had once been growing in the presently urbanized area have become extinct or have been decimated, being unfit to serve as, so to say, door-mats for the crowd. A number of examples are provided to illustrate such processes. But the losses are compensated, in the same cities, by the introduction of new, foreign species. Some of them establish themselves on their own, whereas others will be planted and nursed at high cost in terms of labour, funds and energy, being in a way the pampered kids of urban life. Metropolitan systems tend to eliminate the natural plant cover, replacing in with a foreign flora introduced at great expense. Plant ecology now recommends a less radical solution, encouraging the maintenance of wild grown plants in appropriate spaces whenever possible. This would in the long run lead to a better balance between the native and non-native plant species whose mixture is so characteristic of cities.

Introduction

The history of plants in a city is a story of destruction on the one hand and of overprotection on the other hand. The contradiction in the title reflects the opposite feelings we have when thinking of "urban ecology" and "urban life". For an ecologist nothing is worse in the world than a city: Most plants are eradicated, the remaining ones burdened with all kinds of environmental risks. "Urban life", however, is attractive in its cultural, economical and social values. In most countries city agglomerations grow by the immigration of people from rural areas. The migration in the opposite direction, from cities into agricultural regions, is by no means a movement back to nature. On the contrary: the former villages become urbanized. I want to describe this contradiction in more detail, as to its effects on plant life, using two theoretical considerations and a number of examples.

Historical approach and hemerobiosis

Using a historical approach we have to start with the original natural vegetation. In a city like Berlin, situated in a forest climate, the area formerly was almost completely (except for bogs and open waters) covered by different types of forest, especially bottomland forests with alder, hornbeam or basswood and drier forests with pine and oak (Hueck 1961). These forest communities were changed to a greater or lesser extent during the historical development of the city – some stayed forests, others were converted into a kind of stony desert with very few plants.

First theoretical consideration: To describe the strength of human impact seven degrees of hemerobiosis are used, based on the works of Jalas 1955 and Sukopp 1969. These grades are a semi-quantative means of measuring the alienation from the original natural state ("Naturferne") caused by human activities. The definition of the grades is given in table 1. High degrees of hemerobiosis are working against the natural succession. They therefore need a higher amount of energy input (in the broadest sense, comprising physical engergy, labour, money and material) for their maintenance than lower ones. One of the most important changes within this graduation is the change from 5 to 6: In degree 6 (polyhemerobiosis) no plants are cultivated intentionally. The complete elimination of the occurring plants (city ruderals) did not succeed or did not seem worth while. These are the plants city dwellers are most likely to encounter during a walk along the roads; they are the "door-mats".

Through the action of both natural and anthropogenous factors, the natural mosaic of vegetation units is overlaid by the concentric structures of the city (fig. 1). Since during the development from grade 1 to the middle grades 4 and 5 many new habitats are created, the number of species increases, but decreases again towards the highest degrees of hemerobiosis (fig. 1), The percentage of neophytes (species naturalized later than 1500 A. D.) increases steadily from grade 1 to grade 6 (table 1).

One important conclusion can be drawn for tasks and constraints of nature conservation in cities: Not pure nature is protected but vegetation types of a low grade of hemerobiosis. Historical or present land use is preserved together with flora and fauna.

On the way from grade 1 to grade 7 the original vegetation is more and more altered and finally eliminated. A great number of physical and chemical factors are involved in this process. In the next two chapters the main factors will be briefly discussed.

Table 1: Grades of hemerobiosis in a region with forest climax (C Europe) (compiled from data and definitions given by Jalas 1955, Sukopp 1969, Blume & Sukopp 1976 and Bornkamm 1980)

number	*name*	*cultural influence*	*land use*	*occurrence*	*neophytes (% of flora)*
7	meta-hemerobic	vegetation completely removed	buildings, roads	local (frequent in cities)	—
6	poly-hemerobic	vegetation heavily affected by mechanical and chemical factors	fallow lands, ruderal sites, plant production not intended	local (frequent in settlements)	23
5	α – eu-hemerobic	area deforestated, drainage, heavy fertilizing, use of pesticides	industrial agriculture, gardening	frequent	18—22
4	β-eu-hemerobic	forests replaced by fields, meadows or alien tree plantations	traditional agriculture, intense forestry	frequent	13—17
3	meso-hemerobic	forest vegetation changed or replaced by extensively used heathlands and grasslands	traditional forestry, extensive agriculture	frequent	5—12
2	oligo-hemerobic	low intensity of grazing and cutting in forests, original vegetation ± preserved	extensive forestry	rare	5
1	a-hemerobic	± not influenced	nature conservancy or lacking	nearly extinct	

Physical factors

The most prominent factors of the early reductional stages are physical, like wood cutting and burning. Later an ploughing, digging, weeding, trampling, driving and similar activities play a role. The most extreme effects, resulting in the complete destruction of the biotopes, are exerted by the construction of buildings and paved roads, leading to a sealing of the soil. The percentage of sealing reaches c. 90% in the city center, 50–70% in the suburbs, whereas in the auter city only 10–50% are sealed (Böcker 1984).

Thus most of the physical factors are mechanical influences leading to a reduction of biomass. The consequences of reduced biomass are lower transpiration combined with a lower cooling effect of the vegetation, higher run-off by sealing and by soil compaction which is combined with lower microbial activity and nutrient supply. The destruction of the vegetation enhances some of the typical parameters of the urban climate (Horbert et al. 1983). The urban ecosystems are incomplete in form and function. They share this character with some other ecosystems, like those in extreme deserts (Bornkamm 1987). Especially the food chains are fragmentary.

Chemical factors

The remaining vegetation lives under the stress of chemical pollution. For Berlin (West) three groups of factors shall be discussed here: de-icing salt, heavy metals and noxious gases (most data are taken from Bornkamm, in print; detailed references are given there).

De-icing salt shows two different modes of action: corrosive effects on plant surfaces when spread by the fast-driving traffic, and toxic effects when penetrating the soil and reaching the roots. In 1980 20% (= 40,000) of the Berlin roadside trees had to be regarded as damaged. Under salt stress the trees exhibit a high chloride content, chloroses and finally necroses. The leaves are shed very early, before the retransfer of nutrients to the stem has taken place. The premature loss of leaves, therefore, is not just a sign of early senescence but a pathological process. In Berlin this problem is not so serious

Fig. 1. Zones of similar floristic composition in the city of Berlin (West) (map and data from Kunick 1974). – Zone 1 = heavily built areas; zone 2 = lightly built areas; zone 3 = inner city boundaries; zone 4 = outer city boundaries; hem. = degree of hemerobiosis (see table 1); sp./km^2 = average number of species of spermatophytes observed in six 1 km^2 sized experimental plots. ▷

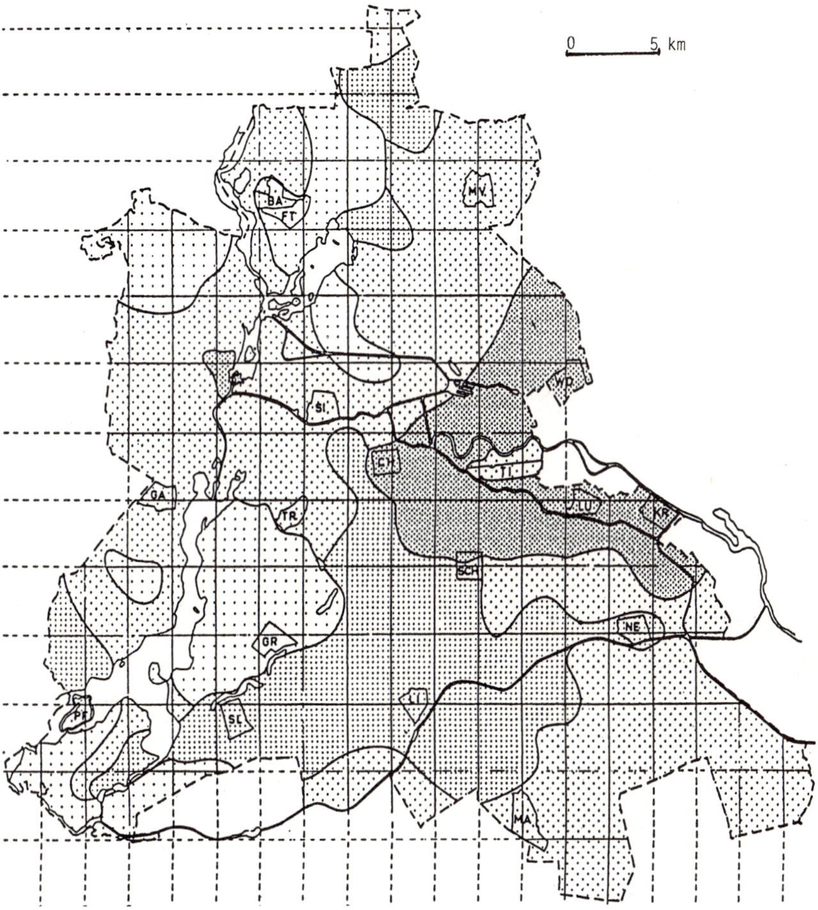

= Zone 1:		380 sp./km^2	hem. 6-7
= Zone 2:		424 sp./km^2	hem. 5-7
= Zone 3:		415 sp./km^2	hem. 5-6
= Zone 4:	open land	480 sp./km^2	hem. 3-4
	forest	234 sp./km^2	hem. 3

any more, because since 1981 the application of de-icing salt has been restricted to some main roads – but it is good to keep in mind the reason why this change in policy was necessary.

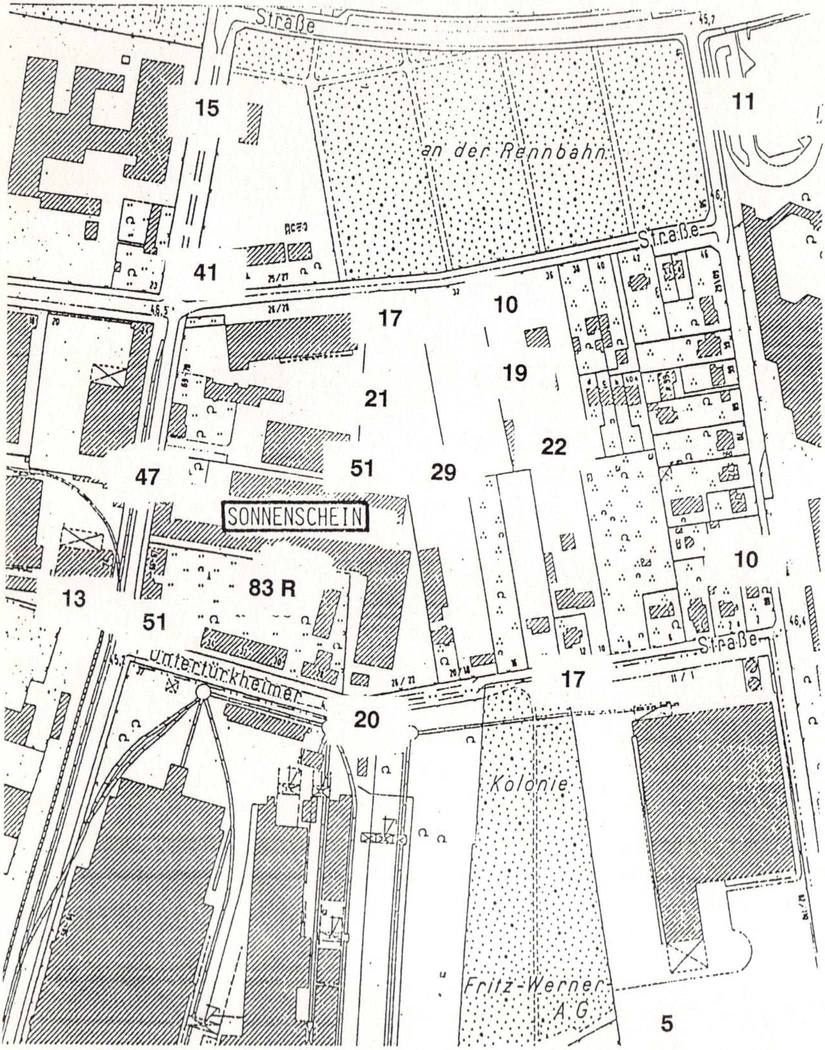

Fig. 2. Lead content (in µg Pb/g d.m.) in leaves of *Solidago canadensis* and *Robinia pseudacacia* (R) in the vicinity of the accumulator factory "Sonnenschein" in Berlin-Mariendorf. Date: 10. 9. 85 (from Rebele 1986).

Heavy metals, mostly stemming from combustion processes, predominantly occur in particulate form. It has been shown repeatedly that the concentration of heavy metals such as lead, bound to large particles, decreases rapidly with the distance from roads or other sources (see fig. 2). The pollution levels measured in certain areas of Berlin reach values capable of affecting sensitive plants. It is, however, unlikely that heavy metals cause excessive damage to plants (this is also true for forest trees), and no real heavy metal vegetation exists except for some species of lower plants such as the lichen *Stereocaulon nanodes,* which can even grow on rusty screws (U. Mezger, pers. comm.). The plants, therefore, mostly play the role of bioindicators, monitoring risks for man and larger animals with their complicated physiology. The evaluation of heavy metal contents of plants is made by calculating possible dietary risks for man and other members of the food chain.

Second theoretical consideration: The discussion on plant reactions to noxious substances makes it necessary to clarify the definitions of some terms involved. In fig. 3 the gradual increase in plant reaction (strain) is shown as related to increasing concentrations (or doses) of the damaging substance (stress). The threshold value can be defined as the concentration (or dosis) at which the most sensitive plant shows its first reaction, *i.e.* the first deviation from the normal state or range of normal states. From an ecological viewpoint aiming at nature conservancy, strain can be called a damage – *e.g.* irrigating a desert can be looked upon as a damage to desert vegetation, because it causes changes in plant composition. The threshold value arises from the conflict between ecological and economical interests, the question being what amount of strain (damage) is publicly accepted, and what amount of effort is accepted to keep the stress below the corresponding value.

Noxious gases are especially deleterious to plants. As open systems they have to transport large gas volumes through their tissues in order to meet their carbon requirement. The ratio carbon dioxide: pollutant, in some cases, is not higher than 1000 : 1. Plants are therefore predisposed for air pollution stress.

Let us focus again on the situation in Berlin (West). Sulphur dioxide can be assumed to contribute to the forest damage (mainly in pine needles) observed in the Berlin forests, taking into consideration the following facts (Cornelius et al. 1984):

- pine needles show twice the sulphur concentration (2800 ppm) as would be normal (1500 ppm, see Jasiek et al. 1984);
- sulphur dioxide sensitive lichens have died out in Berlin;
- in winter we register SO_2 concentrations in the air which are known to be harmful to plants;

- *Lolium perenne* as a bioindicator accumulates S in amounts that could cause damage to more sensitive plants.

It is unlikely that ozone plays a major role in forest damage in Berlin. Ambient concentrations of c. 100 ppb (30 min-value, summer) are however of an order of magnitude that is known to be detrimental to sensitive plants. The very sensitive bioindicator tobacco $BeCW_3$ regularly shows spot necroses caused by ozone (Cornelius et. al. 1985).

A great number of other chemical pollutants will not be discussed here, because they either act locally or have not yet been sufficiently analyzed as to their effects (Bornkamm, in print).

Gardens

Bearing in mind all the facts discussed above, the question arises as to how green plants can possibly survive in a city. They evidently do. When walking in springtime through my residence area I enjoy the view of an ample bloom of

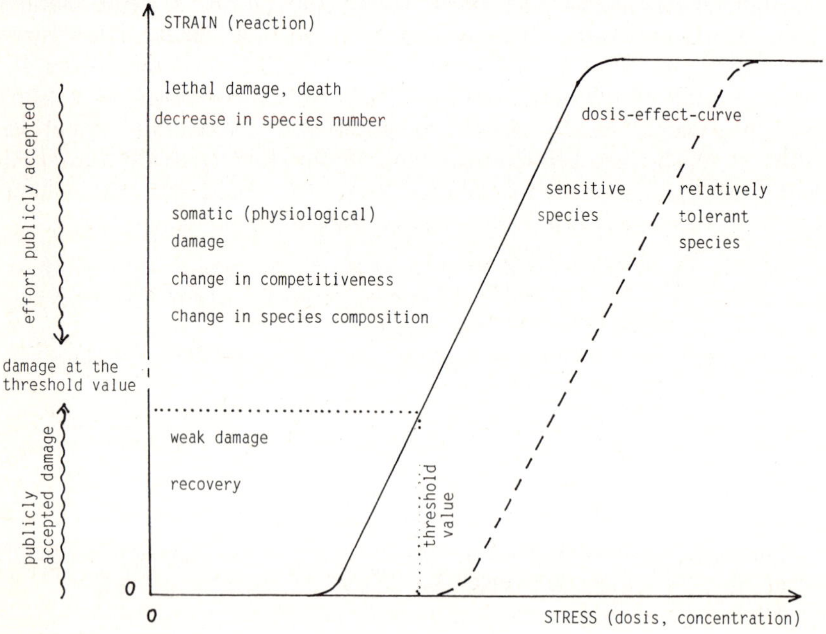

Fig. 3. Relation between reaction of plants (strain) and concentration (or dosis) of noxious substances (stress). For further explanation see text.

variegated flowers. At the same time the woods still looks brownish-green. This miracle is explained by the concentration of exotic plants brough together from all suited climatic regions, which are chosen for their ornamental value and require a lot of work for their maintenance. Gardening covers a wide range of effects, from the simple sowing, watering, fertilizing and weeding to a situation in which plants in autumn are covered every night by plastic sheets, a lawn is mown regularly (children must not tread on this green carpet, they have to play on the paved road), and an iron fence is ornamented with gold-leaf. Here at the uppermost fringe of hemerobiosis degree 5 we encounter the "pampered kids". Taking into account the amount of energy input, again in the broadest sense, we can speak of high-energy horticulture, where only the seed germinates and grows by natural forces while, everything else-such as plant containers, soil, fertilizer, water, herbicides or even illumination – is of anthropic origin. It is evident that both the degradation of natural vegetation down to door-mats and the costly bringing up of plants to pampered kids are two sides of the same coin – namely the increase in hemerobiosis.

The conclusions to be drawn from a plant ecologist's view must be based on the desire to have more plant life (and therefore more animal life) in the city. This cannot be achieved by converting Berlin into a garden, but by using the natural forces of succession. The necessity for a city to have structures like buildings, road systems and railways, well organized gardens and many other things is not questioned. But despite the needs of urban life a considerable amount of space is available for spontaneous and low-energy plant growth, *i.e.* for lowering the degree of hemerobiosis, for creating (at least small grades of) wilderness. Small steps in this direction are:

- covering of walls and roofs with plants;
- using small mosaic pavements instead of complete sealing;
- using road verges and areas around factories;
- reducing the management intensity in some parks and playgrounds;
- and favouring meadows instead of lawns.

Closing this incomplete enumeration I would like to point out that Berlin in some of these respects is well ahead of other cities.

This article emphasises the need for a more natural development of vegetation, but does not claim that by this process a paradise can be created. The environmental burdens described above are still relevant and affect both spontaneously growing and cultivated plants. Therefore, two postulates are to be raised:

- a clean environment and
- more space for the development of natural vegetation.

References

Blume, H.-P. & Sukopp, H. 1976: Ökologische Bedeutung anthropogener Bodenveränderungen. – Schriftenreihe Vegetationsk. **10:** 75–89.

Böcker, R. 1984: Bodenversiegelung. — In: Karten zur Ökologie des Stadtgebietes von Berlin (West). – TU Berlin & Technische Fachhochschule, Berlin.

Bornkamm, R. 1980: Hemerobie und Landschaftsplanung. – Landschaft & Stadt **12:** 49–55.

– 1987: Allochthonous ecosystems. – In: Landscape Ecol. 1987: **1:** 119–122.

– (in print): Belastung der Vegetation. – In: Sukopp, H. (ed.), Stadtökologie von Berlin (West).

Cornelius, R., Faensen-Thiebes, A., Fischer, U. & Markan, K. 1984: Wirkungskataster der Immissionsbelastungen für die Berliner Vegetation. – Landschaftsentw. & Umweltforsch. **26:** 1–81.

– , – & Meyer, G. 1985: Der Einsatz von *Nicotiana tabacum* BEL W-3 zur Immissions- und Wirkungserfasung in einem Berliner Bioindikationsprogramm. – Staub Reinh. Luft **45:** 59–61.

Horbert, M., Kirchgeorg, A. & Stülpnagel, A. von 1983: Ergebnisse stadtklimatischer Untersuchungen als Beitrag zur Freiraumplanung. – Umweltbundesamt Texte **1983** (18): 1–187.

Hueck, K. 1961: Karte der Vegetation der Urlandschaft. – In: Atlas von Berlin. – Akademie für Raumforschung und Landesplanung, Berlin.

Jalas, J. 1955: Hemerobe und hemerochore Pflanzenarten. Ein terminologischer Reformversuch. – Acta Soc. Fauna Fl. & Fenn. **72** (11).

Jasiek, J., Faensen-Thiebes, A. & Cornelius, R. 1984: Schwefelbealstung, Benadelungsgrad und Nährstoffgehalte Berliner Waldkiefern. – Berliner Naturschutzbl. **28:** 21–27.

Kunick, W. 1974: Veränderungen von Flora und Vegetation einer Großstadt, dargestellt am Beispiel von Berlin (West). – Thesis, TU Berlin.

Rebele, F. 1986: Die Ruderalvegetation Westberliner Industriegebiete und deren Immissionsbelastung. – Landschaftsentw. & Umweltforsch. **43:** 1–224.

Sukopp, H. 1969: Der Einfluß des Menschen auf die Vegetation. – Vegetatio **17:** 360–371.

Address of the author: Prof. Dr. R. Bornkamm, Institut für Ökologie, Technische Universität, Rothenburgstraße 12, D-1000 Berlin 41.

Table of Contents